URANIUM SERIES DISEQUILIBRIUM:
APPLICATIONS TO ENVIRONMENTAL PROBLEMS

Uranium Series Disequilibrium:

Applications to Environmental Problems

EDITED BY

M. IVANOVICH AND R. S. HARMON

CLARENDON PRESS · OXFORD
1982

Oxford University Press, Walton Street, Oxford OX2 6DP

London Glasgow New York Toronto
Delhi Bombay Calcutta Madras Karachi
Kuala Lumpur Singapore Hong Kong Tokyo
Nairobi Dar es Salaam Cape Town
Melbourne Auckland
and associates in
Beirut Berlin Ibadan Mexico City Nicosia

Published in the United States by Oxford University
Press, New York.

British Library Cataloguing in Publication Data

Uranium series disequilibrium.
1. Uranium – Decay – Environmental aspects
I. Ivanovich, M.   II. Harmon, R. S.
546'.431   QD181.U7

ISBN 0–19–854423–5

Photo Typeset by Macmillan India Ltd.,
Bangalore
Printed in Great Britain
at the University Press, Oxford
by Eric Buckley
Printer to the University

DEDICATION

E. D. Goldberg

Professor of Oceanography, Scripps Institute of Oceanography,
La Jolla, California

This volume is dedicated to V. V. Cherdynstev, a scholar and creative scientist. His pioneering investigations into disequilibrium within the U-decay series for over 30 years influenced two generations of environmental scientists throughout the international community.

Cherdynstev discovered that ^{234}U was out of radioactive equilibrium with its parent ^{238}U in some natural systems during the early 1950s. The first publication of these results in English was in 1961, and this announcement was greeted sceptically by many western scientists. However, David Thurber, then at the Lamont-Doherty Geological Observatory, seized upon this concept in the early 1960s and was the first scientist outside the U.S.S.R. to exploit the phenomenon in geochemistry. Subsequent work by a number of investigators confirmed and extended the disequilibrium concept, thus laying the foundation for this book.

Today, use of the naturally occurring disequilibrium between ^{234}U and ^{238}U, and between other U-series nuclides, has become a standard 'tool of the trade' for geochemists throughout the world. In geochronology it provides various techniques for dating Recent and Quaternary events over the past million or so years of earth history. It can also be used to trace U, Th, and daughter-nuclide behaviour patterns in both igneous and sedimentary processes. The wide spectrum of investigations employing U-series disequilibrium discussed within this volume attests to the importance of the phenomenon and in particular, of Cherdynstev's pioneering work in the field.

I should also like to offer a further dedication of this volume to those scientists who will further develop some of the radical findings of Cherdynstev and who will enhance our knowledge of radioactivity in natural systems. For example, in his volume on ^{234}U (*Uranium-234*, translated into English by J. Schmorak. Israel Programme for Scientific Translations, Jerusalem, 1971, page 43) Cherdynstev notes:

'Recent studies carried out in the Soviet Union showed a transuranic isotope with an alpha decay energy of 4.4 to 4.7 MeV exists in nature. Its complicated alpha-line spectrum indicates that its atomic weight is odd. This is confirmed by the fact that it is accompanied by Pu-239, which is probably also its decay product. The new alpha emitter, its fissionable products, and plutonium were found in derivatives of basic magmas and vulcanogenic rocks, i.e. this transuranic element tends to accumulate in abyssal rocks as distinct from the elements of the actinium group, which, like rare earth elements, are typically lithophilic and occur in the Earth's crust. . . . it seems to resemble osmium and differs from all other known heavy radioactive elements in that its oxides are volatile. It appears to be a far transuranide with $Z \approx 108$.'

To my knowledge, the identification of this nuclide has not been made yet. An assessment of these observations may provide an investigator with new vistas in environmental radioactivity.

Dr Cherdynstev has argued that studies of 'superheavies' in nature are one of the most important areas for research in geochemistry. He has studied the environment and interpreted his results in the tradition of solid scientific inquiry. He has mastered the concepts and techniques of chemistry and physics relevant to his interests, and his results have provided his colleagues with a deeper understanding of natural radioactivity and geochemistry. For each worker in the field to follow this example in his or her own scientific work is the challenge he has laid before us.

PREFACE

The discovery of the phenomenon of radioactivity occurred less than 100 years ago and became the basis for a revolution in the earth sciences, a revolution that has continued to the present day. The early work in the field, during the first quarter of this century, established that the decay series of the long-lived nuclides ^{238}U, ^{235}U, and ^{232}Th contain radioactive isotopes of several chemically and physically different elements. However, only during the past 25 years has it been recognized that disruption of these decay chains, and separation of parent and daughter nuclides occur as a result of geochemical processes acting in the near-surface environment. Since this realization the application of the naturally occurring radioactive disequilibria has progressed rapidly and has diversified to touch upon a wide spectrum of geological, oceanographic, hydrological, paleoclimatic, and archaeological problems. The first book published on the subject (Cherdynstev 1971) attempted to synthesize the description of the fundamental physical laws and the geochemical behaviour of U and Th nuclides in the near-surface environment responsible for the natural occurrence of radioactive disequilibrium, and was largely based upon the work of Cherdynstev and co-workers in the USSR. The distinguishing feature of Cherdynstev's book, however, is the author's imaginative treatment of, and insight into, a number of problems in the field. During the last decade the field of U-series disequilibrium has expanded rapidly, and hence, we feel the time is appropriate for a book which reviews the general progress of this quarter-century of research, and more specifically, a book which attempts to present an up-to-date summary of recent developments and applications.

Our book concerns itself with naturally occurring radioactive disequilibrium within the ^{238}U and ^{235}U decay series and is addressed to the scientists and engineers who share an interest in uranium in the natural environment. The book is intended for both the specialist already engaged in research in the field and the student wishing to enter a particular area of study, as well as the non-practioner with a general interest in the field. Thus, it is a compromise between an exhaustive treatise and a generalized summary. Although a modest knowledge of fundamental physics, chemistry, and geology is assumed, the necessary background to understand and use the more specialized sections of the book is presented and developed where appropriate.

The purpose of this book is to present a comprehensive discussion of the theory, analytical methodology, and broad application of U-series disequilibrium to problems in the environmental sciences. Our main aim is to emphasize variety, especially the variety of different disciplines and problems to which U-series disequilibrium is, and can be, applied. In order to accomplish this aim we have adopted an integrated, multi-author approach. By choosing this method we have drawn upon the experience and expertise of leading practioners within specialized areas, while hopefully, at the same time, we avoided the inherent problems of a symposium style volume in which individual contributions are often disjointed and only superficially related. Thus, we hope that the regularity in format and style throughout the book will make it more readable and at the same time not destroy the variety of viewpoints and perspectives presented. We sincerely hope that readers will be able to use this book both as a reference text and as a manual for the subject, and that they will be sufficiently stimulated to attempt to apply U-series disequilibrium studies to problems of their own interest.

Harwell   M. I.
East Kilbride   R. S. H.
January 1981

ACKNOWLEDGEMENTS

The preparation of this book would not have been possible without the support and gracious assistance of many people too numerous to name here, but to whom we collectively offer our thanks. However, we should like to express our gratitude and special thanks to the following individuals for their selfless and invaluable help: Misses Jackie Major, Christine Scowen, and Janet Poole of the Nuclear Physics Division, AERE Harwell, and Miss J. McLean and Mrs E. Yuill of SURRC for typing; Miss Rosalind Rees and Mrs Christine Davis of the Engineering Division Tracing Office, AERE Harwell for excellent tracing of numerous diagrams; to Mrs Jackie Hummel of the AERE Library for untiring literature searches and amazingly efficient help in providing the missing bibliographic information.

We are grateful to AERE Harwell Management for granting the senior editor permission to carry out this project and for providing the secretarial, library, and communications assistance in executing the project. Thanks are due to Prof. H. H. Wilson of SURRC for enthusiastic support of the project. Financial support of the Isotope Geology Unit of the SURRC is in part funded by a grant from the NERC.

We should also like to thank all our colleagues for their enthusiasm, for timely submission of their manuscripts, scholarly presentation of their work, and general agreeability and wholehearted support of the editors.

CONTENTS

LIST OF CONTRIBUTORS

C. J. Bland,
Department of Physics,
University of Calgary,
Calgary,
Alberta,
Canada

W. C. Burnett,
Department of Oceanography,
Florida State University,
Tallahassee,
Florida 32306,
USA

J. K. Cochran,
Woods Hole Oceanographic Institution,
Woods Hole,
Massachusetts 02543,
USA

J. B. Cowart,
Geology Department,
Florida State University,
Tallahassee,
Florida 32306,
USA

M. Gascoyne,
Department of Geology,
McMaster University,
1280 Main Street West,
Hamilton,
Ontario L8S 4MI,
Canada

R. S. Harmon,
Isotope Geology Unit,
Scottish Universities Research and Reactor Centre,
East Kilbride G75 0QU,
Scotland
UK

M. Ivanovich,
Nuclear Physics Division,
Building 7,
AERE Harwell,
Oxon OX11 0RA,
UK

T. -L. Ku,
Department of Geological Sciences,
University of Southern California,
Los Angeles,
California 90007,
USA

A. E. Lally,
Environmental and Medical Sciences Division,
Building 364,
AERE Harwell,
Oxon OX11 0RA,
UK

C. Lalou,
Centres des Faibles Radioactivities,
CNRS – CEA,
91190 Gif-sur-Yvette,
France

A. A. Levinson,
Department of Geology,
University of Calgary,
Calgary,
Alberta,
Canada

R. L. Lively,
Minnesota Geological Survey,
1633 Eustis Street,
St. Paul,
Minnesota 55108,
USA

W. S. Moore,
Department of Geology,
University of South Carolina,
Columbia,
South Carolina 29208,
USA

J. K. Osmond,
Geology Department,
Florida State University,
Talahassee,
Florida 32306,
USA

J. N. Rosholt,
US Geological Survey,
Federal Center,
Denver,
Colorado 00225,
USA

H. P. Schwarcz,
Department of Geology,
McMaster University,
1280 Main Street West,
Hamilton,
Ontario L8S 4MI,
Canada

M. Scott,
Department of Oceanography,
Texas A & M University,
College Station,
Texas 77343,
USA

B. J. Szabo,
US Geological Survey,
Federal Center,
Denver,
Colorado 00225,
USA

H. H. Veeh,
School of Physical Sciences,
The Flinders University,
Bedford Park,
South Australia,
Australia SA 5042

NOTATION

A	mass number, total number of nucleons in a nucleus of an atom
Z	atomic number, number of protons in a nucleus of an atom, charge of a nucleus
N	neutron number, equal to $(A - Z)$
λ	decay constant, represents the fraction of radioactive atoms that decays per unit time
$t_{\frac{1}{2}}$	half-life of a radioactive nuclide
t	time
B	binding energy of a nucleus
B/A	average binding energy per nucleon
N_A	Avogadro's number (6.023×10^{23})
M_H	mass of hydrogen atom (1.008 142 amu)
M_n	mass of neutron (1.008 982 amu)
M_e	electron rest mass (0.5488×10^{-3} amu $\equiv 9.108 \times 10^{-31}$ kg)
T	kinetic energy
c	speed of light ($2.997\,76 \times 10^8$ m/s)
ln	logarithm to the base e
e	Naperian constant (2.718 281 828)
log	logarithm to the base 10
π	pi constant (3.141 592 654)
Σ	denotes summation
Δ	incremental change
Eh	oxidation potential
pH	negative logarithm of the chemical activity of the hydrogen ion
M	mole or molar
STP	standard temperature and pressure
δ	deviation in per mil of an isotopic ratio from a defined standard
subscript 0	denotes initial state
superscript 0	denotes standard state

superscript *nnn*   associated with a nuclide symbol identifies the particular isotope of that nuclide (e.g. ^{234}U denotes the nuclide mass number of 234)

subscript *nnn*   associated with λ denotes the nuclide mass number (e.g. λ_{234} represents ^{234}U decay constant)

superscript *n* +   denotes valency of an ion (e.g. U^{6+} denotes the hexavalent state of an U ion)

subscripts   represent phases liquid, vapour, solid, and melt,
l, v, s, m   respectively

UNITS

Energy

J	basic unit of energy
ev	electron volt (1 ev $= 1.6 \times 10^{-19}$J)
KeV	kiloelectron volt
MeV	millionelectron volts (1 MeV $= 0.001\ 074$ amu)

Mass

amu	atomic mass units (1 amu $= 931.56$ MeV)
kg	kilogram (1 kg $= 10^3$g $= 10^6$mg $= 10^9 \mu$g)
g	gram
mg	milligram
μg	microgram

Liquid volume

l	litre (1l $= 10^3$ml $= 10^6 \mu$l)
ml	millilitre (or cc = cubic centimetre)
μl	microlitre

Distance

km	kilometre (1 km $= 10^3$m)
m	metre (1 m $= 10^2$cm $= 10^3$mm $= 10^6 \mu$m)
cm	centimetre
mm	millimetre
μm	micron
Å	angstrom (1Å $= 10^{-10}$m)

Mass concentration

ppm	parts per million (e.g. μg/g, mg/kg)
ppb	parts per billion (e.g. μg/kg, $\sim \mu$g/l)

Time

my	million years (10^6 years B.P.)
ky	kilo years (10^3 years B.P.)
y	year
d	day
h	hour
min	minutes
s	second

Electrical

V	volt, unit of electrical potential, V
amp	unit measure of current, I
Ω	ohm, unit of electrical resistance $R = V/I$
ohm m^2	unit of resistivity

Temperature

°C	unit of temperature expressed as degree Celcius

Radiation

dpm	disintegrations per minute
cpm	counts per minute
Ci	curie is the basic unit of radiation equivalent to 2.22×10^{12} dpm (it has been superceded by the new basic unit called Becquerel) 1 Becquerel (Bq) = 1 nuclear transformation/second, therefore 1 Ci = 3.7×10^{10} Bq
mCi	millicurie (1 mCi = 10^{-3} Ci)
μCi	microcurie (1μCi = 10^{-6} Ci)
nCi	nanocurie (1nCi = 10^{-9}Ci)
pCi	picocurie (1pCi = 10^{-12} Ci)

Concentration activity

μCi/l	microcurie per litre
pCi/l	picocurie per litre
dpm/g	disintegrations per minute per gram
dpm/l	disintegrations per minute per litre

MINERALS CONTAINING U AS AN ESSENTIAL CONSTITUENT
(After Rogers and Adams 1970.)

Oxides

Becquerelite	$7UO_3 \cdot H_2O$
Billietite	$BaO \cdot 6UO_2 \cdot 11H_2O$
Clarkeite	$(Na, K, Ca, Pb)_2U_2O_7 \cdot nH_2O$
Compreignacite	$K_2O \cdot 6UO_2 \cdot 11H_2O$ (PROTAS, 1964)
Curite	$3PbO \cdot 8UO_2 \cdot 4H_2O$
Fourmarierite	$PbO \cdot 4UO_2 \cdot 7H_2O$
Tanthinite	$UO_2 \cdot 5UO_3 \cdot 10.56H_2O$
Gummite	UO_3
Schoepite	$2UO_3 \cdot 5H_2O$
Uraninite	UO_2
Uranospherite	$(BiO) (UO_2) (OH)_3$ (?)
Vandenbrandeite	$Cu (UO_2)O_2 \cdot 2H_2O$
Vandendriesscheite	$PbO \cdot 7UO_3 \cdot 12H_2O$

Phosphates-Arsenates

Abernathyite	$K_2(UO_2)_2(AsO_4)_2 \cdot 8H_2O$
Arsenuranylite	$Ca(UO_2)_4(AsO_4)_2(OH)_4 \cdot 6H_2O$
Autunite	$Ca(UO_2)_2(PO_4)_2 \cdot 8-12H_2O$
Bassetite	$Fe(UO_2)_2(PO_4)_2 \cdot 8H_2O$
Bergenite	$Ba(UO_2)_4(PO_4)_2(OH)_4 \cdot 8H_2O$
Dewindtite	Doubtful Pb, uranyl phosphate
Dumontite	$Pb_2(UO_2)_3(PO_4)_2(OH)_4 \cdot 3H_2O$
Hallimondite	$Pb_2(UO_2)(AsO_4)_2 \cdot nH_2O$
Heinrichite	$Ba(UO_2)_2(AsO_4)_2 \cdot 10-12H_2O$
Hügelite	$Pb_2(UO_2)_3(AsO_4)_2(OH)_4 \cdot 3H_2O$
Kahlerite	$Fe(UO_2)_2(AsO_4)_2 \cdot 8H_2O$
Lermontovite	$(U, Ca, TR)_3(PO_4)_4 \cdot 6H_2O$
Meta-autunite	$Ca(UO_2)_2(PO_4)_2 \cdot 6-8H_2O$
Metaheinrichite	$Ba(UO_2)_2(AsO_4)_2 \cdot 8H_2O$
Metakahlerite	$Fe(UO_2)_2(AsO_4)_2 \cdot 8H_2O$
Meta-kircheimerite	$Co(UO_2)_2(AsO_4)_2 \cdot 8H_2O$
Metanovacekite	$Mg(UO_2)_2 (AsO_4)_2 \cdot 8H_2O$
Metatorbernite	$Cu(UO_2)_2(PO_4)_2 \cdot 6-8H_2O$
Meta-uranocircite	$Ba(UO_2)_2(PO_4)_2 \cdot 8H_2O$
Meta-uranospinite	$Ca(UO_2)_2(AsO_4)_2 \cdot 8H_2O$
Metazeunerite	$Cu(UO_2)_2(AsO_4)_2 \cdot 8H_2O$
Novacekite	$Mg(UO_2)_2(AsO_4)_2 \cdot 8-10H_2O$
Parsonite	$Pb_2(UO_2)_3(PO_4)_2(OH)_4 \cdot 7H_2O$
Phosphuranylite	$Ca(UO_2)_4(PO_4)_2(OH)_4 \cdot 7H_2O$
Przhevalskite	$Pb(UO_2)_2(PO_4)_2 \cdot 2H_2O$
Renardite	$Pb(UO_2)_4(PO_4)_2(OH)_4 \cdot 7H_2O$
Sabugalite	$HAl(UO_2)_4(PO_4)_4 \cdot 16H_2O$
Salceite	$Mg(UO_2)_2(PO_4)_2 \cdot 8-10H_2O$
Torbernite	$Cu(UO_2)_2(PO_4)_2 \cdot 12H_2O$
Troegerite	$H_2(UO_2)_2(AsO_4)_2 \cdot 8H_2O$
Uramphite	$NH_4(UO_2)(PO_4) \cdot 3H_2O$
Uranocircite	$Ba(UO_2)_2(PO_4)_2 \cdot 8-10H_2O$
Uranospathite	$Cu(UO_2)_2(AsO_4, PO_4)_2 \cdot 16H_2O$ (?)
Uranospinite	$Ca(UO_2)_2(AsO_4)_2 \cdot 10H_2O$
Walpurgite	$Bi_4(UO_2)(AsO_4)_2 \cdot 3H_2O$
Zeunerite	$Cu(UO_2)_2(AsO_4)_2 \cdot 8-10H_2O$

Minerals containing U as an essential constituent. (*contd.*)

Carbonates	
Andersonite	$Na_2Ca (UO_2) (CO_3)_2 \cdot 6H_2O$
Baylcyite	$Mg_2 (UO_2) (CO_3)_3 \cdot 18H_2O$
Liebigite	$Ca_2 (UO_2) (CO_3)_2 \cdot 10-11H_2O$
Rabbittite	$Ca_3Mg_3(UO_2)_2(CO_3)_6(OH)_4 \cdot 18H_2O$
Rutherfordine	$(UO_2)(CO_3)$
Schroeckingerite	$NaCa_3(UO_2)(CO_3)_3(SO_4)F \cdot 10H_2O$
Sharpite	$(UO_2)(CO_3) \cdot H_2O(?)$
Swartzite	$CaMg(UO_2) (CO_3)_3 \cdot 12H_2O$
Wyartite	$3CaO \cdot UO_2 \cdot 6UO_3 \cdot 2CO_2 \cdot 12-14H_2O$
Sulfates	
Johannite	$Cu(UO_2)_2(SO_4)_2(OH)_2 \cdot 6H_2O$
Uranopilite	$(UO_2)_6(SO_4)(OH)_{10} \cdot 12H_2O$
Zippeite	Near $2UO_3 \cdot SO_3 \cdot 5H_2O$
Vanadates	
Carnotite	$K_2(UO_2)_2(VO_4)_2 \cdot 1-3H_2O$
Ferghanite	$U_3(VO_4)_2 \cdot 6H_2O$
Francevillite	$(Ba, Pb)(UO_2)_2(VO_4)_2 \cdot 5H_2O$
Metatyuyamunite	$Ca(UO_2)_2(VO_4)_2 \cdot 5-7H_2O$
Rauvite	$CaO \cdot 2UO_3 \cdot 2V_2O_5 \cdot 16H_2O(?)$
Sengierite	$Cu(UO_2)(VO_4)(OH) \cdot 4-5H_2O$
Tyuyamunite	$Ca(UO_2)_2(VO_4)_2 \cdot 7-11H_2O$
Uvanite	$U_2V_6O_{21} \cdot 15H_2O(?)$
Vanuralite	$Al_2O_3 \cdot 2V_2O_5 \cdot 4UO_2 \cdot 17H_2O$
Silicates	
Beta-uranophane	$Ca(UO_2)_2(SiO_3)_2(OH)_2 \cdot 5H_2O$
Boltwoodite	$K_2(UO_2)_2(SiO_3)_2(OH)_2 \cdot 5H_2O$
Coffinite	$U(SiO_4)_{1-x}(OH)_{4x}$
Cuprosklodowskite	$Cu(UO_3)_2(SiO_3)_2(OH)_2 \cdot 5H_2O$
Gastunite	Ca, Pb, uranyl silicate
Kasolite	$Pb(UO_2)(SiO_3)(OH)_2$
Orlite	$3PbO \cdot 3UO_3 \cdot 4SiO_2 \cdot 6H_2O$
Pilbarite	$UO_3 \cdot PbO \cdot ThO_2 \cdot 2SiO_2 \cdot 4H_2O(?)$
Ranquilite	$1.5CaO \cdot 2UO_3 \cdot 5SiO_2 \cdot 12H_2O$
Sklodowskite	$Mg(UO_2)_2(SiO_3)_2(OH)_2 \cdot 5H_2O$
Soddyite	$(UO_2)_5(SiO_4)_2(OH)_2 \cdot 5H_2O$
Uranophane	$Ca(UO_2)_2(SiO_3)_2(OH)_2 \cdot 5H_2O$
Ursilite	$2(Ca, Mg) \cdot O.2UO_3 \cdot 5SiO_2 \cdot 9-10H_2O$
Weeksite	$K_2(UO_2)_2(Si_2O_5)_3 \cdot 4H_2O$
Niobates—Tantalates—Titanates	
Betafite	$(U, Ca)(Nb, Ta, Ti)_2O_9 \cdot nH_2O$
Brannerite	UTi_2O_6
Lodochnikovite (Lodochnikite)	$2(U, Th)O_2 \cdot 3UO_3 \cdot 14TiO_3$
Molybdates	
Calcurmolite	$Ca(UO_2)_3(MoO_4)_3(OH)_2 \cdot 11H_2O$
Iriginite	$UO_3 \cdot 2MoO_3 \cdot 4H_2O$
Moluranite	$UO_2 \cdot 3UO_3 \cdot 7MoO_3 \cdot 20H_2O$
Umohoite	$(UO_2)(MoO_4) \cdot 4H_2O$

MINERALS CONTAINING Th AS AN ESSENTIAL CONSTITUENT
(After Rogers and Adams 1970.)

Brockite	$Ca_{0.42}Sr_{0.03}Ba_{0.01}Th_{0.41}RE_{0.11}(PO_4)_{0.02}(CO_3)_{0.17} \cdot 0.9H_2O$
Cheralite	$(Th, Ca, Ce) (PO_4, SiO_4)$
Ekanite	$(Th, U) (Ca, Fe, Pb)_2Si_2O_{20}$
Huttonite	$ThSiO_4$
Monazite	$(Ce, Y, La, Th) PO_4$
Thorbastnaesite	$Th(Ca_{0.3}RE_{0.2})(CO_3)_2F_2 \cdot 3H_2O$
Thorianite	ThO_2
Thorite	$ThSiO_4$
Thorogummite	$Th(SiO_4)_{1-x}(OH)_{4x}$
Thorosteenstrupine	$(Ca, Th, Mn)_2Si_4(O_{11.24}F_{0.05})_{12.5} \cdot 5H_2O$
Thorutite (Smirnovite)	$ThTi_2O_6$
Uranothorianite	$(Th, U)O_2$
Uranothorite	$(Th, U)SiO_4$

INTRODUCTION

To assist the reader, a standard notation has been adopted throughout the book. A list of symbols and notation, a list of units, and a table of U- and Th-bearing mineral names are included. In general, the notation and units used are a compromise between the current usage and that recommended by the International Union of Pure and Applied Chemistry. The references cited in each chapter are presented in a combined bibliography section at the end of the book. The bibliography section also doubles up as an index to cited papers because each reference is labelled by the section number(s) in which that particular reference has been cited.

Contributions to the book were received between March 1980 and January 1981. A total of 19 authors prepared 19 chapters which can be grouped broadly into five areas or parts. Chapters 1, 2, and 3 present the nuclear theory, geochemistry, and geochronology which form the scientific basis essential to the subject matter pursued in the remainder of the book. Chapters 4 and 5 deal with analytical and measurement techniques and combine to provide a manual for non-practioners wishing to enter the field. The two major parts which follow discuss the application of U-series disequilibrium to the terrestrial (chapters 6 to 14), and marine (chapters 15 to 18) environments. In each part, the first chapters set the appropriate geochemical background, the intermediate chapters discuss the relevant processes affecting nuclide relationships within each particular environment, and the final chapters highlight applications of U-series disequilibrium to specific problems associated with that environment. The final chapter, chapter 19, is a review of the current state of the art and a perspective look into the future. Five appendices presented at the end of the book deal with diverse subjects in some mathematical detail: Appendix A presents a derivation of the basic radioactive decay equations discussed in chapter 1; Appendix B is an illustration of nuclear statistics discussed in section 5.4; Appendices C and D deal with age computations describing in some detail computer codes for the ^{230}Th/^{234}U and ^{231}Pa/^{235}U dating methods, respectively; Appendix E presents the relevant equations used in U-trend empirical model for dating igneous rocks discussed in chapter 10.

The individual contributors' acknowledgements follow each appropriate chapter. The editors, who have much more to be grateful for, list their acknowledgements after the preface.

1 THE PHENOMENON OF RADIOACTIVITY
M. Ivanovich

1.1 An introduction to nuclear physics

The matter of our physical world is made up of some hundred elements, each chemically and physically distinct. Atoms are the smallest entities into which these elements can be divided and still retain their chemical identity. An atom consists of a central *nucleus*, relatively heavy and carrying a positive electrical charge, around which move light negatively charged particles called *electrons*. The positions of the atomic electrons cannot be known precisely and they are best pictured as ill-defined clouds moving in orbits far away from the nucleus. The outer electrons define the chemical properties of the atom. The complete atom is electrically neutral, that is to say, the number of electrons is such that their charge is equal in size but opposite in sign to the charge in the nucleus.

The nucleus itself is made up of particles collectively called *nucleons*. There are two kinds of nucleons: *protons* and *neutrons*. They have nearly equal masses, but protons carry a positive electrical charge equal in size but opposite in sign to the charge of the electrons, and neutrons possess no charge. The total number of nucleons A, determines the mass of the atom and is referred to as the *mass number*. The number of protons Z, determines the charge of the nucleus and is referred to as the *atomic number*. The *neutron number* N, is defined as $(A - Z)$. Since in a neutral atom the number of protons determines the number of extra-nuclear electrons and this number decides the chemical properties of the atom, it follows that the number of protons determines the chemical element to which the atom belongs. Thus, atoms with different numbers of neutrons in their nuclei but the same number of protons will have the same chemical properties. Therefore, they will be atoms of the same element although their atomic masses will be different. They are called *isotopes* of the element. All isotopes of an element have practically the same chemical properties but their physical properties may differ considerably. The word *nuclide* has been proposed by Kohman (1947) as a generic term to represent any particular kind of atom or nucleus. Thus the word or its plural may be substituted for *nucleus, nuclear species,* or *nuclei.* The word isotope is also often used in this generic sense.

The various nuclides are denoted by symbols such as $^{238}U_{92}$, the subscript denotes the number of protons Z, in this particular nucleus, the superscript the total number of nucleons in the nucleus, and the letter the chemical symbol for the element, appropriate to Z. The number of neutrons of this isotope is obtained by subtraction. The subscript is often omitted and the elemental symbol accompanied by the superscript is sufficient to identify any nuclide.

Several other terms are defined here for convenience as they are used in subsequent sections of this chapter. An *isotone* is one of several nuclides having the same number of neutrons in their nuclei. An *isobar* is one of several nuclides having the same number of nucleons in their nuclei. Some nuclei are able to exist in an excited state for a time long compared with nuclear reaction times; these excited nuclei are called *isomers* of their corresponding ground state counterparts.

Of the approximately thousand known nuclides, most that occur are stable and do not change with time, but there are many which are radioactive. A *radioactive nucleus* is one which at any instant may change spontaneously into a different nuclear type. This process, called *radioactive decay*, is a statistical process in which the decay rate is proportional to the number of radioactive nuclei of a particular type present at any time t. The constant of proportionality, λ termed the decay constant, is the probability of decay per unit time interval. It is related to the *half-life*, $t_{\frac{1}{2}}$ of a *radionuclide* (radioactive nuclide) by $\lambda = 0.693/t_{\frac{1}{2}}$. The half-life of a radionuclide is the time required for the decay of exactly one half of the original number of its nuclei present.

Radioactive decay is effected by two alternative types of particle emission: either the moderately heavy *alpha particle* or the light *beta particle*. The particles are released with great energy and are often accompanied by *gamma rays*, penetrating electromagnetic radiation of a similar nature to X-rays. Alpha particles, nuclei of He atoms, are tightly bound assemblies of two protons and two neutrons, have a positive charge of two units, and a mass of four units. By emitting an alpha particle a nucleus of atomic number Z loses two protons and is thus transmuted into an element of atomic number $Z - 2$; for example, $^{238}U_{92}$ decays by alpha particle emission to become $^{234}Th_{90}$. Beta particles are electrons, having negligible mass in comparison with a nucleon, and a negative charge of one unit. Therefore, a nucleus of atomic number Z which emits a beta particle loses a unit of negative charge. This is equivalent to gaining one positive charge with no change in mass, and so becoming another element of atomic number $Z + 1$. For example, $^{234}Th_{90}$ emits a beta particle to become $^{234}Pa_{91}$. Another form of beta particle is the positive electron, or *positron*. Such decay also changes the value of the nuclear Z but in the opposite sense to that for negative electron emission. Thus, $^{22}Na_{11}$ yields $^{22}Ne_{10}$ upon emission of a positron. It may appear surprising that electrons can be emitted from nuclei believed to consist of protons and neutrons only, but the emission of a negative and positive electron in fact represents a nuclear process known as the weak interaction, in which a neutron in the nucleus spontaneously transforms into a proton with the creation and emission of an electron and also a massless particle called a *neutrino*. Alternatively, a proton in the nucleus transforms into a neutron with the creation and emission of a positive electron and a neutrino. Such decays can only take place if the mass of the initial nucleus sufficiently exceeds that of the final nucleus.

It is customary in nuclear physics to express energy in terms of *electron volts* (eV), *kilo electron volts* (keV), or *million electron volts* (MeV), and other larger units of eV. An electron volt is the kinetic energy acquired by an electron falling through a potential difference of 1 volt and is equal to 1.6×10^{-19} J (1.6×10^{-12} erg). Alpha particles emitted from natural radionuclides usually have discrete energies of several MeV; beta particles from a radioactive nuclide, being accompanied by a neutrino, have a continuous range of energies ranging from fractions of an eV up to several MeV.

Most of the long-lived, heavy radioactive elements decay into daughter products that are radioactive themselves. These decay in turn and thus form series (or chains) which end when a stable daughter nuclide is formed. A decay series is said to be in *radioactive equilibrium* when, on average, for each decaying parent atom one of each intermediate daughter atom also decays. The number of atoms of each intermediate daughter product will be in direct proportion to that product's half-life, or in inverse proportion to its respective decay constant. The amount of stable daughter product at the end of the series will continue to increase with time but at a decreasing rate. In most geological environments the natural radioactive series are often in a state of near equilibrium, but instances where they are not are common, and frequently very interesting. As radioactive parent and daughter are chemically different, it follows that they can be separated naturally by ordinary geochemical processes. Thus, the nature and degree of *disequilibrium* which exists in a sample frequently offers a clue to its geological history.

The following sections deal in some detail with the phenomenon of natural radioactivity as a prelude to the main subject of the natural disequilibrium within the U and Th decay series.

1.2 Limits of nuclear stability

Nuclei are composed of protons and neutrons only, bound together by very strong, short-range forces. The shape of the nucleus is essentially spherical, because this shape offers the greatest efficiency for the short-range binding forces between nucleons. There is also substantial evidence that the nuclear volume is essentially proportional to the number of nucleons in a given nucleus, that is to say, nuclear matter is incompressible, and therefore, has a constant density for all nuclei.

For the nucleus consisting exclusively of protons and neutrons the binding energy B, can be expressed as

$$B = ZM_H + NM_n - M, \tag{1.1}$$

where Z and N are the respective atomic and neutron numbers, M_H and M are the neutral atomic masses of H and of the nuclide in question, and M_n is the mass of a neutron. Physically, the binding energy is defined as the energy liberated when Z protons and N neutrons combine to form a nucleus, that is to say, the positive work necessary to break up a nucleus into its constituent nucleons.

Total binding energy in eqn (1.1) may be expressed in terms of A and Z as

$$B = AM_n - Z(M_n - M_H) - M. \tag{1.2}$$

Thus, the average binding energy per nucleon, B/A can be derived from eqn (1.2) after the rearrangement of terms

$$B/A = (M_n - 1) - Z/A(M_n - M_H) - P \tag{1.3}$$

where $P = (M/A - 1)$, is the packing fraction expressed in atomic mass units per nucleon.* The neutron mass excess $(M_n - 1)$ is the predominant term in eqn (1.3). To a first approximation $(M_n - 1)$ is the average binding energy, B/A of the nucleus in all nuclei except the very lightest.

Figure 1.1 shows the variation of average binding energy B/A, with atomic mass A, for $1 < A < 238$ (Evans 1955). One of the features of Fig. 1.1 is that the binding energy for nuclei is positive which means that any nucleus is more stable than an unconnected assembly of its constituent nucleons. This is consistent with the idea that within the nuclear volume the nuclear force between the nucleons is attractive. Since nuclei do not collapse these forces are also capable of becoming repulsive for very close distances of approach between the nucleons.

For $A < 28$ there is a prominent cyclic recurrence in Fig. 1.1, corresponding to strongest binding for nuclides in which A is a multiple of four. Each of these most tightly bound nuclides have Z even, N even, and $N = Z$ (^{4}He, ^{8}Be, ^{12}C,

---

* The binding energy can be expressed either in mass or energy units. This equivalence stems from Einstein's special relativity theory. Two scales for mass measurement are recognized: the absolute scale related to the kilogram; and the atomic mass scale, defined by setting the mass of one atom of the nuclide ^{12}C equal to 12.000 atomic mass units (amu). The absolute value of the atomic mass unit is obtained by noting that for 1 mole of ^{12}C (0.012 kg)

$$0.012 \text{kg} = N_A \times 12 \text{ amu}$$

where N_A is Avogadro's number, the number of atoms in one mole on the physical scale. The above considerations give

$$1 \text{ amu} = 0.001/N_A \text{ kg} = 1.661 \times 10^{-27} \text{ kg} = 931.502 \text{ MeV}/c^2.$$

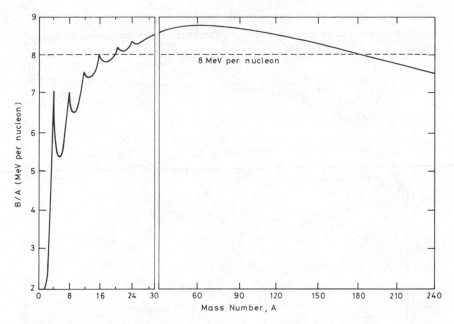

Fig. 1.1. Average binding energy per nucleon of the stable nuclei as a function of mass number (Evans 1955).

^{16}O, ^{20}Ne). This reflects the peculiar stability of the alpha particle.

Several other significant features are displayed in the region of $30 < A < 240$. Pairs of stable isobars first appear with ^{36}S$_{16}$ and ^{36}Ar$_{18}$ and become frequent as A increases, making B/A no longer a single-valued function of A. There is a broad maximum near $A \simeq 60$ (Fe, Ni, Co) where $B/A \simeq 8.7$ MeV per nucleon. Above this region the mean B/A values decrease steadily and B/A declines to a low of 7.3 MeV per nucleon for ^{238}U. This low value of B/A approaches, but does not equal the $B/A = 7.07$ MeV per nucleon exhibited in the alpha particle itself.

Thus B/A is essentially constant and to a good approximation, the total binding energy, B is proportional to the number of nucleons

$$B \simeq \text{constant} \times A \tag{1.4}$$

The nuclear unit which shows saturation contains four particles. This is evident from the binding energy, which reaches its first maximum for the alpha particle, an additional neutron or proton being less tightly bound. As Z increases, the disruptive forces due to Coulomb repulsion between all the protons should prohibit the formation of stable nuclides were it not for the existence of an extra attractive nuclear force present in the nuclear structure provided by neutrons whose number N exceeds Z by a larger and larger amount as Z increases. Thus, it is reasonable to presume that the major role for

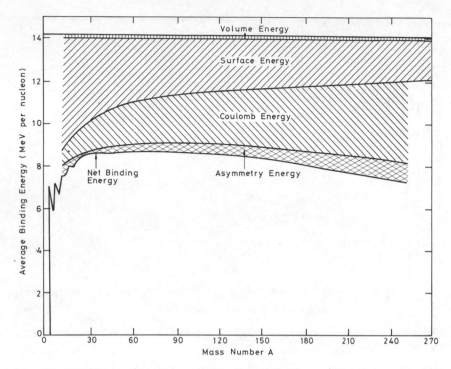

Fig. 1.2. Relative contributions to the net average binding energy per nucleon as a function of the mass number A (Evans 1955).

the excess neutrons $(N - Z)$ is to neutralize the Coulomb repulsion energy.

To conclude, in nuclei of any atomic mass, the forces between nucleons are non-additive. Only nucleons which have the same spatial position (share the same quantum state) are strongly bound to one another. These forces become saturated when, at most, two protons and two neutrons are near each other. Coulomb repulsion between protons is a long-range force and becomes increasingly important with respect to the short-range nuclear forces as the nuclear charge increases. The repulsive effect has to be balanced in a stable nucleus by the presence of extra neutrons to provide extra attractive interactions.

Many attempts have been made to express analytically the main features of Fig. 1.1. The proposed mass formulae proved useful in predicting the binding energy of unstable nuclei and in predicting the energy release in nuclear processes such as alpha particle emission or fission.

The mass M of a neutral atom whose nucleus contains Z protons and N neutrons has been given by eqn (1.2) where the binding energy B, is made up of a number of terms $B_0, B_1, B_2 \ldots$, each of which represents some general characteristic of nuclei. The largest term, the volume energy B_0, is identified as

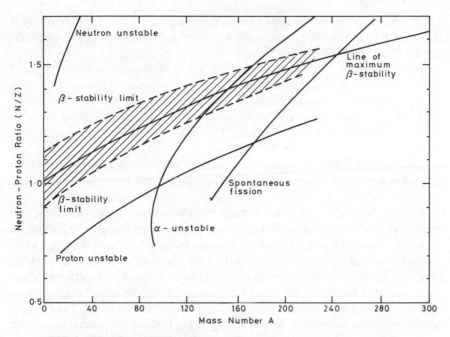

Fig. 1.3. Nuclear stability limits predicted by eqn (1.5) (Segré 1959).

being due to the saturated exchange forces. The second term, the surface energy B_1, is a negative correction term representing the loss of binding energy by the nucleons of the nuclear surface. The only known long range force in nuclei is the Coulomb repulsion between protons. If the total nuclear charge Ze is spread uniformly throughout the nuclear volume, and if the assumption of a constant-density nuclear radius is valid, then the loss of binding energy due to the disruptive Coulomb energy is given by the third term B_2. Another deficit in the binding energy arises from the neutron excess $(N - Z)$. This asymmetry energy is given by B_3. For nuclei with A even, N and Z may be both even or both odd. If both are even, nucleons may be grouped into stable pairs, and the nucleus will be correspondingly more stable than in the case of N, Z odd. The term B_4 corrects for this effect: B_4 is negative for A even, Z even, positive for A even, Z odd, and zero for A odd. Therefore, eqn (1.2) may be rewritten as:

$$M = ZM_H + (A - Z)M_n - (B_0 + B_1 + B_2 + B_3 + B_4)$$

or

$$M = ZM_H + (A - Z)M_n - a_v A + a_s A^{\frac{2}{3}} + a_c Z^2 / A^{\frac{1}{3}} + a_a (A - 2Z)^2 / A$$
$$+ \delta(A, Z) \tag{1.5}$$

where the coefficients a_v, a_s, a_c, and a_a can be determined empirically by fitting the equation to known masses (see for example Evans 1955 or Burcham 1963). The relative contributions to the net average binding energy per nucleon are depicted in Fig. 1.2.

When the known nuclei are arranged in a plot of neutron number N against proton number Z, lines of constant $A = N + Z$ intersect the axes at $45°$. The stable nuclei cluster about the line $N = Z = A/2$ for small A, but the medium and heavy nuclei have many excess neutrons forming a wide band of stable nuclei above the $N = Z = A/2$ line. The unstable nuclei form a fringe for many elements.

The stability limits predicted by eqn (1.5) for various types of decay are shown in Fig. 1.3. The diagram is a plot of neutron–proton ratio N/Z against the mass number A.

As it can be seen from Fig. 1.3, nuclear stability among the lighter elements is dependent exclusively upon the beta decay process. The nuclei with too large a neutron–proton ratio emit negative beta particles, while those with too low a value of this ratio emit positrons or decay by capture of an orbital electron. Among the heavy elements, however, there are nuclides of both beta-stable and beta-unstable variety. In the heavy elements other modes of radioactive decay operating quite independently of beta-decay become important. Thus, alpha-decay becomes important as a stability limit for $Z > 83$, where the nuclei expected to be beta-stable are alpha-active.

1.3 Modes of radioactive decay

1.3.1 Alpha decay

Alpha decay is the ejection from the nucleus of two neutrons and two protons bound together as a He nucleus. Measurements of atomic masses reveal that all nuclei above $A \sim 90$ have the energetic potential for emitting alpha particles. The rate for this process has a highly sensitive dependence upon the available energy. There is a general trend of increasing energy with atomic number, but it is only for elements above Pb that the process becomes dominant. There are however, perturbations in this general trend which have a profound effect upon alpha decay rates.

Empirical regularities in the variation of total disintegration energy, $E_0 = E_\alpha + E_{recoil}$, with mass number, atomic number, and half-life have furnished many guides for the development of nuclear structure theories (Perlman, Ghiorso, and Seaborg 1950). The regularities of the ground state alpha decay energies in the heavy element region are illustrated in Fig. 1.4 where the decay energy is plotted against mass number. Points of equal Z have been connected by straight lines. The resulting shapes exhibit a systematic decrease of alpha decay energy with increasing A among each set of isotopes (for constant Z). Generally, alpha decay energy increases with Z for constant A.

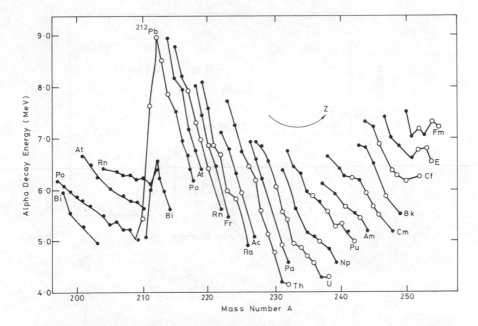

Fig. 1.4. Energy released in the alpha-decay of the heavy elements as a function of mass number A. The ringed nuclides are beta-stable (Burcham 1963).

However, if A is decreased until the neutron number N, falls below 128 the variation of alpha decay energy with A reverses abruptly over a limited region.

Alpha decay spectra are characterized by the presence of fine structure of two types. The first type is due to the excitation of levels of the residual nucleus and is illustrated by Fig. 1.5 where an energy spectrum of the alpha particles emitted in the transition $^{228}\text{Th} \rightarrow {}^{224}\text{Ra}$ ($E_0 = 5.516\,\text{MeV}$) is shown. The excited residual nucleus ^{224}Ra may lose its energy by subsequent gamma ray emission. The exact association between charged particle groups and the subsequent de-excitation spectra is used extensively in the investigation of nuclear reactions and level schemes.

The second type of structure in alpha decay spectra is clearly distinguished from the first type by much larger energy differences and by extremely small intensities, perhaps 1 in 10^5 compared to small energy differences and comparable intensities for the main fine-structure groups. The so-called long-range alpha particles are associated with disintegrations of an excited state of the initial nucleus. Often, as in the case of ^{212}Bi, discussed in some detail by Evans (1955), a beta particle process can populate highly excited states of its product nucleus. The excited states may decay to the ground state by gamma emission but the excited states are also unstable to alpha emission and since the energy release is large, alpha decay can compete with radiative emission.

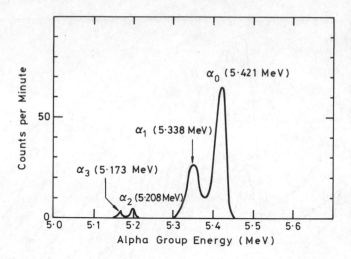

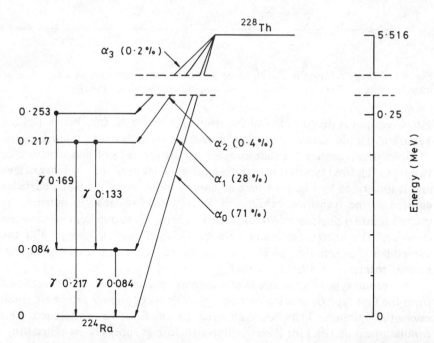

Fig. 1.5. An example of fine structure of alpha particle spectra.

Rutherford (1927) attempted an *ad hoc* theory to reconcile the conflicting conclusions about the dimensions of the U nucleus derived from the energy of the emitted alpha particle and from the alpha scattering experiments. The theory was able to provide a plausible mechanism for the escape of an alpha particle from the nucleus without having to surmount a huge potential barrier due to the Coulomb force of the nucleus, but it failed to account for some of the better known features observed in alpha decay. A much more satisfactory solution to the problem was given by Gamow (1928) and Gamow and Hantermans (1928) and independently by Gurney and Condon (1928 and 1929) when they applied wave mechanics to the alpha decay.

In their theories it is assumed that a preformed alpha particle is moving to and fro inside a spherical nucleus confined to the interior by the Coulomb potential barrier. The equation describing the motion of the alpha particle in the neighbourhood of this potential barrier is then solved with the aid of a number of assumptions. Thus, by applying wave mechanics to the barrier penetration problems it has been shown that there is a finite probability of the particle leaking through the barrier although its kinetic energy is less than the potential energy represented by the height of the barrier. Thus, most particles striking the barrier will be reflected but a certain number will 'tunnel' through and by continuing their forward motion leave the nucleus. Detailed accounts of the solution of the one-body theory of alpha decay have been given by Gamow and Critchfield (1949), Bethe (1937), Preston (1947), Perlman and Rasmunssen (1957), Hanna (1959), Preston (1962), and others.

1.3.2 Beta decay

The emission of the positive or negative electrons from nuclei, and the alternative process of electron capture determine the stability limits for nuclei throughout the periodic system as already discussed in section 1.2. The process changes the atomic number, Z by ± 1. The electrons and positrons that arise in a typical beta decay have a continuous spectrum of energies up to a definite limit determined by the mass change as illustrated in Fig. 1.6, but one beta particle only is emitted per nuclear disintegration. The homogeneous lines observed in such spectra are due to internal conversion of gamma rays and are not connected directly with the beta decay process. However, when a beta transition leads to an excited state of its product nucleus, internal conversion lines may be used to identify the product, since they give accurate values of the X-ray energies of this body. The electrons emitted in beta decay have the same charge/mass ratio as the ordinary atomic electrons.

The first attempt to classify beta-emitting bodies was made by Sargent (1933) who plotted the decay constant logarithmically against the maximum kinetic energy in the beta spectrum for a number of naturally occurring nuclei (see for example Burcham 1963, Fig. 16.8, p. 592). The Sargent diagram demonstrates grouping of points about two lines which are interpreted as 'allowed' and 'forbidden' lines, corresponding to different changes of nuclear spin.

According to the Fermi theory of beta decay, the decay constant λ can be related to a statistical function $f(X, W_0)$ where W_0 is the maximum total energy

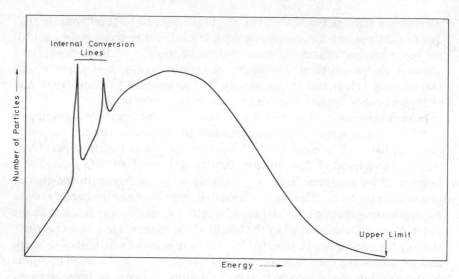

Fig. 1.6. Schematic representation of the beta decay spectrum for a radioactive source.

of the emitted beta particle, including its rest mass energy; W_0 is expressed in units of mc^2, m being the electron mass, so that $(W_0 - 1)mc^2$ represents the maximum kinetic energy of the beta particle. A plot of $\log \lambda$ against $\log (W_0 - 1)mc^2$ gives a straight line only over a limited range of values of Z and W_0, for which $f(Z,W_0)$ may be considered nearly independent of Z and proportional to some power of $(W_0 - 1)mc^2$, the kinetic energy. Recent classifications of beta decays are based on the comparative half-life $f(Z,W_0)t_{\frac{1}{2}}$, (or ft) for which use must be made of tabulated values of the function $f(Z,W_0)$ (see Wapstra and Nijgh 1959 for example). The result of this classification is grouping of nuclides into 'superallowed' or favoured beta-transitions ($\log ft \approx 3$ to 4), allowed transitions ($\log ft \approx 4$ to 5), and forbidden transitions ($\log ft \approx 6$ to 10).

Negatron and positron beta decay can be represented as the transformation of one nucleon in the nucleus into another according to

$$n \rightarrow p + \beta^- + \bar{v} + 780\,\text{keV}$$
$$p \rightarrow n + \beta^+ + v - 1800\,\text{keV} \qquad (1.6)$$

where v is a neutrino and $\bar{v}$ antineutrino. The properties required of the neutrino are that it should have zero charge, zero or nearly zero rest mass, half-integral angular momentum, and extremely small interaction with matter. Many attempts have been made to demonstrate directly the existence of this particle, the first successful one being by Cowan, Reines, Harrison, Kruse, and McGuire (1956).

A process equivalent to positron decay is that of electron capture: whenever it is energetically allowed by the mass difference between two neighbouring isobars, a nucleus with atomic number Z may capture one of its own atomic electrons, transforming to the isobar of atomic number $(Z-1)$. Usually the electron capture (EC) transition involves an electron from the K-shell, because these have the greatest probability density of being in the nucleus. The L-shell electrons may be involved also when energetically possible.

In EC transitions, a nuclear proton changes to a neutron, and the disintegration energy is carried away by a neutrino; thus

$$p + e^- \rightarrow n + v \tag{1.7}$$

The residual nucleus may be left in either its ground state or an excited state. The fundamental difference between EC and its competing positron emission is that in EC transitions the electron participating in the weak interaction is not observed; it disappears along with the proton. Only a neutrino emerges. The Auger electron, if it is observed, is a secondary particle. Following K-capture, there is an electron vacancy in the K-shell. This is filled by a transition of an L-electron into the K-shell and the emission of either an X-ray photon (especially in heavy elements) or an Auger electron (especially in light elements).

1.3.3 Gamma ray emission

An alpha or a beta decay process may leave the product nucleus either in its ground state or, more frequently, in an excited state. A nucleus in an excited state may give up its excitation energy and return to the ground state in a variety of ways. The most common transition is by the emission of electromagnetic radiation called gamma radiation. Gamma rays have a frequency determined by their energy, $E = hv$. Often, the transition proceeds from an upper state to the ground state in several steps involving intermediate excited states. Gamma rays with energies ranging between 10 keV and 7 MeV have been observed following radioactive decays. A gamma transition is characterized by a change in energy of the nucleus without change in Z or A.

In a number of instances a nuclide in an excited state may decay predominantly by alpha or beta decay rather than by gamma ray emission. The long range alpha particles of ^{212}Po and ^{214}Po arise from the decay of excited states in these nuclei so unstable with respect to alpha emission that alpha decay can compete with gamma emission. Usually the half-life of an excited state for gamma emission is much shorter than that for beta or alpha decay. In most cases the gamma decays occur so quickly that the half-lives have not been measured. Indirect evidence from competition between gamma and alpha decay and from the energy widths of gamma emitting levels indicates that many of the half-lives are as short as 10^{-13} s. The theoretical expectations and the experimental evidence agree that the gamma-transition half-lives depend on the transition energy, on nuclear spin change, $\Delta I = |I_i - I_f|$, and on the nuclear mass number, A. The total half-life for de-excitation by gamma emission increases with decreasing transition energy, increasing ΔI, and decreasing A. The selection rules which govern gamma decay and lead to these results are discussed by many authors (see for example Marmier and Sheldon 1969, or Friedlander, Kennedy, and Miller 1964).

1.4 The law of radioactivity

If a large number of radioactive atoms N is subject to a probability that any particular atom will disintegrate in unit time, then this probability is the decay constant, λ. The activity of these atoms, that is to say, the total number of disintegrations per unit time will be simply $N\lambda$. The rate of depletion, dN/dt is equal to the activity, a as long as there is no new supply of radioactive atoms. Because N decreases with increasing time

$$dN/dt = -N\lambda = a. \tag{1.8}$$

The solution of eqn (1.8) can be expressed as

$$N = N_0 e^{-\lambda t} \tag{1.9}$$

provided λ is regarded as time-independent, the initial number of the radioactive atoms is N_0 at $t = 0$, and the number of remaining radioactive atoms is N at some later time t. Recalling that $N_0 \lambda$ is the activity at $t = 0$, a_0, eqn (1.9) can be rewritten in terms of activity ratios

$$a/a_0 = N\lambda/N_0\lambda = e^{-\lambda t} \tag{1.10}$$

in agreement with the empirical law of radioactive decay and the associated disintegration hypothesis of Rutherford and Soddy (1902). The disintegration law of eqn (1.10) applies universally to all radioactive nuclides and the decay constant, λ as already pointed out in section 1.3, is a most important characteristic of each radioactive nuclide. The known radionuclides have decay constants whose values extend between $\lambda = 3 \times 10^6$ s^{-1} (for ^{212}Po) and $\lambda = 1.58 \times 10^{-18}$ s^{-1} (for ^{232}Th), a range of over 10^{24}. The decay constant is essentially independent of all physical and chemical conditions such as temperature, pressure, concentration, or age of the radioactive atoms (however, there are some exceptions, for example ^{7}Be half-life has been found to vary slightly with particular chemical compounds but this only applies to K-capture). If the competing modes of decay of any nuclide have probabilities λ_1, $\lambda_2, \lambda_3, \ldots$ per unit time, then the total probability of decay is represented by the total decay constant λ, which is the linear sum of all the partial decay constants. The 'partial activity' of a sample of N nuclei, if measured by a particular mode of decay characterized by λ_i, is then

$$dN_i/dt = \lambda_i N = \lambda_i N_0 e^{-\lambda t} \tag{1.11}$$

and the total activity is

$$dN/dt = \sum_i dN_i/dt = N \sum_i \lambda_i = \lambda N_0 e^{-\lambda t}. \tag{1.12}$$

Each partial activity falls off with time as $e^{-\lambda t}$, not as $e^{-\lambda_i t}$. Physically, this is because the decrease of activity with time is due to the depletion of the stock of atoms N, and this depletion is accomplished by the combined action of all the competing modes of decay.

The half-life $t_{\frac{1}{2}}$, is the time interval over which the initial number of radioactive atoms N_0, is exactly halved, that is to say $N = \frac{1}{2} N_0$ at $t = t_{\frac{1}{2}}$. The half-life is conveniently determined from a plot of $\log a$ against time t, and is related to the decay constant in eqn (1.10) by

$$t_{\frac{1}{2}} = \ln 2/\lambda = 0.693/\lambda. \tag{1.13}$$

In many cases parts of radioactive decay series where a parent radioactive nuclide 1, decays into a daughter radioactive nuclide 2, which is itself radioactive can be represented by

$$1 \xrightarrow{\lambda_1} 2 \xrightarrow{\lambda_2} 3$$

where λ_1 and λ_2 are the decay constants of type 1 and type 2, respectively. Since the rate of loss by decay is proportional to the amount of radioelement present (see eqn (1.8)) the rate of decay of the daughter atoms is given by the difference between its loss rate and its production rate from its parent

$$dN_2/dt = N_1 \lambda_1 - N_2 \lambda_2 \tag{1.14}$$

where N_1 and N_2 represent the number of atoms of each species present at any time t. If the only source of the parent atoms is an initial supply $N_1 = N_1^0$ at $t = 0$, then $N_1 = N_1^0 e^{-\lambda_1 t}$, and with these initial conditions, eqn (1.14) becomes

$$dN_2/dt = N_1^0 \lambda_1 e^{-\lambda_1 t} - N_2 \lambda_2. \tag{1.15}$$

This differential equation can be solved explicitly for N_2 as a function of time by evaluating the general coefficient for a specific set of initial conditions. The solution for the initial conditions $N_1 = N_1^0$, and $N_2 = N_2^0$ at $t = 0$ is given by

$$N_2 = N_1^0 [\lambda_1/(\lambda_2 - \lambda_1)][e^{-\lambda_1 t} - e^{-\lambda_2 t}] + N_2^0 e^{-\lambda_2 t}. \tag{1.16}$$

Then, the daughter activity is $N_2 \lambda_2$ (not dN_2/dt, see eqn (1.14)), where

$$N_2 \lambda_2 = N_1^0 \lambda_1 [\lambda_2/(\lambda_2 - \lambda_1)][e^{-\lambda_1 t} - e^{-\lambda_2 t}] + N_2^0 \lambda_2 e^{-\lambda_2 t}, \tag{1.17}$$

or since the parent activity at t is $N_1 \lambda_1 = N_1^0 \lambda_1 e^{-\lambda_1 t}$,

$$N_2 \lambda_2 = (N_1 \lambda_1)[\lambda_2/(\lambda_2 - \lambda_1)][1 - e^{-(\lambda_2 - \lambda_1)t}] + N_2^0 \lambda_2 e^{-\lambda_2 t}. \tag{1.18}$$

The daughter/parent activity ratio can then be expressed as

$$\left[\left(\frac{N_2\lambda_2}{N_1\lambda_1} \right)_{t=t} - \frac{\lambda_2}{\lambda_2 - \lambda_1} \right] = \left[\left(\frac{N_2^0\lambda_2}{N_1^0\lambda_1} \right)_{t=0} - \frac{\lambda_2}{\lambda_2 - \lambda_1} \right] e^{-(\lambda_2 - \lambda_1)t}. \quad (1.19)$$

From eqn (1.16), for the initial conditions of $N_1 = N_1^0$ and $N_2 = 0$, the amount of the daughter atoms is zero both at $t = 0$ and $t = \infty$, when all the atoms of both parent and daughter have decayed. At some intermediate time, t_m, the amount of the daughter, and hence its activity $N_2\lambda_2$, passes through a maximum value. The value of t_m is obtained by recognizing that at t_m $dN_2/dt = 0$. The differential of N_2 with respect to time is zero when

$$\lambda_1 e^{-\lambda_1 t_m} = \lambda_2 e^{-\lambda_2 t_m} \quad (1.20)$$

from which it follows that the time t_m of maximum activity of daughter nuclei is

$$t_m = \ln (\lambda_2/\lambda_1)/(\lambda_2 - \lambda_1). \quad (1.21)$$

From eqn (1.14), it is seen that at t_m the activities of the parent and daughter are equal and this condition is referred to as 'ideal equilibrium'. From $t = 0$ to $t = t_m$ the activity of the parent always exceeds the activity of the daughter, dN_2/dt is positive. Conversely, from $t = t_m$ to $t = \infty$ the activity of the daughter exceeds continuously the activity of its parent, dN_2/dt is negative.

The ratio of the activities of the parent and daughter under the initial conditions $N_2^0 = 0$ at $t = 0$, is given by

$$(N_2\lambda_2/N_1\lambda_1) = [\lambda_2/(\lambda_2 - \lambda_1)][1 - e^{-(\lambda_2 - \lambda_1)t}]. \quad (1.22)$$

This activity ratio is zero at $t = 0$, unity for $t = t_m$ and reaches the maximum value at large values of t. Several distinct cases arise, depending on the relative half-lives of the parent and daughter nuclei.

If the half-life of the parent nuclide is smaller than that of the daughter nuclide then the activity ratio increases continuously as t increases and a permanent state of 'disequilibrium' exists. In the extreme case when the half-life of the parent nuclide is much shorter than that of the daughter nuclide, the activity of the daughter substance becomes effectively independent of the residual activity of the parent. Then for $t \gg (t_{\frac{1}{2}})_1$, eqn (1.22) becomes $N_2\lambda_2 \simeq N_1^0\lambda_1 e^{-\lambda_2 t}$.

Another case of radioactive equilibrium in which $\lambda_1 \ll \lambda_2$ and in which the parent activity does not decrease measurably during many daughter half-lives is known as secular equilibrium. In this special case eqn (1.22) takes a particularly simple form

$$N_2\lambda_2 = N_1\lambda_1 (1 - e^{-\lambda_2 t}). \quad (1.23)$$

The daughter activity increases according to the simple exponential growth curve governed by its own decay constant λ_2. Thus, in the case of secular equilibrium for values of t much greater than the daughter half-life, the activity ratio becomes essentially unity, or $N_2\lambda_2 = N_1\lambda_1$.

In general, for a radioactive decay series, secular equilibrium implies that

$$N_1\lambda_1 = N_2\lambda_2 = N_3\lambda_3 = \ldots = N_n\lambda_n. \tag{1.24}$$

If one or more of the daughter products has been lost from a geologic sample by any process other than radioactive decay, then eqn (1.24) is no longer valid and the state of disequilibrium exists.

By analogy with eqn (1.14) the rate of growth of a grand-daughter nuclide in a decay series is given by

$$dN_3/dt = N_2\lambda_2 - N_3\lambda_3. \tag{1.25}$$

Then, extending the analogy with the case of two radioactive nuclides above, the activity $N_3\lambda_3$, subject to the initial conditions $N_1 = N_1^0$, $N_2 = 0$ at $t = 0$, is

$$N_3\lambda_3 = N_1^0\lambda_1[\lambda_2/(\lambda_2 - \lambda_1)\lambda_3/(\lambda_3 - \lambda_1)e^{-\lambda_1 t} + \lambda_2/(\lambda_1 - \lambda_2)$$
$$\lambda_3/(\lambda_3 - \lambda_2)e^{-\lambda_2 t} + \lambda_2/(\lambda_2 - \lambda_3)\lambda_3/(\lambda_1 - \lambda_3)e^{-\lambda_3 t}]. \tag{1.26}$$

The residual activity of the parent radioactive substance is $N_1\lambda_1 = N_1^0\lambda_1 e^{-\lambda_1 t}$.

Equations (1.8), (1.16), and (1.26) which refer to one, two and three radioactive nuclides decaying in a series have been generalized and put into a symmetrical form for any number of products by Bateman (1910): if, at time $t = 0$, there are N_1^0 atoms of the parent nuclide and no atoms of its series of decay products 2, 3, ... M, N, and if the corresponding decay constants are $\lambda_1, \lambda_2, \lambda_3, \ldots \lambda_M, \lambda_N$, then at any time t, the number N_N of N present will be given by the integral of the expression

$$dN_N/dt = N_M\lambda_M - N_N\lambda_N \tag{1.27}$$

where N_M is evaluated from a series of equations similar to eqn (1.27) for the amounts of the preceeding products. Then, the integration will yield

$$N_N = N_1^0(C_1 e^{-\lambda_1 t} + C_2 e^{-\lambda_2 t} + \ldots + C_M e^{-\lambda_M t} + C_N e^{-\lambda_N t}) \tag{1.28a}$$

in which the coefficients are dimensionless functions of the decay constants and have the following systematic values:

$$C_1 = \frac{\lambda_1}{\lambda_N - \lambda_1}\frac{\lambda_2}{\lambda_2 - \lambda_1}\frac{\lambda_3}{\lambda_3 - \lambda_1} \cdots \frac{\lambda_M}{\lambda_M - \lambda_1},$$

$$C_2 = \frac{\lambda_1}{\lambda_1 - \lambda_2}\frac{\lambda_2}{\lambda_N - \lambda_2}\frac{\lambda_3}{\lambda_3 - \lambda_2} \cdots \frac{\lambda_M}{\lambda_M - \lambda_2},$$

$$C_M = \frac{\lambda_1}{\lambda_1 - \lambda_M}\frac{\lambda_2}{\lambda_2 - \lambda_M}\frac{\lambda_3}{\lambda_3 - \lambda_M} \cdots \frac{\lambda_M}{\lambda_N - \lambda_M},$$

$$C_N = \frac{\lambda_1}{\lambda_1 - \lambda_N}\frac{\lambda_2}{\lambda_2 - \lambda_N}\frac{\lambda_3}{\lambda_3 - \lambda_N} \cdots \frac{\lambda_M}{\lambda_M - \lambda_N}. \tag{1.28b}$$

The initial condition that $N_N = 0$ at $t = 0$, requires the sum of the coefficients to be zero. Thus,

$$C_1 + C_2 + C_3 + \ldots + C_M + C_N = 0. \tag{1.28c}$$

In the case of a single radionuclide, eqn (1.27) becomes $dN/dt = -N\lambda$ and its solution is eqn (1.9).

In the case of radioactive parent and daughter series eqn (1.27) becomes

$$dN_2/dt = N_1\lambda_1 - N_2\lambda_2$$

that is to say, eqn (1.14). For initial conditions at $t = 0$, $N_1 = N_1^0$ and $N_2 = N_2^0$ eqn (1.28a) becomes

$$N_2 = N_1^0 (C_1 e^{-\lambda_1 t} + C_2 e^{-\lambda_2 t}) + N_2^0 e^{-\lambda_2 t} \tag{1.29}$$

where C_1 and C_2 are given by eqn (1.28b) and the condition in eqn (1.28c): $C_1 = -C_2$, and thus $C_1 = \lambda_1/(\lambda_2 - \lambda_1)$ and $C_2 = -\lambda_1/(\lambda_2 - \lambda_1)$. On substitution and rearrangement, eqn (1.29) yields the form of eqn (1.19) for the case of a pair of radionuclides, one decaying into the other.

Equation (1.26) for the case of the growth of a grand-daughter nuclide in a decay series can be derived from Bateman's equations too. More specifically, an expression for the grand-daughter activity in a decay series is derived in Appendix A for the case of the ^{230}Th/^{234}U activity ratio in the U-series.

The amount and activity of each radioactive product due to a radioactive accumulation of any duration t, have been obtained through eqns (1.28). It is only necessary to note that the amount N_N remaining at any later time is made up of (i) supply from 1, 2, . . . each acting independently as an originally pure source of 1, 2, . . . and producing the nuclide N in accord with eqns (1.28); and (ii) the residual of the original amount N present, which decays exponentially. Analytically,

$$N_N\lambda_N = (\text{activity of } N) = (\text{growth from } N_1^0) + (\text{growth from } N_2^0) + \ldots$$
$$+ (\text{growth from } N_M^0) + (\text{residue of } N_N^0). \tag{1.30}$$

Thus, no matter how complicated the conditions of accumulation or decay, eqns (1.28) may be applied because eqn (1.27) is always the basic principle governing the radioactive decay.

The general theory of the accummulation of a radioactive product applies also to the accummulation of a stable end-product. For any stable nuclide $\lambda = 0$. Then, if nuclide 1 is radioactive and decays into nuclide 2 which is stable eqn (1.16) becomes

$$N_2 = N_1^0 (1 - e^{-\lambda_1 t})$$
$$N_2 = N_1^0 - N_1 \tag{1.31}$$
$$N_2/N_1 = (e^{\lambda_1 t} - 1)$$

providing the initial condition at $t = 0$ is $N_1 = N_1^0$ and $N_2^0 = 0$. The third expression in eqns (1.31) is useful when, for example, t is to be computed from measurements of the residual amount of type 1 and of the amount of its decay product, type 2 which has accummulated. This is the principle of those geological age measurements which are based on the accummulation of Pb in U and Th minerals.

1.5 The radioactive decay series

All the alpha-emitting nuclides may be classified into four genetically independent decay series as shown in Table 1.1. The Th-series includes alpha-

Table 1.1. Geneology of alpha-emitting nuclides

Name of series	Type	Final nucleus (stable)	Longest lived member Nucleus	Half-life (y)
thorium	$4n$*	^{208}Pb	^{232}Th	1.39×10^{10}
neptunium	$4n+1$	^{209}Bi	^{237}Np	2.20×10^6
uranium	$4n+2$	^{206}Pb	^{238}U	4.47×10^9
actinium	$4n+3$	^{207}Pb	^{235}U	7.13×10^8

* n is an integer

and beta-emitting radionuclides with mass numbers which are multiples of 4, that is, $A = 4n$. The longest-lived member of the $4n$ series is ^{232}Th ($t_{\frac{1}{2}} = 1.39 \times 10^{10}$ y). Its half-life is about three times longer than the currently estimated age of the earth permitting it and its decay products to occur in nature. The immediate parent of ^{232}Th is the alpha-emitter ^{236}U but its half-life is only 2.4×10^7 y, and is, therefore, no longer found in nature. Geophysical evidence has indicated clearly that ^{236}U was one of the isotopes present when the universe was very young (Meyer and Urlich 1923; Riss 1924). Along with ^{236}U a number of heavy nuclei having $A = 4n$ have been produced artificially in a variety of nuclear reactions and they all join the series above ^{232}Th.

The Np-series does not occur in nature because the half-life of its longest-lived member is three orders of magnitude shorter than the age of the universe. The entire series has been produced artificially (Lederer and Shirley 1978).

The U-series is the longest known series. It begins with the nuclide ^{238}U and passes a second time through $Z = 92$ as a consequence of an alpha–beta decay sequence. This is followed by a sequence of five successive alpha transitions. At the end of the series, the alpha–beta–beta sequence is repeated twice (a characteristic of both $4n$ and $4n+2$ series) so that the U-series terminates on the lightest of the radiogenic Pb isotopes, ^{206}Pb.

The Ac-series begins in nature with its longest-lived nuclide ^{235}U and ends with the stable Pb isotope, ^{207}Pb.

In every series there is branching, by competition between alpha decay and beta decay, the relative abundance of the alpha decay branch is always such that the even A series ($4n$ and $4n+2$) exhibit the sequence alpha–beta–beta–alpha in the main branch, whereas the odd-A series ($4n+1$ and $4n+3$) show the decay sequence alpha–beta–alpha–beta in the main branch. A detailed description of the naturally occurring decay series follows.

A schematic diagram of the U-series is shown in Fig 1.7 and Table 1.2 lists essential data for each nuclide (from Lederer and Shirley 1978).

Uranium-238 is the chief constituent of natural U (99.27 per cent abundance) and is the progenitor of the ($4n+2$) series of radioelements. It is an alpha-emitter with a half-life of 4.468×10^9 y corresponding to a specific activity of 0.747 disintegrations per minute per microgramme. The energy of the main group of alpha particles is 4.195 Mev (77 per cent abundance). Conversion electrons of 48 keV gamma ray have been observed indicating that the alpha decay is complex; another two alpha groups have been observed with respective energies of 4.147 MeV (23 per cent) and 4.038 MeV (0.23 per cent).

Pure U emits no beta particles but energetic beta and gamma radiations are soon introduced by the ingrowth of the immediate daughters ^{234}Th and ^{234}Pa. This beta activity reaches equilibrium in undisturbed samples several months after the last purification of the U sample. Thorium-234 ($t_{\frac{1}{2}} = 24.1$ d) is a beta-emitter (Knight and Machlin 1948). The main group of beta particles in its spectrum has an end-point energy of 191 keV, but there is another group complex of about 103 keV energy. Gamma transitions with energies of 30, 60 and 93 keV have been reported.

In the beta decay of ^{234}Th, the principal daughter product is the 1.18 min ^{234}Pa. However, in 0.14 per cent of its disintegrations an isomeric form of ^{234}Pa with a 6.66 h half-life is produced. The 1.18 min ^{234}Pa decays directly to the ground state of ^{234}U in 98 per cent of its transitions. The continuous beta spectrum accompanying this decay has the high end-point energy of 2290 keV. The isomer ^{234}Pa* has an extremely complex beta, gamma, and conversion-electron spectrum.

The fourth member of the U ($4n+2$) decay series is ^{234}U ($t_{\frac{1}{2}} = 2.48 \times 10^5$ y). It is a beta-stable, alpha-emitter. The energy of the first group is 4.768 MeV (72 per cent), and the second group is 4.717 MeV (28 per cent). An additional group with the energy of 4.6 MeV (0.3 per cent) has been observed and gamma ray groups of 53 and 118 keV have been measured. The observations are summarized in the decay scheme of Fig 1.8 (Hyde, Perlman, and Seaborg 1964). As can be seen from Fig. 1.8, the immediate daughter resulting from the alpha decay of ^{234}U is ^{230}Th. This nuclide is another alpha-emitter with a half-life of 7.52×10^4 y (Artree, Cabell, Cushing, and Pieroni 1962). At least four alpha groups have been identified with energies and intensities as follows:

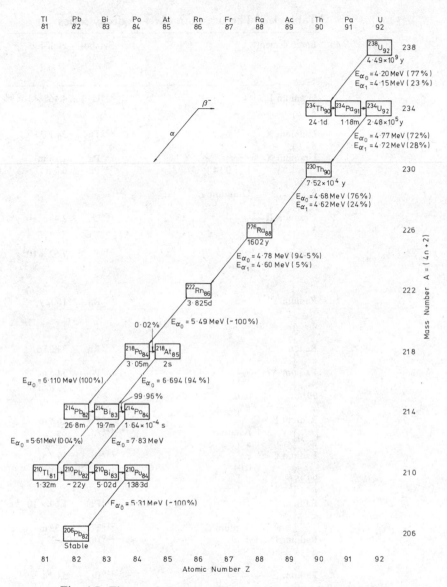

Fig. 1.7. The uranium $(4n+2)$ decay series (after Evans 1955).

4.682 MeV (76 per cent), 4.615 MeV (24 per cent), 4.476 Mev (0.12 per cent), and 4.437 MeV (0.03 per cent). Numerous gamma rays have been observed (see Stephens, Asaro, and Perlman 1957).

Radium-226 ($t_{\frac{1}{2}} = 1.602 \times 10^3$ y) is the immediate daughter of ^{230}Th. It is an alpha-emitter with four known alpha groups: 4.781 MeV (94 per cent), 4.598

Table 1.2. The uranium $(4n + 2)$ decay series

Radioelement	Symbol	Half-life

Uranium I
α

| | ^{238}U | 4.468×10^9 y |

Uranium X_1
β^-

| | ^{234}Th | 24.1 d |

Uranium X_2 ——————┐   I.T. 0.14%

93.86%
β^- Uranium Z

| | ^{234}Pa | 1.18 m |
| | ^{234}Pa | 6.7 h |

Uranium II ◄————
α

| | ^{234}U | 2.48×10^5 y |

Ionium
α

| | ^{230}Th | 7.52×10^4 y |

Radium

α

| | ^{226}Ra | 1602 y |

Radon
α

Radium A ——————┐

| | ^{222}Rn | 3.825 d |
| | ^{218}Po | 3.05 m |

99.98%   0.02%
α β^-

Radium B   Astatine^{218}
β^- α

| | ^{214}Pb | 26.8 m |
| | ^{218}At | 2 s |

Radium C ◄————
          0.04%

| | ^{214}Bi | 19.7 m |

99.96%
β^- α

Radium C'
α

| | ^{214}Po | 1.64×10^{-4} s |

          Radium C''
Radium D ◄ β^-
β^-

          α

| | ^{210}Tl | 1.32 m |
| | ^{210}Pb | ~22 y |

Radium E ——————

| | ^{210}Bi | 5.02 d |

~100%   ~10^{-5}% 1.8×10^{-6}%

β^- ^{206}Hg
Radium F β^-
α Thallium^{206} ◄————
          β^-

	^{206}Hg	8.6 m
	^{210}Po	138.3 d
	^{206}Tl	4.19 m

Radium G ◄————

| | ^{206}Pb | — |

Radiation	Alpha energy (MeV)	Abundance (per cent)	Beta end energy (keV)	Abundance (per cent)
$\alpha(\beta$ stable)	4.195 (α_0)	77		
	4.147 (α_1)	23		
	4.038 (α_2)	0.23		
β^-			191	81
			~ 103	19
β^-			2290	98
β^-			530	66
			1130	13
			1300	$\leqslant 2$
$\alpha(\beta$ stable)	4.768 (α_0)	72		
	4.717 (α_1)	28		
	4.600 (α_2)	0.3		
$\alpha(\beta$ stable)	4.682 (α_0)	76		
	4.615 (α_1)	24		
	4.476 (α_2)	0.12		
	4.437 (α_3)	0.03		
$\alpha(\beta$ stable)	4.781 (α_0)	94.5		
	4.598 (α_1)	5.5		
	4.340 (α_2)	7×10^{-3}		
	4.191 (α_3)	1×10^{-3}		
α	5.486 (α_0)	~ 100		
	4.983 (α_1)	$\sim 8 \times 10^{-2}$		
α and β^-	6.110 (α_0)	(100)	330	(100)
β^-			1030	6
α	6.70 (α_0)	94	670	94
	6.65 (α_1)	6		
α and β^-	5.61 (α_0)	100	3260	(100)
α	7.83 (α_0)	(100)		
β^-			2300	(100)
$\beta^- (\alpha)$	(3.7)	(1.8×10^{-8})	17	85
			64	15
β^- and α	4.93 (α_1)	60	1155	(100)
	4.89 (α_2)	34		
	4.59 (α_3)	5		
β^-			1300	(100)
α	5.305	(100)		
β^-			1520	(100)
stable				

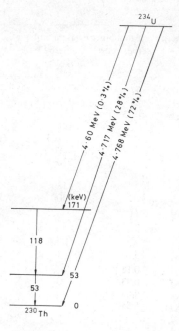

Fig. 1.8. Decay scheme of ^{234}U. (The slight population of levels at 508 and 634 keV is not shown.) (After Hyde *et al.* 1964.)

MeV (5.1 per cent), 4.340 MeV (7×10^{-3} per cent) and 4.191 MeV (1×10^{-3} per cent). Several gammas are observed in the energy region between 186 and 610 keV.

The immediate daughter of ^{226}Ra is ^{222}Rn ($t_{\frac{1}{2}}$ = 3.825 d). Radon-222 also emits only alpha particles: the predominant group of 5.486 MeV (approximately 100 per cent) and a much weaker one of 4.983 MeV ($\sim 8 \times 10^{-2}$ per cent). An accompanying gamma ray of 510 keV with a 0.07 per cent intensity was used by Madansky and Rosetti (1956) to predict the existence of the second group.

The next series of decay products ^{218}Po, ^{214}Pb, ^{214}Bi, ^{214}Po, and ^{210}Tl are quite short-lived and rapidly come to transient equilibrium with Rn. This group is a mixture of alpha- and beta-emitters as illustrated in Fig. 1.7 (also see Lederer and Shirley 1978). The last group of radionuclides belonging to the U-series is headed by ^{210}Pb ($t_{\frac{1}{2}}$ = 22 y) and the series ends with the stable end-product ^{206}Pb. The number of complexities in the alpha and beta spectra associated with the end part of the series is too great to list here and readers are referred to Hyde *et al.* (1964) for full details. Suffice it to add that the isotopic abundance of the stable end-product, ^{206}Pb when compared to other Pb isotopes, had considerable historical importance in the understanding of natural radioactivity and continues to have great significance for dating of minerals.

The second of the three naturally-occurring series of radioactive elements is the Th $(4n)$ decay series. The series originates with ^{232}Th and terminates with the stable nuclide ^{208}Pb. The Th-series is illustrated in Fig. 1.9 and Table 1.3 lists essential details for each nuclide (from Lederer and Shirley 1978).

The Th series could be extended indefinitely to elements above ^{232}Th as follows:

$$^{244}\text{Cm} \xrightarrow[17.9\,\text{y}]{\alpha} \ ^{240}\text{Pu} \xrightarrow[6580\,\text{y}]{\alpha} \ ^{236}\text{U} \xrightarrow[2.39 \times 10^7\,\text{y}]{\alpha} \ ^{232}\text{Th}$$

It is quite possible that several of these higher members of the series were

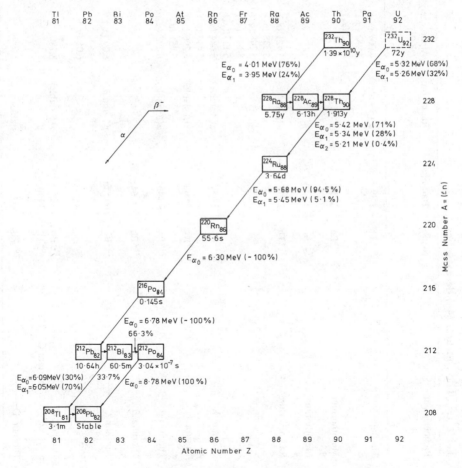

Fig. 1.9. The thorium $(4n)$decay series (after Evans 1955).

Table 1.3. The thorium (4n) decay series

Radioelement	Symbol	Half-life	Radiation	Alpha energy (MeV)	Abundance (per cent)	Beta end energy (KeV)	Abundance (per cent)
Thorium	^{232}Th	1.39×10^{10} y	α (β stable)	4.007 (α_0)	76		100
				3.952 (α_1)	24		
				3.882 (α_2)	0.2		
Mesothorium 1	^{228}Ra	5.75 y	β^-			55	(100)
Mesothorium 2	^{228}Ac	6.13 h	β^-			2110	(100)
Radiothorium	^{228}Th	1.913 y	α (β stable)	5.421 (α_0)	71		
				5.338 (α_1)	28		
				5.208 (α_2)	0.4		
				5.173 (α_3)	0.2		
				5.137 (α_4)	0.03		
Thorium X	^{224}Ra	3.64 d	α (β stable)	5.684 (α_0)	94.5		
				5.447 (α_1)	5.5		
Thoron	^{220}Rn	55.6 s	α (β stable)	6.296 (α_0)	~100		
				5.761 (α_1)	~0.3		
Thorium A	^{216}Po	0.145 s	α (β stable)	6.777 (α_0)	100		
Thorium B	^{212}Pb	10.64 h	β^-			580	(complex)
Thorium C	^{212}Bi	60.5 m	β^- and α	6.09 (α_0)	30	2250	(100)
				6.05 (α_1)	70		
Thorium C'	^{212}Po	3.04×10^{-7} s	α (β stable)	8.780	100		
Thorium C"	^{208}Tl	3.1 m	β^-			1800	(100)
Thorium D	^{208}Pb	—	stable				

Thorium
↓ α
Mesothorium 1
↓ β⁻
Mesothorium 2
↓ β⁻
Radiothorium
↓ α
Thorium X
↓ α
Thoron
↓ α
Thorium A
↓ α
Thorium B
↓ β
Thorium C ──33.7% α──→ Thorium C"
↓ 66.3% β⁻   ↓ β⁻
Thorium C'
↓ α
Thorium D

present at the time the elements were formed, but 4.5×10^9 y later these precursors of ^{232}Th have decayed completely to the much longer lived ^{232}Th. The accepted half-life of ^{232}Th is 1.39×10^{10} y. The main alpha particle group is 4.007 MeV (76 per cent) and a weaker group has been detected with an energy of 3.952 MeV (24 per cent). A third alpha group has been observed with 0.2 per cent intensity at an energy 125 keV below the main group.

The daughter product of ^{232}Th decay is the 5.75 y beta-emitter ^{228}Ra. Since the total decay energy is only 55 keV, the radiations of ^{228}Ra are extremely weak and are difficult to detect. The difficulty in the measurement of this nuclide is caused by the presence of its short-lived daughter ^{228}Ac ($t_{\frac{1}{2}} = 6.13$ h). The latter nuclide has a total beta decay energy of 2110 keV and decays with the emission of a complex mixture of electrons and gamma rays.

The daughter product of the ^{228}Ac decay is ^{228}Th. Thorium-228 is also the product of the alpha decay of ^{232}U and the electron capture decay of ^{228}Pa. The former of these two precursors of ^{228}Th is of particular interest to the techniques of measurement of U-Th disequilibria as a radiochemical tracer. Uranium-232 ($t_{\frac{1}{2}} = 72$ y) is a beta stable, alpha particle emitter. It was first identified (Gofman and Seaborg 1954) following its growth from the shorter-lived beta-emitter ^{232}Pa which had been prepared by the deuteron bombardment of Th:

$$^{232}\text{Th}(d, 2n)\ ^{232}\text{Pa} \xrightarrow[1.32\ \text{d}]{\beta^-}\ ^{232}\text{U}.$$

It has also been produced by other reactions, notably

$$^{230}\text{Th}\,(n, \gamma)\,^{231}\text{Th} \xrightarrow{\beta^-}\ ^{231}\text{Pa}\,(n,\gamma)\,^{232}\text{Pa} \xrightarrow{\beta^-}\ ^{232}\text{U}.$$

The alpha spectrum of ^{232}U consists of three observable groups (Asaro and Perlman 1955): 5.318 MeV (68 per cent), 5.261 MeV (32 per cent), and 5.132 MeV (0.32 per cent). A fourth group with energy 4.998 MeV (0.01 per cent) is deduced from gamma ray measurements.

The alpha-spectrum of ^{228}Th ($t_{\frac{1}{2}} = 1.913$ y) has been investigated by Asaro et al. (1953) who reported five energy groups: 5.421 MeV (71 per cent), 5.338 MeV (28 per cent), 5.208 MeV (0.4 per cent), 5.173 MeV (0.2 per cent), and 5.137 MeV (0.03 per cent) (see Fig. 1.5).

The immediate daughter of ^{228}Th is ^{224}Ra ($t_{\frac{1}{2}} = 3.64$ d). It is an alpha-emitter with two prominent alpha-particle groups: 5.684 MeV (94 per cent) and 5.447 MeV (5.1 per cent). There are three others of very low intensity: 5.159, 5.049, and 5.032 MeV. The second of the two main groups (5.447 MeV) is of particular importance in the analysis of alpha spectra containing ^{228}Th groups because it appears hidden under the ^{228}Th alpha peak.

The remaining members of the (4n) series are all very short-lived, the longest half-life being that of ^{212}Pb, 10.64 h. An overall view of the complex nature of the remainder of the Th decay series is given by Hyde *et al.* (1964). The complexity of the beta decay of ^{212}Pb, ^{212}Bi, and ^{208}Tl may be noted. Additional features of interest are the long-range alpha particles originating from the excited levels of ^{212}Po, the prominent gamma ray with the high energy of 2.615 MeV which occurs in 100 per cent of the disintegrations of ^{208}Tl, and the branching decay of ^{212}Bi. The stable end-product of the $4n$ decay series is ^{208}Pb.

The Ac, or $(4n + 3)$ series is summarized in Fig. 1.10 and in Table 1.4 (from Lederer and Shirley 1978). The primary nuclide from which all the other nuclides in this series are derived is ^{235}U. It is conceivable that higher-mass nuclides in the $(4n + 3)$ series line ^{239}Pu, ^{243}Cm, ^{243}Am, ^{247}Bk, etc. were also present when the elements were first formed, but of all the nuclides of the $(4n + 3)$ type above Pb, only ^{235}U ($t_{\frac{1}{2}} = 7.13 \times 10^8$ y) is sufficiently long-lived to have persisted throughout geologic time. Any sample of natural U no matter what its source, contains ^{235}U in a constant amount relative to ^{238}U, namely 0.72 atom per cent (notable exception are the samples from a fossil, natural ^{235}U fission reactor site at Oklo, Gabon (see Lancelot, Vitrac, and Allegre 1975)). Thus, one milligramme of pure natural U emits exactly 1501 alpha particles per minute. Of these, 733.6 are emitted by ^{238}U, an equal number by ^{234}U in equilibrium with ^{238}U, and 33.7 are emitted by ^{235}U.

Uranium-235 is an alpha-emitter with a complex alpha spectrum and a correspondingly complex gamma spectrum. About ten alpha groups have been observed, all in the energy region between 4.1 and 4.6 MeV. Of these, the two most prominent groups have energies of 4.391 MeV (57 per cent) and 4.361 MeV (18 per cent).

The immediate decay product of ^{235}U is ^{231}Th ($t_{\frac{1}{2}} = 25.64$ h). Thorium-231 is a beta-emitter with a complicated decay scheme. The most energetic beta group has an end-point energy of 300 keV and the total decay energy is 383 keV.

The next nuclide in the $(4n + 3)$ decay chain is ^{231}Pa ($t_{\frac{1}{2}} = 3.43 \times 10^4$ y) (Kirby 1961). It is also an alpha-emitter and the longest-lived of the Pa isotopes. The alpha spectrum of ^{231}Pa is complex with the highest intensity alpha-group energies of 5.05 MeV (11 per cent), 5.016 MeV ($\leqslant$ 20 per cent), 4.999 MeV (25.4 per cent), 4.938 MeV (22.8 per cent), and 4.724 (8.4 per cent). Its daughter is ^{227}Ac ($t_{\frac{1}{2}} = 22$ y). Actinium-227 is 98.8 per cent beta-emitter with an end-point energy of about 45 keV. The alpha decay branching occurs with only 1.2 per cent intensity. The most prominent alpha group energies are 4.949 MeV (48.7 per cent), 4.937 MeV (36.1 per cent), 4.866 MeV (6.9 per cent), and 4.849 MeV (5.5 per cent).

The immediate decay product of ^{227}Ac is ^{227}Th ($t_{\frac{1}{2}} = 18.17$ d) which is an alpha-emitter. This isotope has been subject to considerable interest because of the extraordinary complexity of its alpha and gamma spectra. It has fifteen recognized alpha groups of which the most prominent are at 6.036 MeV

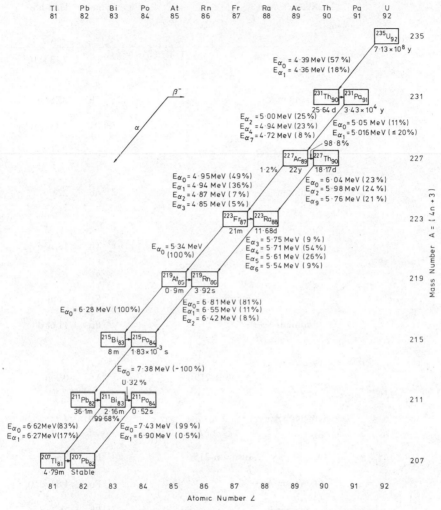

Fig. 1.10. The actinium $(4n + 3)$ decay series (after Evans 1955).

(23 per cent), 5.976 MeV (24 per cent), 5.755 MeV (21 per cent), 5.712 MeV (5 per cent), 5.708 MeV (8.7 per cent), and 5.699 MeV (4.0 per cent). This nuclide is of particular interest in the measurement of $^{231}Pa/^{235}U$ activity ratios and is discussed in some detail in Chapter 5.

The alpha-beta branching observed first in ^{227}Ac continues down the Ac-series in the form of a double chain as illustrated in Fig. 1.10 until ^{215}Bi ($t_{\frac{1}{2}} = 8$ min) is reached. This is a pure beta-emitter which decays to a pure alpha-emitter ^{215}Po. Another double branching occurs at ^{211}Bi, the series ending in stable ^{207}Pb. The ratio $^{207}Pb/^{235}U$ in a U mineral

Table 1.4. The actinium $(4n + 3)$ decay series

Radioelement	Symbol	Half-life
Actinouranium	^{235}U	7.13×10^8 y
$\alpha \downarrow$		
Uranium Y	^{231}Th	25.64 h
$\beta \downarrow$		
Protactinium	^{231}Pa	3.43×10^4 y
$\alpha \downarrow$		
Actinium 98.8% 1.2%	^{227}Ac	22 y
$\beta^- \downarrow$ α		
Radioactinium	^{227}Th	18.17 d
α		
Actinium K β^-	^{223}Fr	21 m
Actinium X ←	^{223}Ra	11.68 d
α		
Astatine-219 β^- 3.0% α 97.0%	^{219}At	0.9 m
Actinon ←	^{219}Rn	3.92 s
α		
Bismuth-215 β^-	^{215}Bi	8 m
Actinium A ←	^{215}Po	1.83×10^{-3} s
$\alpha \downarrow$		
Actinium B	^{211}Pb	36.1 m
β^-		
Actinium C 98.68% 0.32% β^-	^{211}Bi	2.16 m
$\alpha \downarrow$ Actinium C'	^{211}Po	0.52 s
Actinium C'' β^-	^{207}Tl	4.79 m
$\beta^- \downarrow$ Actinium D ←	^{207}Pb	stable

Radiation	Alpha energy (MeV)	Abundance (per cent)	Beta end energy (keV)	Abundance (per cent)
$\alpha(\beta$ stable)	4.391 (α_0)	57		
	4.361 (α_1)	18		
	4.1–4.6	the rest		
β^- (complex spectrum)			300	(~ 100)
$\alpha(\beta$ stable) (complex spectrum)	5.050 (α_0)	11.0		
	5.016 (α_1)	<20		
	4.999 (α_2)	25.4		
	4.938 (α_4)	22.8		
	4.724 (α_7)	8.4		
β^- and α	4.949 (α_0)	48.7	46	(100)
	4.937 (α_1)	36.1		
	4.866 (α_2)	6.9		
	4.849 (α_3)	5.5		
$\alpha(\beta$ stable)	6.036 (α_0)	23		
	5.976 (α_2)	24		
	5.755 (α_9)	21		
	9 other	32		
$\beta^-(\alpha)$	5.340	5×10^{-3}	1150	(100)
α	5.745 (α_3)	9.1		
	5.714 (α_4)	53.7		
	5.605 (α_5)	26.0		
	5.538 (α_6)	9.1		
$\alpha(\beta^-)$	6.28	100	energy known	
α	6.813 (α_0)	81		
	6.547 (α_1)	11		
	6.419 (α_3)	8		
β^-			energy unknown	
α	7.384 (α_0)	100		
β^-			1355	92.4
			951	1.4
			525	5.5
			251	0.7
$\alpha + \beta^-$	6.617 (α_0)	83	energy unknown	
	6.273 (α_1)	17	(calc $Q_{\beta-}$ = 610 keV)	
α	7.434 (α_0)	99		
	6.895 (α_1)	0.5		
β^-			1440	(100)
—				

which contains no primary Pb and which has remained unaltered since its formation is a direct measure of the mineral age through a straight forward application of the laws of radioactive decay. These methods are discussed briefly in Chapter 3.

Acknowledgements

The author should like to thank Dr Joan Freeman for her friendship, selfless help, and invaluable advice. Special thanks are due to Mr Tony Parsons of Nuclear Physics Division, AERE Harwell for proof-reading the manuscript. Drs Emilio Segre and Glenn Seaborg, Longman's Group, McGraw-Hill Book Company, John Wiley and Sons, and Prentice Hall Inc. are gratefully acknowledged for granting their permissions to reproduce some of their diagrams.

2 GEOCHEMISTRY OF THE ACTINIDES AND THEIR DAUGHTERS
M. Gascoyne

2.1 Introduction

2.1.1 Properties of the elements

The actinides are the heaviest naturally-occuring elements in the cosmosphere. Isotopes of the lower members (Ac, Pa, Th, and U) are derived from the three parent nuclides ^{238}U, ^{235}U, and ^{232}Th, which are sufficiently long-lived to survive in abundance at the present time (see Chapter 1 for details). The transuranium elements ($Z > 92$) are too unstable to have persisted as primordial elements, but trace amounts of the elements immediately above U (Np, Pu) have been found in U-rich ores due to neutron bombardment by fission of ^{235}U (Myers and Lindner 1971).

The actinide group takes its name from actinium, the first in a series of elements which are characterized by infilling of the 5f electronic shell (see Table 2.1). Although, formally, fourteen electrons are added to the 5f level between Ac ($Z = 89$) and Nb ($Z = 102$), in fact filling at this level does not begin at least until Pa ($Z = 91$) because it is energetically more favourable for electrons to enter initially the 6d level. The actinides can be compared to the lanthanides in which the 4f shell is filled progressively, but the groups differ in that the 5f electrons have relatively lower binding energies and less effective shielding than 4f electrons. This leads to differences in their chemical properties, such as the tendency of actinides to form complexes rather than strict ionic bonds (Cotton and Wilkinson 1972). Electronic configurations for the more important radioactive daughters of Th and U are also shown in Table 2.1 (other daughters such as Fr, At, and Tl are transient species formed in low frequency branching decays whose half-lives are minutes or less).

The actinides and Ra are electropositive elements and tend to form strong ionic bonds (see Table 2.1). Polonium, Bi, and Pb are amphoteric and in the natural environment may form ionic bonds with some covalent character. An estimate of the amount of covalency is given by the electronegativities for these elements (see Table 2.1). The higher the value for a particular oxidation state, the greater the covalent character. This covalency is reflected in the geochemistry of these elements in that the actinides and Ra are almost

Table 2.1. Physical properties of the naturally-occuring actinides and their daughter elements

Element	Z	Group	Electronic configuration (Hg inner shell)[1]	Oxidation and reduction states[2]	Ionic radii[3] (Å)	Electronegativity[4]
U	92	IIIa	$6p^6 5f^3 6d^1 7s^2$	3+, 4+, 5+, <u>6+</u>	(4+) 0.93 / 1.05	1.22
Pa	91	IIIa	$6p^6$ $5f^2 6d^1 7s^2$ / $5f^1 6d^2 7s^2$	3+, 4+, <u>5+</u>	(4+) 0.96	1.14
Th	90	IIIa	$6p^6 6d^2 7s^2$	3+, <u>4+</u>	(4+) 0.99 / 1.10	1.11
Ac	89	IIIa	$6p^6 6d^1 7s^2$	3+	1.11	1.00
Ra	88	IIa	$6p^6 7s^2$	2+	1.52	0.97
Rn	86	0	$6p^6$	0	–	–
Po	84	VIb	$6p^4$	4+, 2–	(4+) 1.02	1.76
Bi	83	Vb	$6p^3$	3+, 5+, 3–	(3+) 1.09	1.67
Pb	82	IVb	$6p^2$	2+, 4+	(2+) 1.32 / (4+) 0.98	1.55

Notes: (1) [Hg] = $1s^2 2s^2 2p^6 3s^2 3p^6 4s^2 3d^{10} 4p^6 5s^2 4d^{10} 5p^6 6s^2 4f^{14} 5d^{10}$
(2) the most stable oxidation states are underlined
(3) from Cotton and Wilkinson (1972) and Goldschmidt (1954)
(4) Allred and Rochow (1958)

exclusively lithophile elements whereas the stable isotopes of Bi and Pb also exhibit chalcophile properties. This leads to difficulties in predicting the behaviour of Bi, Pb, and perhaps Po, in natural waters of high ionic strength (sea water) or in the presence of organic compounds.

In aqueous media, Ra^{2+}, Ac^{3+}, Th^{4+}, and Pa^{5+} generally form colourless solutions but, because of the presence of one or more electrons in the outer shells, U^{4+} and U^{6+} solutions are often coloured (green and yellow respectively). In aqueous solutions, all the above ions hydrolyse to various extent, depending on the pH and their charge/ionic radius ratio (an index of their reactivity, termed ionic potential, Goldschmidt 1954). Thorium and Pa ions are particularly prone to hydrolysis; Th^{4+} hydrolyses above pH = 3, forming a variety of hydroxide species, and Pa^{5+} appears to hydrolyse more readily than Th^{4+}, a property which caused major problems in early determinations of its chemistry. The Pa^{5+} and U^{6+} ions are not thought to exist in solution simply as hydrated ions but instead as 'oxo ions' of the form MO_2^{n+}. Various data suggest that these ions are linear and form planar complexes with ligands both in solution and in crystal form (Evans 1963). The coordination chemistry in crystalline compounds of U and Th has been summarized by Pertlik (1969) and Bayer (1969) respectively.

2.1.2 General chemical properties and geochemical associations

Thorium and U are concentrated in crustal rocks in an average Th/U ratio of about 3.5 (Rogers and Adams 1969). The constancy of this value among many different igneous rock types indicates the general lack of fractionation of the two elements during magmatic processes. Examples of Th and U abundances are given in Table 2.2. Uranium and Th are preferentially incorporated into late crystallizing magmas and residual solutions because their large ionic radii preclude them from early crystallizing silicates such as olivine and pyroxenes (Goldschmidt 1954). They are therefore found associated mainly with granites and pegmatites. Radium, when fractionated from parent [230]Th, is found in hydrothermal precipitates such as barite and in association with Pb deposits. Most Ra salts are insoluble, particularly the sulphate and carbonate. Bismuth has only one stable isotope ([209]Bi) and is generally found concentrated in apatites, sulphides, pegmatites, and in rare earth minerals. Its salts are generally insoluble in water or, in the case of the chloride, sulphate, and nitrate, hydrolyse to the insoluble oxy-salt. Little is known about the chemical properties of Po because it has no stable isotope and [210]Po is intensely radioactive with a short half-life. Its aqueous chemistry is likely to be dominated by a tendency to hydrolyse to insoluble species or form anionic complexes. Stable Pb isotopes are usually found in greatest abundance in association with sulphides but also occur in silicates (particularly potassium feldspar; the ionic radii of K^+ and Pb^{2+} are almost identical) and in apatite. With the exception of nitrate and acetate, Pb salts are either insoluble or

Table 2.2. Normal range of U and Th concentrations and Th/U ratios in various rock types

Rock type	Name	U (ppm)	Th (ppm)	Th/U
Igneous	granites granodiorites rhyolites dacites	2.2–6.1	8–33	3.5–6.3
	gabbros	0.8	3.8	4.3
	basalts	0.1–1	0.2–5	1–5
	ultramafics	< 0.015	< 0.05	variable
Metamorphic	eclogites	0.3–3	0.2–0.5	2–4.3
	granulites	4.9	21	4.3
	gneiss	2.0	5–27	1–30
	schist	2.5	7.5–19	⩾ 3
	phyllite	1.9	5.5	2.8
	slate	2.7	7.5	2.8
Sedimentary	orthoquartzite	0.45–3.2	1.5–9	1.6–3.8
	greywackes	0.5–2.0	1–7	~ 2
	shales: grey-green red-yellow	2–4	10–13	2.7–7
	black	3–1250	—	low
	bauxite	11.4	49	~ 5
	limestones	~ 2	0–2.4	< 1
	dolomites	0.03–2	—	—
	phosphates	50–300	1–5	< 0.1
	evaporites	< 0.1	< 1	—
	speleothem	< 0.03–100	0–10	—
	living molluscs	< 0.01–0.5	low	—
	fossil molluscs	0.5–8	low	—
	coral	2–4	low	—
	Mn nodules	2–8	10–130	~ 7
	oceanic sands and clays	0.7–4	1–30	0.4–10
	peat	1–12	1–5	⩽ 1
	lignite	< 50–80	—	—
	coal	< 10– < 6000	—	—
	asphalt	10–3760	—	—
	oil	4–77	—	—

(From: Rogers and Adams 1969; Kaufman *et al.* 1971; Harmon *et al.* 1975; Kunzendorf and Friedrich 1976).

sparingly soluble in water, often associated with partial hydrolysis to monomeric and polymeric hydroxide species (for example, $PbOH^+$).

2.1.3 *Influence of weathering*

In the oxidized zone of the terrestrial near-surface environment, U and Th may both be mobilized, but in different ways. Thorium is almost entirely transported bound in insoluble resistate minerals or is adsorbed on the surface of clay minerals. By contrast, U may either move in solution as a complex ion, or like Th, in a detrital, resistate phase. Both elements occur in the $4+$ oxidation state in primary igneous rocks and minerals, but U, unlike Th, can be oxidized to $5+$ and $6+$ states in the near-surface environment. The $6+$ oxidation state is the most stable and forms soluble uranyl complex ions (UO_2^{2+}) which play the most important role in U transport during weathering.

In a closed system, daughter nuclides of Th and U are present in concentrations determined by the concentration of parent U or Th isotope and the time since the system became closed to nuclide migration. Because of the relatively short half-lives of these daughters, their concentration is several orders of magnitude less than that of their parent. If a system has been closed for a time which is long relative to the half-life of the parent nuclide, (less than 10^7 y), the activities of all daughter nuclides should be equal to the activity of their respective parent; this state is known as radioactive secular equilibrium (see Chapter 1). However, most surface and near-surface geological environments are subject to migration of nuclides due either to physical or chemical processes. For instance, Rn gas is an intermediate product in all three decay chains and may diffuse out of a deposit if the structure is sufficiently porous. Radioactive disequilibrium is, therefore, induced between the nuclides above and below in the decay series. Similarly, loss of an intermediate nuclide due to its greater solubility in water (for example, Ra) may also be a cause of such disequilibrium. Many daughters of Th and U, however, are too short-lived to become appreciably fractionated from their immediate parents; for instance, if diffusion of ^{222}Rn is negligible ($t_{\frac{1}{2}} = 3.82$ d), ^{210}Pb is likely to be in equilibrium with its progenitor ^{226}Ra because the other four intermediates all have half-lives of only seconds or minutes. Also, longer-lived intermediate nuclides are found to fractionate from their parents in the near-surface environment, and may concentrate in geological or hydrological bodies. Because they are unsupported, their activity decays away exponentially and its value at any time after formation may be used to determine the age of the body. An example of this is the use of the progressive decrease of ^{230}Th, ^{231}Pa, and ^{226}Ra activities downward in deep sea sediment cores for determination of sedimentation rate. Similarly, deposition of a parent in absence of its daughter, or nuclide migration from a deposit, will leave a system daughter-deficient with respect to its parent. The age of the deposit, or time elapsed since migration, may then be determined from the activity of ingrown daughter (see Chapter 3).

The remainder of this chapter will describe the aqueous geochemistry of Th and U, and examine the various processes by which these elements may be fractionated from each other and from daughter nuclides in the near-surface environment.

2.2 Sources of actinides

2.2.1 Igneous rocks

The upper part of Table 2.2 summarizes the Th and U contents of igneous rock types based on data from several different continental and oceanic island sites (from Rogers and Adams 1969). Uranium and Th are highly concentrated in continental igneous rocks and, in particular, silicic rocks such as granite are enriched in U and Th by up to two orders of magnitude over oceanic basalts. The very low concentrations found in ultramafic rocks further indicate that both elements are strongly fractionated into early-melting, silica-rich phases. Little information is presently available on the partitioning of U and Th between crystals and silicate melt, although data for pyroxenes at 1100 to 1400 ° C and 10 to 25 Kbar (summarized by Irving 1978) show that both U and Th are strongly partitioned into the liquid phase (partition coefficients are typically much less than 0.01).

Although most crustal rocks in Table 2.2 show a Th/U ratio of 3 to 4, analyses of primitive basalts, which best approximate undifferentiated mantle material, show a lower Th/U ratio, in the range 1 to 2 (Tatsumoto, Hedge, and Engel 1965). This suggests that some enrichment of Th over U does occur during crustal evolution.

Uranium and Th are distributed in an igneous rock in three ways: (i) by direct cation substitution in the silicate lattice of the major rock-forming minerals; (ii) as minor or major components of accessory minerals such as apatite, zircon, sphene and monazite; and (iii) by adsorption in lattice defects or onto crystal and grain boundaries. The importance of the latter has been demonstrated by dilute acid leaching of crushed granite samples in which between 15 to 35 per cent of the total U for seven different granites was leached by dilute acid (Szalay and Samsoni 1969). Fission track mapping of U distribution in many types of igneous rocks has also indicated that U is present adsorbed onto grain boundaries, and often concentrated in interstitial secondary minerals (Fleischer, Price, and Walker 1975). The normal ranges of U and Th contents in various minerals are given in Table 6.2. Other than the enriched accessory and resistate minerals, Th is evenly distributed throughout the rock whereas in felsic rocks, U is concentrated in the micas and amphiboles. Uranium is distributed more evenly through the minerals of mafic rocks.

2.2.2 Sedimentary and metasedimentary rocks

Most clastic sediments (sandstones, greywackes, and red/green/grey shales) contain U levels in the range 0.5 to 4 ppm whilst organic-rich 'black' shales and marine phosphates contain from 3 to 1200 ppm U (see Table 2.2). The extremely low solubility of Th in natural waters generally precludes the concentration of Th in sediments by low-temperature chemical processes. Most clastic deposits, therefore, contain Th in concentrations similar to the source rock. High Th concentrations are found in some sands, largely due to mechanical sorting processes which concentrate minerals according to their density (for example, monazite).

Limestones typically contain about 2 ppm U but little or no Th (see Table 2.2). The close correlation of Th concentration and detrital content (Adams and Weaver 1958) indicates that the Th is largely associated with clays and the heavy mineral fraction of the limestone. Dolomites also contain little Th but they generally contain less U than limestones, probably because of U loss during the dolomitization process.

The abundance of Th and U in metamorphic rocks (see Table 2.2) depends mainly on initial rock composition and the effects of migration of nuclides during metamorphism. Although not conclusively demonstrated, it is probable that high grade metamorphic rocks are depleted in U and Th relative to low grade rocks, due to the fluid loss with increased metamorphism and increased mobility of these elements at high pressure and temperature (Heier and Adams 1965).

2.2.3 Recent sediments

Concentrations of U and Th in recent sediments are summarized in Table 2.2. Uranium is found to be strongly enriched in certain organic sediments, particularly those formed from humic substances such as peat, lignite, and coal (Breger and Deul 1956, Szalay 1958). Humic substances* are particularly important in adsorption of the cations of U and Th from water. Organic deposits formed from bituminous and sapropelic materials (hydrocarbons such as resins, algae, spores, and lipids) contain little U (Vine, Swanson, and Bell 1958). Fixation of U as uranyl humate or fulvate by cation exchange is thought to be the concentrating process in humic substances, and this often occurs after death of the plant. Once complexed by organics the uranyl ion may be subsequently reduced to the uranous state because of a decrease in oxidation potential (Eh) associated with organic matter on burial. Alternatively, reduction may be brought about by H_2S produced by

---

* Humic and fulvic acids are the breakdown products of the cellulose of vascular plants. They are mainly composed of a polyaromatic molecular skeleton to which are linked functional groups such as $-COOH$ and $-OH$. These groups give them their acidic character and ion exchange properties. Humic acids are defined as the alkaline-soluble portion of organic material (humus) which precipitates from solution at low pH and are generally of high molecular weight. Fulvic acids are the alkaline-soluble portion which remains in solution at low pH and is of lower molecular weight. The residual insoluble organic material is known as humin.

decomposition of the organics themselves (Langmuir 1978):

$$4(UO_2(CO_3)_3{}^{4-}) + HS^- + 15H^+ = 4UO_2 + SO_4{}^{2-} + 12CO_2 + 8H_2O$$
(2.1)

Doi, Hironi, and Sakamaki (1975) have found U to be concentrated by limonite, clay minerals, zeolite, calcite, and apatite within the pH range 4 to 8. Comparison between naturally-occurring deposits and laboratory experiments led these authors to suggest that the higher U concentrations of geologic deposits were obtained by repetition of the leaching–reduction–precipitation cycle, rather than one single depositional event.

The Th content of recent authigenic sediments (see Table 2.2) such as peat, lignite, and carbonate deposits (muds, coral, limestones, oölites) is generally low. One exception is marine manganese nodules in which Th is enriched with respect to U (see Table 2.2, Kunzendorf and Friedrich 1976).

2.2.4 The hydrosphere

The solubility of uraninite in distilled water is very low (less than 0.01 ppb between pH 2 and 7, Langmuir 1978). However, when oxidized to the uranyl ion, and in the presence of complexing ions, its solubility increases by several orders of magnitude. Uranium contents of seawaters and freshwaters have been tabulated by Rogers and Adams (1969). Seawater averages about 3.3 ppb (Ku, Knauss, and Mathieu 1974). Freshwaters are more variable depending mainly on factors such as contact time with the U-bearing horizon, U content of the horizon, amount of evaporation and availability of complexing ions. Surface waters typically contain 0.01 to 5 ppb whilst groundwaters are somewhat more enriched: 0.1 to 50 ppb; and up to 500 ppb in mineralized areas. Lopatkina (1964) derived an empirical relationship between U in solution (S) and in the bedrock (R), and total dissolved solids in the water (TDS), from a study of over 1000 waters in the USSR:

$$U_S \text{(ppb)} = 0.002 \times U_R \text{(ppm)} \times TDS \text{(ppm)}$$
(2.2)

Large variations of U concentration within the same aquifer have been observed by Cowart and Osmond (1974) and are interpreted as due to Eh–pH changes which cause precipitation of U from solution along the flow direction (see chapter 9).

The Th content of natural waters is extremely low. Various workers have determined the [232]Th content of sea waters to be from 0.0002 to 0.1 ppb (in Miyake, Sugimura, and Yasujima 1970). In a careful evaluation of analyses of 24 samples of ocean water Kaufman (1969) finds the [232]Th content to be less than 8×10^{-6} ppb.

2.2.5 Ore deposits

In spite of their ubiquitous nature as trace elements in continental rocks, U and Th are sometimes found concentrated in a variety of geological environments.

Much of the information of U and Th geochemistry has come from the study of these ore deposits and brief mention is made here of their nature and origin. They are considered in greater detail in chapter 14.

The principal economic sources of U are the uranous oxides, uraninite (UO_2), and pitchblende (UO_2 to U_3O_8), and the silicate, coffinite ($USiO_4$), which are thought to have been formed by precipitation from hydrothermal solutions or ground waters in a reducing environment. Uranyl ores such as the vanadates (carnotite and tyuyamunite), phosphates (the autunite group), and the silicate (uranophane) are generally formed by chemical alteration of the oxide ores although primary deposits of carnotite and autunite have been recently reported (Dall'Aglio, Gragnani, and Locardi 1974). Another important U ore is thucholite, a Th-U-C-H complex often found in black shales, coals and asphalts (Breger and Deul 1956). Although its mechanism of formation is unclear, it appears to involve U adsorption from solution possibly coupled with polymerization induced by radioactivity.

Economic sources of Th are the silicates, thorite ($ThSiO_4$) and thorogummite ($Th(SiO_4)_{1-x}(OH)_{4x}$), and the oxide, thorianite (ThO_2). However, the rare earth orthophosphate, monazite ((Ce, Y, La, Th) PO_4)) is the most important economic source because of its frequent occurence as a placer deposit in detrital sands.

2.3 Geochemical cycles

2.3.1 Mobilization and transport in solution

In the $4+$ oxidation state both U and Th are almost chemically immobile in the near-surface environment at low temperatures. Uranium, however, may be mobilized by oxidation to the $6+$ state (Langmuir 1978):

$$U^{4+} + 2H_2O = UO_2^{2+} + 4H^+ + 2e^-, E° = 0.27 \text{ V}. \tag{2.3}$$

Further complexing of the uranyl ion may then occur depending on pH and presence of other ions.

To determine the speciation of U in natural waters, Hostetler and Garrels (1962) used available free energy data to calculate equilibrium constants for several U species within environmental Eh–pH conditions and thereby define stability fields for the dominant species. They used solubility data for four U minerals forming seven uranyl species. Langmuir (1978) has re-evaluated and extended Hostetler and Garrel's results by using data for 30 U compounds in the formation of 42 species.

The only uranous ($4+$) species found to have appreciable solubility in natural waters are fluoride and hydroxyl complexes at low Eh conditions (up to 1 ppb U for 0.2 ppm total F^- at pH $= 2$; Langmuir 1978, amended by Tripathi 1979). Langmuir's investigation has indicated that the species UO_2^+

(where U is present in the 5 + state) has a greater stability field than previously thought. It may become an important species in reduced waters of less than pH 7.

Uranyl complexes are far more soluble than uranous species. The dominating species present in a water will depend on the Eh–pH conditions, the concentration and availability of complexing ions and the temperature. Figure 2.1 shows the uranyl hydroxyl complexes present in pure water at 25 °C for total $U = 2.4$ ppb (10^{-8} M). It can be seen that in the absence of other complexing ions UO_2^{2+}, UO_2OH^+, and $(UO_2)_3(OH)_5^+$ are the important uranyl species. However, their significance diminishes in the presence of low concentrations of CO_2 and above $pH = 5$ (see Fig. 2.2) such that for $p_{CO_2} = 10^{-2}$ atm at 25 °C various uranyl carbonate complexes are the dominant species.

In addition to carbonate, soluble complexes are also formed with phosphate, sulphate, fluoride, and silicate ions. Figure 2.3 shows the main uranyl complexes that may exist in a groundwater containing small concentrations of these ions. It can be seen that the most important uranyl complexes are the free uranyl ion below $pH = 3$, fluoride between $pH = 3$ to 4, phosphate between $pH = 4$ to 7.5, and the carbonate complexes above $pH = 7.5$. Silicate complexes form at about $pH = 6$ but are insignificant in natural waters. Sulphates are only formed in low concentrations below $pH = 4$.

The importance of the carbonate complexes $UO_2(CO_3)_3^{4-}$ and

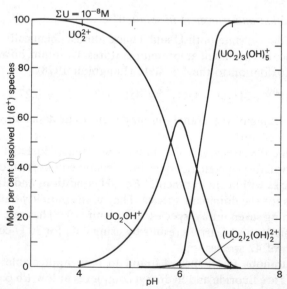

Fig. 2.1. Distribution of uranyl-hydroxy complexes plotted against pH for a total U concentration of 10^{-8} M (2.38 ppb), in pure water at 25 °C (from Langmuir 1978).

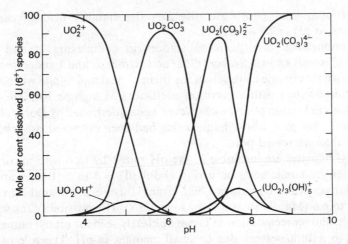

Fig. 2.2. Distribution of uranyl-hydroxy and carbonate complexes plotted against pH for $p_{CO_2} = 10^{-2}$ atm, and a total U concentration of 10^{-8} M (2.38 ppb) at 25 °C (from Langmuir 1978).

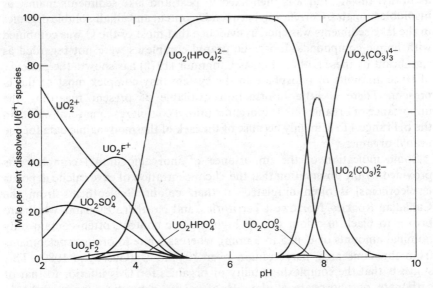

Fig. 2.3. Distribution of uranyl complexes plotted against pH at 25 °C for a total U concentration of 10^{-8} M, in the presence of other ions: $\Sigma F = 0.3$ ppm, $\Sigma Cl = 10$ ppm, $\Sigma SO_4 = 100$ ppm, $\Sigma PO_4 = 0.1$ ppm, and $\Sigma SiO_2 = 30$ ppm (from Tripathi 1979).

$UO_2(CO_3)_2^{2-}$ has been recognized for some time (Hostetler and Garrels 1962) with the dicarbonate species being the most stable over the pH range 4 to 10 depending on the p_{CO_2}. As can be seen from Fig. 2.3, however, the presence of only 0.1 ppm total phosphate displaces the stability field of carbonate complexes to pH = 7.5 or greater.

The importance of organic substances in complexing U and in ore formation is well known. Szalay (1958) and Armands and Landergren (1960) have shown that humic materials in the form of peat and lignite will adsorb U from groundwaters with a partition coefficient as high as 10^4. The role of humic acids in U adsorption has, however, been questioned by Doi et al. (1975) who found that peat, whose humic acids had been extracted, adsorbed even more U than untreated peat.

Uranyl humates are insoluble in the pH range 2.2 to 6 with maximum U adsorption by humic acids occurring about pH = 4 to 5 (Breger and Deul 1956, Manskaya and Drozdova 1968). Uranyl fulvates are insoluble in the pH range 6 to 6.6 (Manskaya and Drozdova 1968). The possibility for solution, transport, and reprecipitation of U can be clearly seen for uranyl humates and fulvates in natural waters, due to small changes in pH. The role of humic substances in mobilizing and depositing U has been examined by Halbach, Von Borstel, and Gunderman (1980) for a granite area overlain by peat. Uranium in the weathered granite was found to be removed in solution mainly as uranyl fulvate, but was then fixed in peat and lake sediments mainly as insoluble humates. Correlation between U content and alkali-soluble organics in the lake sediments was cited as evidence that most of the U was combined with humic compounds. Inorganic uranyl complexes were not regarded as significant by these authors but, as Langmuir (1978) has shown, the presence of trace amounts of phosphate in the system may complex most of the U present. There is little information available at present regarding the importance of organic versus inorganic uranyl complexes in natural waters in the pH range 4 to 8, mainly because of the lack of thermodynamic data for the uranyl organics.

Some indication of the importance of inorganic uranyl complexes is provided by the observation that the U concentration of cave calcite deposits (speleothems) is often unrelated to their colour. Speleothems from the Canadian Rockies, Northwest Territories, and Northern England, which are brown to black in colour due to high organic content, often contain only nominal amounts of U (0.5 to 5 ppm), whereas white to orange speleothems from the same area, may contain over 10 ppm U (Gascoyne 1980). This suggests that the complexing ability of organics for U is inferior to that of carbonate or phosphate at the pH of calcite deposition (approximately, pH = 7 to 8).

2.3.2 Mobilization and transport in particulate matter

The movement of U and Th as particulate matter is controlled only by the physical properties and flow velocity of the transporting medium, except where chemical interaction occurs across the phase boundary. Because of its extremely low solubility in most natural waters, Th is almost entirely transported in particulate matter. Even when Th is generated in solution by radioactive decay of U it rapidly hydrolyses and adsorbs on to the nearest solid surface. For this reason, the upper layers of deep sea sediments are found to contain high activities of unsupported ^{230}Th due to adsorption of the insoluble nuclide onto descending particles, whereas the ^{230}Th/^{234}U activity ratio of ocean water is less than 0.01 (Ku 1976). The ocean water origin of this excess ^{230}Th is further demonstrated by the high activity ratios of ^{230}Th/^{232}Th (up to 158) of recent pelagic clays in the South Pacific Ocean (Holland and Kulp 1954). The decay product, ^{231}Pa (from ^{235}U), is relatively insoluble and it is also found in excess of its parent in recent sediments. Excess ^{226}Ra in sediments is mainly due to the excess of parent ^{230}Th, and upward migration of the nuclide in the sediment column rather than its adsorption on particulates in seawater.

Not all the U present in igneous rocks is taken into solution during the weathering process. Some is tightly bound into accessory minerals which are very resistant to chemical attack. For instance, zircon may contain up to 6000 ppm U (see Table 6.2) little of which is lost during weathering, transportation, and deposition. The suitability of zircon grains for U/Pb dating also indicates the tightness of the system with respect to migration of the more mobile daughters Ra, Rn, Po, Bi, and Pb.

Particulate matter, therefore, assumes little importance in the study of U and Th geochemistry in the near-surface environment except where migration from or adsorption onto the particulate surface occurs. The importance of insoluble humic matter in adsorbing U before deposition as a lake sediment has been described by Halbach et al. (1980) (see above).

2.3.3 Mobilization and transport in the gaseous phase

Two products of radioactive decay in the U and Th series are the gases, Rn and He. Three Rn isotopes are naturally occuring, the longest-lived and most abundant being ^{222}Rn ($t_{\frac{1}{2}} = 3.8$ d). Helium is formed during alpha decay, when the alpha particle loses its positive charge. Loss of He from a mineral will not cause disequilibrium between members of the decay chain, whereas loss of Rn will. However if He is able to diffuse out of the lattice, then Rn may also be lost, but more slowly because of the differences in atomic size.

Radon has an appreciable solubility in water (about 0.5 g/l at STP) and is often found in concentrations far in excess of parent ^{226}Ra. Mazor (1962) has analysed many brines, hot springs, and groundwaters in Israel and finds waters containing high levels of excess Rn (up to 0.02 μCi/l) but with ^{226}Ra contents often below detection level. This accumulation of unsupported Rn is due to dissolution of the gas produced by decay of ^{226}Ra present in the bedrock. Mazor (1962) associates extreme Rn excesses found in waters of the Jordan

Rift Valley with leakage to groundwater flow systems of gas trapped above brine and oil reservoirs.

Radon activities up to 608 pCi/l with a mean $^{222}Rn/^{226}Ra$ activity ratio of 450 have been observed for groundwaters from central England (Andrews and Lee 1979). These authors infer that because of the short half-life of ^{222}Rn, the groundwaters must have been in contact with the Rn source less than about 25 days before measurement at the surface. Mazor (1962) suggests instead that the Rn measured may only be a small fraction of that initially present, implying that a high partial pressure of the gas exists in reservoirs at depth. Andrews and Lee (1979) ascribe the variability of Rn content in groundwater not to time spent underground but to the U content of the rock, its porosity, and grain size, and whether intergranular or fissure flow predominates. High U content, small grain sizes, and high primary porosity lead to high Rn concentrations.

Helium is less soluble in water than Rn (0.0017 g/l at STP) and its solubility

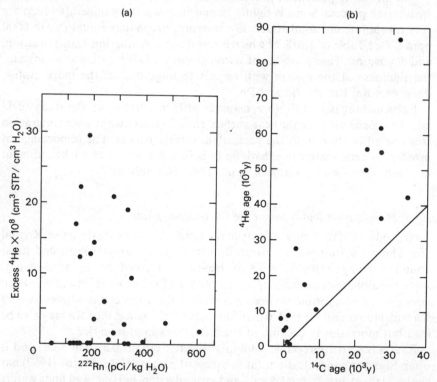

Fig. 2.4. (a) Correlation of excess ^4He with ^{222}Rn concentration in groundwaters from central England (from Andrews and Lee 1979). The ^4He excess is the amount of dissolved ^4He minus the amount dissolved from the atmosphere during recharge. (b) Comparison of ages calculated from excess ^4He and ^{14}C contents of groundwater from central England (after Andrews and Lee 1979).

is almost invariant with temperature over range 0 to 70 °C. Atom for atom, He is generated at a far greater rate because of the frequency of alpha particle decay in all the three decay series. Furthermore, He is stable and will accumulate in groundwater until its maximum solubility is reached. The independent nature of the factors controlling He and Rn content in groundwater is clearly shown by the lack of correlation between ^{222}Rn and ^{4}He concentrations in the groundwaters of central England (see Fig. 2.4a, taken from Andrews and Lee 1979). Helium content may, therefore, be used to determine the age of a groundwater if it is first corrected for He dissolved from the atmosphere before entering the bedrock, and if the U and Th content of the rock is known. Andrews and Lee (1979) compared ^{14}C ages with those determined from a model describing the rate of production of He in the groundwater system using mean Th and U contents of the enclosing rock. They found He ages to be approximately twice ^{14}C ages (see Fig. 2.4b), and ascribed this to inadequate knowledge of rock Th and U contents and He contribution from the upper and lower confining beds of the aquifer.

2.4 Depositional processes in the near-surface environment

2.4.1 Inorganic and biogenic precipitation

If U is present in solution as the carbonate species, reduction of the partial pressure of CO_2 by degassing to the atmosphere will cause the U to precipitate, usually in the form of a coprecipitate with other minerals. For instance, calcite or aragonite deposition in fresh water generally occurs by CO_2 loss from groundwater and U is coprecipitated along with other dissolved trace elements (Mg, Sr, Ra, Ac, etc.). In seawater, carbonate mineral precipitation is almost always a biogenic process in which U is incorporated into the skeleton of an organism via an organic substrate (see Gustavsson and Högberg 1973). Uranium concentrations vary widely between marine animals. Live mollusc shells typically contain an order of magnitude less U than corals (see Table 2.2). After death, however, molluscs absorb U from their surroundings, often attaining concentrations greater than 10 ppm U, whilst aragonite corals remain essentially closed to U or Th migration (Kaufman, Broecker, Ku, and Thurber 1971). Large variations in U concentration in sections of individual living corals and molluscs have been observed by fission track analysis (Schroeder, Miller, and Friedman 1970). High concentrations are thought to be indicative of low growth rates and vice versa.

The U/Ca ratio in corals and oölites is similar to that in ocean water (8×10^{-6}) indicating that the partition coefficient of U between aragonite and seawater is about 1.0. Analyses of groundwater seepages in limestone caves, and their associated speleothem deposits suggest that for calcite, the partition coefficient is somewhat larger, between 2 and 10 (Gascoyne, unpublished results).

Several other ions will coprecipitate U from solution. In oxidizing conditions, and in the presence of potassium and vanadium ions (VO^{2+}, etc.), the uranyl mineral, carnotite, will precipitate at neutral pH, particularly where p_{CO_2} is low (Hostetler and Garrels 1962; Langmuir 1978). Similar conditions, with V absent and phosphate present, will cause precipitation of U as autunite. Only with very high dissolved silica and low phosphate will the uranyl silicate, uranophane, precipitate. At more acidic or alkaline conditions U becomes more soluble either as UO_2^{2+} or as an anion complex, and so precipitation will not occur without decreasing the oxidation potential of the system.

Langmuir (1978) has shown that the reduction of uranyl ions to uraninite can be accomplished by any of the concurrent oxidation reactions of HS^- to SO_4^{2-} (eqn 2.1), FeS_2 to Fe^{2+}, Fe^{2+} to $Fe(OH)_3$, and CH_4 to CO_2 within the region of naturally-occuring Eh conditions (-0.35 to -0.2 V). Hydroxy-apatite and carbonate fluorapatite will form in the marine environment under reducing conditions and concentrate U to levels between 50 and 300 ppm. Various lines of evidence suggest that U^{4+} directly replaces Ca in the apatite structure after initially being adsorbed onto its surface (McKelvey, Everhart, and Garrels 1956). In low phosphate organic-rich reducing environments, U will adsorb onto humic material as described in section 2.3.1. Uranium will also precipitate from reducing solutions which are high in dissolved silica (20 to 120 ppm SiO_2) to form coffinite (Langmuir 1978). At lower SiO_2 concentrations, uraninite, (UO_2), alone will form.

Because all the above processes involve U in solution at some stage, the deposited U is generally formed in absence of other actinides and its more insoluble radioactive daughters (especially Th and Pa). The ability of U to undergo inorganic dissolution and reprecipitation is probably the most important process in the natural environment to cause disequilibrium between radionuclides.

2.4.2 Adsorption

The more insoluble U-series nuclides are removed from solution by adsorption onto the surface of detrital particulates (inorganic and organic), insoluble hydroxides and oxides, and clay minerals. These nuclides form by radioactive decay of U in seawater and either form hydrolyzates or adsorb directly onto particulates before descending to the ocean floor. The excess of ^{230}Th and ^{231}Pa compared to U in the upper layers of deep sea sediments is well documented (Ku 1976). Radium excess in sediments (and ocean water), over parents ^{230}Th and ^{231}Pa is also well known but is due instead to its greater solubility and ability to migrate upwards through the sediments and into the ocean water (Kröll 1953). In freshwaters and anoxic ocean bottom waters, U will also be removed from solution by reduction to the less soluble U^{4+} species. Deposition of U with organic materials occurs either directly, by surface adsorption of U^{4+}, or indirectly, by precipitation of insoluble uranyl compounds (humates, fulvates, etc.) followed eventually by reduction to U^{4+}.

2.4.3 Sedimentation

In lakes, estuaries, and continental shelf environments, Th and U are present in unweathered detrital minerals transported from continental areas by rivers and tidal currents. Mechanical sorting of sediments, particularly in tropical areas where the rate of erosion is high, often leads to the concentration of heavy resistates such as monazite and zircon in beach sands. Coastal sands of Florida, Brazil, and India are known to contain up to 5 per cent monazite (Bell 1954).

In deeper waters, more distant from fluvial inputs, terrigenous clays are diluted with biogenic sediments such as coccoliths, foraminiferal tests, diatoms, and radiolaria. These deposits are depleted in detrital Th nuclides but will contain U-series nuclides in amounts determined by their concentration in seawater, biological fractionation mechanisms, and rate of descent through the water column after death. An enormous amount of literature has accumulated on measurements of concentrations of radionuclides in marine sediments and this is described in detail in chapter 16.

Another form of sedimentation of U-series nuclides is precipitation from the atmosphere, as in the case of ^{210}Pb, derived from ^{222}Rn. In the nearshore environment ^{210}Pb has additional inputs from rivers and decay of ^{226}Ra in seawater. However being largely insoluble, it is rapidly scavenged by descending particulates such that its residence time in coastal regions may be less than one month if biological activity is high (Bruland, Koide, and Goldberg 1974).

2.5 Influence of ageing and diagenesis on isotopic disequilibrium

2.5.1 Fractionation of U isotopes

The parents of the two U decay series, ^{238}U and ^{235}U, are chemically equivalent and, with one important exception, their isotopic abundance ratio in naturally-occuring materials has not been found to deviate from $^{238}U/^{235}U$ $= 137.5 \pm 0.5$ (Rogers and Adams 1969). The exception is that of the Oklo uraninite deposit at Gabon, West Africa where considerably lower ^{235}U contents have been ascribed to a natural fission reaction occuring at a time when ^{235}U was about 3 per cent abundant (some 1.8×10^9 y ago, Lancelot et al. 1975).

Normally the fractionation of heavy nuclides in chemical processes is negligible, particularly in the natural environment. However, Cherdyntsev (1955) observed large fractionations of the isotopes ^{238}U and ^{234}U in rocks and their leach solutions. Subsequently, Thurber (1962) found that even ocean water contained a small excess of ^{234}U. Since then, disequilibrium between ^{238}U and ^{234}U in natural waters and sediments has been found to be the rule rather than the exception, and ^{234}U/^{238}U activity ratios in waters have been observed to vary from 0.5 to 20 (Osmond and Cowart 1976a). Typical ranges of ^{234}U/^{238}U ratio and U concentrations of meteoric waters are shown in Fig. 2.5.

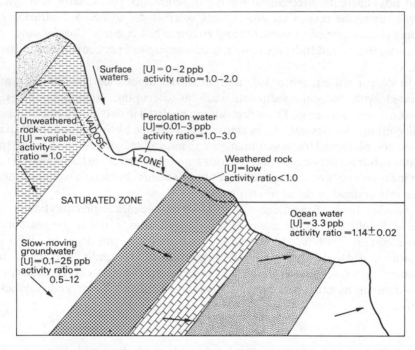

Fig. 2.5. Typical ranges of variation of ^{234}U/^{238}U activity ratios and U content of waters in the hydrological cycle.

Several reasons for this disequilibrium have been cited and some have been substantiated by laboratory experiments.

(i) *Leaching from radiation-damaged sites.* Cherdyntsev (1955) first proposed the mechanism of damaged-site leaching (also known as the 'hot atom' or 'Szilard–Chalmers' effect) to account for small excesses of ^{234}U in river water. Radioactive decay of ^{238}U to ^{234}U involves one alpha and two beta particle emissions, the ejection of which has the effect of damaging the crystal lattice around the parent ^{238}U atom. The atom of ^{234}U produced is often dislocated by recoil from its original site and, as a result, is more accessible and easier to remove by corrosive fluids (rainwater, groundwater, hydrothermal solutions, and others). Laboratory experiments using weak acids and ion-exchange solutions successfully leached ^{234}U preferentially from U-bearing rocks (Baranov, Surkov, and Vilinski 1958) thus supporting the lattice damage theory.

(ii) *Oxidation.* An additional effect thought to enhance the preferential leaching of ^{234}U from a damaged site is oxidation during decay. Either by removal of orbital electrons in the emission of alpha and beta particles, or by change in energy levels between original and dislocated sites, there may be an increase in oxidation state, from U^{4+} to U^{6+}. The uranyl ion so produced is more soluble than its uranous predecessor. A higher proportion of ^{234}U in U^{6+} than in U^{4+} leachates of a mineral was cited as evidence to substantiate this theory, but in fact, oxidation after decay to ^{234}U, may occur via normal environmental processes, and this would be indistinguishable from the proposals above (discussed in Osmond and Cowart 1976a).Numerous other leaching experiments have been performed on ores, igneous rocks, and sediments and these generally show preferential fractionation of ^{234}U into the leachate.

(iii) *Alpha recoil into solution.* As described above, the crystal lattice may be damaged by radioactive decay and recoil of the parent atom. In certain cases, the ejection of an alpha particle into the body of a mineral by a ^{238}U atom situated near the mineral's surface, may recoil the daughter nuclide (^{234}Th) into the medium surrounding the grain. Subsequent decay to ^{234}U would then increase the $^{234}U/^{238}U$ ratio of the groundwater. The reality of this process was demonstrated by Kigoshi (1971) who showed that there was a time-related increase of ^{234}Th in water mixed with powdered zircon, in agreement with the increase calculated from decay constants, U content, and alpha-recoil range in zircon. This mechanism was proposed by Kronfeld (1974) and Cowart and Osmond (1977) to explain the high activity ratios found in some deep-circulation groundwaters in south-west U.S. The enrichment of ^{234}U by the alpha-recoil mechanism is countered by the fact that ^{234}Th has a mean lifetime of 35 days and is likely to hydrolyse and adsorb onto a solid surface before decaying to ^{234}U. Enrichment of ^{234}Th has in fact been observed in the uppermost layers of sediments in near-shore environments (Bhat, Krishnaswami, Lal, Rama, and Moore 1969). However, ^{234}Th soon decays to ^{234}U which would then be in a favourable locus for dissolution.

2.5.2 *Natural variations of $^{234}U/^{238}U$ activity ratio*

In their study of U disequilibrium in groundwaters, Osmond and Cowart (1976a) found a general inverse correlation between $^{234}U/^{238}U$ ratio and U concentration such that, for a given area, higher activity ratios were often associated with lower U concentrations. They suggested that, as a ground-water becomes more removed from the recharge zone, increased reducing conditions are encountered and loss of U from solution to the aquifer walls occurs. This is accompanied by alpha-recoil of ^{234}Th into solution, thereby increasing the $^{234}U/^{238}U$ ratio. Osmond and Cowart developed a classification of groundwaters based on activity ratios and U concentrations, and they proposed that areas of U mineralization might be identified in this way (Cowart and Osmond 1977) (see chapter 9).

Sediments have also been investigated in detail, particularly with respect to the effect on $^{234}U/^{238}U$ ratio caused by reduction–oxidation reactions, and U mobilization and precipitation. River sediments are characteristically deficient in ^{234}U (activity ratio = 0.94, Osmond and Cowart 1976a) as are unoxidized sedimentary ore bodies (Rosholt, Butler, Garner, and Shields 1965). After oxidation, however, these ores may contain excess ^{234}U due to local groundwater movement causing ^{234}U migration and reprecipitation as a secondary mineral.

Excess ^{234}U in ocean water has been ascribed to input by river water (average $^{234}U/^{238}U$ is approximately 1.20), and to diffusion of ^{234}U from deep sea sediments (about 0.3 $dpm/cm^2/10^3$ y, Ku 1965). The importance of variations of $^{234}U/^{238}U$ in natural waters is further considered in chapters 7 to 9.

2.5.3 Fractionation of other actinides and their daughters

(i) *The Th isotopes*. The isotopes ^{238}U and ^{234}U are fractionated in natural processes because of alpha recoil effects and differences in bonding and oxidation state. Similar processes probably apply to the fractionation of Th isotopes, but any observed disequilibrium is better explained by the chemical properties and long half-lives of intermediate nuclides (for example, the formation of ^{228}Th from ^{232}Th involves the intermediate ^{228}Ra $(t_{\frac{1}{2}} = 5.75$ y)). High activities of unsupported ^{228}Th are found in ocean waters ($^{228}Th/^{232}Th$ ratios range from 5 to 30) whereas bottom sediments older than ten years or so are found to have activity ratios less than unity due to ^{228}Ra migration and decay of inital excess of this isotope (Koide, Bruland, and Goldberg 1973; Ku 1976). Lake sediments deposited within the last decade, however, have $^{228}Th/^{232}Th$ ratios greater than unity because of the presence of unsupported ^{228}Th scavenged from the water column by rapidly settling particulates. The excess ^{228}Th activity reduces to zero within 10 years.

It is difficult to make an analogy to U disequilibrium mechanisms when attempting to account for disequilibrium between Th isotopes because: (i) Th is generally insoluble in seawater and is rapidly removed on particulates, and (ii) upward diffusion of Ra from bottom sediments forms the main source of excess ^{228}Th in the water column.

The activity of ^{230}Th in seawater is also negligible; most ^{230}Th is rapidly removed by adsorption onto sediments. It is present as an unsupported nuclide in surface layers of deep sea sediments in concentrations largely determined by the ^{234}U content of seawater and rate of sedimentation, and its activity decreases exponentially with depth. The ^{234}Th activity in ocean water is higher than that of all other Th isotopes because of its production in solution by decay of ^{238}U and short lifetime before further decay to ^{234}U. In shallow waters with high sedimentation rates ^{234}Th may be removed rapidly by particulates and the decay of unsupported ^{234}Th in the uppermost layers of sediment can then be used to examine processes of bioturbation and diagenesis which take place within periods of days or weeks (Bhat *et al.* 1969; Aller and Cochran 1976).

Generally for both seawater and freshwaters, the total activity of Th isotopes decreases in the order ^{234}Th $\gg$ ^{228}Th $>$ ^{230}Th $\geqslant$ ^{232}Th.

(ii) *The Pa isotopes.* Only one isotope of Pa, ^{231}Pa, has any importance in disequilibrium processes because the other naturally-occuring isotope, ^{234}Pa, is too short-lived ($t_{\frac{1}{2}} = 1.18$ min) to migrate appreciably before decaying. The ^{231}Pa is rapidly removed from ocean water before decaying to ^{227}Ac because of its tendency, like Th, to hydrolyse and adsorb onto sinking particulates. It accumulates in marine sediments, and its activity decreases exponentially with depth in the sediment. The ^{231}Pa/^{230}Th activity ratio of recent pelagic sediments can be predicted from the U isotope activities in ocean water (0.091), assuming complete and rapid removal of both isotopes from seawater. This was substantiated by measurements in early studies of deep sea core dating (see Ku 1976). However, many lower values have recently been found indicating either ^{230}Th excess or ^{231}Pa deficiency. Manganese nodules have since been found to have ^{231}Pa/^{230}Th ratios greater than 0.091, therefore, suggesting a chemical fractionation process is occuring to incorporate preferentially ^{231}Pa in manganese nodules, and ^{230}Th in pelagic sediments.

(iii) *The Ra isotopes.* Two of the four naturally occuring isotopes of Ra are useful in disequilibrium studies. The ^{226}Ra, ($t_{\frac{1}{2}} = 1602$ y), the daughter of ^{230}Th, is generally found in excess of its parent in most natural waters due to the greater solubility of Ra over Th and the resulting diffusion into the water column from sediments. It is scavenged mainly by siliceous particulates (phytoplankton) in the water column but redissolves at depth. Initial work suggested that the contribution of Ra to the oceans from near-shore sediments was negligible, but flux calculations now suggest that at least one-quarter of ^{226}Ra in ocean water is derived from these sediments (Li, Ku, Mathieu, and Wolgemuth 1973; Li, Mathieu, Biscaye, and Simpson 1977). In freshwaters, Ra is found in highest concentrations in limestone regions where it is more soluble in HCO_3^- waters (greater than 5 dpm/l of ^{226}Ra, Barker and Johnson 1964). In soils and sediments ^{226}Ra is generally in excess of ^{230}Th and greatly in excess of ^{238}U due either to deposition by ion exchange onto clays and organics from percolating soil waters or to preferential leaching of ^{238}U from the soil or precursor rock (Titayeva, Taskayev, Orchenkov, Aleksakhin, and Shuktomova 1977). Inorganic freshwater carbonates and molluscs have been investigated in dating studies by Kaufman and Broecker (1965a). Excess and deficiency of ^{226}Ra in several of the samples was found, and ^{230}Th/^{234}U and ^{14}C ages of these specimens were often in disagreement showing that either ^{230}Th or ^{226}Ra migration had occured since deposition. Although ideally suited for study, little ^{226}Ra analysis of cave travertine (speleothem) has been done so far.

The ^{228}Ra (from ^{232}Th) has a shorter half-life (5.75 y), but, like ^{226}Ra, it is also found in excess of its parent in natural waters. Again, this excess is attributed mainly to upward diffusion from sediments (Moore 1969b).

The other Ra isotopes (^{224}Ra and ^{223}Ra) are fairly short-lived (periods of days) and are unlikely to migrate sufficiently to cause disequilibrium.

(iv) *The Rn isotopes.* The mobility of Rn isotopes is regarded as one of the main causes of disequilibrium in the U and Th decay series. Gaseous diffusion or dissolution in water are the chief modes of migration, as described earlier in this chapter. The Rn concentrations and isotopic content are used extensively in U ore prospecting methods but relatively few studies of dissolved Rn in ocean water have been made so far. Because of the solubility and short half-lives of Rn isotopes, they are regarded as conservative at all depths in ocean water, and are, therefore, in equilibrium with parent Ra (in contrast to the near-surface continental situation).

(v) *The Pb isotopes.* The isotope ^{210}Pb ($t_{\frac{1}{2}} = 22.6$ y) is the only naturally-occuring radioactive Pb isotope that is stable for periods longer than minutes or hours. The atmospheric flux of ^{210}Pb (from the decay of ^{222}Rn) has been estimated at 15 atoms/min/cm^2 (Rama, Koide, and Goldberg 1961) and this has been substantiated by aerosol activity measurements (Santschi, Li, and Bell 1979). In the oceans and near-shore environments, decay of ^{226}Ra in solution, provides another source of ^{210}Pb. Except for surface ocean waters, dissolved ^{210}Pb is found to be deficient in the entire water column (Craig, Krishnaswami, and Somayajulu 1973; Somayajulu and Craig 1976). Its activity ranges from 15 to 80 percent of ^{226}Ra activity and generally decreases with depth. This indicates that ^{210}Pb is rapidly removed by adsorption onto sinking particulates and may coprecipitate with ferric hydroxide and manganese oxides or form the insoluble lead sulphide in anoxic waters (Bacon, Brewer, Spencer, Murray, and Goddard 1980). Estimates of residence time of ^{210}Pb in ocean water, therefore, vary widely, depending on location (less than one month near land, to over 50 y in the deep ocean.

(vi) *The Po isotopes.* Of the six naturally-occuring Po isotopes, only ^{210}Po is sufficiently long-lived ($t_{\frac{1}{2}} = 138$ d) to appreciably influence the state of equilibrium in the ^{238}U decay series. In the hydrological cycle ^{210}Po generally follows its precursor ^{210}Pb, but it is often found out of radioactive equilibrium with ^{210}Pb in the marine environment. The ^{210}Po is more readily adsorbed than ^{210}Pb onto particulate matter (mainly phytoplankton) in the surface layer of the oceans. However, because of differing chemical properties at greater depths, Po is easily recycled, whereas Pb tends to remain with the sinking particulates (Bacon, Spencer, and Brewer 1976). Additional sources of ^{210}Po have been proposed to account for the overall Po excess at greater depth in ocean water, including coastal waters that are enriched in ^{210}Po and upwelling of deep water (with re-solution of ^{210}Po from particulates).

3 URANIUM SERIES DISEQUILIBRIA APPLICATIONS IN GEOCHRONOLOGY
M. Ivanovich

3.1 Radiometric age determination

The principle aim of geochronology is to classify the earth's history in terms of a sequence of events of known duration (for an excellent review see Fitch, Foster, and Miller 1974). The main available matrix of evidence is contained within the terrestrial rocks. Precise physical measurements, however, capable of providing an estimate of the absolute age of specific rocks and minerals became possible only with the discovery of radioactive decay. Virtually all quantitative determinations of geological age are based on the phenomenon of radioactivity. Thus, determination of absolute age is based on the fact that a given radionuclide decays at a uniform rate, and forms a *geological clock*. In principle, any radionuclide can be used as a radioactive clock provided its half-life, which determines its decay rate, is of chronologically useful duration (see Holmes and Holmes 1978).

A summary of the radiometric dating methods is given in Table 3.1. The listed methods may be subdivided into three basic groups. The first group involves decrease in concentration of a radionuclide from an initial level, or a build up of a stable daughter product (U/Pb, Th/Pb, K/Ar, Rb/Sr, I/Xe, Re/Os, ^{4}He/U, ^{14}C, ^{10}Be, ^{26}Al, ^{32}Si, ^{36}Cl, and ^{41}Ca dating methods). The second group is based on the measurement of the degree of restoration of radioactive equilibrium in a radioactive decay series following an initial external perturbation (U-series disequilibrium dating methods). The third group involves the integration of a local radioactive process in a sample and subsequent comparison with the local flux due to the presence of radionuclides remaining approximately constant in time (thermoluminescence for example). Clearly, age determination made using methods of the first and second groups depend on acccurate measurements of concentration of parent nuclides and their daughter products, while those based upon the third category depend on exploitation of a reliable integrating process. In all cases, certain assumptions must be made about initial conditions, whether the system remains closed, and some other aspects of the subsequent geological history. Often, the only way to

Table 3.1. Summary of radiometric dating methods currently in use in Geochronology and Archaeology

Method	Description	$t_{\frac{1}{2}}(y)$	Effective dating range	Datable materials	References
U/Pb Th/Pb	$^{238}U/^{206}Pb$ $^{235}U/^{207}Pb$ $^{232}Th/^{208}Pb$ $(^{207}Pb/^{206}Pb)$.	4.5×10^9 0.71×10^9 13.9×10^9	From 10^7 y to the age of the earth	zircon, uraninite, pitchblende, monozite, some whole rocks, lava flows, sedimentary rocks, intrusive igneous rocks, metamorphic rocks	Catanzero and Kulp (1964); Tilton (1951); Tilton, Patterson, and Davis (1954); Patterson (1951); Marshall and Hess (1960); Hamilton (1965); Catanzero (1968); Wetherill (1956); York and Farquhar (1972); Ludwig (1980); Ludwig, Lindsey, Zielinski, and Simmons (1980); Ewing (1979); Wasserburg and Steiger (1967).
K/Ar	(1) total degassing (2) $^{40}K/^{39}Ar$ age spectra analysis	1.31×10^9	From 10^5 y to the age of the earth	any K-bearing mineral or rock, primarily applicable to volcanic rocks. (tuff or pumice)	Dalrymple and Lanphere (1969); Curtis (1975); Miller (1972); Fitch Hooker, and Miller (1976); Hurford, Gleadow, and Naeser (1976).
Rb/Sr	$^{87}Rb/^{87}Sr$ (1) mineral ages (2) isochron analysis	47.0×19^9 48.5×10^9 (or 50.0×10^9)	From 10^7 y to the age of the earth	Rb-rich minerals or rocks such as muscovite, biotite, microcline, granite, gneiss, etc.	Zartman (1964); Lambert (1971); Pankhurst (1970); Schreiner (1958); Compston and Jeffery (1959); Nicolaysen (1961).
I/Xe	^{129}I decay to ^{129}Xe	16.4×10^5	up to 10^8 y	meteorites, lunar material, ocean sediments	Raynolds (1960); York and Farquhar (1972); Elmore, Gove, Ferraro, Kilius, Lee, Chang, Beukens, Litherland, Russo, Purser, Murrell, and Finkel (1980).

Table 3.1 (Contd)

Method	Description	$t_{\frac{1}{2}}$ (y)	Effective dating range	Datable materials	References
Re/Os	^{187}Re decays to ^{187}Os (Re does not occur in nature, method limited)	43×10^6	up to 10^8 y	Re occurs mainly in molybdenite and the rare earth minerals. Method can date Mo mineralization, meteorites	Hamilton (1965); Herr, Wolff Eberhardt, and Kopp (1967)
^{10}Be	positron decay and accelerator	1.5×10^6	up to 10^7 y	deep sea sediments	Amin, Lal, and Somayajulu (1975); Inoue and Tanaka (1979).
^{14}C	^{14}C beta-decay	5730	$0-4 \times 10^4$ y	wood, charcoal, peat, grain, tissue, charred bone, cloth, shells, tufa, groundwater	Aitken (1974) and many others.
^{26}Al	positron decay and accelerator	7.3×10^5	up to 3×10^6 y	sediments	Wasson, Adler, and Oeschger (1967); Raisbeck, Yion, and Stephen (1979).
^{32}Si	beta-decay	~ 100	up to 10^3 y	sileceous sediments, ocean sediments (sedimentation rates); groundwaters	Kharkar, Turekian, and Scott (1969); Somayajulu, Lal, and Craig (1973); Elmore, Anoutareman, Fulbright, Gove, Hans, Nishizumi, Murrell, and Honda (1980); Kutschera, Henning, Paul, Smither, Stephenson, Yutima, Alberger, Cumming, and Harbottle (1980).

Method	Process	Half-life	Effective range	Materials/applications	References
^{36}Cl	beta-decay	3.1×10^5	up to 10^6 y	solution geochemistry of Cl argillaceous sediments, evaporites, speleothems, geological event dating (volcanic, glacial erratic, earthquakes), groundwater closed systems	Davis and Schaeffer (1955); Schaeffer, Thompson, and Lark (1960); Elmore, Fulton, Clover, Marsden, Gove, Naylor, Purser, Kilius, Beukens, and Litherland (1979); Nishizumi, Arnold, Elmore, Ferraro, Gove, Finkel, Beukens, Chang, and Kilius (1979).
^{41}Ca	beta decay high energy mass spectroscopy	1.3×10^5	5×10^5 y	dating exposure of formerly buried rock; date of burial of formerly exposed surface, dating of Ca deposits; bone dating	Raisbeck and Yion (1979).
U-series disequilibria	^{230}Th/^{234}U ^{231}Pa/^{235}U etc.	7.5×10^4 3.4×10^4	3.5×10^5 y 2×10^5 y	corals, speleothems, bones, teeth, deep sea cores, shells, volcanic rocks, etc.	Cherdyntsev (1971); Ku (1976); Schwarcz (1980); many others (see this chapter).
^{4}He/U			effective range 0–10^5 y	corals, groundwater, fossils	Fanale and Schaeffer (1965); Thurber, Broecker Blanchard, and Portraz (1965); Schaeffer (1967); Broecker and Bender (1972); Marine (1979); Andrews and Lee (1979).
Fission track ^{238}U	spontaneous fission rate	10^{-16}	from 6 months to the age of the earth	glass, apatite, sphene zircon, epidote, allanite, hornblende, garnet, pyroxene, feldspar, mica, etc.	Fleischer (1975).
Thermoluminescence	integration of alpha, beta, and gamma radiation			pottery, flint, hearths	Aitken (1978)

test these assumptions is by employing an alternative dating method based on unrelated assumptions. Several criteria common to most types of radiometric clocks must be satisfied before a particular method can be considered a geologically useful chronometer. These are: (1) decay constants (or half-lives) of the radionuclides must be known accurately; (2) the analysed sample must be representative of the geological system for which an age is desired; (3) in some cases, the intermediary decay products and final daughter product either must have not been present when the system was formed, or amounts initially present must be taken into consideration in the age calculation; and (4) no gain or loss of the parent nuclide or daughter product can have occured from the system since the time of formation. Radiometric clocks have dating limits which depend on the half-lives and the relative quantities of the radionuclides present in the geological system. Table 3.1 also contains a list of review references for different dating methods should the reader be interested in a more detailed discussion of a particular method. An excellent short general review is given by Hedges (1979). The remainder of this chapter will consider only U-series disequilibria dating methods.

3.2 The U-Series disequilibria dating methods

Mass spectrometric techniques have been instrumental in an early development of dating methods using long lived U-series radionuclides (see Table 3.1). However, because of the long half-lives of the present nuclides, their applicable range is for samples older than a million years. Thus, the dating methods for Quaternary deposits remained unavailable until relatively recently. During the last 25 years much effort has been made to develop the shorter term U-series methods for application to the fields of archaeology, Quaternary geology, and geomorphology. By dating deep-sea carbonates, ice cores, estuarine and lacustrine sediments, recent volcanic events, fossil organic remains, and combining these with paleoclimatic inferences, a detailed picture of the timing and intensity of late Pleistocene climatic events (ice ages) has been constructed. However, very few radiometric techniques are as yet capable of dating events in the 300 to 1000 ky period.

The U-series dating methods are based on the measurement of the activity of U and its various daughter nuclides. In any naturally occurring material which contains U and which has remained undisturbed for several million years, a state of secular equilibrium between the parent and the daughter nuclide in the radioactive series will have been established. However, when a sedimentary deposit is formed, for example, various geochemical processes occur which cause isotopic and elemental fractionation initiating a state of disequilibrium between parent and daughter nuclides (see Chapter 2). If no diagenetic changes or other migratory mechanisms occur after the initial formation of the deposit in this state of disequilibrium it is possible, in principle, to determine the time of the original event by measuring the extent to which the radionuclide system has returned to the state of secular equilibrium.

The U-series dating methods encompass a spectrum of techniques involving many different nuclides and these will be discussed in some detail in the following sections. These methods can be divided into two distinct groups: (i) methods based on accumulation of decay products of U (or daughter-deficiency methods); and (ii) methods based on the decay of unsupported intermediate nuclides in the series (or daughter-excess methods). In the case of the first group, a parent nuclide may be deposited free of its daughters or a daughter deficiency of known extent may be established, so that at some subsequent time, the age of the deposit can be determined from the extent of growth of the daughter(s) into secular equilibrium with its (their) parent(s). Examples of this group of dating methods are the ^{230}Th/^{234}U and ^{231}Pa/^{235}U dating of carbonates. In the case of the second group, the daughter nuclide is initially present in excess of its parent, and the sample is dated by measuring the decay of this excess since the formation of the sample. Examples of this latter group are the ^{230}Th- and ^{231}Pa-excess dating of deep sea sediments. All the U-series disequilibria related dating methods discussed in the following sections are listed in Table 3.2 together with the relevant radionuclide half-lives, dating range, and applicability.

3.3 Dating methods based on accumulation of U-daughter products

Historically, the first dating method based on the accumulation of decay products of U in which U-series disequilibrium was employed, was the dating of U minerals by the accumulation of Ra. Khlapin (1926) showed that the Ra accumulation is controlled by the accumulation of its parent ^{230}Th according to the relationship:

$$^{226}\text{Ra}/^{238}\text{U} = 1 - \frac{\lambda_{226}}{\lambda_{226} - \lambda_{230}} e^{-\lambda_{230}t} + \frac{\lambda_{230}}{\lambda_{226} - \lambda_{230}} e^{-\lambda_{226}t}. \qquad (3.1)$$

The author also used a simpler form of the ^{230}Th growth equation

$$^{230}\text{Th}/^{238}\text{U} = 1 - e^{-\lambda_{230}t}. \qquad (3.2)$$

In both eqns (3.1) and (3.2), it is assumed that ^{234}U is in secular equilibrium with its parent ^{238}U (i.e. the ^{234}U/^{238}U activity ratio is unity). However, Cherdyntsev (1955) discovered that the ^{234}U/^{238}U activity ratio in most terrestrial waters is not unity, and that the ^{234}U anomaly occurs commonly in the natural materials rendering eqn (3.2) invalid. Thus, Broecker (1963),

Table 3.2. A summary of U-series disequilibria dating methods

Dating Method	Nuclide	$t_{\frac{1}{2}}$	Dating range	Application
^{230}Th/^{234}U (accumulation of ^{230}Th)	^{230}Th	7.52×10^4 y	$\leqslant 350$ ky	marine and terrestrial carbonates (fossil corals, shells, speleothems, bones, travertines), volcanic rocks
^{231}Pa/^{235}U (accumulation of ^{231}Pa)	^{231}Pa	3.43×10^4 y	$\leqslant 150$ ky	as above
^{231}Pa/^{230}Th (accumulation of ^{231}Pa and ^{230}Th)	^{231}Pa, ^{230}Th		$\leqslant 200$ ky	
^{226}Ra/^{238}U (accumulation of ^{226}Ra)	^{226}Ra	1.602×10^3 y	$\leqslant 10$ ky	limited application as above, mostly as a check for closed-system
^{234}U/^{238}U (^{234}U-excess decay)	^{234}U	2.48×10^5 y	$\leqslant 1250$ ky	limited, some fossil corals and some waters
^{230}Th-excess decay	^{230}Th	7.52×10^4 y	$\leqslant 300$ ky	deep sea sedimentation rates, Mn nodule formation rate

Method	Half-life	Range	Application
^{231}Pa-excess decay	3.43×10^4 y	$\leqslant 150$ ky	as above
^{230}Th/^{232}Th (^{230}Th-excess decay)	7.52×10^4 y	$\leqslant 300$ ky	deep sea sedimentation rates.
^{231}Pa/^{230}Th (excess decay)	^{231}Pa 3.43×10^4 y ^{230}Th 7.52×10^4 y average $t_{\frac{1}{2}}$ 5.2×10^4 y	$\leqslant 150$ ky	deep sea sedimentation rates
^{234}Th-excess decay	24.1 d	100 d	rapid sedimentation rates in shallow waters, particle reworking and diagenesis study
^{228}Th/^{232}Th (^{228}Th-excess decay)	1.913 y	0.01 ky	shallow water sedimentation rate (as below)
^{210}Pb-excess decay	22.3 y	0.1 ky	sedimentation rates in lakes, estuaries, and coastal marine environment, geochemical tracer, settling rates
He/U (He accumulation)	all alpha emitting nuclides of U-series	1000 ky	fossil corals, groundwaters.

Kaufman (1964), and Cherdyntsev (1971) quote more complex expressions of the type:

$$^{226}Ra/^{238}U = \frac{^{230}Th}{^{238}U} - \frac{\lambda_{230}}{\lambda_{226} - \lambda_{230}} (e^{-\lambda_{230}t} - e^{-\lambda_{226}t}) - \left(\frac{^{234}U}{^{238}U} - 1\right)$$

$$\times \left[\frac{\lambda_{230}\lambda_{230}e^{-\lambda_{230}t}}{(\lambda_{230} - \lambda_{234})(\lambda_{226} - \lambda_{234})} - \frac{\lambda_{230}\lambda_{234}e^{-\lambda_{234}t}}{(\lambda_{230} - \lambda_{234})(\lambda_{226} - \lambda_{234})} \right.$$

$$\left. - \frac{\lambda_{226}\lambda_{230}e^{-\lambda_{226}t}}{(\lambda_{226} - \lambda_{234})(\lambda_{226} - \lambda_{230})} \right]. \tag{3.3}$$

As a dating tool this method is limited in range to only 10 ky. Furthermore, the implicit assumption that the sample started with zero ^{226}Ra content at the time of original U deposition is potentially invalid because in some respects the geochemistry of Ra is similar to that of U. Hence, the likelihood of Ra being present in the original groundwaters and being deposited in the samples is high.

3.3.1 The $^{230}Th/^{234}U$ dating method

If a sample contains no ^{230}Th at the time of formation, then at any later time, the $^{230}Th/^{234}U$ ratio is given by the relationship:

$$^{230}Th/^{234}U = \frac{1 - e^{-\lambda_{230}t}}{^{234}U/^{238}U}$$

$$+ \left(1 - \frac{1}{^{234}U/^{238}U}\right) \frac{\lambda_{230}}{\lambda_{230} - \lambda_{234}} (1 - e^{-(\lambda_{230} - \lambda_{234})t}). \tag{3.4}$$

A derivation of eqn (3.4) is given in Appendix A. Figure 3.1 is an isochron plot showing a graphical solution of eqn (3.4). It illustrates the relationship between $^{230}Th/^{234}U$ and $^{234}U/^{238}U$ activity ratios for closed systems of varying initial $^{234}U/^{238}U$ ratio. The methods used for the measurement of the activity ratios and the calculation of age from eqn (3.4) are discussed in detail in chapter 5. In practice, the dating range of this method is considered to be about 350 ky. As can be seen from Fig. 3.1, in the range of less than 3×10^4 y, the relation between the $^{230}Th/^{234}U$ activity ratio and age is virtually independent of the $^{234}U/^{238}U$ activity ratio in the sample. In this range eqn (3.4) is reduced to the form $^{230}Th/^{234}U = (1 - e^{-\lambda_{230}t})$ as a very good approximation.

3.3.2 The $^{231}Pa/^{235}U$ dating method

The $^{231}Pa/^{235}U$ dating method is based on the decay of ^{235}U to ^{231}Pa. The relation between ^{231}Pa and its grandparent ^{235}U is simple because the half-life

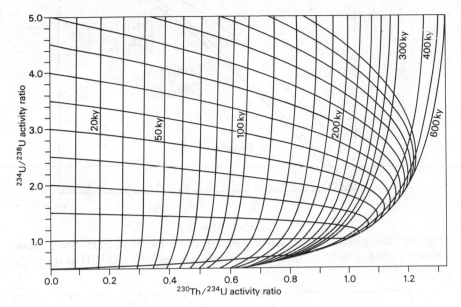

Fig. 3.1. Variation of $^{234}\text{U}/^{238}\text{U}$ and $^{230}\text{Th}/^{234}\text{U}$ activity ratios with time in a closed system with no initial ^{230}Th. The near vertical lines are isochrons (lines of constant age obtained from eqn (3.4) but different $^{234}\text{U}/^{238}\text{U}$ activity ratio). The near horizontal lines show change in the nuclide activity ratios as age increases for different initial $^{234}\text{U}/^{238}\text{U}$ activity ratios as indicated. (After H. P. Schwarcz 1979.)

of the intermediate nuclide ^{231}Th ($t_{\frac{1}{2}} = 25.6$ h) is short enough to consider it always in secular equilibrium with its parent. Thus, the age relationship is:

$$^{231}\text{Pa}/^{235}\text{U} = 1 - e^{-\lambda_{231}t}. \tag{3.5}$$

The ^{235}U activity is usually calculated from the ^{238}U activity because the $^{235}\text{U}/^{238}\text{U}$ activity ratio in nature is considered to be constant at $1/21.7$, and ^{231}Pa is usually measured indirectly by monitoring one of its two daughters, ^{227}Ac or ^{227}Th, and assuming secular equilibrium with ^{231}Pa (see chapter 5). The $^{231}\text{Pa}/^{235}\text{U}$ method has a range of up to 200 ky. Figure 3.2 shows the calculated growth curves of ^{230}Th and ^{231}Pa.

If, however eqns (3.4) and (3.5) are combined to calculate the $^{231}\text{Pa}/^{230}\text{Th}$ activity ratio the following expression is obtained:

$$^{231}\text{Pa}/^{230}\text{Th} = \frac{1 - e^{-\lambda_{231}t}}{21.7\left[\left(1 - e^{-\lambda_{230}t}\right) + \left(^{234}\text{U}/^{238}\text{U} - 1\right)\dfrac{\lambda_{230}}{\lambda_{230} - \lambda_{234}}(1 - e^{-(\lambda_{230} - \lambda_{234})t})\right]} \tag{3.6}$$

Figure 3.3 presents a graphical solution of eqn (3.6), clearly demonstrating that the age range of the age equation written in this form extends to about 300 ky.

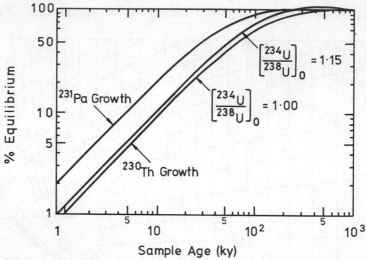

Fig. 3.2. Comparison of the theoretical growth curves for ^{230}Th and ^{231}Pa nuclides. Note the effect of the 15 per cent disequilibrium between ^{234}U and ^{238}U on the ^{230}Th growth curve.

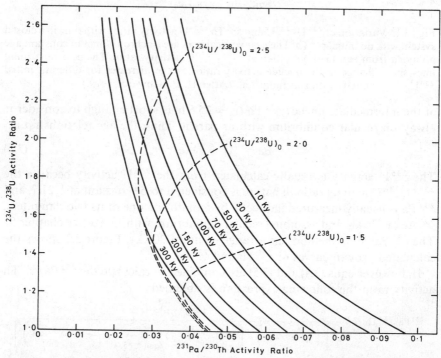

Fig. 3.3. Variation of ^{234}U/^{238}U and ^{231}Pa/^{230}Th activity ratios with time. The solid lines are isochrons representing different ages, as indicated in thousands of years, and the broken lines represent the change of the activity ratios with time for three different initial ^{234}U/^{238}U activity ratios as indicated.

3.3.3 The He/U dating method

Although the He/U dating method is not strictly a U-series disequilibrium dating technique, it represents an accumulation clock associated with the U-series with a dating range of at least 10^6 y, and therefore, it is included here. The potential of He derived from the alpha decay of nuclides in the radioactive series was recognized as a method for dating geological events when alpha particles were identified as He nuclei. However, it was soon realized He could frequently leak out of U and Th bearing minerals, thus creating uncertainty in ages based on the assumption of its total retention. Thus, in cases of age discrepancies between He/U and other dating methods it is usually assumed that the lower ages are due to the He loss by diffusion.

Fanale and Schaeffer (1965) and Schaeffer (1967) applied the method to dating Pleistocene and Tertiary fossil aragonites with some success, particularly in obtaining ages concordant with ^{230}Th/^{234}U ages for the same samples. Bender (1973) determined He/U ages of fossil corals and applied rigorous tests for the reliability of the method. Bender, Fairbanks, Taylor, Matthews, Goddard, and Broecker (1979) applied corrections to the method for the open-system conditions in dating Pleistocene reef tracts of Barbados. The age of a fossil several million years old can be calculated from its He/U ratio with the assumption that the He production rate is constant and equal to the alpha flux derived from the U content, and that the alpha-emitting daughters are in secular equilibrium with each other. For fossil ages less than 10^6 y, initial disequilibrium conditions must be taken into account. In this latter case the age can be calculated by summing the integrated He production from the time when the sample became closed to He loss (the radioactive clock was set) to the present time, from three separate He sources identified by Fanale and Schaeffer (1965) as: (i) the ^{234}U-excess decayed from its initial value to the ^{238}U-supported value; (ii) the contribution from ^{230}Th and its daughters as they grow into equilibrium with the parent ^{234}U; and (iii) the contribution from the constant ^{238}U content. The underlying assumption in this approach is that initially, the fossil is free of Th isotopes and remains so in the case of ^{232}Th. The He contribution from ^{235}U and its daughter is regarded as minor.

Andrews and Lee (1979) and Marine (1979) have applied the method to dating groundwaters and this is discussed in chapter 9 in more detail.

3.4 Dating methods based on the decay of excess U-daughter products

Natural fractionation of U-series daughter nuclides from their parents in the hydrological cycle leads to insoluble daughter nuclide excesses in the bottom sediments (see chapters 2, 15, and 16). This excess decays with time at a rate determined by the decay constant of the particular nuclide, providing a useful dating method if the initial amount of excess is known. Thus, the age or

accumulation rate of a deposit can be determined by a general decay equation
of the form:

$$C = C_0 e^{-\lambda t} \tag{3.7}$$

where λ is the decay constant of the unsupported daughter nuclide, C_0 the
initial excess at the time of deposition, and C, the excess measured at time, t.
Equation (3.7) is valid provided the parent nuclide activity is regarded as
constant (this is quite a reasonable assumption because the parent half-lives
are nearly always much longer than those of their daughters in excess). Clearly
the unknown quantity in eqn (3.7) is the value of C_0, the knowledge of which
depends on the particular environment of the sample.

3.4.1 The $^{234}U/^{238}U$ dating method

The decay of excess ^{234}U towards the secular equilibrium with ^{238}U is
described by eqn (A.7) which is derived from the law of radioactivity in
Appendix A:

$$[(^{234}U/^{238}U)_t - 1] = [(^{234}U/^{238}U)_0 - 1]e^{-\lambda_{234} t} \tag{3.8}$$

where $(^{234}U/^{238}U)_t$ and $(^{234}U/^{238}U)_0$ are the present and initial activity
ratios, respectively. It should be noted that eqn (3.8) can be derived from eqn
(3.7) if the present and initial ^{234}U-excess are set equal to C and C_0 in eqn (3.7),
respectively. In principle, this method should date Pleistocene samples up to at
least 10^6 y. The main problem of this dating method is the inherent difficulty
associated with establishing the value of the initial ^{234}U-excess (C_0). This is not
surprising considering the variability of the $^{234}U/^{238}U$ activity ratio in
terrestrial waters (see chapters 2, 8, and 9). In general, no reliable *a priori*
assumption about the initial $^{234}U/^{238}U$ ratio can be made, thus limiting the
applicability of the method (Cherdyntsev 1955; Cherdyntsev, Kazachevskii,
and Kuz'mina 1963; Thurber 1963; Chalov, Tuzova, and Musin 1964;
Thompson, Ford, and Schwarcz 1975; Osmond and Cowart 1976a). Some
notable exceptions do exist both in terrestrial and in marine environments. In
the former case, fresh water carbonate formed in caves (speleothems) can, in
some very favourable conditions, be dated once the constancy of $(^{234}U/^{238}U)_0$
ratio is established for a group of samples over the length of a specimen.
However, Harmon, Thompson, Schwarcz, and Ford (1978) have observed that
$^{234}U/^{238}U$ ratios often change after interruptions in the growth of the
speleothem rendering the method inapplicable. The situation appears to be
somewhat more favourable in the marine environment where the $^{234}U/^{238}U$
activity ratio is uniform in seawater (Ku, Knauss, and Mathieu 1974). Thus,
unaltered fossil corals have been dated successfully using the $^{234}U/^{238}U$
dating method (Thurber, Broecker, Blanchard, and Potratz 1965; Veeh 1966).

 Chalov, Tuzova, and Musin (1964); Chalov, Merkulova, and Tuzova
(1966a) and Chalov, Svetlichnaya, and Tuzova (1970a, b) estimated the age of

some major Kazakhstan lakes by a residence-time analysis, having observed that the $^{234}U/^{238}U$ activity ratios in the lake waters were lower than in the incoming rivers. These authors attributed the difference to the decay of excess ^{234}U during the U residence in the 'no-outflow' lakes. Thus, by treating the river water activity ratios as the $(^{234}U/^{238}U)_0$ values, Chalov et al. (1964) calculated the ages of the lakes from the present $^{234}U/^{238}U$ ratios of the lake water using the following equation (also see Komura and Sakanoue 1967):

$$[(^{234}U/^{238}U) - 1] = [(^{234}U/^{238}U)_0 - 1][1 - e^{-\lambda_{234}t}]/\lambda_{234}t \qquad (3.9)$$

As Ku (1976) observes, the underlying assumption of constant $^{234}U/^{238}U$ activity ratios in the feedwaters has had to be extrapolated over 2×10^5 y introducing an element of risk in such application, although the ages obtained by Chalov et al. (1964) were in agreement with those derived from chloride accumulation and other geological considerations.

3.4.2 The ^{230}Th-excess dating method

The ^{230}Th-excess dating method was first used by Piggot and Urry (1939, 1942) who measured ^{226}Ra in sediments as an index of ^{230}Th. The sedimentation rate, s (cm/unit time) at a depth in a core, x (cm) may be related to the age, t of the sediment by $s = x/t$. Then, substituting this expression for t in eqn (3.7) yields

$$\ln C = \ln C_0 - (\lambda_{230}/s)x. \qquad (3.10)$$

For a constant C_0 and s, a plot of $\ln C$ against x yields a straight line. Such relation validates the assumption of constant C_0 and s, while the slope of such a straight line is the measure of sedimentation rate, s. To obtain C in eqn (3.10) (i.e. the ^{230}Th-excess) the measured ^{234}U activity is subtracted from the measured ^{230}Th activity (i.e. ^{230}Th-excess $= ^{230}Th - ^{234}U$). Ku (1965) justifies the 'correction' on the grounds that U in a typical pelagic sediment is associated mostly with detrital clays in which only U-supported ^{230}Th is present. This correction becomes particularly significant in deeper strata where ^{230}Th-excess is low. However, Ku (1976) quotes many instances of non-linear plots yielded by eqn (3.10), and suggests that this could be caused by possible variation in initial ^{230}Th concentration due to the change in bulk sediment settling rate. The $^{230}Th/^{232}Th$ and $^{231}Pa/^{230}Th$ dating methods discussed later in this chapter are modifications to the ^{230}Th-excess and ^{231}Pa-excess methods designed to cope with this problem in the marine environment.

3.4.3 The ^{231}Pa-excess dating method

The ^{231}Pa-excess dating method is also based on the decay of unsupported ^{231}Pa in sediment cores and Mn nodules (see chapter 16). The method may be useful as a complementary technique to the ^{230}Th method, although both methods are subject to the same limitations already defined in section 3.4.2.

Sackett (1965) used a different approach in interpreting the ^{231}Pa data. This author assumed that the ^{231}Pa deposition rate was constant for a given area of the ocean floor. Then, by measuring the total ^{231}Pa-excess in the core, and the amount below some depth in the core, the age of that level can be calculated by a modified version of eqn (3.10)

$$t = -\ln(n/n_t)/\lambda_{231}, \qquad (3.11)$$

where n is ^{231}Pa-excess below the depth of age t, and n_t is the total ^{231}Pa-excess in the core. The method depends on the validity of the assumption that the ^{231}Pa deposition rate is constant thus yielding ages independent of variations in the bulk sedimentation rate. A similar approach was applied to ^{230}Th-excess dating method by Kominz, Heath, and Ku (1979) in the measurement of deep sea deposition rates, and by Ku, Omura, and Chen (1979) and Ku and Knauss (1979) in the measurement of Mn nodule growth rates.

3.4.4 The ^{230}Th/^{232}Th dating method

In an effort to minimize the limitations associated with the ^{230}Th-excess dating method, Picciotto and Wilgain (1954) suggested the use of the ^{230}Th/^{232}Th activity ratio for C and C_0 in eqn (3.7) instead of ^{230}Th-excess per total sediment. The reason for this substitution is that the ^{232}Th activity should provide the measure of initial ^{230}Th activity in a sediment because of the chemical identity of these nuclides. However, the method has been critisized on the grounds that although ^{230}Th and ^{232}Th are chemically similar, their geochemical histories in the marine environment differ a great deal (see Ku 1976). The former is produced in the ocean, whereas the latter is mostly an integral part of continental detritus carried to the ocean floor. Indeed, the depth profiles of ^{230}Th/^{232}Th are similar to those of ^{230}Th per non-carbonate. Both methods fail to improve the ^{230}Th per total sediment approach and, in fact, both appear to be vulnerable to fluctuations in carbonate content. However, Ku (1976) concludes that with cautious data interpretation, the ^{230}Th/^{232}Th derived results should be useable.

Isotopic fractionation also occurs between actinides during magmatic processes such that in unweathered Quaternary igneous rocks the ^{234}U/^{238}U ratio is often unity while ^{238}U and ^{235}U are in disequilibrium with their daughters ^{230}Th, ^{231}Pa, ^{226}Ra, and ^{210}Pb (Anestad-Fruth 1963; Somayajulu, Tatsumoto, Rosholt, and Knight 1966; Oversby and Gast 1968; Cherdyntsev, Kislitsina, Kuptsov, Kuz'mina and Zverev 1967; Cherdyntsev, Kuptsov, Kuz'mina, and Zverev 1968). These rocks have been dated using the ^{230}Th/^{238}U disequilibrium relationship (^{230}Th maybe either depleted or in excess with respect to ^{238}U). The basic assumption in this approach (first attempted by Cerrai, Dugnani Logati, Gazzarini, and Tongiorgi 1965) is that various mineralogical phases of solidifying magma inherit different Th/U ratios (hence $^{230}Th/^{238}U$) but identical $^{230}Th/^{232}Th$ and $^{234}U/^{238}U$ activity ratios. Therefore, if a mineral has acted as a closed system for a time t following crystallization, its $^{230}Th/^{232}Th$ activity ratio should be given by the following expression:

$$^{230}Th/^{232}Th = (^{230}Th_0/^{232}Th)e^{-\lambda_{230}t} + (^{238}U/^{232}Th)(1-e^{-\lambda_{230}t}) \quad (3.12)$$

where $^{230}Th_0$ is the initial ^{230}Th activity at the time of crystallization from magma. The two unknowns in eqn (3.12), $^{230}Th_0/^{232}Th$ and t can be determined uniquely for two or more minerals of common origin. This can be expressed analytically for two minerals A and B belonging to the same rock as follows (see Ku 1976):

$$t = \frac{1}{\lambda_{230}} \ln \left\{ 1 - [(^{230}Th/^{232}Th)_A - (^{230}Th/^{232}Th)_B] / \right.$$
$$\left. \times \frac{1}{[(^{238}U/^{232}Th)_A - (^{238}U/^{232}Th)_B]} \right\}. \quad (3.13)$$

Thus, the accuracy of the age measurement depends on the U/Th fractionation between the two minerals. An example of a graphic solution of eqn (3.13) is given in Fig. 6.10. Minerals of the same formation age in a rock form a straight line (isochron) with the slope $(1-e^{-\lambda_{230}t})$, and the $^{230}Th/^{232}Th$ axis intercept $(^{230}Th_0/^{232}Th)e^{-\lambda_{230}t}$. Those not plotted on a defined isochron are products of a different crystallization time and melt $(^{230}Th/^{232}Th)$ composition domain. (For further discussion of the application of this method to volcanic rocks see chapter 6.)

3.4.5 The $^{231}Pa/^{230}Th$ dating method

The initial idea for the $^{231}Pa/^{230}Th$ dating method derives independently from Sackett (1960) and Rosholt, Emiliani, Geiss, Koczy, and Wengersky (1961), and is based on the short residence time of these nuclides in sea water compared to their respective half-lives. As the abundance of their respective parent nuclides ^{238}U, ^{234}U, and ^{235}U, in the ocean is relatively constant (see chapters 15 and 16), and river contribution of unsupported ^{231}Pa and ^{230}Th is negligible (see chapter 8), the $^{231}Pa/^{230}Th$ ratio added to the sediment is expected to be close to their calculated production ratio of 1/11 (or 0.091) provided no fractionation of the two species occurs during precipitation. By anology with eqn (3.7), the age of a sediment with a measured ratio $(^{231}Pa$-excess$/^{230}Th$-excess$)_t$ is given by

$$(^{231}Pa\text{-excess}/^{230}Th\text{-excess})_t = (^{231}Pa\text{-excess}/^{230}Th\text{-excess})_0 e^{-(\lambda_{230}-\lambda_{231})t} \quad (3.14)$$

where $(^{231}Pa$-excess$/^{230}Th$-excess$)_0$ is the ratio of the excesses of the two nuclides in the freshly deposited sediment, having the calculated value of

0.091. The excess values of ^{231}Pa and ^{230}Th are given by ^{231}Pa-excess $=$ ^{231}Pa $- ^{235}$U, and ^{230}Th-excess $= ^{230}$Th $- ^{234}$U, respectively.

In principle, the $(^{231}$Pa-excess$/^{230}$Th-excess) ratio in eqn (3.14) is a function of t only, and thus, being independent of changing oceanic conditions, it should decrease with an average half-life of 6.2×10^4 y. This dating method should, therefore, be applicable to 1.5×10^5 y. However, many authors have documented substantial departures from the calculated value of 0.091 for $(^{231}$Pa-excess$/^{230}$Th-excess)$_0$ implying that the basic assumption of no fractionation between ^{231}Pa and ^{230}Th is invalid (see Ku 1976). On the other hand, since eqn (3.10) often yields compatible results for ^{231}Pa- and ^{230}Th-excess measured in deep sea sediments, $(^{231}$Pa-excess$/^{230}$Th-excess) ratios can be plotted either through eqn (3.10), thus yielding average sedimentation rates (Sackett 1964; Ku and Broecker 1967a; Broecker and Ku 1969), or through eqn (3.14) by substituting the value of 0.091 for the measured $(^{231}$Pa-excess$/^{230}$Th-excess)$_0$ ratio. The former approach is preferred (Ku 1976) because the initial ratio cannot be expected to be constant.

3.4.6 The ^{234}Th dating method

The ^{234}Th is formed in sea and groundwaters by decay of its immediate parent ^{238}U. In common with other Th isotopes, this nuclide is adsorbed rapidly on particulate matter after the ^{238}U decay (see Hussain and Krishnaswami 1980, for example). In the shallow marine environment ^{234}Th is found in excess in sediments in spite of its short half-life. Rather rapid sedimentation rates encountered in shallow water regimes can be obtained through the application of eqn (3.10) after substituting the ^{234}Th decay constant for λ_{230} in that equation. The importance of this nuclide has been established in the studies of particle reworkings and diagenesis in shallow water regimes (see chapter 16 for details).

3.4.7 The ^{228}Th/^{232}Th dating method

The ^{228}Th/^{232}Th activity ratio varies between 5 and 30 (averaging about 15) in surface and deep waters of the ocean (see references in Ku 1976 for example, and for more detail chapters 15 and 16). Thus, only a small proportion of the measured ^{228}Th in these waters derives from ^{232}Th in water, and confirms that the rest must derive from the ^{228}Ra input to the water from the bottom sediments (Koczy, Tomic, and Hecht 1957; Moore 1969b). The existence of disequilibrium between the two nuclides observed in lacustrine environment (for example, Koide et al. 1973; Murray 1975) suggests the possibility of dating modern sediment using eqn (3.7). Because of a relatively short ^{228}Th half-life of only 1.9 y, the range of the method only extends to approximately 10 y.

3.4.8 The ^{210}Pb dating method

The ^{210}Pb dating method has been applied to the measurement of sedimentation rates in lakes, estuaries, and coastal marine sediments. The isolation of ^{210}Pb ($t_{\frac{1}{2}} = 22.3$ y) is ascribed to its precursor ^{222}Rn which escapes from the earth's surface into the atmosphere. Most of ^{222}Rn remains in the troposphere where it decays to ^{210}Pb through a series of short-lived intermediaries. The ^{210}Pb residence time in the troposphere is estimated to range from days to a month before it is removed by precipitation and dry fallout (Francis, Chester, and Hoskin 1970; Peirson, Cambray, and Spicer 1966). This atmospheric flux of unsupported ^{210}Pb is presumed to have remained constant at a given locality, and so Goldberg (1963) first proposed to date permanent snow fields by the measured ^{210}Pb activity.

The age of different horizons in a core can be calculated using eqn (3.7) from the unsupported ^{210}Pb activity (^{210}Pb-excess = ^{210}Pb $- {}^{226}$Ra) assuming that ^{210}Pb is immobile in the core. Two different age calculations can be carried out depending on the choice of two assumptions: (i) constant initial concentration, or (ii) constant rate of supply.

The first method assumes that the ^{210}Pb-excess remains constant with time at a particular location. Then, the difference in age between the surface sediment and sediment at depth, n is given by

$$t_n = \ln \left[(^{210}\text{Pb-excess})_0 / (^{210}\text{Pb-excess})_n \right] \lambda_{210} \qquad (3.15)$$

where $(^{210}\text{Pb-excess})_0$ and $(^{210}\text{Pb-excess})_n$ are excess activities of ^{210}Pb at the two positions in the core in units of dpm/g of dry matter, and λ_{210} is the ^{210}Pb decay constant ($0.031\,08$ y^{-1}). When the accumulation rate is constant in time, then eqn (3.10) applies:

$$\ln (^{210}\text{Pb})_n = \ln (^{210}\text{Pb})_0 - (\lambda_{210}/R) M \qquad (3.16)$$

where M is mass depth in (g/cm^2) dry matter, and R is the accumulation rate of sediment in (g/cm^2)y^{-1} dry matter.

The second method assumes that the flux of ^{210}Pb-excess supply to the sediment is constant with time at a particular location. Then

$$\Sigma (^{210}\text{Pb-excess})_n = F \int_0^{t_n} e^{-\lambda_{210}t} \mathrm{d}t \qquad (3.17)$$

where $\Sigma (^{210}\text{Pb-excess})_n$ is the total excess activity of ^{210}Pb from surface to depth, n in units of (dpm/cm^2), and F is the flux of ^{210}Pb-excess in (dpm/cm^2)y^{-1}. The flux F, is calculated from eqn (3.17) for the total value of ^{210}Pb-excess in the core down to a deposition horizon much older than 100 y.

The most frequently used approach is the one based on the constant initial concentration (Krishnaswami, Lal, Martin, and Meybeck 1971; Robbins and Edgington 1975; Goldberg, Gamble, Griffin, and Koide 1977; Pennington,

Cambrey, Eakins, and Harkness 1976). Oldfield, Appleby, and Butterbee (1978) concluded, however, that when sedimentation rate is constant the two methods yield identical results.

3.5 Criteria for dating suitability

In principle, all radiometric dating methods must be based on a set of assumptions, and criteria for their validity must be formulated. In the case of U-series disequilibria dating methods, the choice of basic assumptions and criteria depends on the kind of sample being dated and the environment from which the samples are derived. There are, however, certain general criteria which apply to U-series dating methods. (1) There must be a measurable quantity of U in the sample (for practical reasons the minimum U content required is 0.05 ppm or an equivalent activity of a daughter nuclide). (2) In the case of dating methods based on accumulation of U daughter products, no daughter nuclide can be present in the sample at the time of deposition of the parent nuclide nor introduced into the sample during the lifetime, fossilization or diagenesis; otherwise appropriate corrections should be implemented to correct for the original presence of the daughter nuclides in the sample. Similarly, in the case of the dating methods based on the decay of daughter excess, the initial nuclide content must be known reliably. (3) The dated sample must be a closed system to post-depositional migration or addition of any radionuclides used for dating. This means radioactive decay and growth must be the only processes responsible for changes in isotopic abundances since formation of the system.

The criteria for ascertaining the extent to which the closed system assumption applies vary for different types of samples, and are discussed in detail for different materials by Gascoyne et al. (1978) (speleothems), Komura and Sakanoue (1967) (corals and shells), Thurber et al. (1965) (coral), Schwarcz (1980) (travertines), and many other authors. These criteria are listed below.
(1) The samples should be impermeable to sea or groundwaters because the flow of water through the system may cause alteration in U or Th content.
(2) There should be no evidence of weathering in the sample.
(3) Deposits known to have been initially composed of aragonite (for example, coral) should show no inversion to calcite, since U or its daughters may have been mobilized during the inversion and recrystallization.
(4) Samples partially replaced by secondary materials should not be dated.
(5) In the case of carbonate samples, there should be little (or no) HNO_3-insoluble residue .
(6) In the case of terrestrial carbonate, there should be no (or little) ^{232}Th present in the sample as this nuclide is usually indicative of the presence of detrital material in the sample.

(7) Apparent radiometric ages must be in agreement with stratigraphic sequence.

(8) Agreement should be obtained for ages determined by several independent methods such as $^{230}Th/^{234}U$, $^{231}Pa/^{235}U$, $^{231}Pa/^{230}Th$, He/U, and ^{14}C.

3.6 Correction techniques for detrital contamination

The presence of ^{232}Th in many marine and terrestrial authigenic carbonates implies a detrital component which indicates that a certain amount of non-radiogenic ^{230}Th is present in the sample. For example, in the case of travertines, Schwarcz (1980) recognized two possible means of correcting for this non-authigenic Th component: (i) an analysis of only HNO_3-soluble fraction, and (ii) analyses of both the soluble and insoluble fractions (either separately or combined). Both approaches assume that some Th was introduced at the time of deposition, incorporated in detrital component with a specific $^{230}Th/^{232}Th$ activity ratio and require multiple sample analysis in order to determine relative contribution of detrital and authigenic nuclide components by some statistical procedure. Thus, it is implied that ^{232}Th activity is a measure of the degree of contamination. For example, Kaufman and Broecker (1965a), observed an average value of 1.7 for the initial $^{230}Th/^{232}Th$ ratio in the detrital component of Lake Lahontan lacustrine sediments. By assuming that ^{232}Th was added initially, and that its associated ^{230}Th has decayed by an equivalent amount to the sample age (obtained by ^{14}C method), these authors obtained the correct radiogenic ^{230}Th component (^{230}Th radiogenic $= {}^{230}Th_{total} - 1.70\ {}^{232}Th\ e^{-\lambda_{230}t}$) which yielded the maximum age for the sediments.

Ku et al. (1979), in dating of caliche, assumed (i) negligible amounts of ^{230}Th and ^{232}Th are initially present in the carbonate phase, (ii) secular equilibrium in the detrital phase, and (iii) $^{230}Th/^{232}Th$ activity ratio in the HNO_3-insoluble residue is the same as that in the initial detritus. These authors analyse the two phases separately, and calculate the age from the corrected ^{230}Th, ^{234}U, and ^{238}U activities.

Schwarcz (1980) quotes a number of examples of the partial dissolution (or leaching) approach in lacustrine carbonates (Kaufman 1971), travertine deposits (Turekian and Nelson 1976), and suggests a general correction procedure for dating travertines (see chapter 12). The author remarks, however, that the great disadvantage of the partial dissolution method is that it does not take into account differences in the solubility of the various nuclides of Th and U due to the difference in diagenetic history, and suggests that this uncertainty can be eliminated by dissolving the HNO_3-insoluble component using other means, and taking its isotopic contribution into account.

Schwarcz (1979) suggests another approach to the analysis of totally dissolved samples by assuming that the samples are a mixture of two phases, the carbonate (soluble) phase in which all ^{230}Th is of *in situ* radiogenic origin (no ^{232}Th is present), and the detrital phase containing some ^{230}Th, ^{232}Th, ^{234}U, and ^{238}U nuclides with daughters not necessarily in secular equilibrium with their respective parents. Then, ^{232}Th is treated as an index of the proportion of detrital material present in the mixture. Thus, a set of linear equations is obtained such that the intercepts at ^{232}Th $= 0$ yield the activities of ^{230}Th, ^{234}U, and ^{238}U in the carbonate component:

$$\left. \begin{aligned} ^{238}U_s &= \left[(^{238}U_d - {}^{238}U_c)/^{232}Th_d \right]^{232}Th_s + {}^{238}U_c \\ ^{234}U_s &= \left[(^{234}U_d - {}^{234}U_c)/^{232}Th_d \right]^{232}Th_s + {}^{234}U_c \\ ^{230}Th_s &= \left[(^{230}Th_d - {}^{230}Th_c)/^{232}Th_d \right]^{232}Th_s + {}^{230}Th_c \end{aligned} \right\} \quad (3.18)$$

where subscripts c, d, and s denote carbonate, detrital, and whole sample activity, respectively.

The general question of the presence of a detrital component in deep sea cores and its effect on the interpretation of data in terms of age and sedimentation rate, has been discussed at length by Ku (1976) and authors quoted therein. It is apparent that an underlying theme of most of the dating methods based on decay of daughter-excess is the effort to minimize the error due to the presence of a detrital component which is essentially of continental, and not marine, origin.

3.7 Open system dating methods

One of the fundamental criteria for datability of any sample is that the sample remains closed to nuclide migration from the time of formation to the time of measurement. Thus, unrecrystallized aragonite corals and pure calcite speleothems are regarded as ideal closed systems (Sakanoue, Konishi, and Komura 1967; Ku 1968; James, Mountjoy, and Omura 1971; Ku, Kimmel, Easton, and O'Neil 1974; Konishi, Omura, and Nakamichi 1974; Szabo and Tracey 1977; and many others). In contrast, open-system examples can be found in marine shells (Rosholt 1967; Szabo and Rosholt 1969; Szabo and Vedder 1971; Kaufman *et al.* 1971), groundwater systems (see chapter 9), and a variety of sediment and soils (see chapter 10). In an effort to date such systems, several authors have reported their attempts to modify the decay equations and formalize disequilibria data mathemetically in terms of an open-system model.

One of these is an open system dating model proposed by Rosholt (1967) and Szabo and Rosholt (1969). The S–R open system dating model was initially proposed for dating fossil shells, and its basis is an assumed two-stage history. In the first stage, shortly after the death of the organism, the shells take up U from the seawater, and distribute it uniformly throughout the shell (see Szabo, Dooley, Taylor, and Rosholt 1970 for the evidence obtained by fission track method). This initially assimilated U and its daughters (formed *in situ*) are assumed by the model to remain in the shell. However, the subsequent weathering of the shell is expected to result in additional uptake of ^{230}Th, ^{231}Pa, and ^{234}U. The ultimate observable outcome of this second stage is an apparent increase in the activity ratios ^{234}U/^{238}U, ^{230}Th/^{234}U, and ^{231}Pa/^{235}U leading to discordance in the ^{230}Th/^{234}U and ^{231}Pa/^{235}U ages. By introducing three types of U, Th, and Pa activities, (i) a group representing the nuclides available to the sample from its environment, (ii) a group which contributes to daughter production but is not retained by the sample, and (iii) a group of measured quantities, the authors construct a set of equations describing the open system which lead to a modified version of eqn (3.4). Thus,

$$(^{230}\text{Th}/^{234}\text{U})_M = (^{231}\text{Pa}/^{234}\text{U})_M \left\{ \frac{(1 - e^{-\lambda_{230}t}) - 1.4352\,(1 - e^{-(\lambda_{230} - \lambda_{234})t}}{1 - e^{-\lambda_{231}t}} \right\}$$

$$+ 1.4352\,(1 - e^{-(\lambda_{230} - \lambda_{234})t} \left(\frac{(^{231}\text{Pa}/^{238}\text{U})_M}{1 - e^{-\lambda_{231}t}} - 1 \right)$$

$$\times \left[1 - (^{231}\text{Pa}/^{234}\text{U})_M \frac{1 - e^{-\lambda_{234}t}}{1 - e^{-\lambda_{231}t}} \right]$$

$$+ 1.4352\,(1 - e^{-(\lambda_{230} - \lambda_{234})t}), \tag{3.19}$$

where subscript M denotes measured activity ratios. Equation (3.19) can be solved for t (open system age) by the method of successive approximations.

Kaufman et al. (1971) critisized the model arguing that the open system ages could not be relied upon without comparison with another independent dating method. In addition, the isotopic migration pattern observed in many mollusc shells cannot be as uniform as the model requires. This latter argument was confirmed by Szabo and Vedder (1971). However, although Szabo (1979a) applied the model successfully to some unrecrystallized open system fossil corals, the author suggests that the measurement of the ^{230}Th/^{234}U and ^{231}Pa/^{235}U ages, and the analysis of several samples from each deposit are essential for the validation of the open system model.

Hille (1979) has proposed another open system dating model based on the assumptions that all ^{238}U daughters remain chemically stable within the system, and that chemically less stable U, having originally migrated into the system shortly after the death of the organism, can change with time according to an exponential law such that

$$^{234,\,235,\,238}\text{U}(t) = {}^{234,\,235,\,238}\text{U}(0)e^{-at} \tag{3.20}$$

where a (greater than zero) is a 'chemical decay constant', a model parameter to be deduced from the experimental data. The first assumption required by Hille's model is justified for ^{230}Th and ^{231}Pa nuclides because of the known slow migration of these nuclides into shells and bones (Szabo, Malde, and Irwin-Williams 1969), but not so for ^{234}U because of its well documented mobility. To obtain the age of a sample, a set of decay equations incorporating the chemical decay constant for U isotopes as well as measured activities of ^{238}U, ^{234}U, ^{230}Th, ^{231}Pa, and ^{235}U is set up and solved by iteration for age t. A graphical solution of the age equations is obtained by plotting $^{230}Th/^{234}U$ against $^{231}Pa/^{235}U$. The model yields isochrons for constant t as well as isolines for discrete values of parameter a. The model has been applied successfully to dating of marine terraces in Southern California with some data of Szabo and Rosholt (1969). However, the author suggests that the success achieved so far is no proof of the validity of the proposed model which will remain unproven until it is shown that radiogenic ^{234}U is indeed more stable against preferential loss from a sample than the initial U at the formation of the sample; and until the time dependence of U loss is investigated to establish whether it is exponential as originally assumed.

The U-trend dating model of Rosholt (1980b) is an empirical open system model which has been used for dating alluvial, glacial, and aolian units. The model does not use rigorous mathematical decay equations to describe the growth and decay of longer-lived radionuclides in U-series, but rather requires time calibration obtained from deposits of known age ranging from 4 to 730 ky. An isochron is determined from analysis of a number of samples representative of soil horizons in a depositional unit (three to ten samples are required). Whole rock samples are used, and accurate concentrations of ^{238}U, ^{234}U, ^{230}Th, and ^{232}Th are determined. Isochrons are constructed, based on results from the plots of $^{238}U/^{232}Th$ against $^{230}Th/^{232}Th$ yielding 'Th-index', and $(^{238}U - ^{230}Th)/^{238}U$ against $(^{234}U - ^{238}U)/^{238}U$ yielding 'U-trend' isochron. Ideally, the isochrons are linear, and the measured slopes change predictably with the increasing age of the deposit. The rate of change of the slope is determined by the 'half-period' of U flux in the local environment. This half-period is obtained from the calibration line established by U-trend isochrons for the depositional units of known age. The starting point for the U-trend clock is the time of deposition of the sediment, and not the initiation of soil development (for further details see chapter 10 and Appendix E).

Acknowledgements

The author should like to thank Drs Mel Gascoyne and Henry Schwarcz for their comments on the manuscript and for granting their permission to reproduce Figs 3.1 and 3.3. Special thanks are due to Mr Tony Parsons of the Nuclear Physics Division, AERE Harwell for proof-reading the manuscript.

4 CHEMICAL PROCEDURES
A. E. Lally

4.1 Introduction

This chapter presents a review of the general analytical chemistry techniques and the methods available for the determination of a range of elements which are of interest to workers in the field of U-series disequilibrium. The elements discussed are U, Th, Pa, Ra, Rn, Pb, and Po. The matrices in which these elements are found are varied and include water (freshwater, seawater and brines), silicates, oxides, carbonates, sulphates, and phosphates. The concentrations of the elements of interest in the above matrices are usually to be found at the ppb to ppm levels. The preliminary extraction of a single or group of elements from the host matrix can often be performed in a relatively simple procedure. The subsequent purification of that element, from others which are chemically very similar, in order that specific radiometric counting may be performed, often leads to lengthy analytical procedures. The majority of procedures used involve three main steps: (i) a preconcentration or extraction, (ii) a separation scheme; and (iii) preparation of a source for radioactive counting.

4.1.1 Preconcentration or extraction

In the case of solids this step can range from obtaining a solution of the matrix by acid digestion, chemical fusion (or a combination of both) to a simple leach in a mineral acid. Once the sample has been rendered into a liquid form, the application of a preconcentration step such as coprecipitation or the adsorption onto ion exchange resin in either the cationic or anionic form can be carried out as a batch process. Coprecipitation is defined as the precipitation of one substance in conjunction with one or more other substances. The mechanism may be due to compound formation, adsorption, solid solution formation, or simply mechanical inclusion and occlusion. Iron and rare-earth hydroxides are frequently used in such operations.

4.1.2 Separation scheme

For the actinides under consideration, U, Th, and Pa, separation from each other is usually performed by liquid–liquid extraction or column separation using selective adsorption and desorption on anion exchange resins. The liquid–liquid extraction, often called solvent extraction, may be accomplished either by extraction of the element or elements to be determined into the organic phase, or by extraction of the interfering elements leaving the element or elements to be determined in the aqueous phase.

Ion exchange resins are synthetic high molecular weight organic polymers which contain a large number of functional groups, which in the ionic form act as labile ions. These ions are capable of exchanging with ions in the surrounding medium without any major physical change taking place in the structure. The two most common forms of resin matrix available are cross linked polystyrene and phenolic. Each is available with a variety of functional groups in both the cationic and anionic forms. Standard types of ion exchange resins are listed in Table 4.1. With functional groups of strong electrolytes such as $-SO_3H$ the resin is completely ionised in a wide pH range. Similarly with the quaternary ammonium type anionic resins this wide pH range increases the applicability of these types of resins.

Table 4.1. Standard types of ion exchange resin

Type of resin matrix	Functional groups	Characteristics
Cross-linked polystyrene	$-SO_3H$	Strongly acid cation
Phenolic	$-OH$, $-CH_2SO_3H$, $-SO_3H$	Strongly acid cation
Cross-linked polystyrene	Quarternary ammonium type ($-NR_3^+$)	Strongly basic anion
Cross-linked polystyrene	$-NH_2$, $-NHR$, $-NR_1R_2$	Weakly basic anion
Phenolic	$-OH$, $-NH_2$, $-NHR$, $-NR_1R_2$	Weekly basic anion

The batch process of ion exchange separation is where sufficient resin in the appropriate form is mixed with a known volume of solution and adsorption occurs. The resin is then separated and treated to remove the adsorbed elements. In the column ion exchange separation method a vertical cylinder is filled with the resin and the solution is poured through the column. The column system can be considered as a large number of consecutive batch equilibria where fresh resin is brought into contact with the ion-depleted solution in each equilibrium step. The distribution coefficient (K_d) is defined by the equation

$$\text{distribution coefficient } (K_d) = \frac{C_1/\text{g resin}}{C_2/\text{ml solution}}, \qquad (4.1)$$

where C_1 is the amount of metal ion adsorbed per gram of dry resin, and C_2 is the amount of metal ion which remains in solution (per ml) after equilibrium has been reached. When K_d is much greater than 10 the batch system can be applied. However, when K_d values of less than 10 are found or where elements with similar K_d values are required to be separated column ion exchange is to be preferred.

Of the other elements under consideration Rn and He are separated in the gaseous phase using conventional techniques, whereas Ra, Pb, and Po are initially separated using classical precipitation techniques which will be discussed later in this chapter.

4.1.3 Source preparation

The type of source preparation is dependent on the choice of the spectrometric method used for the assay of the element, usually present in more than one isotopic form (see chapter 5 for the discussion of spectrometric methods). The preparation of the source, however, is very different for each technique.

The main source requirements for alpha spectrometry are that it should be uniformly distributed, as thin as possible, and on a flat disc. A thin deposit is necessary in order that good resolution may be obtained between the alpha energies present. A thin uniform deposit can be obtained by evaporation of an element-organic complex directly onto a metal disc or by electrodeposition from a variety of electrolyte solutions. The conditions for preparing sources for mass spectrometry are dependent on the phase being used, for example gas or solid and the particular type of spectrometer. For ionization chamber and gas scintillation counting, for example, oxygen and water vapour generally have to be removed prior to counting (see chapter 5).

4.2 General review of analytical separation techniques for U, Th, Pa, Ra, Rn, Pb, and Po

There are many references available concerning the analytical separation techniques which may be applied to these elements irrespective of their host matrices. They have been grouped into coprecipitation characteristics, ion exchange behaviour, and liquid–liquid extraction for convenience.

4.2.1 Uranium

(i) *Coprecipitation characteristics.* When traces of an element are being assayed in a complex mixture a preliminary separation is often made using a coprecipitation procedure. Iron and aluminium hydroxide can be used for the isolation of U from acid aqueous solutions by the addition of carbonate-free ammonium hydroxide. It is essential that ammonia solution is free from carbonate as the U will form a carbonate complex which will remain in solution (Urry 1941). Also, U can be coprecipitated from acid solution as the fluoride or phosphate. The host coprecipitant can be Al, Ti, Zr, or La (Korkisch 1969).

(ii) *Ion exchange behaviour.* There is infrequent use made of cation exchange resins for the separation of U from other elements including the actinides. The sulphuric acid type cation resin shows little selectivity to the divalent uranyl ion UO_2^{2+} compared with other divalent metal ions. Separation is thus not achieved readily. In sulphuric acid solution, however, U can be separated more easily than in nitric and hydrochloric acid media. This is because U is one of a few elements which form stable anionic sulphate complexes $UO_2(SO_4)_2^{2-}$ and $UO_2(SO_4)_3^{4-}$, which allow it to be eluted preferentially from a column containing other elements. Many elements in the 2^+ and 3^+ valent states form strong chloride complexes like U and separation from these is difficult due to non-adsorption on the cation resin. The use of a cation resin in a nitric acid media is also of little use since the majority of transition elements and U form no complexes or else very unstable complexes particularly at low acid concentrations. Distribution coefficients of U and other metals of interest between a typical strongly acidic cation resin, Dowex AG50–WX8, and pure acid media have been published by Strehlow (1960) and Strehlow, Rethemeyer, and Bothma (1965). Other acid media such as phosphoric acid and hydrofluoric acid can be used. However, with phosphoric acid, problems may arise from the formation of sparingly soluble phosphates, particularly of the rare earths (Apte, Subbaraman, and Gupta 1965). Specially made cation resins have been prepared for U separations and are reported by Kennedy, Davis and Robinson (1956) and Bayer and Möllinger (1959).

Use has been made of organic compounds for complexing U for separation on cation resins. These include oxalic acid (Ishimori and Okimo 1956) and ethylenediamine tetra-acetic acid (E.D.T.A.), (Kennedy et al. 1956) but these are perhaps of little value in analysis of environmental samples.

With many common anions U forms anionic or neutral complexes that are useful for the anion exchange separation of the element. The anions used include acetate, chloride, fluoride, nitrate, phosphate, and sulphate. The use of the strongly basic cation resins of the quaternary ammonium type will permit a greater selectivity range than will the weak functional group resins.

Strong anionic complexes of the uranyl ion are formed in aqueous hydrochloric acid solutions $(UO_2Cl_3)^-$ and $(UO_2Cl_4)^{2-}$ (Kraus and Nelson 1956). The adsorption of U onto strongly basic anion resin increases rapidly with increase in the strength of the hydrochloric acid. Many metals form anionic chloride complexes but, by varying the strength of the acid, separation from some of these metals can be obtained. An example of sequential elution with varying concentration of hydrochloric acid is the separation of U, Th, and Pa. Bunney, Ballon, Pascual, and Foti (1959) showed that when a solution containing these three elements was loaded onto an anion column in 12 M hydrochloric acid, Th was not adsorbed and passed straight through. Protactinium was eluted from the column with 4 to 5 M hydrochloric acid, and finally U was eluted with 0.1 M hydrochloric acid. Uranium can be separated from Fe on a chloride column if it is first reduced to the divalent state with ascorbic acid, hydrogen iodide, or ammonium iodide when it will not be retained on the column.

Uranium is only weakly adsorbed onto strong base anion resins in the nitrate form. The distribution coefficient (K_d) rises to about 20 at 8 M nitric acid but then falls with increasing acid strength. This is to be compared with a K_d of about 10^3 in 8 to 10 M hydrochloric acid. Separation of U from Pa and Th can be achieved in the nitrate form, but separation from Fe is only obtained with care. It is for this reason that separation schemes for U based on nitric acid media are not very selective or numerous.

In dilute aqueous sulphuric acid solutions U forms an anionic sulphate complex $UO_2(SO_4)_2{}^{2-}$ which is strongly adsorbed onto strong base anion resins. A K_d value about 10^4 in 0.05 M sulphuric acid rapidly decreases with increasing acid strength. Many other metals form complexes at low acidities and are partly or completely co-adsorbed with U. The adsorption of the sulphate complex is reduced by the presence of nitrate and chloride ions. These can be removed by fuming the sample with concentrated sulphuric acid prior to dilution and loading onto the column.

In the anionic complexes listed above the selectivity of adsorption and separation from other elements can be altered by the use of aqueous systems. An extensive review of these methods and applications has been published by Korkisch (1969), who has also included some useful data on liquid–liquid extraction for a whole range of rarer metal ions.

(iii) *Liquid liquid extraction.* The technique of liquid–liquid extraction has been widely applied to the separation and extraction of U from a wide range of accompanying elements. Organic acids, ketones, ethers, esters, alcohols, and organic derivatives of phosphoric acid have been used for the extraction of U and other actinides. In the early days of U chemistry diethyl ether was used for the extraction of uranyl nitrate but more recently the use of less volatile and flammable ketones and phosphates has become more popular. Methyl isobutyl ketone (MIBK) has been used extensively in the nuclear industry for the large scale separation of U and Pu from spent nuclear fuels (Redox process, see Culler 1956). Ethyl acetate has been used in a nitric acid media to extract U from solutions containing large amounts of salting out agents, for example aluminium nitrate. Callahan (1961) has reported that the extraction is not very pH dependent but that the efficiency of extraction varies according to the concentration of aluminium nitrate present. Iron is co-extracted in small amounts but Th can be retained in the aqueous phase by the addition of sodium phosphate. Uranium can be recovered by back extraction of the organic phase with water or by evaporation of the solvent. A common organic phosphate used is tri-butyl-phosphate (TBP), which is widely used for the separation of U from many accompanying elements in a nitric acid media. The distribution coefficient is very large over an acid range from a pH of 3 to 6 M nitric acid. The advantages of TBP are its non-volatility (boiling point of

289 °C) and its stability to concentrated nitric acid. Iron, Th, and Pa are all co-extracted with U in nitric acid, but Fe can be removed by washing the organic phase with dilute nitric acid. Thorium can be complexed with EDTA before extracting U with TBP, leaving the Th in the aqueous phase. Many different methods have been published for the recovery of U from the TBP complex. These include such solutions as ammonium carbonate (Nemodruck and Varotimtskaya 1962), sodium carbonate (Bril and Holzer 1961), oxalic acid (Pfeifer and Hecht 1960), and water (Wood and McKenna 1962). TBP is not widely used in hydrochloric acid media. However, a separation of U from both Th and Pa can be achieved since the latter two are both poorly extracted relative to U. Tri-n-octylphosphine oxide (TOPO) in both cyclohexane and carbon tetrachloride has been used for the complete extraction of U from 1 to 7 M nitric acid solutions where seperations from metals such as Fe and Cr can be readily achieved (White and Ross 1961).

Of the β-diketone class of compounds thenoyltrifluoroacetone (TTA) is one of the most widely used. Uranium forms a complex with TTA which can be extracted from aqueous solutions of pH 3 or higher using 0.15 to 0.50 M TTA in benzene. Thorium, Fe, and rare earths are also co-extracted but Khopar and De (1960) have reported that the addition of EDTA and a solution of pH 6 improves the selective extraction of U.

4.2.2 Thorium

(i) *Coprecipitation characteristics.* Thorium forms hydroxide, fluoride, iodate, oxalate, and phosphate precipitates. Of these the hydroxide and fluoride have been used extensively to remove tracer amounts of Th from solution. Thorium can be coprecipitated quantitatively as the hydroxide with Fe, La, and Zr by the addition of ammonium or alkaline hydroxide to an aqueous solution. Lanthanum fluoride is used to coprecipitate Th from strongly acid solutions and this method is useful for the separation of small amounts of Th from U solutions (Ko and Weiller 1962). The insoluble fluoride after separation is converted to the hydroxide by metathesis with alkaline hydroxide.

(ii) *Ion exchange behaviour.* Thorium is adsorbed well onto strongly acidic cation resins in all the three mineral acid media, hydrochloric, nitric, and sulphuric. The distribution coefficients for Th in these acids and Dowex 50 resin are given by Strehlow (1961) and Strehlow et al. (1965). The strongest affinity is in the hydrochloric acid media due to the fact that it does not form an anionic complex in aqueous hydrochloric acid. Good separations of Th from most monovalent and divalent elements can be achieved in 1 to 4 M hydrochloric acid solutions (Sackett, Potratz and Goldberg 1958). There is a problem in recovering Th quantitatively from cation resins but the most successful appear to be 3 M sulphuric acid (Korkisch and Antal 1960) and 0.5

M oxalic acid solution (Miyake and Sugimura 1961; Goldberg and Koide 1962).

Thorium, together with trivalent transuranium elements, is not retained on strongly basic anion resins in aqueous hydrochloric acid solutions. This fact is widely used to effect a separation of Th from such elements as Pa, U, and Fe.

Thorium forms a very stable nitrate complex of the form $Th(NO_3)_6{}^{2-}$. The best K_d value is obtained at 5 to 10 M nitric acid solutions. Reasonable selectivity can be obtained for Th from other elements under these conditions. The presence of phosphate greatly decreases the adsorbance of Th as the nitrate complex (Chow and Carswell 1963). Thorium will form an anionic sulphate complex which can be adsorbed onto a strongly basic anion resin. However, as with U, the K_d value falls with increasing acid concentrations. Also above pH 3 Th tends to precipitate out as the sulphate.

(iii) *Liquid–liquid extraction.* Thorium can be extracted from nitric acid solutions with diethyl ether. The use of a salting-out agent such as zinc nitrate greatly improves the extraction (Bock and Bock 1950). Thorium does not form a stable chloride complex (Jenkins and McKay 1954) and is not extracted from hydrochloric acid solutions with diethyl ether. This fact is used to separate Fe from Th in 6 M hydrochloric acid solutions. Thorium, like U, can be extracted with many organic ethers, ketones, and others. Alberti, Bettanili, Salvetti, and Santoli (1959) reported the use of hexone for the extraction of Th from solutions of rock samples. Mesityl oxide (Ko and Weiller 1962), cyclohexanone (Higashi 1958), and ethyl acetate (Andjelkovic and Rajkovic 1958) have been reported as used in environmental sample analysis.

Organic phosphorus compounds are used to extract Th from a variety of matrix solutions. Tri-n-butylphosphate (TBP) in a range of organic solvents will extract Th together with U, Np, and Pu from nitrate solutions (Hesford, McKay, and Scargill 1957). Separation of Th from U can be achieved by extracting with TBP in hydrochloric acid solution (Ishii and Takenchi 1961). Bis-(2-ethylhexyl)-o-phosphoric acid (HDEHP) dissolved in toluene can be used to extract Th from both nitric and hydrochloric acid solution. Butler (1965) reported that extraction from 2 M hydrochloric acid gave the best result. Of the chelating agents used for the extraction of Th, the β-diketone thenoyltrifluoroacetone (TTA) in benzene or toluene is frequently used. Extraction is best carried out from nitric, hydrochloric, or perchloric acid solution. Extraction at pH 1 to 2 gives a good separation of Th from Al, U, alkaline and rare earths (Ko and Weiller 1962).

4.2.3 Protactinium

(i) *Coprecipitation characteristics.* Protactinium forms hydroxide, carbonate, fluoride, phosphate, iodate, and sulphate precipitates from aqueous solutions. Many of the published methods rely, for the initial concentration step, on the coprecipitation of Pa with a flocculent hydroxide (for example Ca or Fe) or the carbonate or phosphate of Ta, Zr, or Ti (Kirby 1959). Manganese dioxide, produced from a 1 to 5 M nitric acid solution, has been used to carry down Pa from acid solution. Katzin and Stoughton (1956) used this carrier to separate Pa from Th.

(ii) *Ion exchange behaviour.* Few papers have been published on the behaviour of Pa on cation exchange resins. Kirby (1959) reviews some unpublished work, and Nelson, Murase, and Kraus (1964) reported that Pa is not adsorbed on strongly basic cation exchange resins from hydrochloric acid solutions. Katz and Seaborg (1957) report on the characteristics of Pa in nitric acid solutions.

Protactinium is strongly adsorbed as an anionic chloride complex on strong base anion resin from hydrochloric acid solutions greater than 4 M in strength. Effective elution of Pa from these chloride columns is achieved using hydrochloric acid solutions containing hydrofluoric acid (Kraus and Moore 1955). Protactinium is only partially adsorbed on strong base anion resins from a nitric acid solutions and thus this method does not appear in Pa separation schemes.

(iii) *Liquid–liquid extraction.* Protactinium can be extracted from nitric, sulphuric, and hydrochloric acid solutions into a wide range of organic solvents. Long-chain alcohols, ketones, and organic phosphorus compounds are the most effective, and in general aqueous chloride media give better yields than nitrate solutions (Kirby 1959). Low molecular weight ethers are not generally used for Pa extractions. Di-isopropyl ketone is widely used with the yield increasing with concentration of acid and salting-out agent (Golden and Maddock 1956). The use of organic phosphorus compounds for the extraction of Pa has been reviewed by Kimura (1960), who reported that effective separation of Pa, Th, and U can be achieved by variation of the nitric acid–oxalic acid concentrations with tri-n-butyl phosphate (TBP). Protactinium can be separated from U with tri-n-butyl phosphine oxide (TOPO). Both Pa and U are extracted with 1 per cent TOPO in toluene from 6 M hydrochloric acid solution. Washing the organic layer with 4 to 6 M hydrochloric acid solution saturated with sodium fluoride removes the Pa (Ishimori, Watanabe, and Kimura 1960). In strong hydrochloric or nitric acid solutions Pa can be extracted with thenoyltrifluoroacetone (TTA) in benzene. At the optimum acidity of 6 M hydrochloric acid or 4 M nitric acid Pa can be separated from Fe, Th, Po, and the rare earths.

4.2.4 Radium

Radium is an alkaline earth element and its properties are similar to Ba.

(i) *Coprecipitation characteristics.* The most frequently used coprecipitation media for Ra is barium sulphate. Radium forms mixed radium-barium sulphate crystals and precipitation is optimum when carried out in hot 0.5 M sulphuric acid solution by the dropwise addition of 0.1 M barium chloride solution. Lead and Sr also form sparingly soluble sulphates and are coprecipitated at the same time. However the Ba-Ra sulphate can be selectively dissolved in an alkaline solution of ethylenediamine tetra-acetic acid (EDTA) (Golding 1961). Barium nitrate (Kirby and Brodbeck 1954), barium chloride (Owers and Parker 1964), and calcium carbonate (Sugimura and Tsubota 1963) have been used as initial coprecipitating media for Ra.

(ii) *Ion exchange behaviour*. The use of ion exchange methods for the separation of Ra is usually restricted to cation resins. Radium is adsorbed on strongly acidic cation exchange resins from dilute acid, neutral to weakly alkaline solutions. Often a complexing agent, for example EDTA, is used to elute Ra preferentially from Th and the alkaline earth elements (Duyckaerto and Lejeune 1960). Diamond (1955) has reported the effect of hydrochloric acid concentration on the K_d for Ra and other alkaline earth elements on cation resins. The adsorption of the alkaline earth elements on anion exchange resin is usually carried out in an ammonium citrate solution (Nelson and Kraus 1955). Radium does not form an anionic chloride complex and this fact has been used to separate Pb from Ra (Petrow, Neitzel, and DeSosa 1960). Using 1.8 M hydrochloric acid solution Pb is adsorbed onto an anion exchange resin while Ra is not adsorbed and passes through in the load solution.

(iii) *Liquid–liquid extraction*. Most Ra compounds have very low solubilities in organic solvents (Kirby and Salutsky 1964). This fact is often used to separate Ra from elements such as Th, Po, Pb, and others, by extracting these elements into the organic phase and leaving the Ra behind in the aqueous phase. Hagemann (1950) used this principle with TTA and Petrow and Lindström (1961) separated Pb from Ra with Aliquat-336 in benzene.

4.2.5 Radon

Radon is a noble gas and is chemically inert. It is usually separated from a matrix by pumping off the gas and removing hydrogen and oxygen by adsorption on a catalyst. The Rn gas is condensed in a cold trap with liquid nitrogen. For aqeuous samples He is usually recycled through the solution to outgas the Rn. The Rn is then assayed in either a proportional or scintillation counter (Harley 1974).

4.2.6 Lead

(i) *Coprecipitation characteristics*. Lead forms many sparingly soluble salts which may be used initially to separate and/or concentrate the element from a variety of solutions. Sulphate, chloride, chromate, and sulphide have all been used in published methods (Gibson 1961).

(ii) *Ion exchange behaviour*. Lead can be adsorbed on a strong acid cation exchange resin from nitric acid solutions. Strehlow (1960) and Strehlow *et al.* (1965) give K_d values for Pb with Bio-Rad AG50-WX8 resin. Bonner and

Smith (1957) report the results of their studies with Pb adsorption on Dowex 50 cation resin from perchloric acid solutions. The adsorption of Pb as the anionic chloride complex has been studied by Nelson and Kraus (1954), who showed that Pb has a maximum adsorption in 1.5 M hydrochloric acid solution and decreases rapidly with increase in acidity. Separation of Pb, Bi, and Po from Ra can be accomplished using an anion exchange resin in the chloride form (Petrow et al. 1960). Lead can also be separated from Th using a chloride system as Th does not form an anionic chloride complex. There are few reports of anion exchange resin being used in nitric acid media, due to Pb being weakly adsorbed at all nitrate concentrations. However, the addition of EDTA greatly increases the adsorption of Pb from nitric acid solutions (Gibson 1961).

(iii) *Liquid–liquid extraction.* Lead will not extract into organic solvents from hydrochloric or nitric acid solutions (Gibson 1961) and this fact can be used to separate Pb from accompanying elements. Lead forms many organic chelate complexes which can be extracted into organic solvents. Complexes with cupferron, diethyldithiocarbamate, oxine, and dithizone are all well used. However, dithizone in carbon tetrachloride is selective for Pb and is widely used (Gibson 1961). Lead is extracted with 2-thenoyltrifluoroacetone (TTA). This complex has the advantage of permitting extraction of Pb from a strong acid solution, avoiding the instability of some complexes in strong acid and the problem of hydrolysis. Lead has been separated from U and Bi by extraction with acetylacetone at pH 7.4 (Freiser and Morrison 1959).

4.2.7 Polonium

(i) *Coprecipitation techniques.* Polonium is frequently isolated from solution using metallic Te as the carrier. The coprecipitation is carried out from a strong hydrochloric acid solution by the addition of sodium hypophosphite. This will reduce both Te and Po to the metallic state. After filtering off, Te and Po can be dissolved in hydrobromic acid/hydrochloric acid mixture and the Te selectively reduced with hydrazine hydrochloride, leaving the Po in solution. Polonium can also be precipitated onto manganese dioxide (Eakins and Morrison 1978). The addition of potassium permanganate solution to a nitric acid solution containing Po will scavenge it with the manganese dioxide precipitate.

(ii) *Ion exchange behaviour.* The application of cation exchange resins for the separation of Po has not been widely reported except for the separation of the element from Bi using EDTA solutions (Korkisch 1969). Polonium is strongly adsorbed on strong basic anion exchange resins from hydrochloric acid solutions. Ishimori (1955) has used this to separate Po from Pb and Bi. Lead is eluted in 3 M hydrochloric acid and Bi in 12 M hydrochloric acid, leaving Po to be eluted with 8 M nitric acid solution.

(iii) *Liquid–liquid extraction.* Polonium is quantitatively extracted from nitric acid solutions with 0.25 M TTA in benzene. It can be separated, along with Th, from Bi at pH 1.5. Millard (1963) used polonium diethyl-dithiocarbamate to determine Po and Pb in minerals. Separation of Po from Pb and Bi can be effected by extraction of Po from 6 M hydrochloric acid solution with 20 per cent TBP in dibutyl ether (Meinke 1964). Bismuth is co-extracted with Po but is removed by washing the organic phase with 6 M hydrochloric acid, the Po being back extracted by shaking with concentrated nitric acid.

4.3 Sample preparation and pretreatment

Samples for analysis can be conveniently divided into two groups. The first group comprises those samples which do not require analysis on location or soon after collection and thus can be collected and stored for later analysis without pretreatment. This group includes all solid samples: rocks, minerals, sediments, cores and so on. These can be packed separately to avoid contamination and returned for analysis in the base laboratory. The second group comprises the liquid samples. The sheer physical size of the sample, in some cases up to 600 l of water have been collected for one single analytical determination (Reid, Moore, and Sackett 1979) prompt an on-site analysis or preliminary separation for the sake of convenience. In the majority of chemical procedures reviewed the initial dissolution and preconcentration of elements from various matrices varies considerably. The final stages of the methods are often similar for identical isotopes. Solid samples are subjected to chemical fusion, total acid digestion or acid leach prior to chemical separation. Radionuclides in liquid samples are preconcentrated by coprecipitation, adsorption on activated charcoal, adsorption on coated acrylic fibres, and adsorption on cation and anion exchange resins.

4.3.1 Solid samples

For most solid samples a complete dissolution of the sample is the preferred technique. This is usually obtained by acid digestion, chemical fusion or a combination of both. The sample should initially be ground as fine as possible and well mixed in order to homogenize the sample before an aliquot is taken for analysis. Where an internal tracer and inactive carrier are used, to determine isotopic and gravimetric recoveries, these should be added at the start of procedure, before the sample has been subjected to any chemical treatment. Various combinations of nitric, hydrochloric, perchloric, and hydrofluoric acids have to be used to dissolve solid samples. Samples containing silicates require the use of hydrofluoric acid to obtain total dissolution. On repeated evaporation of a hydrofluoric/nitric acid or hydrofluoric/perchloric acid mixture the Si will be removed as the volatile silicon tetrafluoride. A typical procedure to obtain a solution of pelagic sediments prior to analysis for U, Th,

Pa, Ra, and Rn has been used by Ku (1965). The method is shown schematically in Fig. 4.1. Up to 10 g of sample was digested by heating with concentrated hydrochloric acid after the addition of several tracers. The sample was allowed to cool and the residue separated from the solution by centrifuging. The supernate was reserved for subsequent analysis. The residue was leached with 6 M hydrochloric acid and the insoluble material separated off. The supernate was added to the original supernate. Hydrofluoric acid and perchloric acid were then added to the residue and the sample taken to fumes of perchloric acid, to remove any silica present. The residue was again leached in 6 M hydrochloric acid and any solid centrifuged from the solution. The supernate was again combined with the original. The residue was fused with sodium carbonate and the melt dissolved in 4 M hydrochloric acid and all supernates were bulked giving a total solution of the sample.

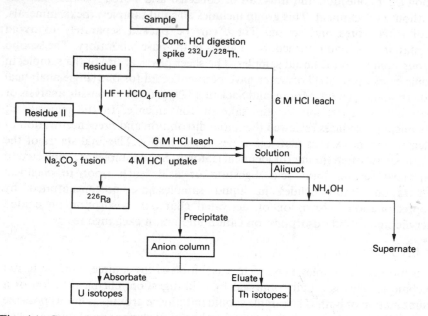

Fig. 4.1. General procedure for sample dissolution for U, Th, and Rn analysis (after Ku 1965).

Allegre and Condomines (1976) used a mixture of nitric and hydrofluoric acids to dissolve volcanic rocks. After heating to dryness to remove silicon tetrafluoride and residual hydrofluoric acid, the residue was dissolved in nitric acid for subsequent analysis. The use of high pressures for the dissolution of samples using a Teflon bomb, has been reported by Krough (1973). Korkisch, Steffan, Arrhenius, Fisk, and Fraser (1977) treated a dried 200 mg sample of manganese nodules with 1 ml of water, 5 ml of perchloric acid and 10 ml of hydrofluoric acid in a sealed Teflon bomb for 48 h at 110 °C. The contents were then transferred to a platinum dish and the hydrofluoric acid removed by evaporation on a steam bath. After further treatment with perchloric acid, hydrobromic acid, and hydrogen peroxide the sample was taken to dryness. The residue was dissolved in hydrochloric acid for subsequent ion exchange separation.

4.3.2 Liquid samples

The effects of storage on water samples collected for subsequent trace element analysis, for radioactive or stable elements, has been the subject of many published papers (for example see Cheeseman and Wilson 1973, Dokiya, Yamazaki, and Fuwa 1974). It is generally recommended that as soon as possible after collection water samples are acidified with nitric or hydrochloric acid to pH 1 approximately and stored in pyrex glass or polyethylene containers. Under these conditions many trace elements are kept in solution and the acid conditions prevent the growth of biological material within the sample. Within the pH range of 6 to 8 typical of the majority of natural waters, many metallic elements will form insoluble hydroxides which can 'plate' onto the walls of the container. These will be difficult to remove without resorting to concentrated acids and/or vigorous agitation. The precipitation of these insoluble hydroxides is seeded by particulate matter in the water and some samples, notably surface waters, are often filtered prior to acidification. Consideration must be given to water samples that have been filtered as to whether the filtered particulate material should be dissolved and added to the filtered water for total analysis (Sill 1976), or simply discarded.

Stable element carriers and radiotracers are also added to the sample as soon as possible after collection. Large volume samples collected in the field or on board ship are often given a preliminary treatment on-site such as a preconcentration step. This can be either a coprecipitation or the passage of the sample through an ion exchange resin bed. These on-site treatments should only be done after several hours, to allow the sample and tracers to become equilibrated. In procedures utilizing a coprecipitation with ferric hydroxide for the determination of U and Th it is usual to allow several days for this equilibration (Briel 1976; Cowart 1974; Kronfeld 1972). Veselesky (1974) also recommends storing the sample for several days, to permit equilibration, before passing the water through a cation exchange resin. Schell (1977) has used alumina filter pads to adsorb Pb from water samples (1000 to 1400 l) prior to freezing to await analysis in a laboratory. Nozaki and Tsunogai (1973a) used carbonate precipitates as a field separation step in their work on Pb and Po in water. Broecker, Kaufman, and Trier (1973) coprecipitated Th and Ra on barium sulphate and ferric hydroxide from 1600 l of water and, after 8 h settling, separated the precipitates from the bulk of the solution for transport back to the laboratory in 20 l bottles. The use of an acrylic fibre coated with ferric hydroxide (Krishnaswami, Lal, Somayajulu, Dixon, Stonecipher, and Craig 1972) has given an alternative method of concentrating such elements as Ra, Th, and Ac. MacKenzie, Baxter, McKinley, Swan, and Jack (1979) have used this technique to preconcentrate ^{228}Ra from in excess of 1000 l of seawater. The water is pumped through 250 g of the manganese dioxide loaded fibre contained in a stainless steel unit. The fibre is then returned to the laboratory for subsequent analysis, which will be described later in this chapter.

4.4 Published procedures for the determination of U, Th, Pb, Po, Pa, Ra, Rn, and He in environmental samples

The measurement of the U-series disequilibria (see chapter 5) can be accomplished in several different ways involving combinations of isotope activity ratios, for example ^{234}U/^{238}U, ^{230}Th/^{234}U, ^{231}Pa/^{235}U, and others. Chemical procedures have been published for the determination of some of these elements either individually or selectively grouped together.

The separation of U and Th from aqueous samples can be accomplished by a variety of extraction procedures. Coprecipitation of the hydroxides with Fe is widely used as an initial step. Typical of such a procedure is that reported by Bhat (1969) and shown diagrammatically in Fig. 4.2. Nitric acid, an Fe carrier, and an equilibrated ^{228}Th/^{232}U spike were added to 30 l of water and the samples allowed to stand for a day. After boiling the sample to remove dissolved carbonates, ferric hydroxide was precipitated by the addition of ammonia and ammonium chloride. The ferric hydroxide precipitate, containing the U and Th was separated from the supernate and washed. These were discarded and the precipitate was dissolved in concentrated hydrochloric acid and loaded onto an anion exchange resin column. The column was washed with hydrochloric acid and the effluent and washings combined for Th determination. Uranium and Fe which were retained on the column were eluted with 0.1 M hydrochloric acid. Ferric hydroxide was precipitated from this solution by the addition of ammonia. After centrifuging, the precipitate was transferred to a laboratory for further processing. Iron was added to the fraction containing the Th and ammonia added to precipitate ferric hydroxide, which was then centrifuged and transferred to the laboratory. The U fraction was further purified from any remaining Th by dissolving the precipitate in concentrated hydrochloric acid and passing through another anion exchange column in the chloride form. The Th-free Fe plus U fraction was evaporated to dryness, dissolved in acetic acid, buffered to pH 4.5 to 5 with ammonium acetate and passed through another anion exchange column. After washing the column with water, the U was eluted with 1 M hydrochloric acid and electrodeposited from an ammonium chloride electrolyte. The Th fraction was purified from any entrained U by dissolution in hydrochloric acid

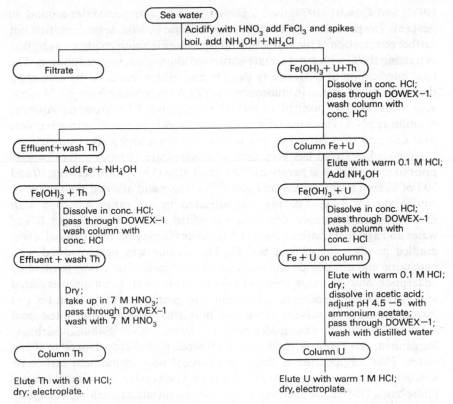

Fig. 4.2. Chemical procedure for the extraction of Th and U from sea water (after Bhat et al. 1969).

and passage through another anion column in the chloride form. The effluent and washes were converted to the nitrate form and loaded onto an anion column in the nitrate form, on which Th is adsorbed. After washing with 8 M nitric acid the column was washed with 6 M hydrochloric acid to remove the Th, which was electrodeposited from an ammonium chloride bath.

Briel (1976) used a similar procedure for surface waters; however, he filtered his samples before analysis. This helped to remove dissolved gases as well as particulate matter. The preliminary ferric hydroxide precipitate was wet oxidized with a mixture of nitric, perchloric, and hydrochloric acids before proceeding with the separation. This oxidation of the tannins present in the sample increased the yield of U by as much as 50 per cent over samples analysed without this treatment. The Fe was removed by solvent extraction with isopropyl ether in 8 M hydrochloric acid prior to ion exchange separation in the chloride and nitrate form. Average recoveries for U of 33 per cent with a range of 1 to 87 per cent were obtained by this technique. Kronfeld

(1972) and Cowart (1974) used a similar method with recoveries around 40 per cent. The procedure of Ku (1965) used a ferric hydroxide precipitation but further purification of the U fraction off the chloride anion column was by first extracting the U with ethyl acetate-saturated aluminium nitrate solution. The final purification of U from any Pa, Th, and others was done by extracting them into 0.4 M thenoyltrifluoroacetone (TTA) in benzene from 0.1 M nitric acid solution. The purified U was extracted into TTA from an aqueous solution at pH 3. This organic phase was evaporated directly onto a stainless steel disc and then flamed to give a thin source for alpha spectrometry.

Activated charcoal has been used to concentrate U from natural waters prior to ion exchange separations. Van and Lalou (1969) took between 10 and 30 l of filtered water, added acid and ^{232}U tracer and allowed the sample to equilibrate for 12 h. The pH was adjusted in the range 4.5 to 5 with hexamethylene-tetramine. Charcoal was added at the rate of 2 g per 10 l of water and agitated continuously for 5 h. After filtering off the charcoal it was muffled in silica for 48 h at 600 °C. The residue was next oxidized with perchloric and nitric acids and evaporated to dryness. The U was purified by adsorption onto an anion column in the chloride form. Uranium was eluted with perchloric/hydrochloric acid mixture to give a separation from Fe, and further purified by solvent extraction into ethyl acetate from nitric acid solution prior to electrodeposition. Spiridonov, Sultankhodzhagv, Surganova, and Tyminskii (1969) used adsorption on charcoal for 60 to 80 l of water. They stripped the U from the charcoal with ammonium carbonate solution. After conversion to the nitrate form U was extracted with tri-n-butyl phosphate (TBP) in the presence of ammonium nitrate as a salting-out agent. The U was back extracted from the organic phase with ammonium carbonate solution and electrodeposited from an ammonium oxalate electrolyte. Recoveries for U averaged 90 per cent.

The use of cation resin to extract U from water has been reported by Veselesky (1974). After acidification of 20 to 30 l of water with hydrochloric acid, 20 mg Fe and ^{232}U tracer were added. The sample was allowed to equilibrate for several days, the pH adjusted to 4, and then passed through a Dowex 50 × 8 resin column. The U and other cations were eluted from the column with 6 M hydrochloric acid. The acidity was adjusted to 9 M with respect to hydrochloric acid and passed through an anion exchange column in the chloride form. Thorium was not retained on the column and after washing the column U and Fe were eluted with 0.1 M hydrochloric acid. The U was purified by extraction into a 10 per cent Alamine-336/xylene solution from a 1 M sulphuric acid solution and then stripped from the organic phase with 0.5 M nitric acid. The purified U fraction was evaporated to low bulk and electrodeposited from an ammonium chloride electrolyte. Average recoveries of 63 per cent were obtained using this method.

The adsorption of U, as a thiocyanate complex, onto an anion resin has given 70 per cent recovery from 10 l of filtered water (Brits 1979). After elution and destruction of the thiocyanate complex the U was further purified by extraction from an ammonium nitrate solution with diethyl ether. The ether fraction was evaporated to dryness, oxidized to remove organic material and then electrodeposited from an ammonium oxalate bath. Lee, Kim, Lee, and Chung (1977) used Chelex-100 chelating ion exchange resin to concentrate U from sea water for subsequent determination by neutron activation analysis. The same resin was used by Hathaway and James (1975) for a U separation scheme terminating in X-ray fluorescence.

The separation of U from waters using the uranyl ion, UO_2^{2+} has been investigated by Burba and Lieser (1979) and Tabushi et al. (1979). Burba and Lieser (1979) used the chelate forming ion exchangers, Hyphan and Salen, which are both based on cellulose and obtained good separations for U over the pH range 5 to 7. Tabushi et al. (1979) used a polymer bound macrocyclic hexaketone dissolved in chloroform to extract the uranyl ion. The adsorbed U was easily removed by treatment with dilute acids. Although these latter methods have been developed for use in extracting U from large volumes of sea water for energy-utilization purposes they could be used as a preliminary separation scheme in geochemical studies.

Many of the methods for U in liquid samples are applicable to soluble matrices after a preliminary dissolution. The method of Ku (1965) is generally accepted as a standard procedure for non-carbonate sediments. This is shown in Fig. 4.1 and has been described in section 4.3.1. This method has more recently been used by Burnett and Veeh (1977) for dissolving phosphorite nodules. Modifications to the method of Ku (1965) can be applied to the analysis of different matrices. Matrices which contain large amounts of organic matter should be ashed in a muffle furnace for several hours, prior to repeated treatment with nitric and perchloric acids to ensure complete oxidation of U and Th. The presence of quantities of phosphate ions in a sample solution can interfere with the ion exchange process and should be removed before this step in the procedure. A controlled hydroxide precipitation of U and Th on Fe, at pH 3 to 3.5, will maintain the majority of the phosphate in solution (Veeh, Calvert, and Price 1974). Harmon, Thompson, Schwarcz, and Ford (1975) used a similar method to that of Ku (1965) for the analysis of calcite speleothems ($CaCO_3$). It is important with carbonate-rich materials to ensure that the sample solution is boiled well to expel dissolved carbon dioxide prior to the precipitation of ferric hydroxide. The analysis of granitic rocks (Manton 1973) can be completed using, after an initial treatment with hydrofluoric and perchloric acid, a fusion with lithium metaborate. He reported that, if the insoluble material was not fused and taken into solution, poor agreement was obtained between duplicate measurements. Sill (1979) has used a pyrosulphate fusion technique for the determination of a range of alpha emitting radionuclides in large environmental and biological samples. This technique ensures the complete dissolution of refractory compounds prior to proceeding with an analytical separation scheme.

The use of an anion exchange system in a magnesium nitrate media has been applied to the analysis of manganese nodules for U and Th (Kuroda and Seki 1980). These elements are adsorbed onto anion exchange resin from a 2.5 M magnesium nitrate and 0.1 M nitric acid solution. Good separation from many elements including Mn, Fe, Cu, Co, and Ni were obtained by washing the column with the load solution. Thorium is eluted with 6 M hydrochloric acid and U with 0.1 M hydrochloric acid.

In analysing sediments for U and Th, Rosholt (1980b) first heated the samples in a muffle furnace at 900 °C to convert calcium carbonate to calcium oxide and to decompose organic matter prior to dissolution in nitric, hydrofluoric, and perchloric acid mixture. The first separation of U and Th was achieved on an anion column in the chloride form. The U fraction was purified by solvent extraction into methyl isobutyl ketone from a nitric acid solution saturated with magnesium nitrate. The U was then purified further on an anion column in the nitrate form to remove Mg and Fe and finally on another chloride anion column, to remove any residual Th, before electrodeposition from an ammonium chloride electrolyte. The Th fraction was first coprecipitated with zirconium pyrophosphate (Rosholt 1957). This was dissolved in oxalic acid and the Th separated from the Zr by coprecipitation with La. The lanthanum oxalate was dissolved in nitric acid and passed through an anion column in the nitrate form to remove the La. The Th was finally purified by solvent extraction into TTA in benzene which was directly evaporated onto a stainless steel disc for alpha counting. Goldberg and Koide (1962) dissolved their sediments in hydrochloric acid, added La and Fe carriers and coprecipitated the Th on the La and Fe, as hydroxides, with ammonia gas. After dissolving the hydroxides in hydrofluoric acid and diluting with water, Th was coprecipitated with lanthanum fluoride. This precipitate was then dissolved in nitric acid and reprecipitated with hydrofluoric acid. The fluoride precipitate was dissolved in nitric acid and the fluoride removed with repeated evaporations with nitric acid. The Th was purified, from the La, by extraction into mesityl oxide from a nitric acid solution containing aluminium nitrate. The Th was back extracted into water, then solvent extracted with TTA in benzene and finally electroplated using an ammonium chloride electrolyte.

In order to measure the U isotope ratios in the surface layers of sediment particles, Joshi and Ganguly (1976) used a saturated ammonium carbonate solution to leach these surface layers, without attacking the mineral core. After leaching the sediment for 8 h the solution was filtered through a 0.22 μm filter and treated with hydrochloric acid and evaporated to dryness to remove ammonium salts. The Fe was extracted into di-isopropyl ether before purification on an anion exchange column in the chloride form. The U was eluted from the column and electrodeposited using an ammonium oxalate bath. Thorium was leached from the surface of sediment particles with 0.5 M ethylenediamine tetra-acetic acid (EDTA) (Joshi and Ganguly 1970). Further purification of the Th was by anion exchange in the nitrate form, followed by a chloride anion column before molecular plating the Th.

The dating of sediments has been attempted using [231]Pa. It can be measured either as the [231]Pa isotope itself, or via its granddaughter [227]Th (Mangini and Sonntag 1977). The measurement of [231]Pa in filtered sea water using [233]Pa as an internal tracer has been reported by Sackett (1960). After coprecipitating the Pa on ferric hydroxide, the Pa was separated from the Fe by coprecipitation with zirconyl iodate $(ZrO(IO_3)_2)$ from a nitric acid solution. The Pa was further purified by solvent extraction into di-isobutyl carbinol from 6 M hydrochloric acid solution. After back extracting the Pa with dilute hydrofluoric acid solution, it was evaporated onto a Pt plate for beta counting to determine the yield of the [233]Pa and then alpha counted for the [231]Pa content. Sediment samples were first dissolved in hydrofluoric/sulphuric acid mixture, and all traces of hydrofluoric acid removed by evaporating down to fumes of sulphuric acid. The Fe was precipitated and the method continued from this step in the procedure for water outlined above. Ku (1968) used a similar purification step for Pa in his method for the simultaneous determination of U, Th, and Pa in corals. The source for alpha counting was prepared by finally extracting the Pa into TTA in benzene and evaporating the organic phase onto a stainless steel planchet. Joshi and Ganguly (1977) have also measured the [231]Pa in the surface layers of sediments by leaching with 1 M oxalic acid. the Pa was adsorbed on an anion exchange column, washed to remove Th, and then Pa eluted with hydrochloric/hydrofluoric acid mixture. The eluate was evaporated onto a Pt tray for counting. Mangini and Sonntag (1977) separated U and Th from deep sea cores and used the granddaughter of [231]Pa, [227]Th to estimate the age. The sediment was fused in sodium peroxide and any insoluble material was treated with nitric acid. Any precipitated silica was filtered off before the U and Th were extracted first with Aliquat-336, and then with TTA in benzene. The purified fractions of U and Th were electrodeposited using an ammonium chloride bath. A general discussion on the separation of U, Th, and Pa by ion exchange and solvent extraction has been reported by Kluge and Lieser (1980).

Lead is often determined in association with Po and Ra. The main isotope of Pb used in dating is [210]Pb. Measurement of this isotope can be made via the daughter [210]Bi, which is a beta emitter, or the granddaughter [210]Po, which is an alpha emitter. Lead has been removed from solution on ferric hydroxide (Craig et al. 1973), as lead chromate (Rama et al. 1961), a mixed ferric hydroxide/lead chromate (Thomson and Turekian 1976), by extraction as a complex with ammonium pyrrolidine dithiocarbamate (APDC) (Bacon et al.

1976), by precipitation as the carbonate (Nozaki and Tsunogai 1973*a*) or adsorption onto alumina filter beds (Schell 1977). Craig *et al.* (1973) coprecipitated Pb on ferric hydroxide by the addition of ammonia. The mixed hydroxides were dissolved in acetic acid and the Pb reprecipitated as the chromate. The lead chromate was dissolved in 1.5 M hydrochloric acid and passed through an anion exchange column in the chloride form to separate off the ^{210}Bi daughter. After elution of Pb from the column with water it was precipitated as the sulphate and mounted to provide a source for beta counting the ^{210}Bi through a 0.9 mg cm^{-2} Mylar window. The source was counted after equilibrium had been reached and at regular intervals over 30 to 35 days. Particulate matter, filtered from the water, was ashed, dissolved in nitric acid, and after passing through an anion exchange column in the nitrate form to remove any Th, Fe was added. Lead was coprecipitated on the ferric hydroxide and processed as for the water. The direct precipitation of lead chromate from natural waters was reported by Rama *et al.* (1961). They purified Pb on an anion exchange column, as did Craig *et al.* (1973), but used a final precipitate of lead chromate as the counting source for the ^{210}Bi daughter.

Thomson and Turekian (1976) used a mixed ferric hydroxide and lead chromate precipitation in their analysis of sea water. A ^{208}Po internal tracer was added before the precipitation by the addition of ammonia gas. After filtering off the lead chromate/ferric hydroxide, it was dissolved in 1 M hydrochloric acid, diluted to about 70 ml and 100 mg ascorbic acid added. The solution was heated to 85 °C and the Po was self-deposited onto an Ag disc. After plating the solution was wet oxidized with nitric and perchloric acids and evaporated to dryness. The residue was dissolved in 8 M hydrochloric acid and Fe, and U removed on a chloride anion exchange column. The Pb fraction was further purified of Bi on an anion exchange column in 1.5 M hydrochloric acid. The Pb was precipitated as the chromate, filtered, and weighed to measure the yield. The precipitate was then dissolved in 6 M hydrochloric acid, some ^{208}Po spike added, and the sample stored for some months to permit ingrowth of the ^{210}Po granddaughter. The ^{210}Po was then self-deposited on Ag as before and counted by alpha spectrometry to give the ^{210}Pb content from the growth of the ^{210}Po. Bacon *et al.* (1976) extracted the Pb from 20 l of filtered sea water by APDC chelate coprecipitation. The precipitate was filtered, heated in a muffle furnace, and Po plated out from a 2 M hydrochloric acid solution. The solution was stored for several months, as above, then Po deposited again onto a Ag disc. Lead carbonate has been used to concentrate Pb from 30 to 50 l of sea water (Nozaki and Tsunogai 1973*a*). After separating off the carbonate precipitate, it was dissolved in nitric acid, and then Po, Pb, and Bi reprecipitated as hydroxides. These were dissolved in 0.5 M hydrochloric acid, hydroxylamine hydrochloride (NH$_2$OH HCl) added, and Po self-deposited onto Ag at 70 to 90 °C. The solution was then made alkaline with ammonia, the hydroxide precipitates filtered off and stored for several months. After ingrowth of ^{210}Po the precipitates were dissolved in 0.5 M hydrochloric acid and the above procedure repeated. Up to 1600 l of water were filtered through alumina filter pads to remove the Pb (Schell 1977). The adsorbed Pb was eluted sequentially with 6 M hydrochloric acid and 6 M nitric acid. After evaporation to low volume, a ^{208}Po tracer was added and the samples were wet ashed with perchloric acid. After dilution to 100 ml with 0.3 M hydrochloric acid, ascorbic acid was added and the ^{210}Po deposited onto Ag in the normal way.

For the determination of ^{210}Pb in lake sediments and other solid matrices Denham, Anderson, and Bacon (1977) either leached the ^{210}Pb from the sediment with concentrated acid or subjected the sample to a total dissolution process with a mixture of nitric, perchloric, and hydrofluoric acids. However, Eakins and Morrison (1978) converted the Po to the chloride and distilled it from the sediment at $500°C$. Excellent agreement was obtained with other methods involving the dissolution of the sample and measurement of the ^{210}Pb via ^{210}Po or ^{210}Bi using conventional techniques.

Mass spectrometry on purified U, Th, and Pb fractions has been used as an alternative to alpha spectrometry (Sinha 1972; Krough 1973; Manton 1973; Allegre and Condomines 1976; Ludwig 1978). The use of delayed neutron counting for the determination of U and Th in rocks and minerals was reported by Gale (1967) and more recently for U preconcentrated onto charcoal from water by Brits and Dias (1978).

Megumi (1979) used gamma ray spectrometry for measuring various isotope ratios on separated fractions of the U-series and Ac-series nuclides in soil. Lead-210 has been measured by gamma ray spectrometry using the 46.52 keV energy peak in lake sediments and air filters (Gäggler, Von Gunten, and Nyffler 1976). Up to 10 g of lake sediments were counted in plastic dishes on a 5 cm^3 planar Ge detector. Air filters representing one months collection of about 25 000 m^3 of air have been counted in a similar fashion. Although the authors state their method is less sensitive than the measurement of the beta activity of ^{210}Bi, no time consuming sample preparation is required.

Two naturally occurring isotopes of Ra that are frequently measured are ^{226}Ra and ^{228}Ra. The former is a member of the ^{238}U decay series and can be measured via its gaseous daughter ^{222}Rn. This technique is known as Rn emanation. It can also be measured via its own alpha emissions following the coprecipitation of Ra onto barium sulphate. Measurement of ^{228}Ra can be accomplished through monitoring its daughter products, the beta emitting ^{228}Ac, or the alpha emitting ^{228}Th (see chapter 5).

Mackenzie et al. (1979) analysed sea water for ^{226}Ra using the emanation method. An outline of the apparatus is shown in Fig. 4.3. Samples of 20 l of water were filtered through a 0.22 μm Millipore filter and placed in the glass equilibration vessel. The ^{222}Rn gas initially present in the water was removed

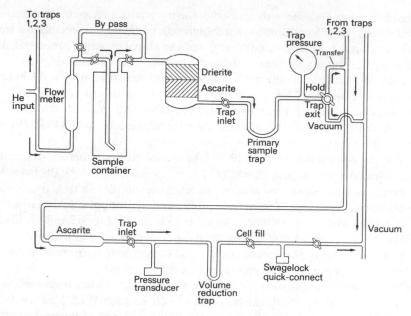

Fig. 4.3. Schematic of Rn stripping and transfer system (after Key, Brewer, Stockwell, Guinasso, and Schink 1979).

by the method described below. After ^{222}Rn removal the sample was sealed and stored for several days during which time the ^{222}Rn grows towards secular equilibrium with the ^{226}Ra (12 days corresponding to 88.7 per cent equilibration). The sample is then returned to the system for evacuating and filling with the gas. The gas is then pumped round the system to strip out the Rn. A solid CO_2/acetone trap is used to remove the water and the Ascarite absorbs any CO_2 in the circulating gas. The Rn is frozen in the glass trap at liquid nitrogen temperature. After isolating the trap and detector from the rest of the system, Rn is quantitatively transferred to a counting cell, coated with zinc sulphide on the inner wall, and isolated. The cell was then placed on the photomultiplier tube of an alpha scintillation counter for counting the ^{222}Rn. If the initial ^{222}Rn concentration in the sample is required this is measured before storing the sample. Key, Brewer, Stockwell, Guinasso, and Schink (1979) used a similar procedure to measure ^{222}Rn and ^{226}Ra in both marine sediments and sea water. Their apparatus was constructed using stainless steel tubing to reduce liquid nitrogen consumption due to the lower thermal conductivity. For sediments the samples were first measured as the total core sample, in the form of the slurry as obtained in the corer. Total Ra content was then determined following a total dissolution of a dried aliquot with hydrochloric, nitric, perchloric, and hydrofluoric acids as described by Ku (1965).

The use of an acrylic fibre loaded with an adsorbent for the in situ removal of Ra, Th, and Pb in sea water was first reported by Krishnaswami *et al.* (1972). They used a fibre loaded with ferric hydroxide and obtained enrichment factors of the order of 1000 from sea water. Isotopic ratios for ^{228}Ra/^{226}Ra activity and Th isotope ratios obtained were similar to those obtained by conventional chemical processing of hundreds of litres of water. Moore and Reid (1973) and subsequently Moore (1976) improved the technique for Ra and Th extraction by impregnating the fibre with manganese dioxide. Knauss, Ku, and Moore (1978) used this improved material for the determination of Ra and Th isotopes in up to 1000 l of sea water with 100 g of the fibre. Relatively large concentrations of ^{226}Ra and ^{234}Th were present in the sampling areas and they used these as natural tracers for other Ra and Th isotopes. Smaller volume samples of 20 l, and fibre samples were collected at the same location so that ^{228}Ra, ^{226}Ra, ^{232}Th, ^{230}Th, and ^{234}Th could also be measured. At stationary sites water was pumped through columns containing the Mn impregnated fibres for 2 to 3 hours at a rate of 5 to 15 l/min. The fibre was also towed behind the ship to give an average value for Ra and Th over tens of kilometres of sea. They reported a fibre extraction efficiency for Ra and Th of about 90 per cent based on the ^{226}Ra and ^{234}Th measured in the small volume sample. The fibres were returned to the laboratory for processing. The Ra and Th was leached with 6 M hydrochloric acid and then coprecipitated on barium sulphate. This was subsequently converted to the chloride and dissolved in dilute hydrochloric acid solution. After adding Fe the Th was coprecipitated as the hydroxide, leaving the Ra in solution. The precipitate was then centrifuged, washed, and dissolved in hydrochloric acid and purified in the usual way. The Th was finally extracted into TTA/benzene and evaporated onto a stainless steel plate for alpha counting. It was counted in an alpha counter for ^{232}Th, ^{230}Th, and ^{228}Th, and in a low background beta counter for the ^{234}Th via ^{234}Pa (see chapter 5 for details). The supernate was taken for Ra analysis. Barium carbonate was precipitated by the addition of sodium carbonate and then dissolved in the minimum volume of hydrochloric acid. Any entrained Th was extracted with TTA in benzene at pH 1 to 2. The pH was then adjusted to between 5.5 and 6, and Ac was extracted again with TTA in benzene and discarded. The solution was acidified to at least pH 1 and stored for around 30 h or longer. After storage, the sample was evaporated to dryness, dissloved in buffer, and diluted to pH 5.7, and Ac extracted with TTA in benzene. The organic phase was evaporated on a stainless steel planchet and counted on a beta counter for ^{228}Ac (see chapter 5). After the ^{228}Ac measurements the Ra solution was analysed for ^{226}Ra using the emanation method. When ^{222}Rn measurements were completed the sample was stored for 1 y for subsequent ^{228}Th measurement. Around the midpoint of the storage period, ^{230}Th spike was added. The Th isotopes were extracted, after storage, with TTA in benzene as before. Mackenzie *et al.* (1979) used a similar procedure for ^{228}Ra analysis. Moore (1969b) used a simple barium sulphate precipitation to remove the Ra from 700 l of filtered sea water. The processing of the barium sulphate was similar to that of Knauss et al. (1978) except that once the solution was free of ^{228}Th, Ra was precipitated with barium carbonate and this precipitate stored for 230 to280 d. During this time the ^{228}Th grew to 18 to 20 per cent of equilibrium with the ^{228}Ra in the sample. The barium carbonate was then dissolved in the minimum volume of hydrochloric acid, a ^{230}Th spike added, and the Th isotope extracted with TTA in benzene and evaporated onto a stainless steel planchet. Broecker et al. (1973) have also measured ^{228}Th/^{228}Ra ratios in sea water using a similar procedure.

Koide and Bruland (1975) developed a procedure for the simultaneous determination of U, Th, Ra, Pb, and Po in waters and sediments. For aqueous samples they coprecipitated the nuclides on aluminium phosphate . If Po was required this was measured, on a 0.5 M hydrochloric solution of the aluminium phosphate, by self-deposition on Ag. The solution was then evaporated to dryness and dissolved in 4 M nitric acid. If Po was not required then the aluminium phosphate was dissolved directly in 4 M nitric acid. Fuming nitric acid was added to coprecipitate Ra with lead nitrate. The supernate then contained U and Th and was purified by the usual techniques of ion exchange and solvent extraction. The lead nitrate was dissolved in hydrochloric acid and separated from Ra on an anion exchange column in 1.5 M hydrochloric acid. Radium, which was not retained on the column, was further purified from alkaline earths by separation on a cation column in dilute hydrochloric acid. The Ra was eluted with 8 M hydrochloric acid, evaporated to dryness and electrodeposited from a chloride bath. The source was flamed to remove ^{222}Rn and alpha counted for ^{226}Ra. The Pb retained on the anion column was eluted with water and precipitated as lead sulphate to provide a source for beta counting of ^{210}Bi.

The measurement of He·in rocks using mass spectrometry has been reported by Fisher (1972) and Schaeffer (1967). Bender (1973) measured the He in corals by mass spectrometry following extraction by acid dissolution in vacuum and equilibration with ^{3}He spike. Clarke and Kugler (1973) measured the dissolved He in groundwaters as a possible method for U and Th prospecting.

4.5 Source preparation

4.5.1 Sources for alpha spectrometry

In order to obtain good resolution with an alpha spectrometer it is necessary to produce a thin, flat, and uniform deposit for a counting source. Ideally the source should have a monoatomic layer of the alpha emitter with no further foreign material above this layer to attenuate alpha radiation. The source should also be capable of being handled, albeit with care, and be stable. All traces of solvent and acid must be removed to prevent damage to detectors and counting chambers when subsequently evacuated prior to counting. Several techniques have been used to obtain an adherent thin film, vacuum evaporation from a hot filament (Jackson 1960), direct evaporation, evaporation from a volatile organic solvent, and electrodeposition. For routine use the method should be simple, quantitative, reproducible, and relatively quick. Vacuum evaporation, although it produces excellent sources, is not readily adaptable to a routine working laboratory environment. Evaporation of an aqueous solution tends to form a ridged source which can result in alpha particle absorption, and therefore decreased resolution. The addition of tetraethylene glycol (TEG) to the solution prior to evaporation has been used but can often result in a poorly adherent deposit (Miller and Brouns 1952).

The evaporation of an organic solvent solution has been widely used, principally the complexes of U and Th with thenoyltrifluoroacetone (TTA) in benzene. Kaufman (1964) and Ku (1965) used this technique after first processing the sample through a separation scheme culminating in reasonably pure fractions of U and Th by ion exchange or solvent extraction separations. The solutions are then evaporated to dryness and treated with perchloric and nitric acids to oxidize residual organic material. After dissolution to give a 0.1 M nitric acid solution the pH is adjusted to 3 for U and 1.5 for Th with sodium hydroxide. The elements are extracted into a 0.4 M TTA solution in benzene. The extraction is carried out from small volumes of 5 ml. Small stoppered centrifuge tubes, or a small Pasteur pipette can be used, to mix the organic and aqueous phases without the introduction of excessive air. The separated organic phase is carefully transferred directly onto a stainless steel disc for evaporation. Even drying can be obtained by placing the steel disc on a heated brass cylinder. After drying the source is flamed to dull red heat to ensure removal of all traces of organic matter. Moore (1969a) used a similar TTA extraction in benzene at pH 6 to remove [228]Ac from solution. The beta activity of this [228]Ra daughter was used to determine the parent activity in the presence of large amounts of another Ra isotope, [226]Ra.

Electrodeposition is well established as a reliable technique for source preparation. The electrodeposition can be carried out from organic or aqueous solutions. Many procedures have been described in the literature, but no individual method appears to have been universally accepted. Electrodeposition from organic media, known as molecular plating, requires high voltages, around 500V, but is reported to be tolerant to the presence of Fe and Al in trace quantities. Rudran (1969) compared the electrodeposition of actinides from isopropyl alcohol and aqueous ammonium sulphate solutions, and concluded that electrodeposition from the organic solution was more rapid and quantitative for a wider range of actinides than the aqueous solution. Although tolerant of Fe and Al, these elements were deposited on the cathode and degraded the resultant alpha spectrum. When these trace elements were present in the aqueous electrolyte they remained in solution thus preventing degradation of the alpha spectrum. Parker, Blidstein, and Getoff (1964) described the molecular plating of Th, U, and rare earths from acetone, butanol, and isopropyl alcohol. Duschner, Born, and Kun (1973) also used molecular plating for the electrodeposition of Pa as fluoride from a range of organic solutions.

The majority of procedures for electrodeposition of actinides use aqueous solutions in either slightly acid or alkaline form. The addition of a complexing agent such as hydrofluoric acid, sodium sulphate, or diethylene tetramine penta-acetic acid (DTPA) have been used. Campbell and Moss (1965) electrodeposited actinides from an alkaline hypochlorite bath. The use of an ammonium sulphate bath is reported by Talvitie (1972) and gives quantitative results for U and Th in 1 and 2 h plating time respectively. Morgan (1971) used a similar bath with the addition of oxalic acid and obtained quantitative platings in 3 h for U and 6 h for Th. Using an ammonium chloride–hydrochloric acid electrolyte Mitchell (1960) reported 100 per cent yields for Th, 95 per cent for Pa, and 98 per cent for U in only 15 min plating. Puphal and Olsen (1972) used a mixed oxalate-chloride bath for the electrodeposition of a range of alpha emitting nuclides. The inclusion of oxalic acid reduced cathodic corrosion which was present with only chloride as the electrolyte. This enabled stainless steel to be used as the cathode instead of platinum. The effect of complexing agents HF and DTPA as well as interfering cations were also studied. Lally and Eakins (1978) used ammonium sulphate with sodium bisulphate and ethylenediamine tetra-acetic acid (EDTA) to obtain quantitative electrodeposition of U and Th in 3 and 5 h respectively. Sodium bisulphate is added to the sample before evaporation to prevent the actinides from baking onto the beaker during the drying procedure. This replaces the fuming with sulphuric acid in the Talvitie (1972) method. Ethylenediamine tetra-acetic acid is added to complex any residual Fe present in the solution and to obtain a clean plate.

In electrodepositions from aqueous media a stainless steel, platinum or copper disc is used for the cathode onto which the actinide is deposited. A stationary or rotating platinum rod is used for the anode. Perspex, PTFE, and polyethylene have all been used to manufacture the electrodeposition cells. Where chloride is incorporated into the electrolyte it may be necessary to provide some exhaust system to remove the chlorine produced during the electrodeposition. A prerequisite of the methods of electrodeposition is a purified solution of the element under consideration, usually obtained after ion exchange or solvent extraction separation. Experiences with the commonly used electrodeposition baths have been reported by Holmes (1967) and Irlweck and Sopantin (1975) reviewed the available electrolytes for the production of calibration sources for alpha spectrometry. Table 4.2 gives a summary of the conditions for the electrodeposition of U and Th given above.

The spontaneous deposition of Po onto Ag discs is widely accepted as the standard method of producing a Po source. Deposition is carried out by immersing a Ag disc in hot hydrochloric acid solution at 80 to 90°C. The method is subject to interference from oxidants, organic materials, and other elements which will self deposit on Ag. Preliminary separation of the Po can be carried out using calcium tannate (Smales, Airey, Woodward, and Mapper 1957) or by co-precipitation with Te (Rundo 1959). Flynn (1968) gives a method for the determination of Po in aqueous effluents and rock samples. The liquid samples are treated with perchloric and nitric acids to oxidize organic material present and finally dissolved in hydrochloric acid. Sodium citrate, hydroxylamine hydrochloride, and 10 mg of Bi carrier are added. The pH is adjusted to 2 with ammonia solution and the sample then warmed to 85 to 90°C for 2 to 3 min to reduce Fe and Cr. The Ag disc is then inserted and, using constant agitation, the deposition is continued for 75 min at 85 to 90°C. The disc is then removed, washed in water and alcohol, and when dried is ready for counting. Rock samples are dissolved in HF/HNO_3 converted to hydrochloric acid solution and treated as for aqueous solutions. Ascorbic acid has also been used to reduce iron prior to the deposition of Po (Turekian 1973).

Table 4.2. Electrodeposition parameters for U and Th

Ref	Electrolyte	Cathode	Current (Amp)	Electrodeposition time (min)	Recovery (%)	Element
(1)	$(NH_4)_2SO_4$	Stainless steel	1.2	60	98	U
				120	100	Th
(2)	NH_4Cl/HCl	Platinum	2	15	94–98	U/Th
(3)	$NH_4Cl/(COOH)_2$	Stainless steel	0.3	180	100	U
				360	100	Th
(4)	$NH_4(COO)_2/NH_4Cl$	Stainless steel	3–6	40	98–100	U/Th
(5)	$(NH_4)_2SO_4/EDTA$	Stainless steel	1	180	100	U
				300	100	Th

References   (1) Talvitie (1972)
              (2) Mitchell (1960)
              (3) Morgan (1971)
              (4) Puphal and Olsen (1972)
              (5) Lally and Eakins (1978)

4.5.2 Sources for gross alpha and beta counting

In certain chemical procedures the final counting source is retained as a precipitate. The production of a uniform source using a precipitate from solution is best obtained with the use of a demountable centrifuge tube or demountable filter stick, incorporating a tared filter paper. In the determination of ^{226}Ra a final mixed Ra–Ba sulphate is counted as an example and the use of a standard amount of carrier to act as precipitant gives reproducible size sources. In order to determine yields and to correct for counting efficiency changes due to self adsorption within a source, the source is dried to constant weight before counting. By reference to a source weight versus counting efficiency calibration chart the correct counting efficiency may be determined. Asikainen and Kahlos (1979) used a special centrifuge tube containing a plate to obtain a uniform source and Harley (1974) described a typical demountable filter.

4.5.3 Sources for mass spectrometry

It is essential that users of mass spectrometers prepare the sources in accordance with manufacturers specific instructions but in general the final purified fraction of the element is evaporated onto a Re or W filament prior to measurement.

4.6 Summary

There is a wide variety of published methods for the determination of U, Th, Pa, Pb, Po, Ra, Rn, and He in environmental samples. The choice of a method to be used will differ according to the combination of elements to be determined. Some methods are designed for the determination of one or two elements, others for six or seven elements. Separation schemes for U and Th are in general modifications to the established procedures published by Ku (1965), and Bhat (1969). For other elements such as Pb and Po the choices are again based on classic methods such as lead sulphate or Po self-deposition on Ag (Craig et al. 1973; Thomson and Turekian 1976). The procedures for Ra and Rn are limited to Rn emanation (Mackenzie et al. 1979; Key, Brewer, Stockwell, Guinasso, and Schink 1979) or ^{226}Ra and ^{228}Ra via barium sulphate (Moore 1969a) or the use of the recently introduced acrylic fibres (Knauss et al. 1978).

Combination methods for a variety of elements have been published by Hussain and Krishnaswami (1980) for U, Th, Pb, Ra, and Rn, and Benninger (1976) for U, Th, Pb, Po, Ra, and Rn. The choice of a counting source for alpha spectrometry is between an evaporated organic phase such as TTA or electrodeposition. There are many options for electrodeposition electrolytes (see Table 4.2) and different authors have their own preferences.

5 SPECTROSCOPIC METHODS
M. Ivanovich

5.1 Interaction of nuclear radiation with matter

The study of the interaction of nuclear radiation with matter contains the clues to the methods of detection of alpha, beta, and gamma radiation. Indeed, in order to consider spectroscopic methods utilized in the U-series disequilibria measurements, it is necessary to consider first some basic principles involved in the interaction of radiation with matter.

The main process by which a charged particle from a radioactive decay loses energy in passing through matter is interaction with atomic electrons through the Coulomb force. When the work necessary to excite these electrons to new levels or to remove them from the atom is small compared with the incident particle energy, this process may be regarded as *elastic*. The nuclei of the stopping medium experience a similar but smaller loss of energy. However, nuclear encounters are of prime importance in determining the scattering of charged particles on their passage through matter. The direct removal of electrons from neutral atoms by the incident particle is the *primary ionization*. The ionization produced by these electrons, referred to as *delta rays*, is known as the *secondary ionization*. Since the delta rays have a short range, the only measurable quantity is the *total ionization*. When the incident particle is no longer able to ionize, it is said to have reached the end of its *range* in the stopping medium and, in the case of particles heavier than the electron, it reverts to a neutral atom.

The interaction of alpha particles with matter is usually much less complicated than the interaction of electrons with matter. Alpha particles lose most of their energy through ionization and excitation of the atoms in the absorber. The paths of these particles tend to be straight. The acquisition of atomic electrons by an alpha particle, on passage through matter, was shown to depend strongly upon its velocity (Henderson 1922; Rutherford 1924; Briggs 1927; Kapitza 1924; and others). These authors have also shown that a regular exchange of electrons between the moving alpha particle and its absorbing medium takes place. About a thousand exchanges occur along the path of a single alpha particle, the interchange becoming most rapid as the alpha particle velocity declines near the end of its range. In each ionizing event an originally neutral atom of the absorbing material is divided into a free negative electron and a residual positive ion. Depending on the nature of the nearby atoms, the liberated electron may remain free or it may become attached to a neutral atom to form a negative ion. The term *ion pair* means the residual positive ion and its negative counterpart, regardless of whether the electron is free or attached. Along the path of an alpha particle in air, some 2000 to 6000 ion pairs per mm are produced, depending on the velocity of the particle at the point of consideration. The number of ion pairs per unit path length is known as the *specific ionization*.

Experimental observations practically always give *total ionization* which includes, in addition to the primary ions all secondary ions due to the absorption of delta rays and the radiation resulting from the primary events. From the specific ionization and the specific energy loss, the average expenditure of energy to form an ion pair (ω) is obtained. Roughly an average expenditure of about 35 eV is required for each ion pair produced in air, the exact value depending upon the alpha particle velocity.

An average alpha particle emitted from a radionuclide makes about 10^6 collisions resulting in small discrete energy transfers to electrons before coming to rest in an absorber. Because of the statistical nature of the energy loss, after passage through a certain thickness of matter, a group of such particles of initially uniform velocity will show a distribution of velocities about a mean value. Therefore, the ranges of the particles will be grouped about a mean value. Since the number of collisions is large, and since they are independent processes, the range (or energy) distribution may be expected to be approximately Gaussian in shape. This phenomenon constitutes the basis of line shapes observed in alpha spectrometry discussed in section 5.3.1.

Electrons penetrating matter lose energy and are scattered. The penetrated matter is also subject to change. The constituent atoms are excited or ionized, and dissociation of molecules, changes in the lattice structure of crystals, changes in the conductivity, and other secondary processes have been observed. For the energy range of the beta-emitting radionuclides, 10 keV to 10 MeV, the deflection of the electrons is due almost entirely to the elastic collisions with the atomic nuclei, while the energy loss results almost exclusively from the inelastic interaction with the atomic electrons. Thus, it is possible to treat the two phenomena separately, although they always occur together. The general behaviour of positrons is the same as that of electrons.

When electrons of a definite energy pass through a foil of matter, a decrease in their energy is observed. This energy loss is due to the inelastic collisions of the incident electrons with the atomic electrons, by which the atoms are excited or ionized. The interaction of the incident electrons with the atomic electrons in the foil is characterized by a very small energy transfer per collision from the electrons to the atoms. Even for very high primary energies excitation is more probable than ionization, and in the latter case the resulting secondary electrons have a mean kinetic energy of only a few eV. The total energy loss after passage through a foil is, therefore, the result of a very large number of small energy losses. It has been shown experimentally for a large range of electron energies (5 keV to 17 MeV), that the energy required to form an ion pair in a gas is nearly independent of the energy of the primary electron and is not very different for different elements. Thus, the kinetic energy of the primary electron can be determined by measuring the number of ions found.

A catalogue of the possible processes by which the electromagnetic field of the gamma ray may interact with matter has been put into a systematic form by Fano (1953). In theory, there are twelve different processes by which gamma rays can be absorbed or scattered. However, in the energy region of 0.01 to 10 MeV, most of the interactions are explainable in terms of just three: Compton effect, photoelectric effect, and pair production. These are depicted diagramatically in Fig. 5.1. If the small effects due to the binding of the atomic electron to the nucleus are neglected, the scattering of an incident energetic photon (gamma ray) by an electron can be described as a simple two-particle collision (Compton 1922). Denoting the frequency of the incident gamma ray, by v_0, the frequency of the scattered gamma ray by v, the incident direction by the angle θ, and the kinetic energy, T, of the electron, coming off at an angle ϕ, Compton scattering can be represented schematically by diagram (a) Fig. 5.1. The wavelength change is independent of the wavelength, and the fractional energy loss in individual Compton processes is quite large for energetic photons.

Below energies of about 0.1 MeV, the predominant mode of gamma ray interaction in all medium—and high—Z absorbing material is the photoelectric process (Evans 1955). As shown in diagram (b) of Fig. 5.1, the primary photon is completely absorbed and a photoelectron is ejected at some angle θ with energy $T = hv_0 - B_e$, where B_e is the binding energy of the ejected atomic electron. Because the presence and participation of the atom are essential, the photoelectric interaction is thought of as one between the incident photon, hv_0

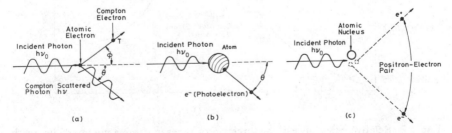

Fig. 5.1. Schematic representation of (a) Compton scattering; (b) the photoelectric process; and (c) pair production.

and the entire atom. The remainder of the energy appears as characteristic X-rays and Auger electrons from the filling of the vacancy in the inner (K or L) shell where the photoelectric reaction mostly takes place.

Above incident photon energies of 1.02 MeV, the third type of interaction becomes increasingly important. In this interaction, known as pair production, the photon is completely absorbed and replaced by a positron and electron pair whose total energy (kinetic plus rest mass) is just equal to $h\nu_0$. The process is shown schematically in diagram (c) of Fig. 5.1. Thus,

$$h\nu_0 = (T_- + m_0 c^2) + (T_+ + m_0 c^2)$$

where T_- and T_+ are the kinetic energies of the electron and positron, respectively, and $m_0 c^2 = 0.51$ MeV is the electron rest energy. The process occurs only in the field of charged particles, mainly in the nuclear field but also to some degree in the field of an electron.

The relative importance of the three processes described above is shown graphically in Fig. 5.2 in terms of linear attenuation coefficients σ (Compton scattering), τ (photoelectric process), and κ (pair production). As can be seen from the diagram, photoelectric collisions are important only for small $h\nu$ and large Z. Pair production is of major importance only for large $h\nu$ and large Z. Compton collisions predominate in the entire domain of intermediate $h\nu$, for all Z.

Thus, the effects produced by photons in matter are almost exclusively due to the secondary electrons. A photon produces *primary ionization* only when it

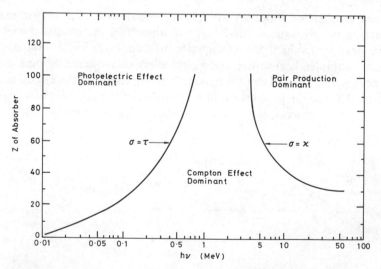

Fig. 5.2. Relative importance of the three major types of gamma ray interaction with matter. The lines show the values of Z and $h\nu$ for which the two neighbouring effects are just equal (Evans 1955).

removes an electron from an atom by a collision (photoelectric or Compton), but from each primary ionizing collision the fast secondary electron which is produced, may have nearly as much kinetic energy as the primary photon. This secondary electron dissipates its energy mainly by producing ionization and excitation of the atoms and molecules in the absorbing matter. If on the average, the electron loses about 32 eV per ion pair produced, then a 1 MeV electron produces of the order of 30 000 ion pairs before being stopped. The one primary ionization is thus completely negligible in comparison with the very large amount of secondary ionization.

General review articles on the subject of interaction of alpha particles with matter have been written by Taylor (1952), Bethe and Ashwin (1953), and Allison and Warshaw (1953), and on the subject of beta particles with matter by Bothe (1932), Bethe and Ashwin (1953), and theoretical principles have been discussed in detail by Seuter (1953), and Mott and Massey (1948). The subject of alpha, beta, and gamma radiation interaction with matter is treated at some length in Evans (1955), Burcham (1963), Siegbahn (1965), Marmier and Sheldon (1969), and many others.

5.2 Detection of nuclear radiation

5.2.1 Radiation detectors

The general principle behind most methods of detection of nuclear radiation (notable exception being electromagnetic spectrometers) is that whatever the form of the radiation, it gives up some or all of its energy to the medium of the detector, either by ionizing it directly, or by causing the emission in it of a charged particle, which in its turn produces ionization in the medium. The ionization thus produced is then detected by one of several methods: (i) the charge may be collected directly by electrical means, as it is in gas ionization and semiconductor detectors; (ii) it may cause the emission of photons, as in scintillation detectors; (iii) it may leave a trail of ionized atoms in the detection medium along which bubbles may condense or chemical change occur, as in cloud or bubble chambers and photographic emulsions; (iv) it may cause permanent damage to the medium which can then be made visible, as in solid state track detectors. Not all of the above types of detection are relevant to the methods of U-series disequilibria measurements, and therefore, the principles of operation of only some of these detectors are discussed below. For general interest, the reader is referred to Staub (1953); Sharpe (1960); Burcham (1963); Siegbahn (1965); Cuninghame (1975).

(i) *Gas ionization detectors.* In a gas ionization detector the counting gas fills a space containing two electrodes, between which there is an electric field. When radiation causes ionization in the gas, the electrons produced are drawn rapidly to the anode while the heavy positive ions (or *positive holes*) move more slowly to the cathode. The result is a current flowing through the circuit which may be used as a measure of the amount of radiation causing the ionization.

The output pulse height from an ionization detector varies with applied voltage as shown in Fig. 5.3. Region I is the ion chamber region. Two main types of ion chamber are recognized: direct current (d.c.) and pulsed. The basic difference between them is the duration of the output pulse, the latter having a much shorter time constant. In a typical pulsed ion chamber the electrons from one ionizing event will be collected in a time of the order of 10^{-6} s, but the positive ions being much heavier, and therefore, less mobile will take about 10^{-3} s. This causes the time between accepted pulses, *the dead time*, to be rather long. In addition, the positive ions left behind after collection of the electrons affect the size of the pulse produced by the latter, and the magnitude of this effect depends on their distance from the anode, and hence on the position of the ionization track in the chamber, thus destroying the proportionality of pulse height and energy. The solution to this problem as suggested by Frisch (1944), consisted of placing a grid at negative potential across the chamber in such a way that the electrons in the region between this grid and the anode

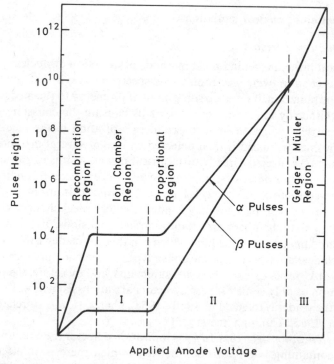

Fig. 5.3. Variation of pulse height in gas ionization detectors as a function of applied high voltage.

would be screened from the positive-ion space charges near the cathode, and that their collection times should, therefore, be independent of position. The counting gas must not give rise to negative ions and the electrons must have as high a drift velocity as possible so that the charge can be collected quickly (argon containing 10 per cent methane is commonly used). Gridded ion chambers are used extensively in alpha pulse-height analysis and Fig. 5.4 shows a design in extensive use at AERE Harwell.

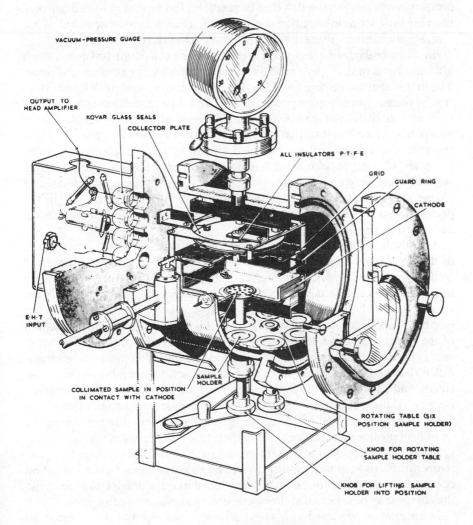

Six position gridded ionization chamber.

Fig. 5.4. Gridded ion chamber used at AERE Harwell for alpha spectrometry (Glover 1956).

In region II of Fig. 5.3, the proportional region, the pulse height at any particular applied voltage is also proportional to the energy of the radiation, provided the detector is large enough to contain the whole of the ionization track from the event. If the voltage is increased to that of region III, the detector becomes a Geiger–Müller (GM) detector. In this case, the *secondary avalanches* expand along the anode under the influence of the emission of photoelectrons, and a large pulse independent of the energy of the original ionizing particle is produced. As can be seen, this last type of gas-filled detector has very little scope in radiation spectrometry. (For detailed discussion of gas ionization chambers readers are referred to Wilkinson (1950).)

(ii) *Semiconductor detectors.* If the gas dielectric in a parallel-plate gas-filled ion chamber is replaced by a solid dielectric two immediate advantages accrue. The first is that as the density of the solid dielectric is very much higher than that of the gas, the stopping power is improved. The second advantage is that in most solid dielectrics the average energy necessary to form an electron–hole pair is very much less than that required to form an ion pair in the gas. Thus, the statistical accuracy of the collected charge is, in principle, much better. However, the electric fields which have to be applied to achieve efficient charge collection are very large because of charge carrier trapping by defects, and charge carrier scattering by thermal phonons in the material. Only in the perfect single crystal is this trapping sufficiently small to allow useful charge collection to be obtained.

As a result of their proximity in crystalline material, the sharp energy levels of the constituent atoms have become broad energy bands. This is illustrated by diagram (a) in Fig. 5.5. Electrons fill up these bands from the lowest energy up to the top of the *valence band*. The band above the valence band is the *conduction band*, and conduction can occur because of the mobility of any electrons which may have been excited into this band. The electron population of the conduction band depends on the temperature and on the difference in energy between the valence and the conduction bands: this energy difference is called the *energy gap*. In the most frequently used semiconductor materials, silicon and germanium, the energy gaps are 1.1 eV and 0.66 eV, respectively. Because of these relatively large energy gaps there are only a few electrons in the conduction bands of these semiconductors, the number being an equilibrium value depending on the temperature, but the much smaller gap for germanium means that detectors made of it, must be operated at low temperatures to keep this number down and so reduce *noise*: when ionization occurs in a semiconductor, electrons are excited into the conduction band, and the charge may be collected at the electrodes by application of an electric field.

In an *intrinsic semiconductor* (a pure crystal described above) there is an equal number of electrons (n) and holes (p) in the conduction and valence bands respectively. If impurity atoms are present in the crystal lattice, the impurity band level usually comes within the energy gap (see diagram (a) of

Fig. 5.5). Impurities of n-type, or donors, donate electrons to the conduction band, while impurities of p-type, or acceptors, accept them from the valence band. Thus, the numbers of holes and electrons in the valence and conduction bands are no longer equal. A detector composed of either pure p-type or pure n-type material is called a *homogeneous conduction detector*, while one which has a combination of the types is a *junction detector*.

The general principle behind a junction detector is that if n- and p-type materials are in contact and a bias voltage is applied such that it is positive to the n-type material, the free charge carriers (electrons from the n-type and holes from the p-type) are drawn away from both sides of the junction, leaving a layer called the *depletion* or *barrier layer* which is an excellent intrinsic semiconductor (see diagram (b) of Fig. 5.5). The thickness of the depletion layer is proportional to $\sqrt{V\rho}$ where V is the applied voltage, and ρ the resistivity in ohm cm. When ionization occurs in the depletion layer the resulting electrons and holes are swept rapidly out by the applied field and the charge is collected.

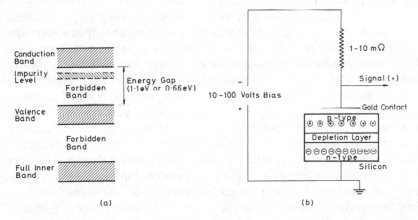

Fig. 5.5. (a) Schematic representation of the electronic band structure and impurity level in semiconductor. (b) Illustration of a junction counter and its circuit arrangement.

Junction detectors are the most commonly used semiconductor detectors. The main types are the diffused junction, the surface barrier, and the lithium drifted. Diffused junction detectors may be prepared with either n- or p-type material. In the case of p-type material, boron is diffused into silicon and the n-layer is made by painting on phosphorous and diffusing it into the material at high temperature simultaneously with diffusion of aluminium from the other face to make an electrical contact. Surface barrier detectors are made from n-type material whose surface is allowed to oxidize. A thin layer of gold is then evaporated to act as a contact. It also influences the oxide film to form a p-layer over a period of about a week. Depletion layer thickness for both types of detectors are limited to about 2 mm only. Lithium drifted detectors are prepared from p-type material into which lithium is diffused at high temperatures with bias applied. By this technique depletion depths of several centimetres can be achieved, and hence, lithium drifted germanium detectors are preferred for gamma spectroscopy. For charged particle spectroscopy, however, silicon detectors are normally used. Because of their higher resolution, the surface barrier and diffused junction detectors are preferred for the charged particle spectrometry over other types of detectors discussed in this section*. For further details on the subject of semiconductor detectors readers are referred to England (1974), Bertolini and Coche (1968), and Dearnaley and Whitehead (1961).

(iii) *Scintillation Detectors.* Scintillation detectors in one form or another have been used for the detection or spectroscopy of almost every form of radiation. As in all other types of detectors, the basic principle is that ionization is caused in the medium of the detector by the slowing down of the particle or secondaries from gamma radiation, but, whereas in gas or semiconductor detectors the ionization is measured electrically, in a scintillation detector it first causes the emission of light which is then either viewed optically or is converted to an electrical pulse by means of a photomultiplier. There exists a group of transparent dielectric materials including many of the noble gases (for example, Xenon), organic and inorganic single crystals, polycrystalline materials or polymers, and organic liquids based on benzene ring compounds, which are transparent to some part of the wavelength spectrum of the photons emitted by the excited atoms and ions along the path of the charged particle (see Garlick 1952). These photons can, therefore, be detected outside the transparent medium, and it is found that their number is proportional to the energy lost within the material in exactly the same way that the number of ions produced within a gas dielectric detector is proportional to the energy lost in the detector. The main use of scintillators is

---

* The resolution of a detector, which represents the ability of the detector to separate lines of different energies, is measured in terms of the *full line width at half maximum* (FWHM) or the *relative line width*. Different causes contribute to the broadening of a mono-energetic particle line. The main causes are statistical fluctuations in the number of carriers created and in the energy losses; influence of the thickness of the source (as well as the entrance window of the detector); various noise sources which are related either to the detector or the associated electronic circuits (see section 5.2.2). The half-width of the line, L is the quadratic sum of the partial widths, L_i introduced by each of the listed contributions:

$$L^2 = \sum_i L_i^2$$

Neglecting the noise contributions from the detector itself and from the amplifier, Engel–Kemeir (1967) has estimated that the intrinsic line width for a 5.5 MeV alpha particle detected in a surface barrier detector should be less than 9 keV.

in the detection of electrons, either directly from beta particle emitting sources, or from the interaction of gamma rays and X-rays with the scintillator. (The best known examples of these detectors are NaI(Tl) and Ge(Li) detectors). Detailed discussions of the theory and application of scintillation detectors can be found in England (1974), Siegbahn (1965), and Burcham (1963).

5.2.2. Ancillary equipment

The essential parts of any simple electronic counting system are shown in block schematic form in Fig. 5.6. The electrical pulses produced by the detector may range in size from about 1 mV for a gas ionization chamber to 50 V for some Geiger–Müller detectors, and therefore, the amplification system will not be the same in all cases. The amplification system is usually split into two stages, a *head* or *preamplifier unit*, and a *main amplifier unit*. A detector is a high impedance device, and the headamplifier, which is placed as close to the detector as possible to avoid pulse distortion, may do some amplifying and pulse shaping but its chief purpose is to match the pulses into low impedance cables so that the main amplifier and other parts of the counting system may be situated more conveniently elsewhere. The head-amplifier is also a convenient point of entry for the voltage supply to the detector. This voltage may range from a few volts for some semiconductor detectors to several kilovolts for gridded ion chambers and proportional detectors.

The main amplifier amplifies and shapes the pulses to the required degree, both gain and time constants usually being variable. The necessity for pulse shaping arises from the existence of long tails on the detector pulses. Pulses with long tails may be subject to *pile up*, that is to say, a new pulse may arrive while the system is still responding to a previous pulse. As pointed out earlier,

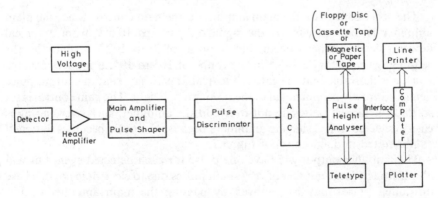

Fig. 5.6. Block diagram of a typical electronic system used with a nuclear radiation detector interfaced to a computer and its peripherals.

one of the aims of a detector design is to collect the information about a counting event as quickly as possible. Thus, the important sections of a detector pulse are its leading edge and its top. Therefore, the object of pulse shaping is to preserve maximum information while reducing the pulse duration as much as possible. This process is termed *clipping*, and is carried out most commonly by means of a resistance–capacitance (RC) circuit. The two types of RC circuit, called differentiating and integrating circuits, are shown in diagrams (a) and (b) of Fig. 5.7, respectively. Differentiation affects mainly the pulse top and causes an overshoot, while integration alters its leading and trailing edges. The total effect of a practical network of these two types is shown in diagram (c) of Fig. 5.7. In this example, equal differentiation and integration time constants of 1 μs were used. (The rectangular shape of the input pulses was chosen for the diagram to show more clearly what effect the two RC circuits have on input pulses.)

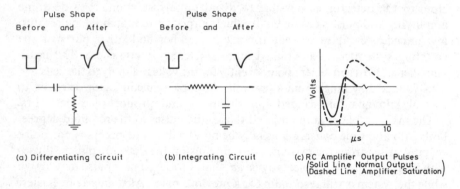

Fig. 5.7. Pulse shaping in RC amplifiers.

The other control on the main amplifier is the gain control. Since the main amplifier is used to convert the signal of perhaps 10 mV from a typical detector-headamplifier system into a pulse of 1 to 100 V amplitude, the required gain factors may be in the range of 10^2 to 10^4. For spectroscopic work it is desirable that the gain is linear, that is to say, that the output pulse height is directly proportional to the input pulse height. The gain control is set so that the largest pulses do not overload the amplifier; such overloading is easily detected in RC shaping amplifiers because the pulse becomes flattened as illustrated in diagram (c) of Fig. 5.7.

The amplifier output will have one pulse for each detected event but will also include a large number of very small pulses due to electronic noise. These unwanted pulses may be removed by passing the main amplifier output through a *discriminator*, a device which generates an output pulse when triggered by an input pulse whose amplitude exceeds some preset value.

For spectroscopic studies such as the measurement of energy spectra of the particles emitted by a radioactive source, a suitable detector and recording system must be chosen. The pulses, whose amplitudes will be proportional to the detected particle energies, are fed to a *pulse height analyser* after amplification and shaping. The pulse height analyser sorts the pulses according to their amplitudes thus forming the required spectrum. The pulse enters an analogue-to-digital converter (ADC) which gives out a digital signal proportional to the pulse height. This signal is then used to add one unit into the memory word or *channel* corresponding to the size of the digital signal. The memory may have a number of channels, 8192 (2^{13}) for example, and usually, it can be split up into smaller number of channel groups from 2^6 to 2^{13}. The rest of the pulse height analyser consists of units performing such functions as automatic timing, spectrum manipulation and analysis, and output of spectral data. These devices usually record data on magnetic or punched paper tapes, as well as control plotters and teletypes for hard copy. Thus, the spectroscopic data can be analysed 'off line' in a computer.

5.3 Methods of detection and measurement of radionuclides

5.3.1 Alpha spectrometry

The choice of the detector for alpha spectrometry application is governed by (i) the detector's high energy resolution; (ii) its high counting efficiency* suitable for low activity measurement; (iii) low background; (iv) speed of response, capable of handling high counting rates; and (v) high tolerance for beta and gamma radiation. Table 5.1 gives the optimal values of energy

Table 5.1. Comparison of resolutions achieved by some alpha spectrometers

Spectrometer	Alpha energy (MeV)	FWHM (keV)	Experimental conditions	References
Surface barrier detector	5.48	11	$7\,mm^2$, $-30\,°C$	Siffert (1966)
Ionization chamber	5.681	14	$Ar + 0.8\% \ C_2H_2$	Vorob'ev, Komer, and Korolev (1962)
Magnetic spectrometer	6.110	3.5	Double-focusing instrument	Baranov, Zelenkov, Schepkin, Beruchlov, and Malov (1959)
Scintillation detector	5.305	95	CsI (Tl)	Martinez and Senftle (1960)

* Counting efficiency here refers to the source-detector geometry, and is given by the factor representing that fraction of the total radiation emitted by the source which is recorded by the detector. In this context, counting efficiency is a function of (i) radiation-sensitive area of detector presented to source; (ii) size of active area of source, and (iii)distance between source and detector.

resolution obtained with the most common types of alpha spectrometers for energies between 5 and 7 MeV (after Siffert *et al.* 1968). Clearly, magnetic spectrometers have the best resolution but suffer from low counting efficiency, and are very expensive and complicated. Resolution of scintillation detectors is not comparable to that of the other three types of detectors and is mostly inadequate for the requirements of analytical applications.

Silicon surface barrier detectors of 60 micron depletion depth and about 100 mm^2 surface area will consistently give a resolution of better than 14 keV (FWHM) for 5 MeV alpha particles but with the relatively low counting efficiency of less than 10 per cent. Resolutions of the order of 20 keV (FWHM) are obtainable with 450 mm^2 area surface barrier detector having counting efficiency of about 20 per cent. The surface barrier detector has the advantage of compact size, simple power supply requirements, low gamma ray sensitivity, and relatively low cost. Their main disadvantage is that their active surfaces can be damaged easily by the effect of oil and grease in the source chamber atmosphere, and have an energy resolution which is inversely proportional to their surface area as shown above.

The gridded ion chamber of the construction shown in Fig. 5.4 (Glover 1956), gives a resolution of 25 keV (FWHM) with a thin source at a counting efficiency of 40 per cent. Although the achievable resolution is not comparable with the best of surface barrier detectors, there are applications for which the gridded ion chamber is the more suitable instrument. Thus, for alpha spectrometry the two detectors should be regarded as complementary. For example, because of its high counting efficiency the gridded ion chamber is the best instrument for determining the constituents of samples where the specific activity is low and the quantity of material is limited. Additionally, the gridded ion chamber can accommodate sources up to 10 cm in diameter so that, for low activity samples, the source thickness can be reduced effectively by spreading the activity over a larger area, or alternatively, the count rate can be increased by increasing the area of the deposit. Figure 5.8(a) shows a typical calibration spectrum obtained with a gridded ion chamber; and Fig. 5.8(b) a similar calibration spectrum obtained with a 450 mm^2, 100 micron depletion, surface barrier detector illustrating the better resolution obtained by the latter device (note the appearance of the individual doublets for each radionuclide in diagram (b), absent in diagram (a)) (Ivanovich 1980, unpublished). Energies of the peaks in an unknown sample can be read off subsequently from the plot of energy (MeV) against peak channel number shown in the diagram. The countrates due to the constituent radionuclides present in a mixture of this type are obtained by integrating the counts under each peak, after correcting for background and low energy tail.

The shape of the spectrum is an important consideration since a non-gaussian tail is almost always present on the low energy side of an alpha peak. The low energy tails make the lower energy lines of weak intensity difficult to

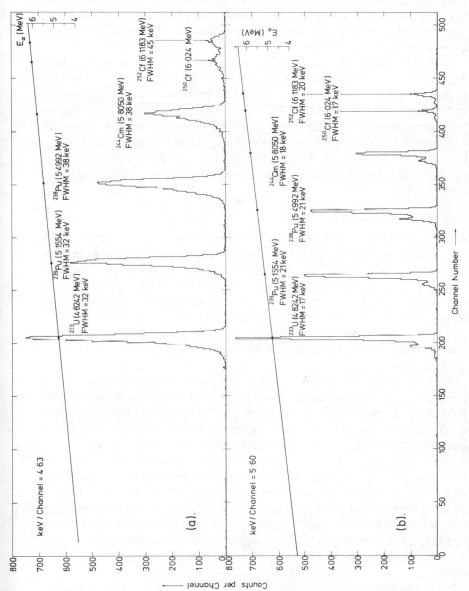

Fig. 5.8. (a) A typical gridded ion chamber calibration spectrum with the alpha energy against channel number plot. (b) A typical 450 mm² silicon surface barrier detector calibration spectrum with the alpha energy against channel number plot.

resolve. The tail extends down to zero energy and its area is of the order of one per cent of the main peak area. Nevertheless, the use of collimators, which eliminate some of the scattered particles, reduces the non-gaussian tail by about 30 per cent (Chetham–Strode, Tarrant, and Silva 1961). Of course, for a thin alpha source the peak shapes are near-gaussian and the tail corrections are negligible.

Typical background countrates for a surface barrier detector are about 0.001 to 0.003 counts per minute (cpm) whereas typical gridded ion chamber background in the energy range of 4 to 8 MeV tends to be of the order of 0.010 to 0.050 cpm. As can be seen, alpha spectrometry utilizing surface barrier detectors and gridded ion chambers is a very precise and sensitive technique with a lower limit of detection of a few disintegrations per hour. Specific alpha spectrometric procedures adopted for the U-series disequilibria measurements by the majority of the laboratories in the field are discussed in section 5.5.

5.3.2 Beta spectrometry

A beta particle has a very much higher velocity than an alpha particle of the same energy because it is much lighter. Consequently it spends less time in the vicinity of an individual atom of the counting medium and thus produces far fewer ion pairs per unit length of track. The net result is that (i) beta particles have a much larger range than alpha particles (a 5 MeV beta particle has a range of about 10 mm in silicon compared with only 0.03 mm for a 5 MeV alpha particle); and (ii) the beta particle tracks are very erratic. However, total *path length* can be much greater than its range because of the numerous large-angle scatterings.

With the exception of the expensive magnetic spectrometers (Siegbahn 1965), the only simple and accurate method of measuring the spectra of beta particles and conversion electrons with good resolution is that based on a semiconductor detector. A 5 mm depletion depth lithium-drifted silicon detector is sufficiently thick to stop beta particles having energies up to about 2.5 MeV. All that is necessary is that the detector and a thin source are mounted in a vaccum chamber with a suitable collimator in front of the detector to prevent the particles from entering it at the edges. Resolutions of 30 keV (FWHM) are easily attained and this can be improved by cooling the detector, or for lower energies, by using thinner detectors. Both surface barrier and lithium-drifted silicon detectors can be used for electron spectrometry. A comparison of their characteristics is given by Bertolini and Rota (1968).

To measure the absolute disintegration rate of a beta source it is usual to employ a 4π counter; Geiger–Müller and proportional, gas flow and filled counters, have been used. The end-window proportional detector, however, is the most important instrument for this purpose today.

5.3.3 Gamma spectrometry

Gamma ray spectrometry utilizing the method of pulse-height analysis makes possible the direct determination of individual radionuclides in a gamma emitting sample. Such determinations, as in the case of alpha and beta spectrometry, are possible because the method provides a basis for the identification of specific nuclear transitions, and these, in turn, are characteristic of specific radionuclides. The method is easy to use, highly sensitive, and fast.

As already seen in the case of alpha particles, the pulse-height spectrum is usually represented by a plot in which events per unit time per pulse-height increment are plotted as a function of pulse-height increment. In the case of gamma rays, such a plot is characterized by a series of approximately gaussian shaped peaks superimposed on a nearly characterless continuum. Both of these features are a direct manifestation of the various interaction processes which have taken place within the detector/shield assembly. Detailed description of a gamma ray spectrum is given by Covell (1975), and specific examples of various gamma spectra obtained by semiconductor detectors by Cappelani and Restelli (1968).

In order to choose an appropriate spectrometer to analyse a particular gamma ray spectrum, it should be considered whether it is only necessary to recognize the presence of a certain gamma ray line, or whether a quantitative determination of the peak intensity is required. In the first case, the only important parameter is the energy resolution. In the second case, if nearby lines which could be overlapped are not present, it is important to evaluate the extent to which an improvement in energy resolution can balance the drawback of a reduced efficiency. Such a consideration is particularly worthwhile when a choice has to be made between a NaI (Tl) scintillator, and a Ge(Li) spectrometer for a specific gamma ray spectrum analysis.

5.3.4 Selected alternative methods of radionuclide analysis

Three alternative methods available for the measurement of the heavy radionuclides from the decay series are: (i) neutron activation analysis; (ii) mass spectrometry; and (iii) fission track technique. They were selected particularly because of their successful recent application to the methods in geochronology utilizing the measurements of trace concentrations of U, Th, and Pb isotopes in natural samples.

(i) *Neutron activation analysis.* In general, activation analysis relies on the production of radionuclides in a sample, and the subsequent detection and measurement of the induced radiation. Sensitivities for neutron activation for about 70 elements require sample quantities ranging from 50 μg to 10^{-6} μg. In practice, the sensitivity and the accuracy of the analysis are freqently based on the ability to distinguish the radiation of the radionuclides of interest from the other activated constituents of a sample.

In the past, neutron activation of U and Th species has been carried out mainly along two approaches: the first approach involved a neutron capture reaction, (n,γ) followed by chemical separation and measurement of the product isotopes, for example determination of U through the reaction* $^{238}U(n,\gamma)^{239}U \xrightarrow{\beta} {}^{239}Np$, or Th through $^{232}Th(n,\gamma)^{233}Th \xrightarrow{\beta} {}^{233}Pa$. The second approach is the case of neutron capture, fission reaction, (n, fission), followed by radioassay of one or more of the longer-lived fission products. Chemical separation can be avoided rarely in the case of determination of U at trace concentrations in complex samples. However, non-destructive methods for U determination use a specific feature of the fission process, such as the emission of delayed neutrons, or the detection of the high-ionization-density fission-fragment track in the matter.

A good example of the first of the above approaches is determination of ^{231}Pa in geological samples using the alpha radiation of ^{232}U formed through the reaction chain $^{231}Pa\,(n, \gamma)\,^{232}Pa\xrightarrow[1.3\,d]{\beta^-}{}^{232}U\xrightarrow[72\,y]{\alpha}{}^{228}Th$ (Rosholt and Szabo 1968). The ^{232}U was preferred over the 1.3 day β–γ emitting ^{232}Pa despite its longer half-life of 72 y. The choice of ^{232}U was, however, guided by the fact that it was not the ^{231}Pa concentration as such which was the quantity of interest, but rather the ratio $^{231}Pa/^{235}U$. The alternative method of determination of this ratio would have been alpha spectrometry since both nuclides are alpha emitters (see section 1.5).

In the case of the second approach, in view of the energy dependence of the fission cross sections, the detection of delayed neutrons following exposure of a sample to thermal neutrons is characteristic of the presence of ^{235}U, while if the exposure was to fast neutrons the presence of ^{232}Th and/or ^{238}U is indicated. If, therefore, two determinations are made for each sample, one in a mixed (fast and slow) neutron flux, one with the slow neutrons screened out by a cadmium shield, both Th and U may be determined.

For detailed discussion of the principles of the neutron activation readers are referred to Bowen (1975), May (1973), Morrison (1973), Gale (1967), Amiel (1961 and 1962), Dyer, Emergy, and Leddicotte (1962), and Keepin, Wimitt, and Zeigler (1957). Interesting examples of application to the measurement of U and Th nuclides can be found in Kulmatov and Kist (1978), Millard (1976), and Aly et al. (1975).

(ii) *Mass spectrometry.* In general, the mass spectrometer consists of a source of ions, the ions containing the element whose isotopic composition is to be determined, an analyser capable of separating the ion beam, and a

* The usual nomenclature used in reaction nuclear physics is as follows:
Target nuclide (projectile, observed radiation or nucleus) Residual nuclide where the observed gamma ray in the above neutron activation reaction originated in the excited nucleus of the residual nuclide following the neutron capture by the target nucleus. This is then followed by a beta decay and the formation of the final nuclide.

collector and electronic circuit to measure the ion currents produced. The components of the ion beam can be separated by means of a magnetic analyser, 'time of flight' techniques, and velocity selector spectrometers. The latter device can bunch ions into discrete bundles according to their charge to mass ratio.

An excellent review of radiogenic isotope research over the period 1970 to 1975 is given by Church (1975), and de Bièvre (1976) gives a good review of the mass spectrometry applications. An interesting example of the application to the determination of U in sea water is given by Rona, Gilpatrick, and Jeffrey (1956).

(iii) *Fission and alpha track technique.* Heavy charged particles produce submicroscopic tracks in many insulating solids including crystals, inorganic glasses, and plastics. It has also been found that many solids contain ancient particle traces which yield information about the radiation history of the sample. Although the narrow, damaged regions, or tracks may be seen at very high magnification in an electron microscope, they are usually enlarged to a size suitable for optical viewing by chemical etching. Fleischer, Price, and Walker (1965) have reviewed the subject of fission track detectors in some detail. Examples of application of this method to geological dating can be found in Fleischer and Price (1964a,b), Wagner (1966), and examples of measurement of low U concentration in natural samples in Price and Walker (1963), Chakarvarti and Nagpaul (1979), and Danis and Valjin (1979). Fission tracks have been used to test the distribution of U in bones and molluscs (Szabo and Rosholt 1969).

A combination of fission and alpha track analysis using plastic and mica detectors has been used to measure the age of deep sea sediments by Fisher (1977a,b, 1978). The technique cannot compete in precision with alpha spectrometry but its great advantage is relative simplicity. Fisher (1978) gives a successful example of deep sea core sedimentation rate measurement using fission/alpha track counting. In principle both alpha and fission track analysis can be used to measure U and Th isotopes plated onto metal discs (replacing alpha spectrometry). A study of $^{234}U/^{238}U$ activity ratios in groundwaters by this method has been carried out by Schwarcz and Mohamad (1981). Finally, for readers interested in the comparison of the achievable precision of various methods of U and Th determination specifically in granitic rocks, Stuckless, Bunker, Bush, Doering, and Scott (1977) give a useful account.

5.4 Nuclear statistics and treatment of data

A physical magnitude can never be measured exactly, that is to say, an error is always associated with the measurement. Progressively more elaborate efforts in measurement result only in reducing the possible error of the determination, never eliminating it completely. Thus, when reporting the result of any measurement it is necessary to quantify the probability that the result is in error by some specified amount. The theory of statistics and fluctuations deals with the mathematical procedures used for estimating uncertainties associated with the reported data.

Systematic error represents reproducible inaccuracy introduced by faulty equipment, calibration, or technique. *Random error* is an uncertainty representing indefiniteness of a result due to finite precision of the measurement. In other words, it is a measure of fluctuation in a result obtained after repeated measurement. *Probable error* is a numerical measure of the extent of error estimated to be associated with a specific set of measurements of same physical quantity. The *accuracy* of an experiment is a measure of how close the result is to the 'true' value. Therefore, it is a measure of the correctness of the result. The *precision* of an experiment is a measure of how reproducible the result is. (For further detail see, for example, Bevington 1969.)

In any series of measurements, the frequency of occurrence of particular values is expected to follow some 'probability distribution law' or 'frequency distribution'. Nuclear decay processes are random in character, and are described mathematically by an exponential decay function derived from probability theory (see Evans 1955 for example). The Binomial Distribution represents the true probability that a particular atom will decay. If the probability of occurrence is very small ($P_x \ll 1$) and the number of observations large, the Poisson Distribution can be used as a very good approximation:

$$P_x = \frac{m^x}{x!} e^{-m} \tag{5.1}$$

where m is the true number of disintegrations whose measured value is x. Both distributions rely on a very short observation time relative to the half-life of the nuclide.

For any frequency distribution, the standard deviation for a large number of observations is given by

$$\sigma^2 = \sum_{x=-\infty}^{x=+\infty} (x-m)^2 P_x. \tag{5.2}$$

In the case of the Poisson Distribution, substitution for P_x from eqn (5.1) yields

$$\sigma^2 = \sum_{x=-\infty}^{x=+\infty} \frac{(x-m)^2 m^x}{x!} e^{-m} \tag{5.3}$$

which on expansion gives

$$\sigma^2 = m \text{ or } \sigma = \pm \sqrt{m}. \tag{5.4}$$

Thus, the standard deviation for the Poisson Distribution has a definite value in terms of the mean value, m, that is to say, the 'plus or minus' error incurred in measuring m disintegrations is simply equal to $\sqrt{m}$ if m is large.

The laws governing the propagation of errors are rigorous for the standard deviation. When a physical magnitude is to be obtained from the summation or the differences of independent observations on two or more physical quantities, the final error, R of the derived magnitude is obtained from

$$R^2 = r_1^2 + r_2^2 + \ldots + r_n^2 \tag{5.5}$$

where $r_1, r_2, \ldots, r_n$ are the absolute values of the errors in the mean values of the several quantities, expressed in the same units. The arithmetic of subtraction may be illustrated by the problem of measuring a counting rate due to a radiation source. When measuring the countrate, s of a source, separate measurement of the associated background rate, b must be carried out. If in time T_b a total of bT_b background counts is recorded, then the average background rate, B is expressed as

$$B = [bT_b \pm \sqrt{(bT_b)}]\frac{1}{T_b}$$

or

$$B = b \pm \sqrt{(b/T_b)} \tag{5.6}$$

where $\pm \sqrt{(b/T_b)}$ is the statistical error associated with the value of b. The magnitude of the error is inversely proportional to the square root of the counting time. Therefore, for a source of activity S, the best estimate of total countrate, $(S + B)$ is given by

$$S + B = [(s+b)T_s \pm \sqrt{((s+b)T_s)}]\frac{1}{T_s}$$

$$= (s+b) \pm \sqrt{(s/T_s + b/T_s)}.$$

Thus, for the average countrate due to the source alone

$$S = s \pm \sqrt{(s/T_s + b/T_s + b/T_b)}. \tag{5.7}$$

It should be noted that background appears twice in eqn (5.7), once for variation in measuring $(s+b)$, and once for variation in measuring b alone.

If a physical magnitude Y is to be obtained by multiplication or division of results of several independent observations on two or more physical magnitudes $y_1, y_2, \ldots y_n$, the fractional probable error R/Y in the resulting value of Y depends upon the fractional probable errors $r_1/y_1, r_2/y_2, \ldots r_n/y_n$ in the measurement of $y_1, y_2, \ldots y_n$, and is given by

$$(R/Y)^2 = (r_1/y_1)^2 + (r_2/y_2)^2 + \ldots + (r_n/y_n)^2 \tag{5.8}$$

Thus to determine errors of quotients or products of the corrected countrates given by eqn (5.7), the composite error R, in the final result can be expressed by

$$R = Y \sqrt{((r_1/y_1)^2 + (r_2/y_2)^2 + \ldots + (r_n/y_n)^2)} \qquad (5.9)$$

Equation (5.9) is used to calculate the error associated with the $^{234}U/^{238}U$ activity ratio in the example given in Appendix B. Most errors quoted with the U-series disequilibria results, isotope concentrations, dates, and other quantities calculated from measured countrates are based on the Poisson distribution model given above and derive from eqns (5.7) and (5.9).

5.5 Measurement of U series disequilibria

The most frequently used method in the measurement of U-series disequilibria is that of alpha spectrometry. The neutron activation method is more complex to apply, dependent on expensive facilities, and therefore, rarely used (see section 5.3.4 for an example of the application of this method to the measurement of ^{231}Pa). Gamma spectrometry (Ostrihansky 1976; Komura, Sakanoue, and Konishi 1978; Megumi 1979; Yokoyama and Nguyen 1980), and more recently, fission/alpha track method (Fisher 1977a,b, 1978), do not offer the accuracy, resolution, and/or sensitivity for all the nuclides required of the disequilibria measurements for the purpose of dating. Therefore, this section is devoted almost exclusively to the alpha spectrometric techniques for the measurement of disequilibria between the nuclides of the ^{238}U and ^{235}U decay series. Other techniques are mentioned and described only when they are linked in some way with the main alpha spectrometric measurement.

5.5.1 Measurement of U concentration and $^{234}U/^{238}U$, $^{230}Th/^{234}U$, and $^{230}Th/^{232}Th$ activity ratios

Because ^{230}Th and ^{234}U emit alpha particles of very similar energy (see Fig. 5.9), their alpha groups cannot be resolved satisfactorily by pulse height analysis, and thus, it is necessary to separate U and Th isotopes chemically. Since the chemistries of U and Th are quite different (see chapter 4), their extraction efficiencies will be different. To measure their respective chemical yields, a radioactive tracer mixture or 'spike' of known concentration of one isotope of each element, is added at an early stage of the extraction procedure, and then determined from the respective U and Th alpha spectra.

(i) *Spike solution and its calibration.* An ideal spike should not contain nuclides that occur naturally in the sample, their alpha particle energies must be sufficiently different from those of the sample-derived nuclides for complete resolution of the resulting peaks in the alpha spectrum; there must be no chemical fractionation between spike and sample nuclides; and the activity ratio of nuclides in the spike must be known accurately. The most frequently used spike which fulfills all of the above requirements is the daughter/parent

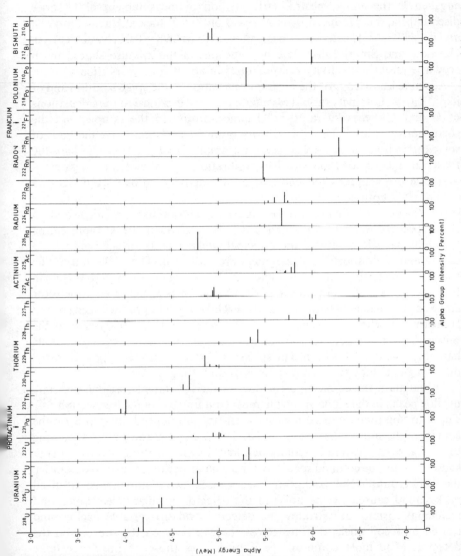

Fig. 5.9. A schematic representation of alpha group energies and their abundances, designed to aid in the interpretation of the U and Th spectra encountered in the measurement of the U-series disequilibrium by alpha spectrometry.

mixture ^{228}Th/^{232}U. (When ^{232}Th peak is found in the Th alpha spectrum, appropriate correction is applied to the spike ^{228}Th countrate for the presence of the naturally occurring ^{228}Th in secular equilibrium with its parent ^{232}Th.) The solution of ^{228}Th/^{232}U spike recently purchased by many laboratories engaged in the alpha spectrometric method of measurement of U series disequilibria, was produced from a 12 year old ^{232}U stock at Harwell in 1979, and its ^{228}Th/^{232}U activity ratio is estimated to be 0.98 ± 0.01 (Ivanovich, Ku, Harmon, and Smart 1981). The two nuclides in the mixture should reach secular equilibrium activity ratio of 1.027 in about four years time.

Accurate knowledge of the ^{232}U concentration in the spike solution and of its ^{228}Th/^{232}U activity ratio is essential because it is critical to the calculations of U and Th recovery rates, ^{238}U concentration in the sample, and the corrections to the ^{230}Th/^{234}U activity ratio. Several alternative methods of spike calibration can be applied (see for example Gascoyne 1977). Perhaps the most satisfactory method involves calibration by measurement against a uraninite ore standard old enough to have activity ratios ^{234}U/^{238}U and ^{230}Th/^{234}U both equal to unity.

(ii) *The analysis of alpha spectra.* Figure 5.10 shows characteristic U and Th alpha spectra obtained from a standard carbonate source. Peaks representing the alpha particle groups of the principal nuclides used for the ^{230}Th/^{234}U dating and their short-lived daughters are indicated. The ^{232}U and ^{228}Th peaks are due to the added spike solution. The energy resolution (FWHM) in Fig. 5.9 is 23 keV for U peaks and 27 keV for Th peaks. The eight-detector alpha spectrometer currently in use at AERE Harwell to produce such spectra is shown in a schematic form in Fig. 5.11. The pulse height analyser sorts the pulses from eight detectors simultaneously by means of two ADC units handling four detectors each. The spectra are printed on a teletype and stored on a magnetic tape for future analysis and plotting off-line.

To maintain the counting error due to nuclear statistics at one per cent, nuclide peaks in the alpha spectrum must contain 10^4 counts (see section 5.4). The counting time required for such statistics will depend, therefore, on the activity of the source, that is to say, U and Th content of the sample. Counting time of at least 10^3 min is usually required for most natural samples. Blank sources should be counted regularly for a similar period of time to determine the background. The background comprises two components: (i) fixed background caused by the noise in the detector, ancillary electronics, and incidental cosmic rays striking the detector; and (ii) variable background caused by recoil nuclei drifting across the gap between the source and the detector. (The most common recoil nuclei are those of the ^{232}U spike daughters, ^{228}Th, ^{224}Ra, ^{220}Rn, and others.) The precise knowledge of this variable background is essential when low activity samples are counted. Because the backgrounds due to embedded recoils differ for U and Th sources, these should be counted on separate, U and Th dedicated detectors if possible.

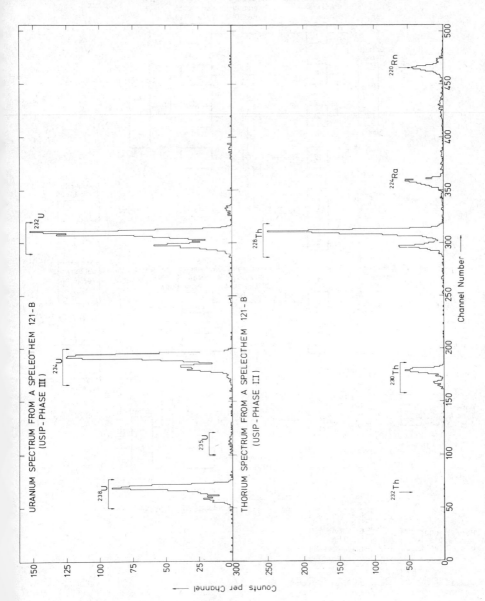

Fig. 5.10. Typical U and Th spectra obtained from a USIP-phase III-speleothem sample 121 B. (The two spectra do not share the same energy/channel calibration.)

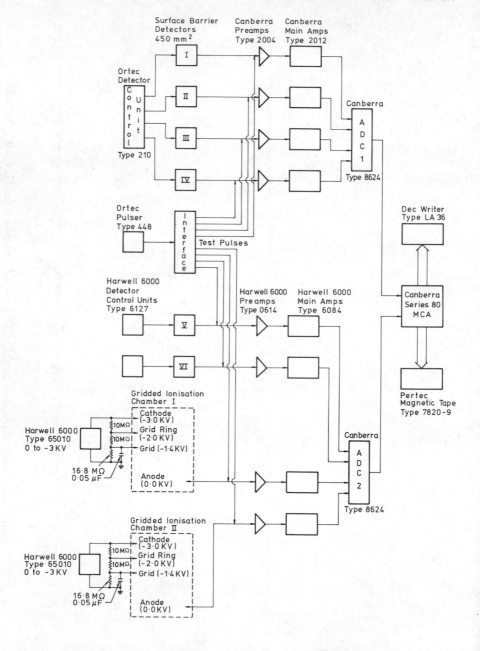

Fig. 5.11. Schematic diagram of the Harwell eight-detector alpha spectrometer.

The interpretation of the alpha spectra leading to the determination of U and Th nuclide countrates is an important, but at the same time, a subjective step. The peak integration is carried out usually over a roughly equal number of channels for all the peaks in the spectrum. The ideal criterion for setting the integration limits is that the counts per channel fall to zero on either side of the peak value. Often this is not the case either because of background spread across the spectrum or because of a combination of a thick source, poor detector resolution, and high activity present in the source. Then, various criteria for integration limits can be adopted. For example, for peaks which do not overlap, integration limits may be defined by the channels containing one per cent of the peak channel counts (Rosholt 1978, personal communication). Alternatively, should the low energy tail encroach on another peak in the spectrum, computer codes can be used to calculate the exponential shape of these tails to channel zero and correct all the overlapping peaks using the same shape for their tails. This latter approach becomes necessary only in the extreme cases of very thick sources and/or very active sources, a rare occurrence in most natural samples encountered. A simple approximation for tail correction can be applied using shapes of tails of more isolated peaks in the spectrum. For example, the tail of ^{216}Po peak is used to correct the ^{228}Th peak in the Th spectrum for the overlap with the ^{224}Ra tail (Gascoyne 1980, personal communication).

The analysis of a typical U alpha spectrum (see Fig. 5.10) is straightforward in the sense that the peaks identified as ^{238}U, ^{235}U, ^{234}U, and ^{232}U are usually well resolved. Once the integration limits are set for each peak, and the integration carried out, the only correction required is that due to the background, provided the sources are counted soon after the last column separation step (or the last TTA step, if TTA is used). Otherwise, the shorter lived daughter nuclides grow in rapidly yielding alpha peaks which, in some cases, cannot be resolved from the peaks of interest. In the case of Th spectra, the treatment is somewhat complicated by rapid ingrowth of a 5.1 per cent ^{224}Ra group under the ^{228}Th peak. The correction of the ^{228}Th countrate due to the ^{224}Ra ingrowth is carried out by first obtaining the 94.5 per cent ^{224}Ra countrate from the main ^{224}Ra peak. After the appropriate background correction, this countrate is used to calculate the 5.1 per cent component in the total background-corrected ^{228}Th countrate.

Figure 5.9 is a schematic diagram of all the alpha group energies and their abundances designed to aid the interpretation of most U and Th alpha spectra encountered in the U-series disequilibria measurements.

(iii) *Chemical yield calculations.* The chemical yields (or extraction efficiencies) of U and Th are obtained from the observed ^{232}U and ^{228}Th countrates in their respective spectra. The counting efficiency of the detector (otherwise referred to as the detector geometry) must be known for the chemical yield calculation. As already mentioned in section 5.3.1, the counting efficiency depends on the area of the detector, the active area of the source, and the distance between the source and the detector. The simplest method of measuring the counting efficiency of a detector system is to use a calibrated source, that is a source counted absolutely in a chamber of known counting efficiency, having an identical active area to the sources used for the disequilibria measurements, and containing a long-lived alpha emitting nuclide such as ^{241}Am. The counting efficiency is given by the expression

$$E(\%) = \frac{A \text{ counted (cpm)}}{A \text{ calibrated (cpm)}} \times 100. \tag{5.10}$$

The percentage chemical yields of U and Th are calculated from

$$Y_U(\%) = \frac{(^{232}U)\,100}{(E/100)W_s[^{232}U]_s},$$

and

$$\tag{5.11}$$

$$Y_{Th}(\%) = \frac{(^{228}Th)\,100}{(E/100)W_s[^{232}U]_s R_s},$$

respectively, where W_s is the weight of added spike solution (g); $[^{232}U]_s$ is the concentration of ^{232}U in the spike, (dpm/g); R_s is the spike $^{228}Th/^{232}U$ activity ratio; and (^{232}U) and (^{228}Th) are the measured countrates from the integration of the ^{232}U and ^{228}Th peaks in the respective U and Th spectra.

(iv) *U-concentration calculation.* For a sample weight, W (g), the U concentration in (ppm) is given by

$$[U](ppm) = \frac{(^{238}U)W_s[^{232}U]_s}{0.747W(^{232}U)} \tag{5.12}$$

where the factor 0.747 is the activity in dpm of 1µg of ^{238}U. The only assumption made so far is that both ^{238}U and ^{232}U are extracted with identical chemical yield, that is to say, no U isotope fractionation took place during the the extraction procedures; and that both U isotopes were subject to an equal probability of measurement of disintegrations by the detector. The detector efficiency is not required for this calculation.

(v) *The $^{230}Th/^{234}U$ activity ratio calculation.* Whereas the activity ratios $^{234}U/^{238}U$ and $^{230}Th/^{232}Th$ are calculated directly from the background corrected countrates derived from their respective spectra, the $^{230}Th/^{234}U$ activity ratio must be corrected because the two countrates are obtained from different spectra. The true $^{230}Th/^{234}U$ activity ratio is given by the following expression

$$[^{230}Th/^{234}U]_{True} = [^{230}Th/^{234}U]_M [^{232}U/^{228}Th]_M [^{228}Th/^{232}U]_s \tag{5.13}$$

where $[^{230}Th/^{234}U]_M$ is the measured activity ratio of ^{230}Th and ^{234}U,

$[^{232}U/^{228}Th]_M$ is the measured activity ratio of ^{232}U and ^{228}Th from their respective U and Th spectra, and $[^{228}Th/^{232}U]_s$ is the calibrated spike activity ratio. Equation (5.13) obviates the need to allow for different counting efficiencies, chemical yields, and concentrations of the U and Th isotopes.

(vi) *Determination of the $^{230}Th/^{234}U$ age.* Given the corrected $^{230}Th/^{234}U$ and $^{234}U/^{238}U$ activity ratios for a sample, an age can be calculated using eqn (3.4) (see chapter 3 for details). Two methods are generally used for solving this equation: (i) graphical; and (ii) iterative solution by computer.

The graphical method makes use of a plot of the $^{234}U/^{238}U$ against $^{230}Th/^{234}U$ activity ratios (see Fig. 3.1). Isochrons are nearly vertical lines, whose slopes decrease with increasing age. Thus, for ages less than 50 ky, the values of the $^{234}U/^{238}U$ activity ratio have little influence on the age. Ages are determined by interpolating between isochrons. The age error limits are determined by drawing an ellipse whose major and minor axes correspond to the $\pm 1\sigma$ error bars for the two activity ratios. The upper and lower age limits are then given by the isochrons touching the ellipse spaced farthest apart.

The iterative method consists of substituting trial values of t into eqn (3.4) and approaching a solution of the equation in the limit. Various methods have been proposed for this procedure, most making use of computers. An example of such a program is UTAGE-3 written originally by Thompson (1973), and modified and corrected extensively by Gascoyne (1977). The Harwell version of this program, UTAGE-3, and its functions are described in detail in Appendix C. The program also corrects the measured countrates for background, natural contamination, growth of daughter nuclides since U/Th separation, and chemical yields. The $^{230}Th/^{234}U$ age is calculated by the Newton–Raphson iterative method described in Appendix C. Countrate errors (1σ) and the corresponding age errors are determined. The output of this program presents corrected present activity ratios for $^{230}Th/^{234}U$, $^{234}U/^{238}U$, and $^{230}Th/^{232}Th$, U concentration, U and Th chemical yields, age data, and the initial $^{234}U/^{238}U$ ratio. A listing of the program, and an example of a typical output are also given in Appendix C.

Three sources of error affecting both precision and accuracy of these measurements can be recognized: (i) analytical errors; (ii) counting errors; and (iii) defects in the physical or chemical nature of the sample. Of these, only counting errors are readily quantifiable and usually quoted as $\pm 1\sigma$ in the activity ratios. Analytical errors, however, include: (i) contamination of reagents by U and Th (blank correction); (ii) failure to separate U from Th in chemical procedures; (iii) 'memory' effects in analytical equipment, either glassware or detectors; (iv) failure to equilibriate the spike U and Th with respective isotopes in sample, resulting in selective loss of sample or spike nuclides during chemical separation; (v) fractional loss of U or Th from spike solution due, for example, to lowering or raising of pH, leading to selective adsorption of Th onto glassware. It should be noted that uncertainties in the concentration of spike solution do not affect age calculation, nor do errors in pipetting of spike or other reagents, as long as the spike isotopic ratio remains unaffected. Counting errors are estimated from the Poisson approximation (see section 5.4) for countrates of all nuclides. Clearly, errors of this type associated with the spike activity ratio should also be included in the quoted age error. Defects in the sample include: (i) absence of the closed system conditions after sample deposition; (ii) contamination of the sample by detritus or organic material, which either introduces excess ^{230}Th leading to anomalous high ages, or adsorbs excess U leading to anomalously low ages (see chapter 12 for details). Various correcting schemes for the effect of detrital components have been proposed, and are summarized in chapter 3. Nevertheless, alternative dating methods are necessary to confirm that single ^{230}Th/^{234}U ages are reliable.

(vii) *Gamma spectrometry method for U/Th activity measurement*. Earlier attempts to apply gamma spectrometry to the measurement of U-series disequilibria (Osmond and Pollard 1967; Yokoyama, Tobailem, Grjebine, and Labeyrie 1968; Reyss, Yokoyama, and Duplessey 1978) utilized NaI (Tl) scintillation detectors. Only two nuclides, ^{226}Ra and ^{228}Th could be measured whereas ^{230}Th, ^{232}Th, ^{234}U, and ^{238}U could not be measured due to insufficient energy resolutions. However, Yokoyama and Nguyen (1980) used high resolution (0.8 to 2.3 keV in the energy range 62 to 1330 keV), high purity Ge detectors and succeeded in measuring directly ^{230}Th from its gamma peak at 67.7 keV. They also measured ^{238}U indirectly from ^{234}Th ($E_y = 63.3, 92.3,$ and 92.8 keV) and ^{234}Pa ($E_y = 1001$ keV), and ^{228}Th from ^{208}Tl ($E_y = 582.3$ keV). (Other successfully measured nuclides will be discussed later on in the chapter where appropriate.) The ^{232}Th was not reported presumably because it was assumed that ^{232}Th was in radioactive equilibrium with its measured daughter ^{228}Th. However, the method failed to measure ^{234}U nuclide ($E_y = 53.3$ keV) due to a superimposed gamma group from ^{214}Pb ($E_y = 53.2$ keV). The great advantage of this approach is that no chemical procedures are required to analyse the samples, while the disadvantages, of course, are the method's inherent inapplicability to the measurement of ^{234}U in natural samples, the low efficiency of the gamma detection (limiting the application of the method to samples with high U content), and the relatively high cost of the counting equipment.

5.5.2 *Measurement of* ^{231}Pa/^{235}U *activity ratio*

The Pa dating method is based on the decay of ^{235}U to ^{231}Pa via a short-lived, intermediate nuclide ^{231}Th, and is an important alternative to the ^{230}Th dating method (see chapter 3). The ^{235}U activity can be obtained from a U alpha spectrum of the type described in section 5.5.1, either by direct integration of the ^{235}U pek or, preferably, from the ^{238}U countrate and their constant activity ratio (^{238}U/^{235}U = 21.7). In contrast, the ^{231}Pa activity can be measured either directly by alpha spectrometry of ^{231}Pa or indirectly, by measuring some of its longer-lived daughters assumed to be in secular equilibrium with their parent in the sample.

(i) *Direct measurement of* ^{231}*Pa.* The ^{231}Pa and added ^{233}Pa $(t_{\frac{1}{2}} = 27.4$ days) spike nuclide are separated from the sample and plated on a platinum disk (see chapter 4 for details). Chemical yield is determined by beta counting of ^{233}Pa. The ^{231}Pa activity in the same source is determined by alpha counting in a low background detector.

Ku (1968) improved the precision of the method (compared to the methods reported by Rosholt and Antal 1962 Sackett 1960; Ku 1966) by preparing thin U and Pa standard sources from very old uraninite. The ^{231}Pa activity in the sample is then estimated by comparing the alpha and beta countrates from ^{231}Pa and ^{233}Pa respectively, the sample source with those of the standard source. Thus, the actual ^{231}Pa activity in the sample is given by Ku (1968) as

$$(^{231}Pa)_{sample}dpm = (^{231}Pa)_{standard}dpm \frac{(^{231}Pa)_{sample}cpm}{(^{231}Pa)_{standard}cpm} \cdot$$

$$\cdot \frac{(^{233}Pa)_{standard}cpm}{(^{233}Pa)_{sample}cpm} f \qquad (5.14)$$

where f is the ratio of the amount of ^{233}Pa spike added to the sample to the amount added to the standard. The obvious advantage of this approach is that there is no need to measure the counting efficiencies of the detectors and that the uncertainties associated with the ^{232}U spike concentration do not enter the calculation of the ^{231}Pa/^{235}U activity ratio.

(ii) *Measurement of* ^{231}*Pa by* ^{227}*Th.* Protactinium is difficult to isolate free of U and Th, and, for that reason, only some laboratories determine ^{231}Pa by direct alpha spectrometry. The alternatives involve either the beta activity measurement of its immediate daughter ^{227}Ac $(t_{\frac{1}{2}} = 21.6$ y), or the alpha activity measurement of the ^{227}Ac immediate daughter ^{227}Th $(t_{\frac{1}{2}} = 18.2$ d). Of course, both alternatives are based on the assumption that the daughter and the granddaughter are both in radioactive equilibrium with their parent, ^{231}Pa (Mangini and Sonntag (1977) show that this is the case for deep sea cores). The former approach is not attractive because of the need to count beta particles and because it requires mounting another set of analytical techniques and equipment for extraction, isolation, and calibration of ^{227}Ac. The advantages of the latter are relative ease of measurement and the fact that the ^{227}Th activity may be compared to ^{230}Th activity using the same spectrum. The two Th isotopes are related, of course, through the natural abundance relationship of their respective parents ^{235}U and ^{238}U. Thus, no spike is needed because correction for chemical yields of Th source becomes unnecessary. In fact, the addition of a ^{228}Th/^{232}U spike would only complicate matters because the short-lived daughters of ^{228}Th grow in very rapidly masking a much weaker ^{227}Th peak in the Th spectrum.

Cherdyntsev, Kazachevskii, and Kuz'mina (1965) and Rosholt (1967) applied the ^{227}Th dating method to travertines and molluscs respectively. Thompson (1973) adopted the method for a limited number of speleothem samples, Moore and Somayajulu (1974) sketch out the principle of the ^{227}Th method and its application to dating of fossil corals, and Mangini and Sonntag (1977) report successful dating of deep sea cores. Gascoyne (1977, 1980) gives a detailed account of the method which forms the basis for the following account. The method requires an adequate U content of at least 1 ppm of U because the natural activity of the ^{231}Pa parent, ^{235}U is only 5 per cent of ^{238}U. If the U content is sufficient, however, the ^{231}Pa/^{235}U method provides an excellent check on the 'closed system' assumption.

The problems associated with the ^{227}Th measurement stem from the fact that it decays rapidly after its separation from ^{227}Ac parent, and its daughter ^{223}Ra ($t_{\frac{1}{2}} = 11.68$ d) grows in equally rapidly during counting, and has a very similar alpha energy. Their growth/decay relationship is illustrated in Fig. 5.12. An additional complication derives from the interference of daughters of ^{228}Th when a small amount of ^{232}Th is present in the sample. To minimize the rapid decay of ^{227}Th after separation from its parent ^{227}Ac note must be taken of the elapsed time between the cation column separation and plating (DSPT(h) in Fig. 5.12), and the source must be counted immediately after plating to reduce the ^{223}Ra ingrowth. Similarly, the delay time between the plating and counting must be taken into account (DCPT (min) in Fig. 5.12). To minimize the background due to the ^{228}Th daughters, a new detector which has never been exposed to ^{228}Th/^{232}U spiked samples is used.

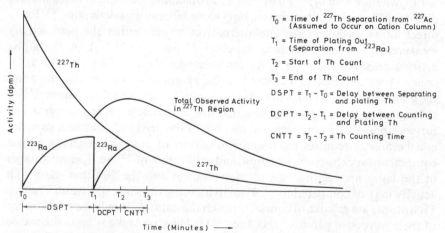

Fig. 5.12. Graphical representation of the decay of ^{227}Th and ingrowth of its daughter, ^{223}Ra (after Gascoyne (1977)).

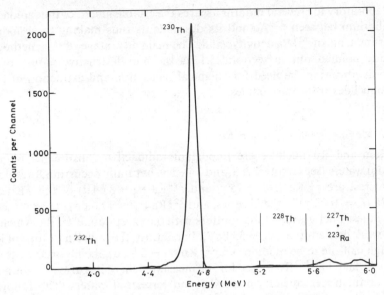

Fig. 5.13. Alpha spectrum of ^{227}Th used in the measurement of ^{231}Pa/^{235}U activity ratio (Gascoyne (1980)).

A typical ^{227}Th alpha spectrum free of ^{228}Th/^{232}U spike is shown in Fig. 5.13.

Processing of data obtained by the ^{227}Th method for ^{231}Pa/^{235}U dating involves subtraction of background from the new countrates, correction for the ingrowth of ^{224}Ra from ^{228}Th over the period of counting (if there is ^{232}Th present in the sample), correction for the ingrowth of ^{223}Ra from ^{227}Th over the period of counting, and correction of this new ^{227}Th activity for decay since separation from the parent at the cation column stage (the exact point at which ^{227}Th becomes unsupported is in some doubt due to the general lack of knowledge of Ac chemistry, but it is probably at the cation exchange step). The age calculation from the corrected countrates involves an iterative procedure similar to the one in the ^{230}Th/^{234}U activity ratio age program given in Appendix C. The full listing of the program, and a typical output is reproduced in Appendix D.

(iii) *Measurement of ^{231}Pa by gamma spectrometry.* Yokoyama and Nguyen (1980) attempted to measure ^{231}Pa activity by direct gamma measurement (E_γ = 299.94, 302.52, and 283.56 keV) in marine samples, but the spectra were too complex for quantitative interpretation in the relevant energy region. Instead, they measured ^{231}Pa activity by gamma rays from ^{227}Th (E_γ = 50.2, 236.0, and 252.2 KeV), ^{223}Ra (E_γ = 154.3 and 269.4 keV), ^{219}Rn (E_γ = 271.0 and 401.7 keV) and ^{211}Pb (E_γ = 404.8 keV). The authors argue that since Mangini and Sonntag (1977) have demonstrated that ^{227}Ac and ^{231}Pa are in radioactive equilibrium in marine sediments, the storage of samples in the laboratory for several months assures the establishment of the radioactive equilibrium between [227]Ac and its descendants thus making these indirect measurements of [231]Pa activity valid. The main advantage of this method, as already pointed out in section 5.5.1, is the non-destructive nature of the analysis obviating the need for chemical separation and extraction of these radionuclides from the samples.

5.5.3 Measurement of Ra and Rn

Radium and Rn nuclides are important radioactive constituents of most natural waters (see chapters 2, 9, and 14). The naturally occurring Ra isotopes of interest are [228]Ra ($t_{\frac{1}{2}} = 5.75$ y) and [224]Ra ($t_{\frac{1}{2}} = 3.64$ d) of the Th series. [226]Ra ($t_{\frac{1}{2}} = 1602$ y) of the U series, and [223]Ra ($t_{\frac{1}{2}} = 11.68$ d) of the Ac series. All of these nuclides are alpha emitters with the exception of [228]Ra, which is a very weak beta emitter ($E_\beta = 55$ keV). In contrast, the only Rn isotope of long enough half-life to be of interest is [222]Rn ($t_{\frac{1}{2}} = 3.82$ d). Its importance derives from the fact that [222]Rn is used for the measurement of its parent [226]Ra, and it is a useful tracer nuclide in marine and terrestrial waters. The following paragraphs only sketch the most important methods of the longer-lived Ra and Rn nuclides mentioned above, and for further details, the reader is urged to follow up the references quoted therein.

(i) *Radium-228.* Radium-228 is usually determined by extraction and beta counting of its immediate daughter [228]Ac($t_{\frac{1}{2}} = 6.13$ h), (Hagemann 1950; Moore 1969; Smith and Mercer 1970; Johnson 1971; Krishnaswami *et al.* 1972; Koide and Bruland 1975). Knauss *et al.* (1978) found those procedures to be plagued by incomplete extraction and interferences and their experiences suggest that the method is applicable only when adequate sensitivity of detection is available. The alternative approaches are used when high detection sensitivity is required. One of them involves an indirect measurement of [228]Ra from the growth of its daughter [228]Th (Moore 1969a, b; Smith and Mercer 1970; Koide and Bruland 1975; Michel and Moore 1980). In principle, the [228]Th method requires that the Ra planchet is allowed to stand 3 to 12 months after the original Ra plating. The planchet is flamed to volatilize [222]Rn and is counted in an alpha spectrometer. The buildup of [228]Th and its short-lived daughter [224]Ra is measured. The [224]Ra (5.68 MeV, 94.5 per cent) group is used to determine the amount of [228]Th buildup. The initial [228]Ra content is then calculated by Bateman's equations (see section 1.4). This technique has obvious advantages over the method involving chemical separation of [228]Th from a stored Ra solution. However, when the latter method is used, halfway through the growing period in solution, a [230]Th spike must be added to act as a chemical yield tracer for the ingrowing [228]Th (Knauss *et al.* 1978). In the gamma-spectrometric method described by Yokoyama and Nguyen (1980) (see sections 5.5.1 and 5.5.2), ^{228}Ra activity is measured indirectly from its daughter ^{228}Ac ($E_\gamma = 911.1$ keV).

(ii) *Radium-226*. The determination of the alpha-emitting Ra isotopes, particularly ^{226}Ra, in various materials has been the object of many studies. Kirby and Salutsky (1964) give a useful review of the Ra radiochemistry and 43 procedures for determination of Ra in many materials are described in detail. Nearly all reported measurements of ^{226}Ra are carried out indirectly by the ^{222}Rn emanation technique (see for example Føyn, Karlik, Petterson, and Rona 1939; Urry and Piggot 1941; Broecker 1965; Mathieu 1977). This method will be described in more detail in paragraph (iv) of this section which is concerned with the ^{222}Rn measurement. Koide and Bruland (1975) assayed Ra in sea water and sediments by isotope dilution with ^{223}Ra, isolating Ra isotopes and electroplating onto platinum planchets. Naturally occurring ^{224}Ra in sediments has also been used as the yield monitor in the ^{226}Ra determination by alpha spectrometry when related to ^{228}Th in the sample. Yokoyama and Nguyen (1980) measured ^{226}Ra ($E_\gamma = 186.1$ keV) directly by gamma spectrometry.

Radium data for both marine and terrestrial waters are usually quoted in terms of the ^{228}Ra/^{226}Ra activity ratios (generally greater than unity) (Moore 1980; others). Then, the ^{228}Ra and ^{226}Ra contents obtained by one of the above quoted methods are used to establish the ^{228}Ra/^{226}Ra activity ratios.

(iii) *Radium-223 and 224*. These two Ra isotopes are too short-lived to be of any great interest to the geochemical or geochronological considerations and are usually expected to be in secular equilibium with their parents in the environmental samples. However, they can be useful in the role of chemical yield tracers either as natural constituents in the sample or as artificial tracers added to the sample. When the ^{228}Ra/^{226}Ra activity ratio is required only, as in the case of Moore (1976) and Knauss *et al.* (1978), the extraction efficiency is not important. However, the extraction efficiency becomes necessary for determination of the specific activities of ^{228}Ra and ^{226}Ra (Moore and Reid 1973; Key, Brewer, Stockwell, Guinasso, and Schink 1979; Key, Guinasso, and Schink 1979). Thus, for example, Reid *et al.* (1979) describe a method for testing manganese oxide coated acrylic fibres for extraction of Ra, Th, and Ac from sea water in which they spike their samples with ^{227}Ac, ^{227}Th, and ^{223}Ra. These authors, then compare gamma spectrum of a small aliquot withdrawn from the sample to a gamma spectrum obtained with a standard and the comparison yields the information about the recovery efficiency of ^{223}Ra, and through this nuclide, using Bateman's equations, the extraction efficiencies for Ac and Th. An alternative application of ^{223}Ra is the measurement of Ra extraction efficiency by isotope dilution method using alpha spectrometry (see Koide and Bruland 1975, for example). The naturally occurring ^{224}Ra can also be used as Ra yield monitor when related to the parallel measurement of naturally occurring ^{228}Th in the same sample (Koide and Bruland 1975).

(iv) *Radon-222*. The mobility of Rn is held responsible for the occurrence of disequilibrium between the short-lived lower members of the U and Th decay series, and the much longer-lived members preceding Rn in the chain (see chapter 2). Radon-222 is the most important of the Rn isotopes because it is frequently used for an indirect measurement of its parent ^{226}Ra by the so-called emanation method (Broecker 1965; Key, Brewer, Stockwell, Guinasso, and Schink 1979; Michel and Moore 1980, for example), and because the direct measurement of the ^{222}Rn content of sea water or groundwater is frequently made as part of the U-series disequilibrium study.

Broecker (1965) describes in detail a simple *in situ* system for extraction of ^{222}Rn from sea water samples. Helium is pumped through the water sample to drive all the gaseous content of the sample into a vacuum system comprising a liquid nitrogen trap where Rn is liquified and He gas remains gaseous and is pumped away. Radium is then mobilized by heating, and is collected quantitatively by repeated pump-throughs with He into an initially evacuated scintillation detector. The ^{222}Rn activity is then measured by a direct application of alpha spectrometry. If the sample is sealed, stored for at least seven days, and the extraction procedure repeated, the amount of Rn recovered is then proportional to ^{226}Ra present in the sample. The ^{222}Rn growth, extraction, and measurement procedure may be repeated as often as required for a reproducible result, yielding the ^{226}Ra content of the sample. A much simpler method for the measurement of ^{222}Rn content of deep sea cores is given by Yokoyama and Nguyen (1980), where they measured it indirectly by gamma spectrometry of one of its descendants, ^{214}Bi ($E_\gamma = 609.2$ keV).

5.5.4 *Measurement of Pb, Bi, and Po*

Both ^{210}Pb$(t_{\frac{1}{2}} = 22.6$ y) and ^{210}Po$(t_{\frac{1}{2}} = 138.3$ d) are recognized tracers of natural processes in the atmosphere and on the earth's surface. They are both radionuclides with the longest half-lives of their respective species (all others have half-lives in minutes or hours as can be seen in chapter 1). Thus, for example, the ^{210}Pb half-life allows determination of sedimentation rates over a period of up to 100 y. On the other hand, ^{210}Bi$(t_{\frac{1}{2}} = 5.02$ d) is of interest mainly because its growth from ^{210}Pb provides an indirect means of ^{210}Pb activity measurement.

There are three basic approaches to the measurement of ^{210}Pb activity: (i) by alpha spectrometry of ^{210}Po , assuming radioactive equilibrium exists between the two nuclides; (ii) by beta spectrometry, observing the growth of its daughter ^{210}Bi; and (iii) by gamma spectrometry, measuring the gammas from ^{210}Pb directly.

Assuming ^{210}Po is in equilibrium with its precursor ^{210}Pb, the latter can be measured indirectly by alpha spectrometry of ^{210}Po (Flynn 1968; Turekian,

Kharkar, and Thompson 1973; Benninger 1976; Santschi *et al.* 1979; Krishnaswami, Benninger, Aller, and Von Damm 1980). For example, Flynn (1968) describes a simple method for plating ^{210}Po onto silver planchets for alpha-spectrometric measurements in a low-background, high geometry chamber (50 per cent) containing argon counting gas. Integration of the ^{210}Po ($E_\alpha = 5.31$ MeV) single peak yields the ^{210}Pb countrate. However, often in natural samples ^{210}Pb and its granddaughter ^{210}Po are in disequilibrium, and then the assumption on which this technique is based is invalid.

Krishnaswami *et al.* (1972) and Craig *et al.* (1973) describe a method originally developed by Rama *et al.* (1961), in which ^{210}Pb and ^{210}Bi are extracted from the sample and then separated from each other by an anion exchange technique (see chapter 4). A pure Pb source is produced in which the growth of ^{210}Bi is monitored by counting its betas regularly over a period of 30 to 35 days in a low background 2π-beta detector (see Lal and Schink 1960). The counting efficiency for ^{210}Bi is determined with a standard ^{210}Pb source. The ^{210}Pb activity is calculated from the final counting rates after a growth period of a month or more. Oversby and Gast (1968) describe a similar ^{210}Bi ingrowth procedure where ^{210}Bi betas are measured in a low background proportional detector. An Al absorber (11 mg/cm^2) is used to eliminate ^{210}Pb ($E_{\beta max} = 64$ keV, 15 per cent) and ^{210}Po($E_\alpha = 5.31$ MeV) activities and thus ensure that only the ^{210}Bi($E_{\beta max} = 1155$ keV) is counted.

The measurement of the beta radiation of the ^{210}Pb–^{210}Bi pair requires a considerable effort in the quantitative separation of these nuclides from complex samples. In order to overcome the complexities of the above mentioned methods, Gäggler *et al.* (1976) determine ^{210}Pb directly, without chemical separation, by measuring the intensity of its 46.52 keV gamma line. This gamma is associated with internal conversion and has an abundance of 4.1 per cent per decay. Like Yokoyama and Nguyen (1980), these authors use a high resolution, high purity Ge detector (resolution of 390 eV at $E_\gamma = 6$ keV and 610 e V at $E_y = 122$ ke V). Gäggler *et al.* (1976) tested their method on ^{210}Pb standards against the beta counting technique and found an excellent agreement. They also determined detector efficiency by adding ^{210}Pb standard solutions to identical samples of sediments and found it to be 4.25 per cent for the 46.5 keV gamma ray. The verdict on the direct ^{210}Pb gamma measurement method is that it is less sensitive than the beta counting of isolated ^{210}Pb–^{210}Bi sources but has the advantage of avoiding complex analysis and sample preparation.

Acknowledgements

The author is grateful to Drs Joan Freeman, Mel Gascoyne, and Henry Schwarcz for their invaluable comments on the manuscript. Special thanks are due to Mr Tony Parsons of Nuclear Physics Division, AERE Harwell, for proof-reading the manuscript. Dr Mel Gascoyne is gratefully acknowledged for granting permission to reproduce diagrams in Figs 5.12 and 5.13 from his PhD thesis, Mrs Kate Glover for providing the diagram in Fig. 5.4, and McGraw-Hill Book Company for granting permission to reproduce the diagram in Fig. 5.2 from the book *The Atomic Nucleus* by R. D. Evans.

6 IGNEOUS ROCKS
R. S. Harmon and J. N. Rosholt

6.1 Concentrations of radioelements in igneous rocks

The systematics of radioactive equilibrium have been discussed in previous chapters in this book. In fresh plutonic extrusive rocks, radioactive equilibrium in the U and Th decay series and the distribution of the actinide elements are controlled by the two parent elements, U and Th. A vast amount of data has been published regarding the distribution and geochemistry of U in crystalline rocks because of the interest shown in U as a source of nuclear energy. Studies related to the heat budget of the earth and the Pb/U, Pb/Th, and U-series dating techniques used in geochronology have generated many summaries of the geology, geochemistry, and distribution of these elements. Summaries in handbook format include Rogers and Adams (1969) and Clark *et al.* (1966).

The ranges of U concentration in the major igneous rock groups (see Fig. 6.1) show that the abundance of U in igneous rocks generally increases with Si content and rarely exceeds 30 ppm. Most silicic igneous rocks contain

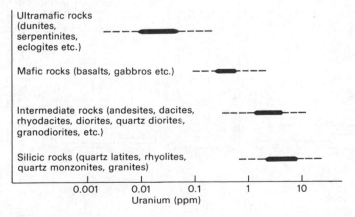

Fig. 6.1. Generalized distribution of U in various broad groups of igneous rocks. (After Rogers and Adams 1969.)

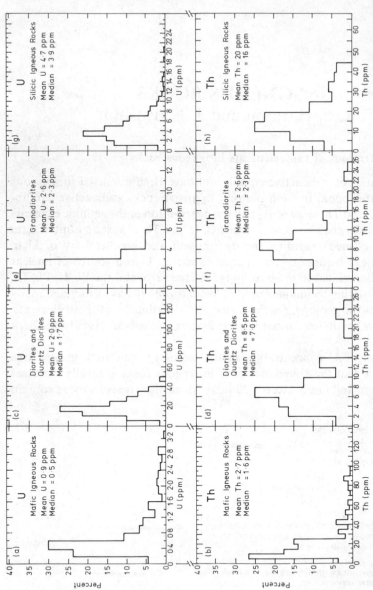

Fig. 6.2. Histograms of U and Th contents of various types of rocks. (After Clark *et al.* 1966, Figs. 22–2 to 22–9). (a) Plot represents 125 individual analyses and 14 average values for volcanic rocks, and 81 individual analyses and 16 average values for plutonic and hypabyssal rocks. (b) Plot represents 107 individual analyses for volcanic rocks, and 53 individual analyses and 9 average values for plutonic and hypabyssal rocks. (c) Plot represents 59 individual analyses and 10 average values. (d) Plot represents 39 individual analyses and 8 average values. (e) Plot represents 126 individual analyses and 30 average values. (f) Plot represents 75 individual analyses and 17 average values. (g) Plot represents 242 individual analyses and 3 average values for volcanic rocks, and 194 individual analyses and 96 average values for plutonic and hypabyssal rocks. (h) Plot represents 111 individual analyses and 45 average values mostly for plutonic rocks.

about 2 to 10 ppm U (Rogers and Adams 1969). Primary igneous U abundances greater than 10 ppm have been documented in some obsidians (Zielinski 1978). A similar correlation of abundance with Si content exists for Th. The Th/U ratios of 3 to 5 are most common in igneous rocks. Abundance data for various igneous rock groups are displayed in histogram form in Fig. 6.2, where separate histograms for U and Th content, each based on analyses of a large number of rocks, are reproduced from Clark *et al.* (1966). The groups of calc-alkaline and calcic rock types are divided into: (i) mafic rocks—gabbro and basalt; (ii) intermediate rocks—diorite, quartz diorite, andesite, and dacite; (iii) intermediate rocks—granodiorite (rhyodacites are not included because of an insufficient number of analyses); and (iv) silicic rocks—quartz monzonite, granite, and rhyolite. No histogram is shown for ultramafic rocks (dunites, periodotites, and eclogites) because of lack of sufficient high-quality analyses of these low U/Th rocks. Clark *et al.* (1966) suggested that Th concentration is more closely related to U concentration in mafic and intermediate rocks (Fig. 6.3, correlation coefficient of 0.93) than in granitic rocks (Fig. 6.4, correlation coefficient of 0.73). Many granitic rocks have a higher Th/U ratio, which suggests that U had been leached subsequent to crystallization. Uranium and Th concentrations and their ratios in a large number of granitic rocks has been compiled recently (J. S. Stuckless 1980, personal comm.). Average data from 1850 granite samples from throughout the United States are shown in Table 6.1. Data on all analyses, which include several samples from the same drill holes, are shown in part A; results based on one average data point for every 10×10 m^2 area are shown in part B. When information regarding composition is available, data on the granites are divided into peraluminous and metaluminous types (Stuckless and Peterman 1977).

Uranium, Th, and K are the major radioelements contained in the crust, and they are responsible for crustal radiogenic heat production. Lachenbruch (1968) discussed the apparent discrepancy between radiogenic heat production and measured heat flow and suggested that this discrepancy may be resolved by postulating that the concentrations of heat producing elements in the crust decrease exponentially with depth. However, based on measured cores sampled to depths of 1.2 km in the Boulder batholith (Tilling and Gottfried 1969) and 0.6 km in the Sierra Nevada batholith, Tilling, Gottfried, and Dodge (1970) suggested that there is no change in radioelement content with depth. One of the few deep boreholes for which there were analyses of U and Th at a large number of equally spaced intervals (94 samples at 30 m intervals) to a depth of 2.98 km drilled in crystalline schists of the Yukon-Tanana Upland, Alaska, was reported by Bunker, Bush, and Forbes (1973). Averages of the 66 samples at depths greater than 1 km had a larger radioelement content (3.52 ± 0.52 ppm U, 16 ± 4 ppm Th, and 2.6 per cent K) than the average of 28 samples at less than 1 km depth (2.06 ± 0.62 ppm U,

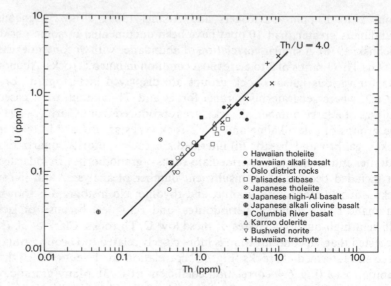

Fig. 6.3. Plot of U against Th in mafic and intermediate rocks. The correlation coefficient is 0.93. (After Clark *et al.* 1966).

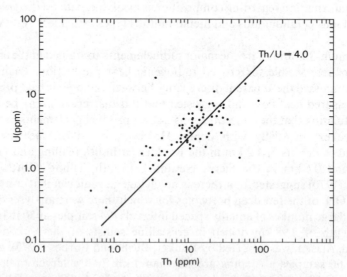

Fig. 6.4. Plot of U against Th in granitic rocks. The correlation coefficient is 0.73. (After Clark *et al.* 1966).

Table 6.1. Compilation of average U and Th concentration and Th/U in granitic rocks from U.S.A.

	U (ppm)	Th (ppm)	Th/U
A			
1850 granites	4.56	22.3	7.3
710 peralumina	4.95	22.5	6.3
250 metalumina	3.97	18.2	5.5
B			
320 granites	4.16	18.9	5.6
210 peralumina	4.49	19.0	5.0
150 metalumina	3.71	16.5	5.1

A is averages of all rocks; B is based on one average data point per $10 \times 10\,m^2$ area.

(J. S. Stuckless 1980, written comm.)

10.9 ± 2.7 ppm Th, and 1.8 per cent K).Thus, no laboratory observations seem to indicate that radioelement content decreases significantly with depth in the crust.

6.2 Distribution of U and Th

In general, high U and Th contents accompany high SiO_2 and K_2O and low CaO. Lyons (1964) stated that enrichment of the residuum in elements rejected by most of the major rock-forming minerals appears to be the simplest and best explanation of the U and Th distribution in igneous rocks. Uranium was thought by Neuerburg (1956) to have six modes of occurrence in the fabric of igneous rocks: (1) in U minerals; (2) in crystallographic sites or structural defects of major rock-forming minerals and minor accessory minerals; (3) in cation-exchange positions; (4) adsorbed on mineral surfaces; (5) dissolved in fluid inclusions; and (6) dissolved in intergranular fluids. Neuerburg apparently was the first to use the term *labile* U as that part of the U content of a rock which is readily dissolved. Although not as well documented, similar igneous geochemistry studies of Th suggest that the above discussion for U distribution also should apply for Th distribution in most igneous rocks.

In crystalline rocks, U and Th are concentrated mainly in minor accessory minerals, such as zircon, sphene, apatite, monazite, allanite, and epidote. Data on the abundances of U and Th in these accessory minerals and in the major rock-forming minerals are summarized by Rogers and Adams (1969). A compilation of contents of Th and U, and Th/U ratios in the minerals, taken from Adams, Osmond, and Rogers (1959) is shown in Table 6.2.

Table 6.2. Uranium, Th, and Th/U ratios in the minerals of igneous rocks

Mineral		U (ppm)	Th (ppm)	Th/U
Accessory minerals				
Allanite	{ accessory	30–700	500–5000	5–10
	{ pegmatite	?–100	1000–20 000	high
Apatite	{ accessory	5–150	20–150	1
	{ coarse aggregate	10–50(?)	50–250(?)	1–5
Epidote		20–50	50–500	2–6
Ilmenite		1–50		
Magnetite (and other opaque minerals		1–30	0.3–20	
Monazite		500–3000	25 000–200 000	25–50
Sphene		100–700	100–600	1–2
Zenotime		500–35 000	low	low
Zircon	accessory	300–3000	100–2500	0.2–1
	pegmatite	100–6000	50–4000	1
Major minerals				
Biotite		1–40	0.5–50	0.5–3
Hornblende		1–30	5–50	2–4
Potassium feldspar		0.2–3	3–7	2–6
Muscovite		2–8		
Olivine		0.01	low	
Plagioclase		0.2–5	0.5–3	1–5
Pyroxene		0.01–40	2–25	4–5
Quartz	rocks	0.1–5	0.5–6	2–5
	beach sands	0.7	2.0	3

After Adams *et al.* (1959)

Zielinski (1978) studied U distribution and abundance in some glassy and crystalline rhyolites (felsites) in silicic volcanic rocks. He found that U distribution is inhomogeneous in felsites, and accessory minerals concentrate U in the order of zircon > sphene > apatite. Unaltered biotite and magnetite phases contain little or no U, excluding possible inclusions of the above accessories. Quartz, crystobalite, alkali feldspar, and sodic plagioclase phases compose the major minerals of felsites; all contain little or no U. Obsidians (non-hydrated glass) and perlites (hydrated glass) are unique among igneous rocks in that they have homogeneous U and Th distributions.

Element mapping is an important technique used to determine the distribution of U and Th in rocks. Using alpha-autoradiographs of the Mesozoic Conway Granite, New Hampshire, Richardson (1964) concluded that minor accessory minerals, thorite, huttonite, and zircon, contained most of the Th; U also was present in uraniferous accessory minerals, but more than half of the U was in a dispersed phase. Fission-track images are much more sensitive, and more precise, and are specific for U or U together with Th, depending on the experimental conditions (Fleischer *et al.* 1975). Using this technique, several studies (Wollenberg and Smith 1968; Bowie, Simpson, and Rice 1973; Berzina, Yeliseyeva, and Popenko 1974; Zielinski 1978; Tieh, Ledger, and Rowe 1980; Zeilinski, Peterman, Stuckless, Rosholt, and Nkomo 1981) have all shown the heterogeneous distribution of U in crystalline igneous rocks. These studies also showed: (i) a small amount of U actually contained in rock-forming minerals; (ii) a major primary component in the uraniferous accessory minerals, zircon, sphene, allanite, apatite, and monazite (resistate U); and (iii) the other major component of U is distributed in grain boundaries, cleavage traces, and microfractures (intergranular U). Evidence of U redistribution is common; usually the redistribution is directed along zones of higher permeability, and the U occurs with hematite, which forms as alteration rims on ferromagnesian minerals. Fisher (1977b) has used both alpha particle tracks and neutron-induced fission tracks (fission/alpha method) for measuring the U and Th contents and their distributions in a geological specimen in which the decay series are in radioactive equilibrium.

6.3 Chemical behaviour of U and Th in the igneous environment

Uranium and Th are strongly lithophilic, in that they have a strong tendency to concentrate in the sialic crust. The chemical behavior of U and Th during igneous processes produces an increase in the concentration of these two elements in late stage silicic and alkali members of the differentiated series (Fig. 6.2). This enrichment of the igneous residuum is explained by the large ionic size and high ionic charge of Th and U ions, which effectively preclude their entry into the crystal sites of the common rock-forming minerals. Increases of both U and Th correlated well with increasing alkali, and especially K contents (Tilling and Gottfried 1969; Killeen and Heier 1975).

One major chemical element for which these two elements could substitute, is Ca, having an ionic radius of 0.99. Apparently U does substitute for Ca in apatite (Altschuler, Clark, and Young 1958), and Th tends to substitute in this mineral to a somewhat lesser extent (see Table 6.2); however, both of these actinides may be controlled more by the scavenging effect of the phosphate anion in apatite. The strong negative correlation between Ca and the two actinide elements in differentiated rock series indicates that substitution for Ca is not extensive in most rock-forming minerals.

In rocks relatively undisturbed by the effects of water penetration, U and Th behave similarly in the tetravalent state, and Th/U ratios remain fairly constant within petrographic provinces (Larsen and Gottfried 1960). However, U^{4+} can be oxidized to the hexavalent state, but Th cannot exist in this valence state. Hexavalent U is considerably more soluble and forms more soluble compounds than tetravalent Th. In crystalline rocks, a significant fraction of U and Th resides along grain boundaries and in minerals that become degraded and unstable because of radiation damage (Z. E. Peterman

1979, personal comm.). Whitfield, Rogers, and Adams (1959) observed an apparent increase of Th relative to U with petrogenic evolution which can be explained by the increase of the fraction of labile U with increase in U and Th content of rocks and subsequent removal of some loosely-bound U. Labile U has been discussed in section 6.1 and in chapter 14.

A study of the behaviour of U and Th under postmagmatic conditions was reported by Zielinski (1978), who investigated the abundance and distribution of U in 11 suites of coexisting rhyolitic obsidians, perlites, and felsites from the western United States. The suites ranged in composition from peralkaline to calc-alkaline, and in age from Pleistocene to Oligocene. Obsidians are non-hydrated glass, perlites are hydrated glass, and felsites are crystallized rhyolite consisting of fine-grained intergrowths of alkali feldspar and cristobalites; felsite also contains abundant small grains of primary Fe-Ti-Mn oxides. Uranium distribution was found to be homogeneous in obsidians and perlites but inhomogeneous in felsites. Obsidians and coexisting perlites have nearly identical concentrations, indicating that little or no U is lost during hydration, whereas rhyolitic felsite shows as much as 80 per cent depletion compared to coexisting obsidian. Some felsites have U abundances indistinguishable from coexisting glass, but most of these samples are among the youngest suites. The abundance of U in felsites normalized to coexisting obsidians varies with age and compositional type, as shown in Fig. 6.5. Relative depletions are most

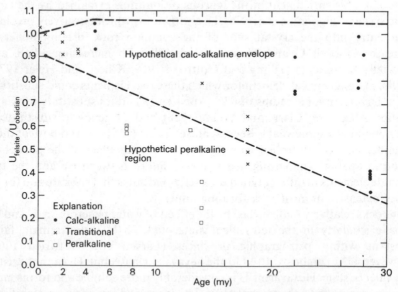

Fig. 6.5. Relative abundances of U in crystallized rhyolites, normalized to coexisting glasses; plotted against age, and coded according to compositional type. (After Zielinski 1978).

noticeable in older suites of the same compositional group, which implies that U loss is not generally controlled by transient, post-crystallization phenomena such as volatile transport and/or high temperature solubility. Rather, U apparently was lost slowly over a long period of time, presumably under the influence of alteration. Felsites having large relative U depletion show signs of U mobilization, as indicated by reprecipitation of U with secondary Fe-Ti-Mn oxides as fissure fillings and grain coatings. Thorium abundance in coexisting glass-felsite pairs shows no significant relative differences (Rosholt, Prijana, and Noble 1971), which reflect the low solubility of Th during alteration.

The U content in surface samples of plutonic rocks may not be the original U content at the time of emplacement of the rock because of secondary redistribution of loosely-bound (labile) U. Labile U is equivalent to inter-granular U prior to secondary redistribution. Considerable evidence shows that this redistribution has occurred, based on the disequilibrium that exists between parent ^{238}U and stable decay product ^{206}Pb (Stuckless and Nkomo 1978; Stuckless and Ferreira 1976; and Rosholt et al. 1973). To determine if there was a loss (or gain) of U and Th, the amount of radiogenic Pb is measured, and from this quantity, the amount of U and Th required, at time of emplacement, to support this Pb, is calculated, where the age is known or can be determined by Pb/Pb isochrons (Rosholt, Zartman, and Nkomo 1973). In cases where Pb immobility can be documented, these results substantiate that labile U exists in plutonic rocks characterized by high contents of Si, Th, alkalis, and biotite. In one study, radiogenic Pb excesses indicated removal of as much as 70 per cent of the U from Archean granite (Stuckless and Nkomo 1978) whereas little or no Th was removed. A more detailed description of this U/Th/Pb system in Precambrian rocks is included in chapter 14.

6.4 Geochronology

6.4.1 Introduction

Study of U-series disequilibrium in late Quaternary volcanic rocks offers a unique opportunity to obtain information on the timing of the geological processes involved in their formation, which is difficult to obtain by other dating techniques. The application of U-series techniques to young volcanic rocks is feasible because chemical fractionations occur during magmatic processes in such a way that disequilibrium commonly occurs between U-parents and some daughter nuclides.

Secular equilibrium will be attained within the entire ^{238}U decay series if a magma source region or a magma in transit within the crust remains closed for a period of more than about 10^6 y: ^{234}U reaching equilibrium with ^{238}U after 1.25×10^6 y, ^{230}Th with ^{234}U after 4.5×10^5 y, ^{226}Ra with ^{230}Th after 9×10^3 y, and ^{210}Pb with ^{226}Ra after only 100 y. Conditions of secular equilibrium will be disturbed by any physical or chemical fractionation of the members of the decay series. Observed disequilibrium can thus be used to determine the age of crystallization, provided the magma reached the surface in a time which is short compared to the half-lives of the nuclides involved.

Although fractionation of ^{234}U from ^{238}U occurs during weathering (see chapter 7), it does not occur during magmatic processes. Thus, fresh igneous rocks show no U-isotope fractionation (Somayajulu *et al.* 1966), and, therefore, ^{234}U is not useful in chronologic studies. Significant differences in time between the formation of a magma at depth and the time it is erupted on the surface may also be likely. Therefore, ^{226}Ra and ^{210}Pb may have approached significantly secular equilibrium, and thus be of only limited value in chronologic studies. As a result, the $^{230}Th/^{238}U$ system is the most appropriate to apply to upper Quaternary volcanic rocks because of its relatively long half-life of 7.52×10^4 y.

6.4.2 Theory

In dealing with young volcanic rocks, ^{230}Th- ^{238}U isotope systematics can provide detailed chronological information in a manner analogous to the $^{87}Rb/^{87}Sr$ isochron method for old igneous rocks (Allegre and Condomines 1976). If a mantle or crust source rock is partially melted, the resultant liquid will be either enriched or depleted in U relative to Th depending upon the pressure-temperature conditions of the source area, the composition of the source rock, and the degree of partial melting. For a system initially at isotopic equilibrium and presuming a short transport time to the surface so that ^{238}U decay is negligible, a plot of $(^{230}Th/^{232}Th)$ against $(^{238}U/^{232}Th)$ for the individual mineral phases (see Fig. 6.6) will define a horizontal line at the time of crystallization ($t = 0$). At this time the minerals will have the same $(^{230}Th/^{232}Th)$ ratios, but different $(^{238}U/^{232}Th)$ ratios because U and Th will not be partitioned to the same extent into the constituent mineral phases (see Table 6.2). Because of this, $(^{230}Th/^{232}Th)$ and $(^{238}U/^{232}Th)$ mineral ratios will define an internal isochron whose slope will vary as a function of time according to the relationship:

$$\left(\frac{^{230}Th}{^{232}Th}\right)_t = \left(\frac{^{230}Th}{^{232}Th}\right)_0 e^{-\lambda_{230}t} + \left(\frac{^{238}U}{^{232}Th}\right)_t (1 - e^{-\lambda_{230}t}). \qquad (6.1)$$

Thus, the ordinate intercept of the horizontal line defined by the mineral phases at $t = 0$ will be the initial $(^{230}Th/^{232}Th)_0$ ratio of the whole system. With time, the $(^{230}Th/^{232}Th) - (^{238}U/^{232}Th)$ activity ratios for the mineral phases will subsequently define a suite of lines (isochrons) whose slopes progressively change according to the time function $(1 - e^{-\lambda_{230}t})$ rotating about a point equal to the initial $(^{230}Th/^{232}Th)_0$ ratio of the melt, and whose

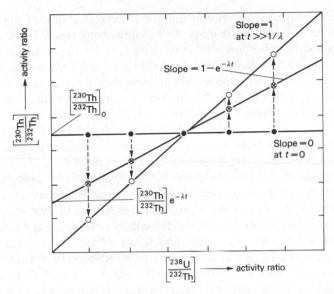

Fig. 6.6. Uranium and Th isotopic behaviour in the system ^{230}Th/^{232}Th–^{238}U/^{232}Th for the separate mineral phases of a volcanic rock initially at isotopic equilibrium with different U/Th ratios, but the same ^{230}Th/^{232}Th ratios. Filled circles represent the system at $t = 0$, the crossed circles at some elapsed time t, and the open circles at secular equilibrium where $t = \infty$. The slope of the line connecting the individual phases is defined by the chronometric function $(1 - e^{-\lambda t})$ and gives the age of the system at any time t. Thus, radioactive decay within the system ^{230}Th–^{238}U–^{232}Th is represented by a vertical vector in the ^{230}Th/^{232}Th–^{238}U/^{232}Th space whereas chemical alteration of either U or Th is represented by a horizontal vector. (After Allegre and Condomines 1976).

ordinate intercepts at any time, t are given by the relationship:

$$(^{230}\text{Th}/^{232}\text{Th})_t = (^{230}\text{Th}/^{232}\text{Th})_0 \cdot e^{-\lambda_{230}t}. \qquad (6.2)$$

At secular equilibrium this line will reach a slope of unity and pass through the origin since, at that time, $(^{230}\text{Th}/^{232}\text{Th}) = (^{238}\text{U}/^{232}\text{Th})$.

In order to obtain the accurate ages by this isochron technique four criteria must be met: (i) the mineral phases must have had identical $(^{230}\text{Th}/^{232}\text{Th})$ ratios but different $(^{238}\text{U}/^{232}\text{Th})$ ratios at the time of crystallization; (ii) the transit time of the magma to the surface must be short with respect to the half-life of ^{230}Th, (iii) the mineral phases must have formed at essentially the same time, and (iv) the system must have remained closed with respect to post-crystallization isotope migration.

If the transfer time of the magma to the surface is short with respect to the half-life of ^{230}Th, then the system behaves as described above; the measured time t being that elapsed since eruption and crystallization of the rock. If, however, the transfer time is long, the isotopic composition of the melt will proceed toward equilibrium before crystallization. Upon reaching the surface and crystallizing, progress toward equilibrium will continue defining an internal isochron which rotates about the point on the equilibrium line corresponding to the $(^{230}Th/^{232}Th)$ ratio at the time of crystallization. Similar reasoning applies to phenocrysts crystallizing from a melt at depth and later carried to the surface by the magma. If the initial $(^{230}Th/^{232}Th)_0$ and $(^{238}U/^{232}Th)_0$ ratios of the melt are known, then some estimate of the transfer time or time of phenocryst formation can be obtained.

Once the initial $(^{230}Th/^{232}Th)$ and $(^{238}U/^{232}Th)$ isotope ratios are set at the time of crystallization, a volcanic rock must remain closed to either inward or outward isotope migration. Otherwise, the measured isotope ratios for the mineral phases in the rock will not define a coherent isochron, and thus the age of the rock will be indeterminable. This is not so much a problem with Th which is generally immobile under normal conditions in the near-surface environment, but can be a problem for U which is both highly mobile in aqueous solution and subject to leaching due to its tendency to be located along mineral grain boundaries (see Chapter 2). The presence of ^{234}U in radiation damaged sites, a feature which increases with the age of a rock due to its continual production from ^{238}U decay, is a second factor which can contribute to enhanced mobility of U. Such effects, if present, will not, however, go unnoticed in an isochron plot unless all mineral phases have been affected to the same extent.

6.4.3 Examples

The dating of young volcanic rocks is a relatively recent application of U-series disequilibrium methods. It is, therefore, difficult at present to predict the ultimate contribution of the method to geochronological problems. It certainly has a great potential for the dating of young igneous rocks, especially basic rocks, which are difficult to date by conventional geochronological techniques. The examples which follow illustrate this point and also indicate some of the problems of the method.

The first attempt to use U-series disequilibrium to date young igneous rocks was by Cerrai, Dugnani Lonati, Gazzarini, and Tongiorgi (1965) who analysed zircon and ilmenite from a beach sand thought to have concentrated these resistate minerals from a nearby volcanic tuff. Shortly thereafter, Taddeucci, Broecker, and Thurber (1967) and Kigoshi (1967) reported isochron techniques for ^{230}Th-dating of young igneous rocks. The results of these studies are shown in Figs 6.7 and 6.8. The data of Taddeucci et al. (1967) were obtained on phases for five rhyolitic tuffs from the Mono Craters of California which had been previously dated by the K/Ar techniques. Their results, based largely on hornblende-glass pairs, gave crystallization ages of between 2 and 10 ky which were, with one exception, concordant with K/Ar ages and geological relationships. However, Allegre (1968) noted that the one sample for which all constituent phases had been analysed gave a hornblende-quartz-olivine mineral isochron which was significantly older than that for hornblende-glass (see Fig. 6.7). He interpreted this older isochron to indicate a two-stage evolution for this rock: crystallization of phenocrysts at depth in a magma chamber at about 25 ky and later eruption of the lava about 1 ky ago. A similar situation was noted by Baranowski and Harmon (1978) in a more detailed study of rhyolitic volcanics in the Long Valley, California. Again ^{230}Th-ages agreed well with K/Ar ages, but in this instance the glass phase was displaced above the biotite/hornblende-quartz-feldspar-whole rock isochron (see Fig. 6.9). Because the glass phases showed no petrographic evidence of alteration, and had equivalent U content and ^{234}U/^{238}U ratios, it was suggested that post-eruption leaching of U was not responsible for the anomalous behaviour of the glass matrix, but rather that U was depleted from the melt prior to eruption, perhaps through vapour-phase fumarolic activity.

By contrast, the data of Kigoshi (1967) was obtained on different leach fractions of the analysed rocks so that it was not possible to more than estimate chronologic information from the samples. In spite of potential problems with

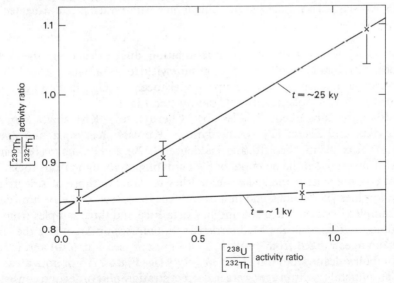

Fig. 6.7. The ^{230}Th/^{232}Th–^{238}U/^{232}Th isochron diagram for a young rhyolite from Mono Craters area, California, USA (data from Taddeucci *et al.* 1967). The rock can be interpreted as having a two-stage evolution; the hornblende-olivine-quartz isochron defining the age of crystallization of the phenocrysts at depth in the magma chamber prior to eruption (~ 25 ky) and the hornblende-glass isochron defining the actual eruption age of ~ 1 ky. (After Allegre 1968.)

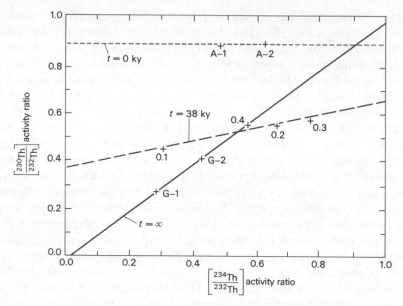

Fig. 6.8. The ^{230}Th/^{232}Th–^{238}U/^{232}Th isochron diagram for a Cretaceous granite, a ~ 35 ky Late Pleistocene pumice, and a historical lava flow. Note that ^{235}U was measured via ^{234}Th assuming secular equilibrium between the parent-daughter pair (After Kigoshi 1967.)

incomplete leaching or isotope fractionation during leaching, the results appear satisfactory, as a Cretaceous granite yielded an infinite age, an historic lava gave a modern age, and an upper Pleistocene pumice had an age which agreed with that obtained by ^{14}C-dating (see Fig. 6.8).

Although subsequent studies by Cherdynstev, Kislitsina, Kuptsov, Kuz'mina, and Zverev (1967) Cherdyntsev, Kuptsov, Kuz'mina, and Zverev (1968), Discendenti, Nicoletti, and Taddeucci (1970), and Fukuoka (1974) have further investigated the potential of ^{230}Th-dating of young igneous rocks, the most comprehensive and successful studies to date are those of Allegre and Condomines (1976) and Condomines and Allegre (1980). These authors dated six samples from the Irazu volcano in Costa Rica and three samples from the Stromboli volcano in the Mediterranean Eolian Archipelago. For the Irazu volcano ages ranged from 100 ± 16 ky to present (see Fig. 6.10) and for the Stromboli volcano from $156 ^{+45}_{-32}$ to 35 ± 9 ky (see Fig. 6.11). For both areas the ^{230}Th-mineral isochron ages were in correct stratigraphic order and consistent with geological evidence. Isotopic equilibrium was observed within individual mineral phases with no significant difference noted between phenocrysts and glass matrix.

Disequilibrium studies can also provide unique information on the transfer time of a magma to the surface. In the ^{238}U decay series, ^{226}Ra will reach

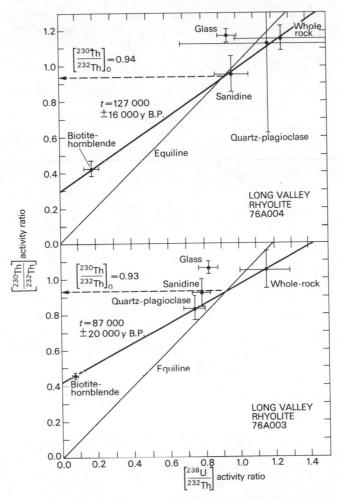

Fig. 6.9. The ^{230}Th/^{232}Th–^{238}U/^{232}Th isochron diagrams for two rhyolites from Long Valley, California. (After Baranowski and Harmon 1978.)

equilibrium with its immediate parent ^{230}Th after only 10^4 y, and ^{210}Pb will be in equilibrium with ^{226}Ra in just over 10^2 y. Thus, if a magma travels to the surface in a time which is short relative to the respective half-lives, one should expect ^{226}Ra/^{230}Th or ^{210}Pb/^{226}Ra disequilibrium, provided no other processes affect these two nuclides.

Additional information regarding magma transfer time can, in theory, be obtained from a ^{230}Th/^{232}Th-^{238}U/^{232}Th isochron diagram. Partial melting in the source region and direct transfer to the surface will result in the mineral phases and total rock initially defining a horizontal line on an isochron

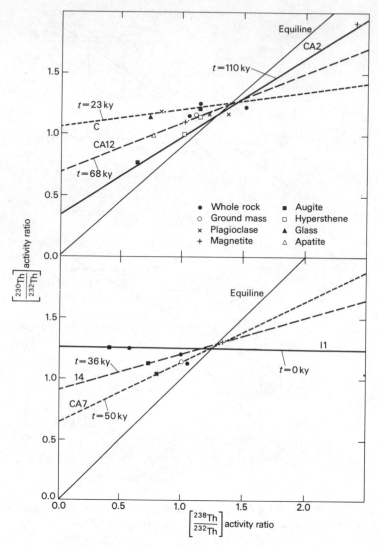

Fig. 6.10. The ^{230}Th/^{232}Th–^{238}U/^{232}Th isochron diagrams for six Quaternary andesite (five lavas and one volcanic bomb) from the Irazu volcano, Costa Rica. (After Allegre and Condomines 1976.)

diagram where the intercept on the y-axis is the initial ^{230}Th/^{232}Th ratio for the bulk system at the time of partial melting. However, if the transfer time is long relative to the half-life of ^{230}Th, then the isotopic composition of the magma will evolve toward equilibrium prior to the crystallization of the rock when the magma is erupted at the surface. For this situation, the mineral

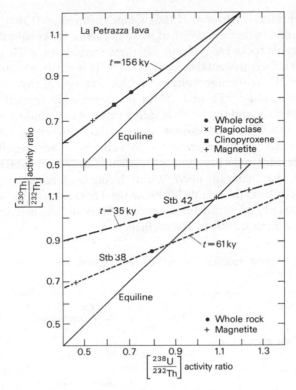

Fig. 6.11. The $^{230}Th/^{232}Th-^{238}U/^{232}Th$ isochron diagram for three lavas from the Stromboli volcano. (After Condomines and Allegre 1980.)

phases and whole rock will initially define a horizontal line on an $(^{230}Th/^{232}Th)_0$ – time diagram offset to a higher $^{230}Th/^{232}Th$ initial ratio. An estimate of the transfer time can be obtained if the $(^{230}Th/^{232}Th)$ and U/Th ratios of the source are known independently (see Fig. 6.12).

6.5 Tracer studies

6.5.1 Introduction

If a magma source region in the mantle or crust has remained a closed system for more than about 10^6 y, then all members of the ^{238}U decay series will be in secular equilibrium. However, chemical processes involving members of the U-decay series which occur during magma genesis or affect a magma during transit to the surface can act to upset the conditions of equilibrium within the U-decay series. Thus, disequilibrium studies can be an important means for identifying and characterizing chemical fractionation during partial melting, contamination of a magma during transit or while in residence in a volcano prior to eruption, and post-eruption alteration of volcanic rocks.

Studies by Somayajulu *et al.* (1966), Cherdynstev *et al.* (1967), Oversby and Gast (1968), and Nishimura (1970) of young volcanic rocks have shown that: (i) oceanic volcanic rocks are typically characterized by Ra > Th > U, whereas Ra $\geqslant$ U > Th for continental volcanics; (ii) ^{234}U is in virtual equilibrium with its parent ^{238}U in oceanic volcanic rocks, but not always in continental volcanic rocks; (iii) ^{230}Th and ^{226}Ra are commonly present in excess of equilibrium values; and (iv) ^{210}Pb is frequently not in equilibrium with ^{226}Ra. Some aspects of U-series isotope relationships in young volcanic rocks illustrating these points are presented in Table 6.3. These results suggest an intrinsic difference in the source area for oceanic and continental volcanics and the presence of a chemically open system during magma genesis, the removal of partial melts from the mantle lowering the Th/U ratio of the source region as well as a variety of post-magma genesis effects such as contamination and volatile loss during transit to the surface.

Table 6.3. U-series nuclide activities for some recent and Quaternary volcanic rocks

Sample description	^{238}U	^{234}U	^{232}Th	^{230}Th	^{226}Ra	^{210}Pb	Ref.
	(			dpm/g		)	
1926 tholeiite flow, Mauna Loa, Hawaii	0.16	0.16	0.15	0.23	—	0.19	(1)
1907 tholeiite flow, Mauna Loa, Hawaii	0.14	0.14	0.12	0.18	—	0.30	(1)
1801 alkalic flow, Hualalai, Hawaii	0.40	0.40	0.46	0.51	—	0.40	(1)
1921 tholeiite flow, Halemaumau, Hawaii	0.33	0.33	0.30	0.37	—	0.23	(1)
Late Pleistocene mugearite, Kohala, Hawaii	1.65	1.67	1.52	1.78	—	—	(1)
1961 trachy andesite flow, Tristan de Cunha	2.39	—	3.22	2.96	3.73	2.31	(2)
1944 nephelene basalt flow, Mt Vesuvius, Italy	3.93	—	3.90	4.27	43.7	34.1	(2)
1958 alkali basalt flow, Faidal, Azores	0.76	—	0.98	1.15	1.88	1.15	(2)

(1) Somayajulu *et al.* (1966)
(2) Oversby and Gast (1968)

6.5.2 Theory

The presence of disequilibrium within the ^{238}U decay series brings about the possibility of using Th/U whole-rock isotope systematics as a tracer to follow the chemical evolution of young volcanic rocks. As detailed by Allegre and Condomines (1976) and Condomines, Berna't, and Allegre (1976), insight into the chemical evolution of a magma can be obtained by considering whole-rock initial $(^{230}\mathrm{Th}/^{232}\mathrm{Th})_0$ ratios as a function of time for specimens dated by the internal isochron method described in section 6.4. The U/Th abundance relationships are also useful in this regard.

If a source rock is partially melted to produce a magma which is periodically tapped and transported to the surface, where it is erupted and crystallizes, the evolution of the magma in the chamber and of the separated fractions will be described by the relationship:

$$(^{230}\mathrm{Th}/^{232}\mathrm{Th})_0 = [(^{230}\mathrm{Th}/^{232}\mathrm{Th})_0^R - (^{238}\mathrm{U}/^{232}\mathrm{Th})_t^R]e^{\lambda_{230}(t - t_R)}$$

$$+ (^{238}\mathrm{U}/^{232}\mathrm{Th})_t^R \qquad\qquad (16.3)$$

where the subscript '0' indicates initial conditions and 't' the conditions at some later time, and R refers to the reservoir. Thus, $(^{230}\mathrm{Th}/^{232}\mathrm{Th})_0$ is a linear function of $e^{\lambda_{230}t}$ (see Fig. 6.12).

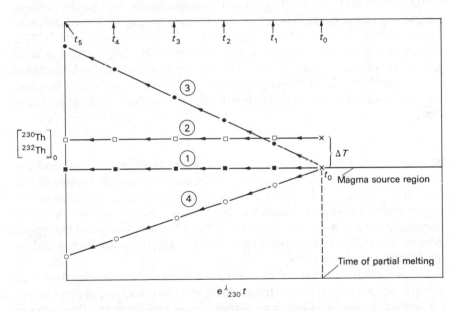

Fig. 6.12. Initial $^{230}\mathrm{Th}/^{232}\mathrm{Th}$ ratios as a function $e^{\lambda_{230}t}$ for an evolving volcano. Six arbitrary points in time $(t_0 \rightarrow t_6)$ are shown for four different models: (1) successive eruptions of a magma derived from partial melting of a source in secular equilibrium with rapid transfer to the surface; (2) successive eruptions of a magma derived from partial melting of a source in secular equilibrium with a long, but constant transfer time, ΔT; (3) successive eruptions of a magma from a differentiating magma chamber with a constant U/Th ratio which is higher than the source ratio; and (4) successive eruptions of a magma from a differentiating magma chamber with a constant U/Th ratio which is lower than the source ratio. The effect of continual variation in the U/Th ratio of the magma chamber would cause the linear evolution lines to become curved in the direction of the change. (After Allegre and Condomines 1976.)

Where successive eruptions of a magma chamber initially in a state of secular equilibrium are characterized by a short transfer time, the evolution of a magma through time is shown by trend (1) and a longer transfer time by trend (2); trend (3) illustrates the conditions for successive tappings of a differentiating magma chamber where the U/Th ratio of the chamber is constant, but greater than the source material; trend (4) denotes the same situation for a U/Th ratio that is constant but less than that of the source. Other more complicated histories can be modelled by allowing the U/Th ratio of the reservoir to change with time in a constant or variable manner, by mixing of single or multiple contaminants into the differentiating magma, or by considering longer or variable transfer times to the surface.

Contamination of a magma during transit to the surface or during residence in a volcano prior to eruption in a single, bulk event (for example, assimilation of U-enriched crust or Th-enriched sediments) would result in either higher $(^{234}U/^{232}Th)$ or $(^{230}Th/^{232}Th)_0$ ratios than expected for bulk partial melting. Simple two-component mixing of a differentiating magma with a contaminant would generate a linear correlation between $(^{230}Th/^{232}Th)$ and $(^{234}U/^{232}Th)$, whereas continuous or multiple-source contamination could result in total disruption of the $^{230}Th-^{234}U-^{232}Th$ isotope systematics. Such contamination would be difficult to recognize unless expressed by differences in initial ratios for modern lavas from a single volcano erupted during a short period of time.

6.5.3 Examples

Because investigation of U-series disequilibrium in young volcanic rocks is relatively recent, its application as a tracer has not yet been fully exploited. The majority of studies to date have been concerned with understanding the causes of observed Pb-isotope variations in igneous rocks (see for example Somayajulu *et al.* 1966; Oversby and Gast 1968). Some indication of the type of information that can be obtained from U-series tracer studies is provided by the following two examples.

The calc-alkaline volcanism of Costa Rica is part of the Central American volcanic belt where presently active andesite volcanoes are built upon a crust of Cenozoic sediment and Cretaceous ultramafic rocks. Allegre and Condomines (1976) have analysed rock and mineral samples from three volcanoes: Poas, Arenal, and Irazu. The $^{230}Th/^{232}Th-^{238}U/^{232}Th$ isochrons for six Irazu samples gave crystallization ages of between 100 ky and the present, with initial $(^{230}Th/^{232}Th)_0$ ratios ranging from 1.01 to 1.32 (see Fig. 6.10). Consideration of the variation of the initial $(^{230}Th/^{232}Th)_0$ ratios with time (see Fig. 6.13) indicates that differentiation in the magma chamber proceeded at constant U/Th activity ratio of about 1.6, and that the magma chamber beneath the Irazu volcano was established at about 140 ky with a $(^{230}Th/^{232}Th)$ ratio of 0.81, presuming a typical mantle Th/U value of 3.7 for the source partial melt (Oversby and Gast 1968; Tatsumoto 1966). Subsequent magmatic evolution occurred primarily through fractional crystallization/differentiation of the largely unmodified mantle partial melt. If it is further assumed that the Poas and Arenal lavas were derived from the same ultimate source, then each volcano must have evolved independently, the respective magma chambers established at 50 ky and 35 ky (Fig. 6.13).

The U contents, $^{234}U/^{238}U$ ratios, and some $^{230}Th/^{234}U$ ratios have been determined by Bacon (1978) and MacDougall, Finkel, Carlson, and Krishnaswami (1979) for oceanic basalts from the Atlantic and Indian Oceans: both studies attributed (see Table 6.4) observed disequilibrium within the U-decay series to low-temperature alteration by, and exchange with, seawater. Enrichment of U in palagonite (a hydrated mafic glassy material formed by the rapid breakup of chilled magma during eruption into a cold environment) rims, relative to unaltered glass and the crystalline interior of the basalt, occurs generally accompanied by leaching of authigenic, decay-generated ^{234}U, to produce low $^{234}U/^{238}U$ ratios. Old, altered crystalline basalts have $^{234}U/^{238}U$ ratios in excess of the equilibrium value, but are deficient in ^{230}Th, as expected from prolonged low-temperature U exchange with seawater. By contrast, the

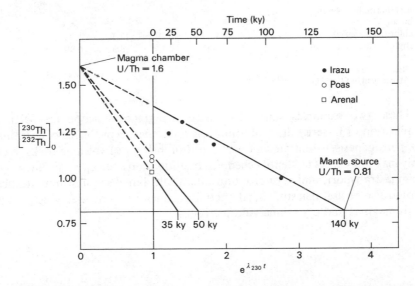

Fig. 6.13. Plot of initial $^{230}Th/^{232}Th$ ratios against age for the six whole-rock samples from the Irazu volcano dated by the $^{230}Th/^{232}Th-^{238}U/^{232}Th$ isochron technique (Allegre and Condomines 1976). If the mantle $^{230}Th/^{232}Th$ value is taken to be 0.81 (i.e. Th/U = 3.7) then an age of ~ 140 ky is derived for the establishment of the Irazu volcano. Data for modern lava flows from two younger volcanoes, Poas and Arenal, indicate ages of ~ 50 ky and ~ 35 ky respectively assuming the same U/Th ratio during differentiation within the magma chambers of these volcanoes.

margins of hydrothermally altered samples contain less U than crystalline interiors, but contain a slight excess of ^{234}U which possibly indicates subsequent low-temperature interaction with seawater.

Table 6.4. U-series elemental abundancies and nuclide activity ratios for oceanic basalts from Atlantic and Indian Oceans

Sample description	Magnetic anomaly age (10^6 y)	U (ppm)	Th (ppm)	$\left[\dfrac{^{234}U}{^{238}U}\right]$	$\left[\dfrac{^{230}Th}{^{234}U}\right]$	Ref
Fresh pillow basalt	<1					(1)
glassy rim		0.07	0.17	1.05	1.06	
crystalline interior		0.05	0.21	1.08	0.77	
Fresh glassy pillow basalt	<1	0.06	—	1.00	—	(2)
Fresh pillow basalt	1.5	0.10	—	0.97	—	(2)
Weathered pillow basalt	5	4	0.38	0.92	—	(1)
Weathered pillow basalt	12	0.16	—	1.19	—	(2)
Weathered pillow basalt	23	0.61	—	1.14	—	(2)
Weathered pillow basalt	46	0.28	—	0.94	—	(1)
Hydrothermally altered basalt	<1					(1)
chloritized margin		0.14	0.19	1.05	1.05	
slightly altered interior		0.19	0.17	1.03	0.98	

(1) Bacon (1978)
(2) Macdougall *et al.* (1979).

These two examples show that there is significant scope for further application of U-series disequilibrium tracer studies to problems concerning the petrogenesis; evolution, and post-eruption history of volcanic rocks. Not only may one expect to identify chemical fractionations during partial melting, magma transport, and post-eruption alteration, but disequilibrium studies should also be useful in studying the actual geochemical processes which affect both magmas and crystalline volcanic rocks.

7 MOBILIZATION AND WEATHERING
J. N. Rosholt

7.1 Introduction

Silicic igneous rocks such as granites and rhyolites are considered to be important sources for U mobilization because of their high U and Th content relative to other igneous rocks (see chapter 6), their consistent spatial association with U deposits, and because they can contain a significant fraction of labile U (see chapter 14). In the cycle of U shown in Fig. 7.1, a small but economically significant fraction of mobilized U is fixed in sedimentary ore deposits, but the major fraction ultimately enters the ocean via river systems where it has a mean residence time in ocean water of about 5×10^5y (Goldberg and Koide 1962). If a constant U concentration and isotopic abundance is maintained in the ocean, its removal from ocean water must

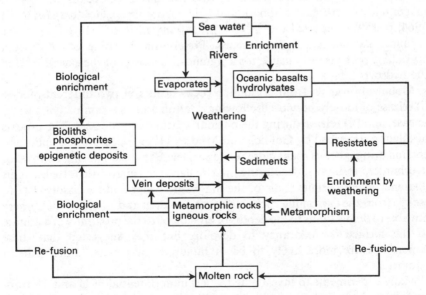

Fig. 7.1. The U cycle. (Modified from Rankama and Sahama 1950, and Hambleton–Jones 1978.)

proceed at the same rate as input from rivers. Organic-rich sediments and coexisting phosphorites remove some of the total supply (Veeh 1967) and several investigations have shown that weathered marine basalts have acquired substantial amounts of U from ocean water (Aumento 1971; Thompson 1973; MacDougall 1973, 1977; Bacon 1978). Bloch (1980b) interpreted these results to show that low temperature alteration of the oceanic crust is a major sink for U supplied to the oceans and may account for about 50 per cent of the estimated present day input of this element. Uranium uptake by organic-rich sediments and coexisting phosphorites on continental margins may remove in excess of 10 per cent of the total supply. High temperature alteration of oceanic basalts, metalliferous sediments, carbonate sediments, and sediments in anoxic basins deeper than 200 m play a relatively minor role in removal of U; each of these sinks is responsible for uptake of less than 5 per cent of the overall input.

The part of the geochemical cycle that encompasses initial mobilization by erosion and weathering of igneous rocks, is discussed in this chapter, along with consideration of mobilization through continental surficial sediments.

7.2 Mobilization and leaching experiments

Natural ground and surface waters have an excess of ^{234}U compared to that required for equilibrium with its parent ^{238}U (Osmond and Cowart 1976a). It has been demonstrated that the large reservoir of U, the oceans, reflects the magnitude of isotopic fractionation of U (Thurber 1962). Goldberg (in Rona 1964, p. 1596), reported the $^{234}U/^{238}U$ activity ratio of 1.14 ± 0.01 in the Atlantic, Pacific, and Indian oceans. Preferential leaching of ^{234}U from crystalline rocks is responsible for a significant fraction of the excess ^{234}U in the hydrosphere.

Mobilization of U from crystalline rocks occurs in two different modes: (i) release of loosely-bound U following dilation and its accompanying water release; and (ii) release during the normal weathering processes. The process of dilation described by Goldich and Mudrey (1969) occurs when uplift and erosion bring igneous rocks close to the surface. Expansion during removal of overburden results in fracturing, particularly of quartz-rich rocks. This dilatancy effect permits some of the magmatic water and its dissolved U to escape from the crystalline rock. However, Goldich and Mudrey (1972) were not able to define limits to the depths of erosion or to the proximity of a granite to the surface for dilatancy to develop, but they suggested that these dimensions are more likely to be in hundreds of meters rather than in kilometers.

Many experiments to investigate the leaching potential of U and Th from igneous rocks have been made. Leaching of crushed rocks with alkaline or acid solutions gives a semi-quantitative estimate of the amount of these elements loosely held in igneous rocks, most likely along grain boundaries, fractures, and in soluble U-bearing minerals (Z. E. Peterman 1962, personal comm.). In addition, loosely-bound Th and especially U have received considerable attention because of the interest in the possibility of obtaining economic quantities of these elements from silicic igneous rocks.

Brown and Silver (1956) reported that in addition to U and Th, a fraction of rare earth elements is also removed by strong acid treatment. Larsen and Gottfried (1960) presented acid leaching data for six samples of various phases of the Southern California batholith and found that, although greater proportions of mafic samples were dissolved, the largest fraction of the total soluble U occurs in the silicic members of the series indicating that the amount and percentage of soluble U are closely related to the U content and compositions of the rocks. Hurley (1950) noted similar relations for four diabases and eight granites treated with dilute hydrochloric acid where averages of 57 and 82 per cent of the alpha particle activity were removed from the diabases and granites, respectively. Neuerburg, Antweiler, and Bieler (1956) leached 250 samples of various igneous rocks and reported that the relation between readily soluble U to total U contents, rock types, and other features is essentially random.

Limited data indicate that comparable amounts of Th in igneous rocks are readily soluble in acid. Tilton, Patterson, Brown, Inghram, Hayden, Hess, and Larsen (1955) reported the removal of 34 per cent of the total U and 42 per cent of the total Th from an Ontario granite leached with 6 M hydrochloric acid. Leaching studies by Larsen (1957) on 16 volcanic and plutonic rocks from the Colorado Front Range with 1:4 hydrochloric acid indicated that between 10 and 70 per cent of the U, and between 20 and 91 per cent of the Th were removed. For all of the samples the Th/U ratio was decreased by leaching and many were reduced by more than half of the original ratios. In a leaching study of fresh Boulder Creek granodiorite from Colorado, Pliler and Adams (1962) demonstrated that as much as 90 per cent of the Th, and 60 per cent of the U could be removed by 2 M hydrochloric acid leaching solution.

It is difficult to compare results of leaching studies because of the use of the different methods and reagents. Tauson (1956) and Leonova and Tauson (1958) used ammonium carbonate solutions. Tauson (1956) noted that only about 70 per cent of the easily soluble U could be dissolved with ammonium carbonate alone, but addition of hydrogen peroxide brought the remaining 30 per cent in solution. Krylov and Atrashenok (1959) performed leaching studies with 0.2 M sodium carbonate solutions where crushed rocks were leached for 5 days at room temperature. These authors suggested that the early leaching studies, especially those in which acids were used, do not give a reliable estimate of the amounts of loosely held U because considerable amounts of the total rock are dissolved.

More recent leaching studies have been made by Szalay and Samsoni (1969) for investigations of U leachability from granite, gabbro, andesite, and basalt using 0.2 per cent sodium bicarbonate solution in aerated water. They found that: (i) approximately one per cent of the total U content could be leached from granite detritus; (ii) leaching from granite was greater than from any of the other rock types investigated; (iii) equilibrium was established within 3 to 5 h after the leachable U was depleted from surfaces of the grains exposed by crushing the sample; and (iv) successive releaching several times with replacement water indicated that equilibrium was reestablished each time with leaching concentration decreasing to about 1/5 of the original value. Using a more concentrated leaching solution, Titayeva and Veksler (1977) made a series of experiments on leaching of Th and U from granite rock samples of massifs in USSR, with 5 per cent sodium carbonate solution at room temperature (solid to liquid ratio of 10). After 24 h of leaching, from 0.9 to 20 per cent of Th (8 per cent average), and from 0.1 to 30 per cent of U (5 per cent average) passed into solution. In most samples the percentage of extracted Th was higher than the percentage of extracted U. They concluded that Th isotopes are mobile on the earth's surface, but the relative instability of their compounds in natural waters and the ease with which they are adsorbed limits their mobility; as a result, Th is redistributed during weathering and often forms secondary accumulations. From Wyoming granites, Harshman (1972) found that a leaching solution (0.002 M each of sodium carbonate and sodium sulfate) removed 0.1 per cent to 1 percent of the U with an average value of 0.5 per cent, from 11 samples of crushed granites.

Zielinski (1979) reported a leaching experiment to determine the rate of U removal from rhyolitic obsidian, perlite, and felsite by oxygenated alkaline carbonate solution. The conclusions of this study were that: (i) under experimental conditions U removal from crushed glassy samples proceeds by a mechanism of glass dissolution in which U and Si are dissolved in approximately equal weight fractions; (ii) rate of U loss correlates positively with the surface area of glass samples; (iii) the rate of U removal from crushed felsite is time dependent with rapid initial loss of a small component approximtely (2.5 per cent) representing mobilization of loosely bound U associated with mineral surfaces and alteration products; and (iv) the fractions of U removal during the experiment (leaching with 0.05 M $Na_2CO_3 + 0.05$ M $NaHCO_3$, at pH 9.9, 25 °C) ranged from 3.2 per cent for felsite to 27 per cent for pearlite. In a similar study on leaching of volcanic glass, Zielinski (1980) also found that U and Si were dissolved in approximately equal percentages, implying that U removal proceeds by a mechanism of glass network destruction. Of the measured variables, temperature had the greatest influence on the removal of glass components; other variables which influence the rate of removal of U, but to a much lesser extent, were pH, total carbonate content of solutions, glass composition, and grain size (surface area). Leaching during a 30 day experiment indicated that about

3 per cent of the total U was dissolved at $60°C$, about 14 per cent dissolved at $90°C$, and about 30 per cent dissolved at $120°C$. Leaching experiments under similar conditions for a 20 hour period on petrographically-fresh granites (Zielinski *et al.* 1981, written comm.) indicated that 2 to 6 per cent of the total U was removed with an average $^{234}U/^{238}U$ activity ratio of 1.72 ± 0.29 in leaching solutions of the 6 drill-core samples which had $^{234}U/^{238}U$ ratios of unity before leaching. No measureable amounts of Th isotopes were found to be leached from the granite at pH 8.5.

All these reports show that there is a wide variation among results of leaching experiments by different investigators using diverse chemical leaching solutions to remove Th and U. To obtain meaningful leaching experiments regarding U and Th removal from rocks, solutions that are similar to those occurring in nature should be used; however, it is not possible to obtain the effects of geological time under the limitations of experimental conditions. Table 7.1, taken from Zielinski (1979), shows some experimental leach solutions and pertinent natural analogues.

7.3 Alteration and weathering

Tuff sequences of Pleistocene age at Lake Tecopa, California (Sheppard and Gude 1968) and of Pliocene age at Keg Mountain, Utah (Zielinski *et al.* 1980), were analysed for U and Th isotopic abundances by Rosholt (1980*a*). Three volcanic ash beds that have been correlated with equivalent dated ash deposits in other areas (G. A. Izett 1979, personal comm.) occur at Lake Tecopa (Izett, Wilcox, Powers, and Desborough 1970): tuff A, Pearlette type O ash (0.6 my old, Naeser *et al.* 1973); tuff B, Bishop Tuff (0.7 my old, Izett and Naeser 1976); and tuff C, Pearlette type B ash (2 my old, Naeser, Izett, and Wilcox 1973) or Huckleberry Ridge ash bed. Samples of tuff A that have been altered to more than 90 per cent zeolite contain about 60 per cent excess ^{234}U and an even greater excess of ^{230}Th; predominantly glassy samples of Pearlette type O have about 20 per cent excess ^{234}U. In statigraphically older ash of Lake Tecopa (tuff B, and tuff C) highly zeolitized samples only have 3 to 9 per cent excess ^{234}U, whereas samples composed essentially of glass are in radioactive equilibrium. Altered tuff from the Keg Mountain area (10 my old) has zeolite components that are 50 per cent deficient in ^{234}U and glass components that are in radioactive equilibrium. In Fig. 7.2, the $^{234}U/^{238}U$ activity ratios in altered tuff and glass samples are plotted against per cent zeolite. Extrapolation of $^{234}U/^{238}U$ trends to 100 per cent zeolite indicate that: (i) zeolites in 0.6 my old Pearlette ash (tuff A) have about 66 per cent excess ^{234}U; (ii) excess ^{234}U in 0.7 my old Bishop Tuff (tuff B), and 2 my old Pearlette ash (tuff C) was diminished to about 10 per cent; and (iii) deficient ^{234}U predominated in 10 my old Keg Mountain area tuff extending to about 50 per cent deficiency in zeolite components.

Table 7.1. Comparison of experimental leach solution and pertinent analogues

Source	Typical concentration of major components	Typical pH	Typical ionic strength
Groundwater draining rhyolite tuff	~ 0.001–0.005 M HCO$_3$ [Na + K]	7–8.5	0.01–0.02
Groundwater draining carbonates	~ 0.001–0.01 M HCO$_3$ [Ca + Mg + Na]	7–8.5	0.01–0.02
Experimental solution	0.05 M Na$_2$CO$_3$ + 0.05 M NaHCO$_3$	9.9	0.2
Seawater	~ 0.5 M NaCl	8.15	0.7
Alkaline lakes, brines	0.5–5 M NaCl + Na$_2$CO$_3$ + NaHCO$_3$	8–10	1–10
Industrial carbonate leach of U ores	0.25–0.5 M Na$_2$CO$_3$ + 0.1–0.5 NaHCO$_3$	10–11	1–2
Industrial in-situ alkaline leach of low-grade U deposits	0.01–0.05 M NH$_4$HCO$_3$	8.0	0.01–0.05

After Zielinski (1980)

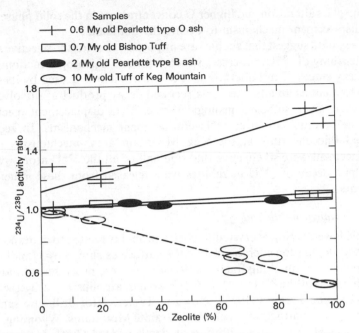

Fig. 7.2. The $^{234}U/^{238}U$ activity ratios plotted against per cent zeolite in glassy zeolitized tuff. (After Rosholt 1980a.)

Interpretation of radioactive disequilibrium in zeolitically altered glass at Lake Tecopa and Keg Mountain area indicate that the predominant initial process was selective emplacement of ^{234}U and ^{230}Th. Two separate mechanisms may be responsible for emplacement of ^{234}U and other daughter products produced from parent atoms in solution: (i) some of the recoiling nuclei ejected at the water–particle interface become imbedded in adjacent grains; and (ii) adsorption of ^{230}Th and the precursors of ^{234}U, ^{234}Th, and ^{234}Pa, on surfaces of particulate matter at this interface. Beta decay of short-lived ^{234}Th and ^{234}Pa would have left some of their daughter ^{234}U atoms implanted on the particulate matter. The ^{234}U emplacement atoms were not bound strongly to the solid surfaces and subsequently, over extended periods of time, significant fractions were leached back into the water phase at a rate greater than their disintegration rate (2.5×10^5 y half-life). Gradual leaching of emplaced ^{234}U was accompanied by the ^{234}U displacement mechanisms summarized in section 7.3.1. Initial emplacement by sorption-recoil processes probably was limited by the concentrations of dissolved species and by the sorptive capacity of the solids. In contrast, displacement by leaching-recoil was controlled by the concentrations in the solid phase and the solubility of leached isotopes. Apparently, continuous exposure of leachable ^{234}U sites during glass alteration and higher U concentrations in the solid phases caused the displacement mechanism to eventually dominate.

These data suggest that zeolites are important sites for the selective sorption and leaching of ^{234}U, processes which influence U isotope fractionation. In the early stages of tuff alteration, ^{234}U enrichment is caused by its selective emplacement as an adsorbed or a recoiled decay product of dissolved ^{238}U. However, given sufficient geological time, ^{234}U displacement mechanisms become predominant over ^{234}U emplacement mechanisms. In zeolitically altered Pliocene tuffs at the Keg Mountain area, mechanisms of ^{234}U displacement were so effective that one half of all the ^{234}U atoms produced by alpha decay of ^{238}U in zeolites were released from these mineral components.

7.3.1 Incipient weathering

Incipient weathering of crystalline rocks, probably related to dilatancy effects, starts at depths considerably below the surface as shown by U mobilization and redistribution along fractures. Results of these processes on radioactive disequilibria in the ^{238}U–^{234}U–^{230}Th system are illustrated by the data of Stuckless and Ferreira (1976), shown in Table 7.2 for drill core samples of fractured and unfractured rock from Granite Mountains, Wyoming. Similar results on U series disequilibrium at depth in Owl Creek Mountains and Laramie Mountains of Wyoming are shown by Nkomo, Stuckless, Thaden and Rosholt (1978) and Nkomo, Rosholt, and Dooley (1979). Investigation of the ^{238}U–^{234}U–^{230}Th system in silicic crystalline rocks includes analyses of 64 whole-rock specimens ranging from deep drill cores to surface and near-surface rocks. This study includes results from a drill core to 400 m depth (Stuckless and Ferreira 1976) and near-surface samples (Rosholt and Bartel 1969) in the Granite Mountains of Wyoming, a drill core to 150 m (Nkomo *et al.* 1978) in the Owl Creek Mountains of Wyoming, a drill core to 58 m (Nkomo *et al.* 1979) in the Laramie Mountains of Wyoming, a drill core to 450 m and surface samples of Pikes Peak granite in Colorado and a drill core to 530 m in the Sierra Nevada of California. The data are shown on a ternary diagram in Fig. 7.3. For this illustration, samples have been divided into three categories: (i) core samples that have abundant microfractures designated by 'V'; (ii) unfractured core samples designated by 'O'; and (iii) surface and near-surface samples designated by '+'. The ternary diagram has apexes for ^{238}U, ^{234}U, and ^{230}Th and the ^{234}U/^{238}U activity ratio for a given area on the diagram is shown by the dotted lines that cross the diagram. The location of the data point on the diagram can be used to indicate which of several processes affected the rock. A number of unfractured core samples and a few surface samples cluster about the radioactive equilibrium point indicating that the abundances of U and/or its long-lived daughter products in these rocks have not been disturbed during the past 300 ky.

Table 7.2. Alpha-spectrometric analyses of granites from the Granite Mountains, Wyoming

Sample	Depth (m)	Sample	U (ppm)	Th (ppm)	Activity ratios				$^{238}U/^{206}Pb*$
					$^{234}U/^{238}U$	$^{230}Th/^{238}U$	$^{230}Th/^{234}U$	$^{226}Ra/^{230}Th$	
MS-1----	0	S	15.4	43.7	0.85	0.89	1.06	0.95	1 <
TCM-2---	0	B	3.81	19.1	0.94	0.97	1.03	0.88	1 <
DDH-7---	0.5	B	3.62	40.4	0.78	0.90	1.15	0.88	1 ≪
GM-1----	20.43	B	12.9	60.1	1.04	1.10	1.05	0.87	1 <
GM-1----	50.11	BF	1380.0	45.8	0.95	0.93	0.98	0.87	≫ 1
GM-1----	50.75	AF	107.5	55.1	0.90	0.82	0.91	1.22	≫ 1
GM-1----	61.63	A	7.28	73.2	1.01	1.00	0.99	0.68	1 <
GM-1----	78.58	B	26.3	65.2	1.01	1.01	1.01	0.88	1 <
GM-1----	87.17	B	56.4	46.3	1.01	1.01	1.00	0.80	> 1
GM-1----	230.95	AF	11.2	6.96	1.23	1.26	1.02	1.46	1 <
GM-1----	248.26	L	23.4	15.1	1.00	1.00	1.00	1.01	> 1
GM-1----	259.66	L	13.7	6.95	1.00	0.98	0.98	0.76	0
GM-1----	389.69	LF	4.54	20.8	1.05	1.05	1.00	1.05	1 ≪
GM-1----	404.29	LF	22.7	25.1	1.28	1.28	1.00	0.64	> 1

Analysts: C. P. Ferreira and J. N. Rosholt.
Sample types are: S = silicified and epidotized granite; B = biotite granite; A = albitized granite; L = leucocratic granite; F = rock with abundant fractures.
$^{206}Pb*$: Radiogenic ^{206}Pb only, calculated by assuming an age of 2640 my and initial $^{206}Pb/^{204}Pb$ ratio of 13.77. $^{238}U/^{206}Pb*$ ratio indicates gain (>1) or loss ($1 <$) of ^{238}U during Cenozoic time.
(Modified from Stuckless and Ferreira 1976.)

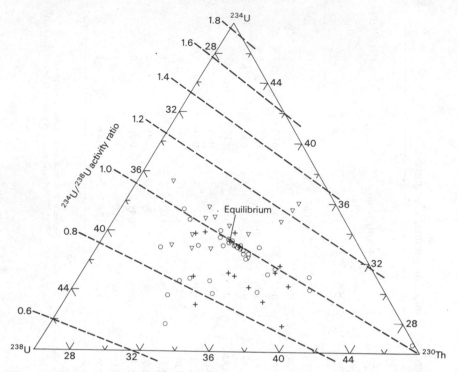

Fig. 7.3. Ternary diagram of relative activites of ^{238}U, ^{234}U, and ^{230}Th for 64 granitic rocks from western United States. V indicates core sample with abundant microfractures, O indicates unfractured core sample, and + indicates surface or near-surface sample.

A description of geochemical processes and the area on the ternary diagram where they are effective is shown on Fig. 7.4. The five most important geochemical processes are: (i) U leaching where ^{238}U and ^{234}U were removed with little or no fractionation: (ii) preferential ^{234}U leaching by alpha-recoil displacement processes (^{234}U recoil loss) with lesser ^{238}U loss; (iii) ^{234}U recoil loss with little or no ^{238}U loss; (iv) U assimilation where both ^{238}U and ^{234}U were deposited with the present-day ^{234}U/^{238}U activity ratios varying from 0.8 to 1.2; and (v) addition of ^{234}U and ^{230}Th by daughter emplacement processes (^{234}U and ^{230}Th recoil gain). The ^{234}U displacement mechanism (see item (iii)) operates when ^{234}U produced by decay of structurally incorporated ^{238}U is selectively displaced from mineral surfaces as a leached or alpha-recoiled decay product and this mechanism has been extensively documented in natural samples (Osmond and Cowart 1976a) and in laboratory experiments (Kigoshi 1971; Fleischer and Raabe 1978). Fleischer (1980) reported experiments which indicate that, in addition to the direct

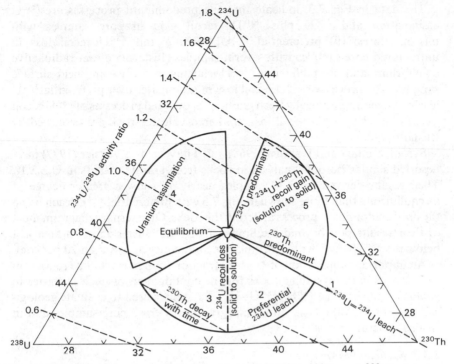

Fig. 7.4. Ternary diagram of relative activities of ^{238}U, ^{234}U, and ^{230}Th showing processes that affect the rocks and the areas where sample data indicate that these processes were effective.

recoil ejection into the liquid interface, a second mechanism of ^{234}U displacement results when many of the recoiling nuclei which are ejected from mineral grains become embedded in adjacent grains, producing alpha-recoil tracks; the subsequent track etching by natural solutions release some of the recoil nuclei. In contrast to ^{234}U displacement, less attention has been given to the complimentary process (item (v)) of daughter emplacement where dissolved ^{238}U and ^{234}U atoms decay by alpha disintegration, and recoiling daughter nuclides of ^{230}Th, ^{234}Th, ^{234}Pa, and ^{234}U are adsorbed or driven into particulate matter at the solid/liquid interface. After sufficient geological time, this mechanism (^{234}U and ^{230}Th recoil gain) results in solids that have ^{234}U/^{238}U and ^{230}Th/^{238}U activity ratios greater than the equilibrium value of unity. This is the only process described that involves transport of ^{230}Th. All other changes of abundance of ^{230}Th with time are in response to decay of ^{230}Th excess that is unsupported by its parent ^{234}U. Excess of ^{230}Th occurs following leaching of U (item (i)), preferential leaching of ^{234}U (item (ii)), and ^{234}U recoil loss processes (item (iii)).

The data in Fig. 7.3, indicate that the predominant processes are (i) U assimilation and ^{234}U plus ^{230}Th recoil gain in core samples with microfractures; (ii) preferential ^{234}U leaching and ^{234}U recoil loss in unfractured core samples with several samples clustering about radioactive equilibrium; and (iii) preferential ^{234}U leaching in surface and near-surface samples. The process of ^{234}U recoil loss predominates over preferential^{234}U leaching in petrographically fresh granites. Sequential processes of ^{234}U recoil gain followed by ^{234}U recoil loss in altered volcanic rock are described by Rosholt (1980).

Syromyatnikov and Ivanova (1968) and Titayeva and Veksler (1977) have reported similar isotopic analyses in rocks from granite massifs in U.S.S.R. Their results derived from 19 samples nearly all show a greater degree of disequilibrium than those shown in Fig. 7.3 with preferential ^{234}U leach being the only predominant process affecting the rocks. On a ternary diagram, most of their results plot beyond the limits of Fig. 7.3 clustering in an area just below the ^{230}Th apex with ^{234}U relative abundance at less than 20 per cent.

An important application of U-series fractionation in crystalline rocks is in the evaluation of crystalline rocks for the containment of nuclear wastes in geological media. The ^{238}U–^{234}U–^{230}Th system is used to evaluate geologically recent U mobility in both petrographically fresh core samples and in 'sealed' fracture zones.

7.3.2 Surficial weathering

Surficial weathering processes affecting U and Th migration and radioactive disequilibrium have been reported by several investigators. Hasen (1965) studied the result of weathering as reflected in residual soils developed on weathered granite in the Sierra Nevada Mountains of California. He found that: (i) both U and Th were concentrated to a greater degree in residual soils than in underlying rocks; (ii) U concentrated more than Th in topsoils with accumulations of Th occurring in B-horizon components; and (iii) both U and Th were most highly concentrated in clay fractions with concentrations of U being as much as ten times greater in clays than in underlying rock. The same conclusions were summarized by Hansen and Stout (1968), and based on patterns of isotopic distribution of ^{230}Th, ^{232}Th, and ^{234}U to describe movement of Th, they postulated that Th complexes readily with organic compounds in topsoils and reprecipitates as it is leached to zones of low organic content. Thorium distributions in a sequence of morainal soils in California were studied by Hansen and Huntington (1969), using gamma ray spectrometry, their data indicating distinct Th accumulation in soil horizons immediately underlying horizons of high organic content. Distributions of ^{226}Ra were explained in terms of U retention by organic matter, mobilization of ^{230}Th by organic matter, plant absorption of Ra, and time. Rosholt, Doe, and Tatsumoto (1966) concluded that ^{234}U was mobilized more readily than

^{238}U in early stages of rock weathering, but some ^{234}U in excess of ^{238}U accummulated in topsoils of high organic content.

Pliler and Adams (1962) made comparative studies of re-exposed weathered material in the Boulder Creek granodiorite of Colorado by gamma ray spectrometry. Their samples showed that, during exposures in the Pennsylvanian Period, weathering agents effectively removed the available U at an early stage, whereas little, if any, Th was removed at that stage. As weathering proceeded in later geological times, U and Th were concentrated to about the same extent; the most weathered material had four times more U and five times more Th than the slightly altered material above the fresh granodiorite. Their leaching studies of the weathered rock indicated that U was present largely in primary resistates, such as zircon, xenotime, and apatite, and Th occurred mainly in or on clays or in secondary resistates (minerals formed during weathering). Tieh et al. (1980) determined the abundance, distribution, and nature of occurrence of U and Th in granitic rocks in central Texas, and in soil profiles and local stream sediments derived from these rocks. Results were obtained by fission track mapping, delayed-neutron counting, and gamma ray spectrometry. In the granites, U occurs primarily in weathering-resistant accessory minerals (called resistate U) and along grain boundaries of major minerals, particularly biotite (called intergranular U). During in situ weathering and initial erosion of the granite, changes in U concentration are controlled by the chemical mobility of intergranular U and dispersal of the resistate U.

The concentration of U-series and Ac-series nuclides were determined by Megumi (1979) for samples of soil derived from weathering of granite at Murho, Japan. Each activity ratio of a radioactive descendant to ^{238}U or ^{235}U in the weathered rock was found to be greater than unity, and adsorption of ^{230}Th and ^{231}Pa on the surfaces of soil particles was responsible for the radioactive disequilibria. Extraction experiments indicated that more than 50 per cent of the ^{238}U, ^{231}Pa, ^{230}Th, ^{226}Ra, and ^{210}Pb resided on the surfaces of soil particles. Moreira–Nordemann (1980) used ^{234}U/^{238}U disequilibrium obtained by alpha spectrometry, to measure the chemical weathering rate of rocks from the Preto River basin, Bahia State, Brazil. The ^{234}U/^{238}U activity ratio measured in various samples of weathered rock was used to determine the U fraction which was removed from rocks during the weathering processes and, based on these results, a mathematical model indicated that the rock underwent a weathering rate of 1 m per 25 ky in that region under existing climatic conditions.

One of the more highly weathered products in continental environments is laterite and especially bauxite deposits that have more than normal U and Th contents. Adams and Richardson (1960) analysed 29 samples of bauxite from different locations around the world by gamma ray spectrometry and found that Th concentrations ranged from 5 to 133 ppm and average 49 ppm, whereas, U concentrations ranged from 3 to 27 ppm and average 11 ppm. The Th/U ratios ranged from 1.5 to 21, with an average value of 5. The actinide content of the bauxites is related to the type of source rock and those derived from nepheline syenites contain more Th and U than bauxites derived from other rock types; bauxites derived from basic igneous rocks have the lowest Th and U concentrations, and bauxites derived from carbonate rocks have the lowest Th/U ratios. Their study indicated that much of the Th and U in bauxites occurs in either primary or secondary resistate minerals. Based on the assumption that no Th is lost during laterization, the authors estimated that approximately 20 per cent of the U is leached from igneous and shale parent rocks, and 75 per cent of the U is leached from carbonate rocks during laterization; however, mass wasting of the matrix rock during laterization results in Th and U content that usually is greater than that in the original unweathered rock.

Additional weathering products that interface in the cycle of U include continental sediments and soils which are discussed in Chapter 10. Fluvial sediments, marine sediments, and marine carbonate and phosphate sediments are discussed in the marine environment chapters.

8 THE CHEMISTRY OF U AND Th SERIES NUCLIDES IN RIVERS

M. R. Scott

8.1 Introduction

The major path by which most products of chemical weathering on land reach the sea is transport by rivers. The concentrations and isotope ratios of U- and Th-series nuclides in river water and sediment are, therefore, very important parameters needed to construct geochemical balances for these elements. But 'average' river water is a rather problematic concept, especially where the more soluble members of the U- and Th-decay series are concerned. The literature on U in river water is preoccupied with attempts to estimate a world average for U content. Such attempts have been complicated by the extreme variability of these parameters for individual rivers, and the fact that chemical reactions which occur during riverine transport continue to affect U and Th and their daughter nuclides in transit to the ocean.

Several types of chemical reactions may take place in the riverine environment. Redox reactions strongly affect the chemistry of U itself and U can be scavenged from solution into reducing river bottom sediments (Sackett and Cook 1969; Lewis 1976). Additions of reduced groundwater or acid mine wastes to river water may be followed by the precipitation of Fe or Mn oxides and oxyhydroxides as oxidation of the dissolved constituents occurs. Such freshly precipitated oxide surfaces make extremely effective scavengers of nuclides like ^{210}Pb (Lewis 1976). Further precipitation of Fe hydroxides and of organic matter in estuarine mixing zones may serve to remove additional amounts of dissolved constituents from river water. Desorption of Ra by the increased salt content in the mixing zone drastically changes the dissolved Ra contributed to the ocean by rivers (Li et al. 1977; Hanor and Chan 1977), and ion exchange reactions between river water and river sediment take place continually during transport to the ocean.

The fractionation of ^{234}U from ^{238}U during chemical weathering, first documented by Cherdyntsev (1955), is conspicuously evident in the $^{234}U/^{238}U$ activity ratios of river water (see Tables 8.1 and 8.2). These activity ratios and those of groundwaters have been used by a number of workers to arrive at mixing ratios for water masses (Osmond and Cowart 1976*a*). Such considerations, discussed in chapter 9, are not included in this chapter. The following discussion of U and Th series nuclides in rivers is based upon the recent scientific literature and on previously published data. The data currently being accumulated by the U.S. National Uranium Resource Evaluation project is not included here.

8.2 Uranium

8.2.1 Uranium in river water

An inspection of the data in Tables 8.1 and 8.2 shows an extremely large range of variation in U content of river water, from 0.02 µg/l for the Amazon to 6.6

Table 8.1. Dissolved uranium content of selected rivers

A. Major rivers

River	^{238}U (μg/l)	Activity ratio ^{234}U/^{238}U	Reference	Location, Date
Amazon	0.04 ± 0.002	1.10 ± 0.05	(11)	
	0.02 ± 0.002	—	(1)	
Congo	0.12 ± 0.001	—	(1)	
	0.08	1.09 ± 0.07	(8)	
Mississippi	0.03	0	(14)	
	1.83	—	(13)	
	1.0 ± 0.05	1.31 ± 0.06	(11)	
	1.3 ± 0.4	—	(6)	
	$<0.4 \pm 0.4$	—	(6)	
	$<0.4 \pm 0.4$	—	(6)	
	0.31 ± 0.01	—	(1)	
	0.10 ± 0.05	—	(1)	
	0.97 ± 0.05	—	(17)	
	0.74 ± 0.02	1.27 ± 0.04	(3)	Delta
	*1.08 (0.59–1.34) mean, 11 samples	1.25 (1.19–1.30)	(3)	Vicksburg
	0.14 ± 0.02	1.37 ± 0.03	(16)	St Francisville, La., July 1979
	0.86 ± 0.03	1.21 ± 0.04	(16)	Delta, SW Pass, Sept. 1980
	0.79 ± 0.03	1.21 ± 0.05	(16)	Delta, SW Pass, Sept. 1980.
Ganges	1.9 ± 0.2	1.03 ± 0.03	(2)	Hardwar, Jan. 1968
	*6.6 ± 0.8	1.04 ± 0.02	(2)	Allahabad, Jan. 1968
	4.1 ± 0.5	1.12 ± 0.03	(2)	Patna, Jan. 1968
	1.6 ± 0.2	1.07 ± 0.03	(2)	Nabdeep, Jan. 1968 (mouth)
Columbia	0.5 ± 1.4	—	(6)	

Table 8.1 (*contd.*)

River	^{238}U (μg/l)	Activity ratio ^{234}U/^{238}U	Reference	Location, Date
MacKenzie	0.51	—	18	Fort Providence, NWT (source)
	0.83	—	18	Arctic Red River, NWT

B. Other rivers

Europe				
Rhone	0.64 ± 0.01		(1)	
Po	0.16 ± 0.01		(1)	
Neckar	1.46 ± 0.04	1.49 ± 0.04	(7)	
Rhein	2.48 ± 0.12	1.29 ± 0.07	(7)	
Lahn	0.71 ± 0.04	1.91 ± 0.1	(7)	
Mosel	1.8 ± 0.2	1.4 ± 0.1	(7)	
Maas	1.2 ± 0.2	1.6 ± 0.2	(7)	
Nahe	3.5 ± 0.2	1.6 ± 0.1	(7)	
Charente	2.0 ± 0.3	1.22 ± 0.05	(9)	
Gironde	0.7 ± 0.1	1.32 ± 0.06	(9)	
Japan	0.57 (0.34–1.23) mean, 10 rivers	—	(10)	
India				
Sabarmati	3.50 ± 0.40	1.50 ± 0.03	(2)	
Godaveri	0.68 ± 0.04	1.35 ± 0.03	(2)	
Krishna	1.08 ± 0.12	1.58 ± 0.04	(2)	
Cauveri	0.58 ± 0.04	1.28 ± 0.03	(2)	
Mutha	0.01 ± 0.001	1.44 ± 0.05	(2)	
Ulhas	0.01 ± 0.001	1.31 ± 0.12	(2)	
Godavari	0.54 ± 0.03	1.42 ± 0.03	(19)	
Narbada	0.84 ± 0.04	1.39 ± 0.04	(19)	
Tapti	0.36 ± 0.02	1.26 ± 0.04	(19)	
Russia	1.2 (1.12–1.29) average		(4)	
South America				
Rio Maipo	0.23 ± 0.01	—	(1)	Chile
Rio Preto	0.11 ± 0.01 *(0.06–0.19)	1.23 ± 0.05 (1.07–1.33)	(12)	Brazil
North America				
Yukon	0.4–1.4	—	(6)	
Klamath	1.22 ± 0.02	—	(1)	California
Mad	0.03 ± 0.002	—	(1)	California
Russian	0.03 ± 0.002	—	(1)	California
Eel	0.06 ± 0.004	—	(1)	California

Table 8.1 *(contd.)*

River	^{238}U (µg/l)	Activity ratio $^{234}U/^{238}U$	Reference	Location, Date
Rio Grande	1.15, 3.49	—	(17)	
	4.50 ± 0.07	1.74 ± 0.03	(16)	Del Rio, Texas, June 1979
Pecos	3.46 ± 0.08	1.74 ± 0.05	(16)	Pandale, Texas, June 1979
Coatzacoalcos	0.29	—	(17)	Northern Mexico
Tonalla	0.40	—	(17)	Northern Mexico
Grijalva	0.18	—	(17)	Northern Mexico
Candelaria	0.70	—	(17)	Northern Mexico
Brazos	1.6 ± 0.2	1.22 ± 0.09	(15)	
	1.11, 0.70	—	(17)	
	1.06 ± 0.01	—	(1)	Hwy 59, Texas
Suwannee	0.064 ± 0.003	2.03 ± 0.08	(16)	Oldtown, Florida, July 1979
	0.154 ± 0.007	1.90 ± 0.11	(16)	Live Oak, Florida, July 1979
Susquehannah	$*0.022 \pm 0.003$	1.79 ± 0.21	(5)	West Branch, Northumberland Pennsylvania, January 1974
	$*0.043 \pm 0.003$	1.31 ± 0.09	(5)	East of Gettysburg, Virginia, Jan. 1974

* See also Table 8.2.
Errors are 1σ from counting statistics
(1) Bertine, Chan, and Turekian (1970)
(2) Bhat and Krishnaswami (1961)
(3) Kaufman, unpublished data
(4) Kazachevskii, Cherdyntsev, Kuzmina, Sulevzhitskii, Mochalova, and Kyaregan (1964)
(5) Lewis, D. M. (1976)
(6) Mallory, E. C., Johnson, J. O., and Scott, R. C. (1969)
(7) Mangini, Sonntag, Bartsch, and Müller (1979)
(8) Martin, Meybeck, and Pusset (1978)
(9) Martin, Nijampurkar, and Salvadori (1978)
(10) Miyake, Sugimura, and Tsubota (1964)
(11) Moore (1967)
(12) Moreira-Nordeman (1980)
(13) Rona, Gilpatrick, and Jeffrey (1956)
(14) Rona and Urry (1952)
(15) Sackett and Cook (1969)
(16) Scott and Salter, unpublished data
(17) Spalding and Sackett (1972)
(18) Turekian and Chan (1971)
(19) Borole, Krishnaswami, and Somayajulu (1977)

µg/l for the Ganges, more than two orders of magnitude difference. The seasonal variations for U in individual rivers are shown in Table 8.2 for cases in which samples were taken at the same sites and analysed by the same laboratories for each data set. Seasonal variations of a factor of 2 or 3 are typical in these examples. Assembling such data into a world average value for the content of river water is obviously a problem. Moreover the average $^{234}U/^{238}U$ activity ratio is also crucial to geochemical balance calculations.

River water ratios are high, about 1.2 to 1.3, and the residence time of U in sea water is about 3×10^5 y, a significant length of time compared to the half-life of ^{234}U. These circumstances combine to make the $^{234}U/^{238}U$ activity ratio of average river water a possible means of isotopically calculating the residence time of U in the ocean. Several different approaches have been taken in the search for a 'best average' concentration and isotope ratio of U in river water.

A number of authors have estimated an average U concentration and $^{234}U/^{238}U$ activity ratio directly from analyses of river water. The analyses of Amazon and Mississippi River water U isotopes by Moore (1967) given in Table 8.1 were widely used as typical values in subsequent works by other authors. Bhat and Krishnaswami (1969) used Moore's data in combination with analyses of water from Indian rivers to evaluate the behavior of U in the ocean. They concluded that an additional source of U with a high $^{234}U/^{238}U$ activity ratio was needed in addition to river water U to achieve a geochemical balance, and suggested that diffusion from sediment pore water (Ku 1965) might provide this extra U. Mallory et al. (1969) analysed water from a large number of rivers and calculated the amount of U added to the ocean annually by rivers. Unfortunately, their analytical techniques did not allow determination of $^{234}U/^{238}U$ ratios, and the detection limit was 0.4 µg/l.

Turekian and Chan (1971), using data from Bertine et al. (1970) and from additional analyses of river water, concluded that 0.3 µg/l was an appropriate average value for U in river water. By using a diffusion rate of 0.3 dpm $^{234}U/cm^2/10^3$ y from deep sea sediments (Ku 1965), Turekian and Chan (1971) then estimated the average $^{234}U/^{238}U$ activity ratio in river water to be about 1.2. Their model resulted in a calculated U residence time in the ocean of 4×10^5 y and a calculated deposition rate of 3 µg $U/cm^2/10^3$ y over the entire sea floor. This model inspired additional efforts by other authors either to re-evaluate the U content of river water or to find the required number of sinks for U in the ocean (see chapter 15).

Sackett, Mo, Spalding, and Exner (1973) considered the marine geochemistry of U and arrived at a minimum residence time of 2×10^5 y. The river water concentration used by Sackett et al. (1973) was 0.6 µg/l, including their corrections for U added by phosphate fertilizers (Spalding and Sackett 1972; Sackett and Cook 1969). The implication of the shorter residence time for U was that a number of as yet unrecognized sinks for U must exist in the ocean if steady state prevails in U geochemistry. In addition to anoxic sediments described earlier by Veeh (1967), Sackett et al. (1973) suggested carbonates, siliceous oozes, and shallow water shelf sediments as possibly significant removal sites; they also suggested that their choice for the U content of river water might be too high.

Another approach to determining the average U content of river water has been to estimate this value by determining the relationship between U and other known geochemical parameters. The link between U in river water and

Table 8.2. Seasonal variation in uranium content and isotope ratios of river water

River	^{238}U ($\mu g/l$)	Activity ratio $^{234}U/^{238}U$	Date	Location	Reference
Mississippi	1.04 ± 0.03	1.21 ± 0.04	12-3-74		
	0.98 ± 0.03	1.26 ± 0.04	12-3-74		
	1.34 ± 0.04	1.24 ± 0.04	7-5-74		
	1.34 ± 0.04	1.25 ± 0.04	7-5-74		
	1.04 ± 0.03	1.19 ± 0.04	9-7-74		
	1.07 ± 0.03	1.21 ± 0.04	9-7-74	Vicksburg, Mississippi	M. I. Kaufman, (unpublished data furnished by J. K. Osmond)
	1.25 ± 0.04	1.27 ± 0.04	6-9-74		
	1.28 ± 0.04	1.28 ± 0.04	6-9-74		
	1.30 ± 0.04	1.30 ± 0.04	8-11-74		
	0.69 ± 0.02	1.29 ± 0.04	14-1-75		
	0.59 ± 0.02	1.29 ± 0.04	14-1-75		
Susquehannah	0.022 ± 0.003	1.79 ± 0.21	1-74		
	0.024 ± 0.001	1.53 ± 0.09	3-74	Northumberland, Pennsylvania	Lewis (1976)
	0.045 ± 0.003	1.68 ± 0.12	6-74		
	0.043 ± 0.003	1.31 ± 0.09	1-74		
	0.032 ± 0.001	1.74 ± 0.10	3-74	Near Gettysburg, Virginia	Lewis (1976)
	0.13 ± 0.01	1.32 ± 0.07	6-74		

River			Date	Location	Reference
Rio Preto	0.141 ± 0.007	1.33 ± 0.05	4-11-73	Teolandia, Brazil	Moreira–Nordeman (1980)
	0.123 ± 0.006	1.26 ± 0.1	12-8-73		
	0.075 ± 0.004	1.28 ± 0.03	17-10-73		
	0.191 ± 0.010	1.19 ± 0.03	1-12-73		
	0.092 ± 0.005	1.07 ± 0.05	7-12-73		
	0.062 ± 0.003	1.27 ± 0.05	11-12-73		
Ganges	7.0 ± 1.00	1.00 ± 0.04	5-67	Allahabad, India	Bhat and Krishnaswami (1969)
	6.6 ± 0.80	1.04 ± 0.02	1-68		
	0.9 ± 0.03	1.03 ± 0.03	7-68		
	5.7 ± 0.40	1.04 ± 0.02	10-68		
	6.1 ± 0.60	1.08 ± 0.03	11-68		
	5.1 ± 0.30	1.01 ± 0.02	1-69		
Sabarmati	2.5 ± 0.40	1.45 ± 0.07	5-67	Ahmedabad, India	Bhat and Krishnaswami (1969)
	3.5 ± 0.40	1.49 ± 0.03	2-68		
	4.2 ± 0.20	1.57 ± 0.04	4-68		
	2.4 ± 0.12	1.58 ± 0.04	9-68		
	3.0 ± 0.11	1.51 ± 0.02	10-68		
	$\simeq 3$	1.51 ± 0.02	12-68		
Cauvery	0.7 ± 0.09	1.30 ± 0.03	2-67	Trichi, India	Bhat and Krishnaswami (1969)
	0.6 ± 0.04	1.28 ± 0.03	10-67		
	0.3 ± 0.01	1.31 ± 0.04	6-68		
	0.2 ± 0.01	1.29 ± 0.09	10-68		

Errors represent 1σ from counting statistics

phosphate fertilizers, described by Spalding and Sackett (1972), has been studied in detail in the Charente River by Martin, Nijampurkar, and Salvadori (1978). The influence of these studies has been generally to cause other workers to use lower value ranges of U content in river water in order to avoid having the geochemical balance skewed by a non-steady state, human influence on rivers.

Broecker (1974) and Mangini *et al.* (1979) suggested that the U concentration in surface waters depends on the concentration of HCO_3^- in river water because of the importance of the uranyl-carbonate ion in the solution chemistry of U. Mangini *et al.* (1979) concluded that high U content in river water reflects high HCO_3^- content, rather than contamination by phosphate fertilizer. Turekian and Cochran (1978) have suggested that the correlation of U with total dissolved solids in river water represents a general control on U content by intensity of weathering, and that the correlation with HCO_3^- is minimal for samples they have examined. Bhat and Krishnaswami (1969) also report a correlation between U and total dissolved solids.

The work by Mangini *et al.* (1979) and by this author suggests that there is a positive correlation between U and both HCO_3^- and total dissolved solids in river water. However, the mutual correlation is inevitable because HCO_3^- is the major anion present in most river water and is, therefore, bound to correlate with total dissolved solids by the constraints of charge balance. It is thus pointless to argue which of these correlations accounts for the U content of river water without additional information. This relationship is illustrated in Fig. 8.1.

On the basis of the relationship between U and total dissolved solids plus an annual total dissolved solid flux to the ocean of 3.5×10^{15} g/y Borole, Krishnaswami, and Somayajulu (1981) derive a mean U content of river water of 0.3 µg/l. Mangini *et al.* (1979) similarly estimate a U content of 0.3 µg/l U in rivers on the basis of the correlation between the HCO_3^- and U content of surface waters. Both sources state a probable $^{234}U/^{238}U$ activity ratio in river water of about 1.2. These two approaches, basically the same, have yielded the same answer.

Ku, Knauss, and Mathieu (1977) have used a steady state geochemical balance model to set limits on the U content and isotope ratio of river water. Their model assumes a diffusional input of ^{234}U from marine sediments of about 0.3 dpm/cm^2/10^{3}y and is based on this input, the riverine input of both ^{234}U and ^{238}U, and the content of U in sea water (3.3 µg/l, $^{234}U/^{238}U = 1.14$). The model results in a range of 0.1 to 0.3 µg/l U content of river water, an activity ratio of 1.2 to 1.3, and a residence time in the ocean of 2×10^5 to 4×10^5 y. Calculations of the residence time of U in sea water by Osmond and Cowart (1976a), made without considering ^{234}U input from sediment pore water, resulted in a river water U content of 0.5 µg/l and an activity ratio of 1.3. Bloch (1980b) re-examined the available geochemical

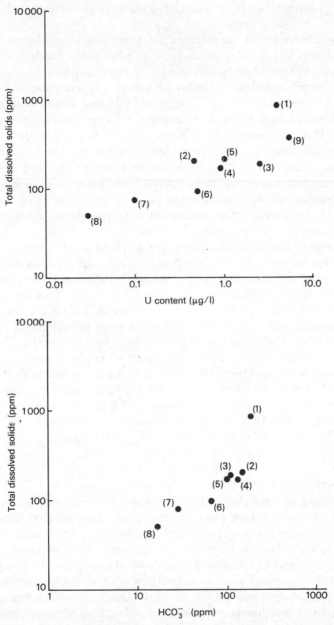

Fig. 8.1.(a) Plot of dissolved U against total dissolved solids in river water. (1) Rio Grande (2) MacKenzie River (3) Rhine River (4) Yukon River (5) Mississippi River (6) Columbia River (7) Congo River (8) Amazon River (9) Ganges River. The data are from Tables 8.1 and 8.2, Martin and Meybeck (1979), and Holland (1978). (b) Plot of total dissolved solids against HCO_3^- in river water. (1) Rio Grande (2) MacKenzie River (3) Rhine River (4) Yukon River (5) Mississippi River (6) Columbia River (7) Congo River (8) Amazon River. The data are from Holland (1978).

sinks for U in the ocean. He included the removal by metalliferous sediments, hydrothermal alteration of ocean crust at ridge crests and low temperature weathering of submarine basalts along with removal processes considered by others. According to Bloch (1980b) the sum of all these removal processes is sufficient to balance a river U input of 0.6 µg/l in a steady state model.

The above discussion shows that there is a consensus about the average $^{234}U/^{238}U$ activity ratio of river water (1.2 to 1.3) and that the most recent estimates of average U concentration in river water vary by a factor of 2 (0.3 to 0.6 µg/l). Future work will undoubtedly be aimed at resolving the disputed average concentration, but the amount of uncertainty involved in most terms of the geochemical balance equation suggests that a factor of two range is reasonably accurate. Further discussion of the geochemical balance of U is given in chapter 15.

The detailed geochemical controls of the U content of individual rivers are poorly understood and deserve much more study. One aspect of the behaviour of U in river water that has been neglected is the considerable evidence for significant seasonal variation of U content in rivers. This can be seen in analyses of Mississippi River water U listed in Tables 8.1 and 8.2. The variation represents a random sampling of what is probably a fluctuation influenced by rainfall, run-off, and soil-leaching phenomena. To date no intensive studies of these temporal variations have been made.

Bhat and Krishnaswami (1969) observed a low U content in rivers draining a basaltic terrain, and Koczy et al. (1957) have suggested that higher U contents are found in rivers crossing sedimentary rocks. However the correlations of U with HCO_3^- and total dissolved solids seem to be generally observable. Turekian and Cochran (1978) have suggested that such correlations reflect dependence on intensity of weathering.

8.2.2 Uranium in estuaries

The addition of trace elements to the ocean by rivers may be significantly affected by events taking place in the salinity gradient that forms during estuarine mixing. Studies of estuaries (for example, Sholkovitz 1976) have shown that the chemical interactions between rivers and the ocean are complex. Iron and Mn dissolved in river water precipitate as oxyhydroxides during mixing with sea water. Precipitation of dissolved organic matter also occurs, and all of these substrates may adsorb dissolved elements from solution. Moreover, the increase in dissolved salts as mixing progresses may cause marked desorption of some elements adsorbed to surfaces of river sediment particles. Thus the contribution of dissolved Ba and Ra to the ocean is greatly increased over river water concentrations by desorption processes discussed below. The situation for U, however, is not so clear.

The adsorption chemistry of U has been studied by a number of workers (Szalay and Samsoni 1969; Van der Weidjen and Langmuir 1976; Muto,

Hirono, and Kurata 1968, Doi *et al.* 1975). As pointed out by Langmuir (1978) if the adsorbed U is not reduced to its sparingly soluble 4^+ state, it can easily be desorbed by an increase in alkalinity at constant pH, or an increase in pH. The desorption of U from particles across a salinity gradient is, therefore, quite possible. Langmuir (1978) has also pointed out the strong association between Fe $(OH)_3$ and adsorbed U. The matter is complicated by a change in the surface charge of Fe $(OH)_3$, surfaces from positive at or below pH 7 to negative at higher pHs (Balustrieri and Murray 1978). If this substrate is controlling the adsorption behaviour of anionic U species, the transfer of particles from the river into the ocean may favour the desorption of U because of an increase in pH. Another process which is at present unevaluated is the possible scavenging of U from solution by the humic substances precipitating in the lower ranges of the salinity gradient (see Chapter 2).

Martin and Meybeck (1979) have described U as being an element which undergoes conservative mixing in estuaries. However, Turekian and Cochran (1978) cite evidence for adsorption of U onto particles in streams.

Borole *et al.* (1977) report conservative mixing of U in the estuarine mixing zones of three Indian rivers, and Borole *et al.* (1981) have found minor amounts of desorption of U in estuaries. Martin, Nijampurkar, and Salvadori (1978) report evidence of U removal during estuarine mixing of two rivers in France. Cochran (Chapter 15) has plotted the data of Borole *et al.* (1977) and Martin, Nijampurkar, and Salvadori (1978) on a $^{234}U/^{238}U$ against $1/U$ diagram and has concluded that the data are not sufficiently precise to determine whether adsorption or desorption of U is occurring.

Figure 8.2 is a plot of U data from the Mississippi River salinity gradient (see Table 8.3). Conservative mixing of river water with sea water (3.3µg U/l, $^{234}U/^{238}U = 1.14$) should yeild a straight line plot. There is enough scatter in these data to suggest the possibility of either U desorption or adsorption, but the amount must be negligible if it exists at all, and would not make an appreciable difference in values now being used for river additions of U to the ocean.

8.2.3 Uranium in river sediment

The U in river sediments may have three possible origins. The majority of the U is inherited from soils in the drainage basin of the river and has isotope ratios reflecting the fractionation of U during weathering. The subject of U chemistry in soils is treated in detail in chapter 7. The U in river sediment may also consist in part of U added by adsorption from river water during transport in the river. A third source of U is removal from water to river sediments during the time the sediments are part of reducing river bottom deposits (Bertine *et al.* 1970; Lewis 1976; Sackett and Cook 1969).

Martin and Meybeck (1979) include U in the list of elements for which more than 90 per cent of the riverine flux to the ocean is contained in the particulate

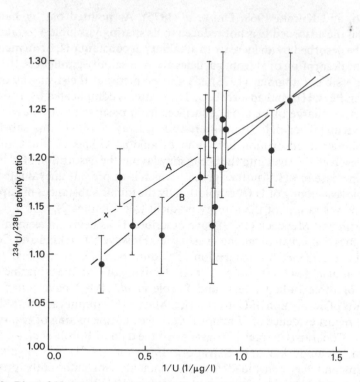

Fig. 8.2. Plot of U content variation in Mississippi River estuarine mixing zone (see Table 8.3). Conservative mixing between average sea water (X) (Ku *et al.* 1977) and river water end member is represented by line A. Line B represents mixing of river water with the apparent composition of coastal sea water at the time of sampling.

load. Their estimate of the average U content of river suspended matter is 3 ppm. Analyses of U in selected river suspended and bottom sediments have been listed in Table 8.4.

The fractionation of ^{234}U from its parent ^{238}U that occurs during chemical weathering is reflected in the $^{234}U/^{238}U$ activity ratios of river sediments. Most river sediments for which U isotopes have been analysed have $^{234}U/^{238}U$ ratio between 0.9 and 1.0, with an average of about 0.95 (Scott 1968; Martin, Nijampurkar, and Salvadori 1978, Moore 1967). Lewis (1976) has found $^{234}U/^{238}U$ activity ratios up to 1.8 in Susquehannah River sediments. As local soil profiles have much lower ratios, the conclusion is that the high U isotope ratios represent removal of U from the river water to the sediment.

An inspection of the data in Table 8.4 shows that the $^{234}U/^{238}U$ activity ratios are at, or slightly below, equilibrium except for a few cases. The Rio Grande, Pecos, and Suwannee Rivers all have U activity ratios greater than

Table 8.3. Uranium in the estuarine mixing zone of
the Mississippi River

Salinity (%)	^{238}U (µg/l)	$^{234}U/^{238}U$ activity ratio
0	0.86 ± 0.03	1.21 ± 0.04
0	0.80 ± 0.03	1.26 ± 0.05
2.34	1.09 ± 0.03	1.23 ± 0.04
2.81	1.11 ± 0.04	1.24 ± 0.07
3.07	1.15 ± 0.04	1.22 ± 0.06
3.15	1.18 ± 0.04	1.13 ± 0.06
3.20	1.11 ± 0.02	1.19 ± 0.03
3.50	1.15 ± 0.04	1.22 ± 0.05
4.17	1.23 ± 0.03	1.22 ± 0.04
4.67	1.20 ± 0.03	1.25 ± 0.05
5.10	1.26 ± 0.03	1.18 ± 0.03
10.58	1.69 ± 0.04	1.12 ± 0.04
21.31	2.34 ± 0.04	1.13 ± 0.03
26.30	2.69 ± 0.05	1.18 ± 0.03
37.20	3.69 ± 0.08	1.09 ± 0.03

Scott and Salter, unpublished data
Errors represent 1σ from counting statistics

unity in their suspended sediment. Some of the chemical properties of the water and sediment from these sites were determined for comparison with the radiometric data. The Suwannee River sample was observed to contain 4 per cent organic C in its suspended sediment. This river drains the Okefenokee Swamp and has the 'blackwater' appearance typical of such streams, although its dissolved organic carbon content was only 3.9 mg/l. Rosholt et al. (1966) have pointed out the occurrence of $^{234}U/^{238}U$ ratios greater than one in poorly drained topsoil layers which are rich in organic matter. Removal of such material to a river would result in a high $^{234}U/^{238}U$ ratio in the suspended sediments. However, it should also be noted from Table 8.1 that the U activity ratio in Suwannee River water at the same sites was very high (2.03,1.90). Adsorption of dissolved U from the water onto sediment particles or incorporation by algae in the water could lead to a high ratio in the suspended particles. Bottom sediment from the Suwannee River has a typical, slightly low, $^{234}U/^{238}U$ ratio of 0.96 (see Table 8.4).

The high $^{234}U/^{238}U$ ratios in the Rio Grande and Pecos River sediments are of less obvious origin. Neither of these rivers contains much organic matter in its suspended sediment (0.3 and 0.1 per cent, respectively). Both of them had an unexpectedly large amount of $CaCO_3$ in the suspended sediment, 24 per cent for the Rio Grande sample and 50 per cent for the Pecos sample. None of the other river samples contained measurable amounts of $CaCO_3$. This material is probably related to the formation of caliche in the soils of arid climates. Uranium is very effectively complexed in solution by carbonate ions

Table 8.4. Uranium and thorium isotopes in rivers draining into the Gulf of Mexico: suspended and bottom sediments

River	Suspended sediment				Bottom sediment				Reference
	^{238}U (ppm)	$^{234}U/^{238}U$ activity ratio	^{232}Th (ppm)	$^{230}Th/^{234}U$ activity ratio	^{238}U (ppm)	$^{234}U/^{238}U$ activity ratio	^{232}Th (ppm)	$^{230}Th/^{234}U$ activity ratio	
Rio Grande	3.39 ± 0.06	1.16 ± 0.03	8.44 ± 0.23	0.88 ± 0.03	2.90 ± 0.05	1.07 ± 0.03	21.16 ± 0.16	1.98 ± 0.04	(1)
Pecos	1.76 ± 0.06	1.23 ± 0.06	5.26 ± 0.16	1.19 ± 0.05	2.05 ± 0.03	1.30 ± 0.03	4.24 ± 0.03	0.78 ± 0.01	(1)
Guadalupe	1.63 ± 0.05	0.96 ± 0.04	8.46 ± 0.19	1.53 ± 0.05	1.25 ± 0.04	0.95 ± 0.04	4.77 ± 0.08	1.34 ± 0.05	(1)
Brazos	2.43 ± 0.05	0.94 ± 0.03	12.17 ± 0.14	1.68 ± 0.04	1.19 ± 0.03	0.98 ± 0.04	4.36 ± 0.11	1.28 ± 0.03	(1)
Mississippi	2.97 ± 0.06	0.95 ± 0.03	13.51 ± 0.26	1.72 ± 0.04	2.59 ± 0.09	0.98 ± 0.05	10.42 ± 0.07	1.38 ± 0.07	(1)
(St Francisville, La.)	2.90 ± 0.10	1.04 ± 0.05	10.86 ± 0.28	1.28 ± 0.05	2.71 ± 0.04	0.97 ± 0.02	11.36 ± 0.07	1.54 ± 0.03	(1)
	2.93 ± 0.12	1.05 ± 0.07	11.15 ± 0.29	1.35 ± 0.07	2.82 ± 0.04	0.91 ± 0.04	9.02 ± 0.52	1.35 ± 0.09	(1)
Tombigbee	4.43 ± 0.06	1.00 ± 0.02	13.98 ± 0.18	1.22 ± 0.02	1.68 ± 0.02	1.00 ± 0.02	8.16 ± 0.11	1.28 ± 0.01	(1)
Alabama	3.89 ± 0.12	1.10 ± 0.05	14.98 ± 0.42	1.21 ± 0.05	2.34 ± 0.03	1.05 ± 0.02	14.83 ± 0.10	1.29 ± 0.02	(1)
Flint	2.64 ± 0.07	1.00 ± 0.04	9.82 ± 0.06	1.03 ± 0.04	2.07 ± 0.03	1.04 ± 0.02	11.36 ± 0.08	1.04 ± 0.02	(1)
Suwannee	3.93 ± 0.32	1.39 ± 0.15	10.92 ± 0.46	1.24 ± 0.10	4.71 ± 0.06	0.96 ± 0.02	11.74 ± 0.10	1.71 ± 0.02	(1)
Mississippi (Delta)	—	—	—	—	3.31 ± 0.06	0.92 ± 0.02	—	—	(2)
Mississippi (Baton Rouge)	4.18 ± 0.08	0.98 ± 0.02	—	—	2.66 ± 0.05	0.92 ± 0.02	—	—	(2)
Mississippi (Vicksburg)									
12-3-74	2.92 ± 0.06	0.93 ± 0.02	—	—	0.74 ± 0.01	0.96 ± 0.02	—	—	(2)
7-5-74	3.22 ± 0.06	0.96 ± 0.02	—	—	0.55 ± 0.01	0.94 ± 0.02	—	—	(2)
9-7-74	3.41 ± 0.07	0.97 ± 0.02	—	—	0.75 ± 0.02	1.00 ± 0.02	—	—	(2)
6-9-74	1.87 ± 0.04	1.00 ± 0.02	—	—	0.57 ± 0.01	0.94 ± 0.02	—	—	(2)
8-11-74	3.08 ± 0.06	0.96 ± 0.02	—	—	0.78 ± 0.02	0.94 ± 0.02	—	—	(2)
14-1-75	3.00 ± 0.06	0.94 ± 0.02	—	—	0.66 ± 0.01	0.88 ± 0.02	—	—	(2)

Errors represent 1σ from counting statistics
(1) Scott and Salter, unpublished data
(2) M. I. Kaufman, unpublished data

(Langmuir 1978) and can be expected to coprecipitate with carbonate minerals forming in the soil profile. High U isotope ratios have recently been reported in a caliche deposit by Rosholt and McKinney (1980).

A comparison of suspended sediment and bottom sediment values for both U and Th (see Table 8.4) shows that the bottom sediments generally have somewhat lower concentrations of both elements. The differences are probably related to grain size of the samples and amount of dilution by quartz. Size fractionated bottom sediment samples from five rivers in the United States were analysed by Scott (1968) for U and Th isotopes. In those samples, higher concentrations of both Th and U were found in the less than 2 μm size fraction in comparison to larger sizes (2 to 20 μm). Increased amount of surface area per gram probably play a role in the concentration of trace elements in the smaller size fractions.

8.3 Thorium and Protactinium

The chemistries of Th and Pa in the environment differ markedly from that of U. Both Th and Pa are essentially insoluble in normal surface waters, and neither element undergoes an oxidation-state change under environmental Eh-pH conditions. The charge for Th is 4^+ and for Pa is 5^+, so that even if a minor amount of solution occurs, the dissolved ion is extremely reactive and tends to form hydroxide complexes and to adsorb or precipitate. Both of these elements are, therefore, almost exclusively transported on particulate matter in rivers.

The concentration of the most abundant Th isotope, ^{232}Th, is listed as 14 ppm for river sediments (Martin and Meybeck 1979), a value 3 to 4 times the concentration of U in the same material. Analyses of Th isotopes in river sediments are listed in Tables 8.4 and 8.5. The ^{232}Th values in Table 8.4 average somewhat lower than 14 ppm, but the ratios to ^{238}U are about 3. In addition to ^{232}Th, the isotopes ^{230}Th, ^{228}Th, and in some cases ^{234}Th are of interest to geochemists.

The ^{230}Th daughter of ^{234}U is present in excess of equilibrium with respect to both ^{234}U and ^{238}U in essentially all of the river suspended sediments in Table 8.4 and in those reported by Scott (1968) and Martin, Nijampurkar, and Salvadori (1978). This disequilibrium comes about by preferential solution of U from soils during the weathering process, or possibly preferential addition of ^{230}Th. The latter suggestion is far less probable, because U is much more readily dissolved than Th. The enhanced ^{234}U content of suspended sediments in the Rio Grande River results in a ^{230}Th/^{234}U value of less than one (0.88). The other samples substantiate the findings that a small amount of unsupported ^{230}Th is added to the ocean along with river suspended sediments (Scott 1968).

The average unsupported ^{230}Th (dpm/g ^{230}Th minus dpm/g ^{234}U) in the

Table 8.5. Thorium isotopes in river sediment and river water

Sample	^{232}Th (ppm)	^{230}Th/^{232}Th activity ratio	^{228}Th/^{232}Th activity ratio	Reference
Charente River suspended sediment				Martin, Nijampurkar, and Salvadori (1978)
CH. TOC. 5	20.4 ± 0.3	1.32 ± 0.03	1.13 ± 0.04	
CH. 04. 18	—	1.22 ± 0.03	0.81 ± 0.04	
CH. 04. 07	—	1.16 ± 0.04	0.90 ± 0.04	
Gironde River suspended sediment				Martin, Nijampurkar, and Salvadori (1978)
G. 76–01	13.0 ± 0.4	1.23 ± 0.05	—	
Charente estuarine sediment ($< 1\% $ Cl$^-$)				Martin, Nijampurkar, and Salvadori (1978)
CH. 1	9.81 ± 0.13	1.20 ± 0.04	0.82	
CH. 2	9.68 ± 0.40	1.06 ± 0.09	0.75	
CH. 3	6.06 ± 0.30	1.46 ± 0.15	1.82	
Amazon river water	$0.096 \pm 0.005 \times 10^{-3}$	0.74 ± 0.04	1.40 ± 0.07	Moore (1967)
Mississippi river water	$0.011 \pm 0.001 \times 10^{-3}$	1.3 ± 0.13	1.2 ± 0.12	Moore (1967)
Amazon River suspended sediment	7.8 ± 0.4	0.66 ± 0.03	—	Moore (1967)
Mississippi River suspended sediment	10 ± 0.5	1.32 ± 0.07	—	Moore (1967)

four Mississippi River suspended sediment samples (see Table 8.4) is 0.93 dpm/g. Assuming even an unreasonably high value of 1 gm/cm^2/10^3 y accumulation rate for land-derived detritus in the deep sea, the unsupported ^{230}Th added to the ocean by rivers would amount to only 0.9 dpm/cm^2/10^3 y, 10 per cent of the ^{230}Th derived from ^{234}U in sea water (3800 m average water depth). This amount of ^{230}Th is not enough to affect the balance of this isotope in the ocean, or to explain its anomalous concentrations and ratios to ^{231}Pa in some marine sediments (see Chapter 15).

Very little work has been done on the ^{231}Pa content of river sediment. This isotope, a daughter of ^{235}U, is believed to behave very much like Th in the environment, and is almost entirely associated with the solid phases. In a system at isotopic equilibrium the activity ratio ^{230}Th/^{231}Pa should be equal to the ^{238}U/^{235}U activity ratio, or 21.7. A few samples of river sediments analysed by Sackett and Cook (1969) all showed ^{230}Th/^{231}Pa activity ratios between 20 and 25. These values are close to the theoretical value, and little or no chemical fractionation is evident in river sediment being transported to the ocean.

Evidence does exist, however, for some remobilization of Th isotopes during soil formation. A correlation between Th/U weight ratios and

^{230}Th/^{234}U activity ratios in the suspended sediments in Table 8.4 is shown in Fig. 8.3, and has also been observed for river bottom sediments (Scott 1968) and soils (Rosholt *et al.* 1966). The ^{232}Th, parent of a separate decay series, must be remobilized to some extent during the weathering process in order to correlate with ^{230}Th, a daughter of ^{234}U.

In a few cases, river sediments have been analysed for ^{228}Th as well as the other Th isotopes (Moore 1967; Martin, Nijampurkar, and Salvadori 1978). The ^{228}Th is in the ^{232}Th decay series and its immediate parent is ^{228}Ra. Unlike its daughter Ra is easily able to go into solution at grain surfaces and undergo ion-exchange reactions. Therefore ^{228}Th/^{232}Th activity ratios measured in river sediments are less than one (see Table 8.5). Martin, Nijampurkar, and Salvadori (1978) report ^{228}Th/^{232}Th activity ratios of 0.8 to 0.9 in sediments of the Charente River.

Moore (1967) has reported ^{228}Th/^{232}Th activity ratios of 1.4 in Amazon River water and 1.2 in Mississippi River water. The excess ^{228}Th has grown in from its parent ^{228}Ra in solution. Any small amounts of Th present in river

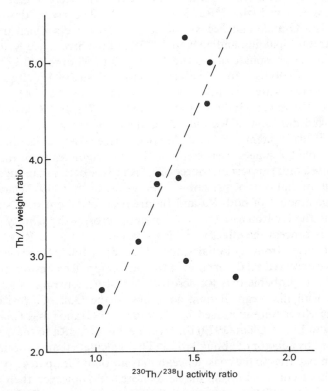

Fig. 8.3. Suspended sediment data from Table 8.4 plotted as Th/U against ^{230}Th/^{238}U ratio.

water are undoubtedly removed very quickly onto the sediment particles. The [228]Th measured in solution in coastal sea water has been generated in the water column by decay of [228]Ra, rather than by addition of Th from rivers. However, the Ra parent may come in part from river water, in addition to diffusion out of the sediments on the sea floor.

8.4 Radium

Two isotopes of Ra, [226]Ra and [228]Ra, have been abundantly used in studies of diffusion and mixing in the ocean (see Chapter 15). The major source of Ra to the ocean appears to be diffusion from sediments on the sea floors; rivers contribute 3 to 10 per cent of the total Ra found in sea water (Turekian 1971; Turekian and Cochran 1978).

Analyses of Ra in river water are given in Table 8.6. Part A of Table 8.6 gives previously published [226]Ra values used in the calculations discussed in chapter 15. Part B shows some additional results on rivers emptying into the Gulf of Mexico, including the Mississippi River. The second data set shows some unexpectedly high [226]Ra concentrations. The rivers draining arid climates (Rio Grande and Pecos), the south Texas rivers which traverse U bearing strata, (Spalding and Sackett 1972) and the Suwannee River, which crosses Florida phosphate deposits all show 20 to 30 dpm/100 l [226]Ra.

Reid *et al.* (1979) have found values of 6 to 11 dpm/100 l [226]Ra in Gulf of Mexico surface waters. Surface water in the Atlantic Ocean has 7.4 dpm/100 kg (Broecker, Goddard and Sarmiento 1976), and Reid *et al.* (1979) have invoked diffusion of Ra out of sediments as the major source of the increased [226]Ra content of Gulf of Mexico surface waters. The high values listed in Table 8.6 suggest that some of the Ra originates from the rivers.

Rivers are a more important source of [226]Ra to the ocean than suggested by the amount of that isotope present in river water. Li *et al.* (1977) have shown that a large amount of both Ra and Ba are desorbed from river suspended sediment in the Hudson estuary as the material crosses the salinity gradient. During this process the effective [226]Ra content of Hudson River water is greatly increased from its initial value of 2.3 dpm/100 l. Hanor and Chan (1977) have observed Ba desorption in the Mississippi River estuarine mixing zone, and it is probable that Ra desorption is also occurring as all of these rivers mix with the ocean. Barium additions to the Gulf of Mexico by the Mississippi River were increased by a factor of 3 (Hanor and Chan 1977). According to Li and Chan (1979) the riverine flux of [226]Ra to the ocean may be increased by a factor of about 9 to 22. They calculate that the contribution of Ra from estuaries to the coastal ocean may account for up to 43 per cent of the total flux to the ocean from the near-shore area; however, their model is based on a 2 dpm/100 l concentration of [226]Ra in river water. The river water values reported in Table 8.6 suggest that the amount of [226]Ra added to the

Table 8.6. Radium in river water

A. Previously published results

River	^{226}Ra (dpm/100 l)	^{228}Ra (dpm/100 l)	References
Amazon	2 ± 0.2	1.2 ± .1	Moore (1967)
Mississippi	7 ± 0.7	—	Moore (1967)
Hudson	7	—	Rona and Urry (1952)
Hudson	0.9 ± 2.3	—	Li and Chan (1979)
St Lawrence	0.5	—	Rona and Urry (1952)
Ganges	20 ± 3	15.6 ± 2.3,	Bhat and Krishnaswami (1969)
		9.8 ± 1.5	Bhat and Krishnaswami (1969)
Godaveri	5.0 ± 0.75	—	Bhat and Krishnaswami (1969)
Krishna	5.0 ± 0.75	—	Bhat and Krishnaswami (1969)
Sabarati	9.0 ± 1.35	30.6 ± 4.6	Bhat and Krishnaswami (1969)
			Bhat and Krishnaswami (1969)
10 Japanese rivers	8.2 ± 30.4 (mean 18.2)	—	Miyake et al. (1964)

B. Radium in rivers entering the Gulf of Mexico*

River	^{226}Ra (dpm/100l)
Rio Grande	25.6 ± .5
Pecos	25.6 ± .5
Nueces	29.6 ± .3
Guadalupe	20.8 ± .7
Colorado	20.1 ± .2
Brazos	18.7 ± .4
Mississippi (La)	7.90 ± .24
Mississippi (Ark)	11.1 + .4
Mississippi (Il)	13.3 ± 1.2
Ohio	7.27 ± .22
Miami (Oh)	2.78 ± 1.11
Tombigbee	6.15 ± .23
Alabama	5.23 ± .2
Flint	3.31 ± .12
Suwannee	28.9 ± .1

* Scott, Key, and Salter, unpublished data

ocean by rivers is larger than previously believed, at least in the Gulf of Mexico.

There are relatively few available river water analyses for Ra, and choosing an average Ra concentration is fraught with the same difficulties already discussed in section 8.2.1 for U in river water. A plot of U against ^{226}Ra content (see Fig. 8.4) shows no correlation between these two elements in river water; their concentrations clearly are controlled by different mechanisms in the environment, so that neither HCO_3^- nor total dissolved solids would be appropriate for estimating average ^{226}Ra in rivers (see section 8.2.1). Furthermore, an estimate of the amount of Ra actually contributed to the ocean must be dependent on the amount of suspended sediment undergoing

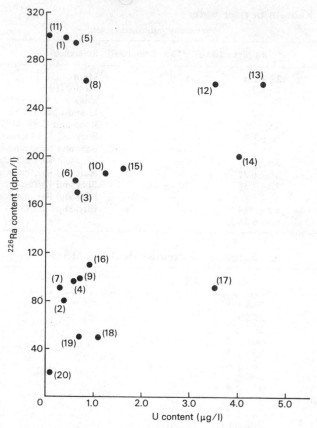

Fig. 8.4. Plot of ^{226}Ra content against U content in river water samples listed in Tables 8.1 and 8.6. (1) to (10) Japanese rivers; (11) Suwannee River; (12) Pecos River; (13) Rio Grande; (14) Guadalupe River; (15) Brazos River; (16) Mississippi River; (17) Sabarmati River; (18) Krishna River; (19) Godavari River; (20) Amazon River.

Ra desorption in the estuarine mixing zone. The ^{228}Ra, a daughter of ^{232}Th, is found in natural waters far in excess of equilibrium with its parent because of its ability to diffuse out of sediments into the overlying water (Moore 1969a). Desorption of ^{228}Ra is also to be expected in estuaries. Abundant analyses for ^{228}Ra have been made in the ocean (see chapter 15), but are scarce for river water (see Table 8.6). Michel and Moore (1980) have recently measured ^{228}Ra and ^{226}Ra in ground water in South Carolina and found ^{226}Ra concentrations as high as 6000 dpm/100 l, several orders of magnitude higher than the river water values in Table 8.6. The ^{228}Ra/^{226}Ra ratios in the ground water samples range from 0.23 to 3.0; the variability appears to depend on the variable distribution of the U parent of ^{226}Ra.

8.5 Lead-210

Lead-210 is unique among the other members of the U-and Th-decay series discussed here, because its major pathway to the ocean and to the top layers of soils is the atmosphere. The ^{222}Rn, parent of ^{210}Pb with four short-lived intervening decay products, is a noble gas and enters the atmosphere by diffusion from soils (Poet, Moore, and Martell 1972; Wilkening, Clements, and Stanley 1975). The non-reactive Rn remains in the atmosphere, but ^{210}Pb is removed to the land surface and the sea surface by both wet and dry fallout.

Because ^{210}Pb is highly reactive in the environment, it is almost entirely associated with particulate matter, and as a consequence, is essentially absent from river water. However in the case of acid mine drainage into the Colorado River (Rama et al. 1961) and into the Susquehannah River (Lewis 1976 and 1977) measurable amounts of ^{210}Pb can be found in solution as long as the pH is 4 or less. In normal river water Lewis (1977) reports dissolved ^{210}Pb to be less than 0.01 dpm/l. By contrast, in acid water dissolved ^{210}Pb was found to reach 14 dpm/l in the Colorado River system (Rama et al. 1961) and 0.17 dpm/l in the West Branch of Susquehannah River (Lewis 1976 and 1977; Benninger, Lewis, and Turekian 1975).

Lewis (1977) reports that precipitation of Fe and Mn occurs as the pH of the acid river water increases during mixing downstream; these freshly precipitated surfaces are excellent scavengers of Pb and other trace metals from solution. Moreover, the presence of soil organic matter in river suspended sediment provides an additional source of Pb scavenging surfaces. Lewis (1977) and Benninger et al. (1975) report the strong positive correlation of ^{210}Pb and organic matter in soil profiles. Organic matter in river bottom sediments may also adsorb Pb from solution, and may result in reducing conditions which provide a continual flux of dissolved Mn^{2+} and Fe^{2+} to the oxidizing river water.

The residence time of ^{210}Pb in soils has been calculated to be 2×10^3 to 5×10^3 y by Benninger et al. (1975) and Lewis (1977). These authors suggest that the behaviour of ^{210}Pb in the environment may be used to indicate the general behaviour of highly reactive trace metals, whether naturally occuring or pollutant. Scott and Salter (1978) have determined a residence time in soils of 3×10^3 y for $^{239-240}Pu$ from weapons testing fallout, a value consistent with the residence time calculated for ^{210}Pb.

Acknowledgements

This work has been supported by DOE Contract EY-76-5-05-3852. Invaluable assistance in laboratory and field work has been given by P.F. Salter and J. R. Foote.

9 GROUND WATER
J. K. Osmond and J. B. Cowart

9.1 Background and theory

9.1.1 Introduction

The most extreme examples of U-series disequilibrium are found in the hydrosphere. Not only are elemental differences in geochemical behaviour exploited by aqueous systems, but a number of radioactivity related isotopic fractionation processes also come into play. For example, the ratio of ^{235}U to ^{238}U is essentially constant in nature, both in U disseminated in minerals and dissolved in the hydrosphere. On the other hand, the ratio of ^{234}U to ^{238}U (and ^{235}U) is found frequently to depart significantly from its equilibrium value in rocks, and almost always in water (Osmond and Cowart 1976a). Fig. 9.1 illustrates the very great scatter in concentration and activity ratio values for U dissolved in ground waters.

The geochemical behaviour of U, and several of its daughters in the hydrosphere, is discussed at length in chapter 2. In this chapter the radiogenic factors causing elemental and isotopic fractionation in the hydrosphere are presented and some of the studies which illustrate these fractionation processes and their applications to hydrological and environmental problems are reviewed.

9.1.2 Fractionation in the hydrosphere

Radioactive disequilibrium results from geochemical sorting or differentiation processes, whereby one decay series daughter is more mobile than another. These differentiation processes may be physical as well as chemical, and isotopic as well as elemental. Fractionation can be observed to occur at liquid/solid, gas/solid, and solid/solid phase boundaries. However, solution and precipitation are by far the most important sorting processes and the locales of disequilibrium production of geological interest are primarily at liquid/solid phase boundaries. Thus, most of the examples of U-series disequilibrium observed in nature have their origin in the hydrosphere. Especially important are the ground water and aquifer environments where water and minerals are intimately mixed, i.e., where porosity, surface area,

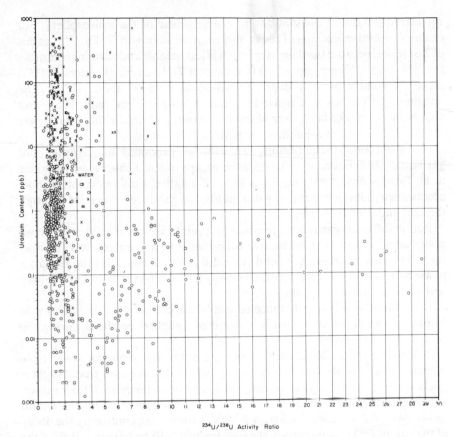

Fig. 9.1. Scattergram showing the range of U content and $^{234}U/^{238}U$ activity ratios in ground water. In mineralized areas U content may exceed one ppm, the maximum shown in this plot, by an order of magnitude. Several samples having activity ratios greater than 30 have also been analysed. (Modified from Osmond and Cowart 1976a.)

and residence times have large values, and where both chemical and physical differentiation processes have opportunity to operate.

(i) Solution and precipitation. A common differentiation process is selective leaching by ground water as it percolates past the solid mineral grains in a aquifer or soil. Some of the daughters of a series are more soluble than others, under given conditions of Eh (oxidation potential), pH, and so on (see chapter 2). The result is a liquid phase with excesses of the soluble daughters, and a solid phase with surfaces which are deficient in them. Subsequently, dissolved species produce daughters by decay, which may be less soluble than their mobile parent. Precipitation or adsorption processes then become responsible for creation of radioactive disequilibrium. Chemical precipitates often exhibit the extreme states of disequilibrium, characteristic of the hydrosphere. For example, selective leaching of rocks may result in soils with $^{230}Th/^{234}U$ ratios of 2 or more, because the daughter ^{230}Th is less soluble and remains behind as the ^{234}U and ^{238}U are carried off (see chapter 7). In oxidized waters, ^{234}U is relatively soluble, with a dissolved residence time of about 10^5 y. The daughter ^{230}Th has a dissolved residence time of less than 10^2 y and accumulates on lake beds, near-shore sediments, and on the sea floor, where $^{230}Th/^{234}U$ ratios of a hundred or more can be found.

(ii) Diffusion. Disequilibrium can also be produced at gas/solid and gas/liquid phase boundaries, by diffusive escape of Rn, a short-lived daughter in all three series. Although Rn is in itself relatively unimportant in most geochemical considerations, any disequilibrium produced is carried through the remainder of the series.

Radon diffusion can readily produce disequilibria even at solid/solid phase boundaries, for example, between small radioactive mineral crystals and host rock. To a lesser extent, the solid state diffusion of other daughters across phase boundaries may also occur, especially if the half-lives are relatively long and the concentration gradients high.

(iii) Alpha recoil. Another mechanism which can produce disequilibrium at phase boundaries of all kinds is recoil of daughter products during the process of alpha decay.

Given that the alpha particles from the naturally occurring series have energies of 4 to 9 MeV, conservation of momentum requires that daughter nuclides recoil with energies inversely proportional to the ratio of their masses to that of alpha particles, i.e. $E_d = E_\alpha(4/M_d)$. In the case of ^{234}Th, this would be $(4/234) \times 4.2 = 72$ KeV, where 4.2 MeV is the energy emitted by the decay of the parent ^{238}U. This is one of the least energetic alpha particles of all of the decay events in the three natural series, but it is energetic enough to cause recoil displacement of the daughter by 200 Å or more in crystalline materials (Kigoshi 1971).

One way to look at recoil displacement is as a special kind of diffusion, which operates in even the most impermeable media, but which occurs only during the process of decay from one nuclide to another. The rate of such 'diffusion' then is a function of (i) the number of decay events in the chain; (ii) the energy of the individual decay events; and (iii) the concentration gradient of the parent nuclides. The distance over which this kind of 'diffusion' operates is limited by the cumulative recoil ranges of all of the daughters of a series (essentially a random walk).

In geochemical systems, recoil displacement may be a significant process in causing measurable disequilibria where phase domains are small in size, as for example in aphanitic (microcrystalline) igneous rocks, in clay soils, in disseminated precipitates in aquifers, or where concentration gradients are large between adjacent phases, as in the case of radioactive mineral grains in contact with ground water.

In Fig. 9.2 the normal effects of recoil on the relative abundance of dissolved radionuclides are shown schematically. Minerals usually have a higher U concentration than do pore waters, so that the transfer of daughter nuclides from solid to liquid phases will exceed the reverse transfer. If only recoil displacement probabilities are considered, the relative activities of successive daughters in the solid (near the interface) are expected to decrease, whereas their relative activities in the liquid should increase. This explains the commonly observed low activity ratios of ^{234}U to ^{238}U in fine-grained rocks and sediments, and the correspondingly high ratios in most ground waters.

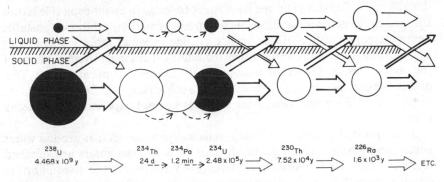

Fig. 9.2. The recoil model of U-series isotopic disequilibrium. Decay chains in the liquid and solid phases from ^{238}U through the first several daughters are shown diagramatically, and involve a series of alpha (solid arrows) and beta (dotted) events. Alpha decay involves recoil displacement and may result in transfer of the atoms from the solid phase to the liquid, or vice versa. Sizes of atoms and arrows represent relative activity. The commonly measured isotope pair, ^{234}U and ^{238}U, are shaded.

The Rn escape process in soils is not just a matter of true diffusion of one short-lived nuclide; rather it is only the last step in a process that may involve recoil 'diffusion' of three or four preceding daughters.

The best documentation of recoil-displacement as a mechanism for causing disequilibrium involves the study of the escape of ^{234}Th from zircon. Kigoshi (1971) noted that water containing zircon grains became enriched in ^{234}Th as a result of decay of ^{234}U in the crystals. No dissolution of zircon was involved.

It is difficult to isolate recoil displacement as a mechanism causing disequilibrium in natural processes because of the role of competing solution mechanisms. However, in one instance, the role of recoil in producing disequilibrium must be important. The disequilibrium of ^{234}U relative to ^{238}U, so universal in the hydrosphere and weathered rocks, is difficult to explain in other terms. The parent and daughter have the same chemical characteristics, and the intervening daughters, ^{234}Th and ^{234}Pa are relatively insoluble and have short half-lives, so that leaching is an unlikely explanation. Furthermore, the disequilibrium in some cases involves solid phases only, or solid phases as the adsorbing or precipitating phase, so that leaching is not involved at all (Osmond and Cowart 1976a).

(iv) *Recoil and vulnerability.* One other process has been suggested as the cause of natural disequilibrium, and it also is especially difficult to isolate and identify: the Szilard–Chalmers effect. Applied to the natural series, this translates to a 'vulnerability to leaching' process. Radioactive decay, especially an alpha event (sometimes a beta event), greatly disrupts the crystalline lattice along the path of recoil and in the neighbourhood of the displaced daughter. The daughter isotope itself is apt to lodge in an inhospitable lattice site, and may, as the result of its nuclear transformation, exhibit an unstable electronic configuration. Thus, such a nuclide becomes more vulnerable to leaching than its neighbouring atoms, including other longer lived members of the same series, and even other isotopes of the same chemical species. This process would be especially significant in such cases, as ^{234}U and ^{238}U or ^{228}Th and ^{232}Th where one of the isotopes has not experienced any decay event.

It is important to note that in any 'milking' process, such as ground water percolating past a mineral grain, the rate at which daughters are supplied relative to the rate at which they are removed determines the disequilibrium effected. Both the recoil displacement and the recoil-vulnerability models depend on alpha decay rates to supply the available daughters.

9.1.3. Mobility factors

To define the pH/Eh fields of stability of various oxidation states and complexes of a given element, and then to determine the corresponding solubility products in laboratory solutions is a challenging endeavour. It is an even greater challenge to extrapolate successfully such data to real ground water conditions, where a myriad of complicating dissolved, colloidal, and solid components are present (see chapter 2). But even if these difficulties can be overcome, the definition of parameters controlling the radioactivity in natural water due to a given U-series nuclide has just begun. This point is illustrated by the many examples of extreme *isotopic* variation exhibited by U-series elements in natural waters.

The fact that a nuclide occurs as part of a series results in an amplification of its potential for dissolution. Even if a nuclide itself is not alpha radiogenic, to have had an ancestor decay in this mode is to increase its chances of residing in an unstable lattice site. Furthermore, if any of its ancestors has been in solution, a daughter will either be formed in the liquid phase, or on a precipitation or adsorption surface, where re-solution is facilitated. Thus, the paucity of observable ^{234}Th in natural water (Hussain and Krishnaswami 1980) does not contradict the recoil model of ^{234}U mobilization.

An important determinant of the level of abundance of a given radionuclide in solution is its half-life. This comes into play in several ways. In the case of unsupported dissolved species, shorter-lived nuclides will decay away while longer-lived sisters will persist. Even though two nuclides have similar chemical properties, the shorter-lived species, or isotope, will be scarce unless there is a continuing supply contributed to the water by the aquifer rock. This aspect of half-life as a control, on mobility of radionuclides has been universally recognized. Figure 9.3 shows qualitatively how the various members of the natural decay series can be ranked according to the concept of 'range'. Range is defined as a log-log product of both half-life and solubility. The U, Ra, and Rn frequently occur in natural waters at the pCi/l level, and are thought of as geochemically soluble. The Th, Pa, and Po are seldom detected in solution, and are essentially insoluble in natural waters. The Pb, Bi, and Ac may be found under certain conditions, in solution. For those nuclides which are readily soluble, the factor which limits their occurrence in natural waters may be their average life times, equal to their half-life divided by the natural log of 2 (0.692). Unless supplied by decay from parents in solution, or in the immediate vicinity in the aquifer rock, short-lived nuclides will not be widely dispersed. Each nuclide, therefore, has its own tendency toward mobilization which depends on its elemental solubility and its mode of

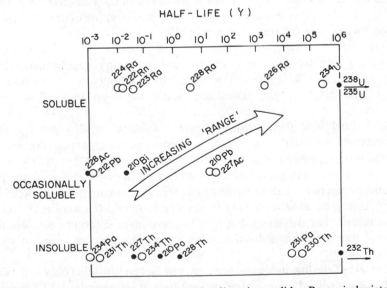

Fig. 9.3. Geochemical mobility and range of radioactive nuclides. Range is depicted as the product of both half-life and solubility of each nuclide. Circled nuclides are products of alpha decay, and are susceptible to mobilization by alpha recoil.

radiogenesis; and its own geochemical range which depends on its elemental solubility and half-life.

Range, as a geochemical property of nuclides, must be applied with caution, inasmuch as the solubility of elemental species varies greatly depending on ground water composition. Nevertheless, even for very soluble nuclides, a short half-life is tantamount to a limited range; conversely, even very insoluble elements may be mobilized by recoil processes.

The second way that half-life exerts a control on the amount of radionuclide in solution is the opposite to the first. Where a dissolved radionuclide is supplied by decay of a parent nuclide in the aquifer rocks, the rate of supply is limited by the growth curve of the daughter. This is especially true if the mobilization process is brought about by the decay event itself, i.e. alpha recoil. Either of two aquifer circumstances will cause this situation: (i) water flow rate through the aquifer which is fast relative to the half-life of the mobile daughter; or (ii) a localized concentration of the parent in the aquifer such that the exposure of water to the zone of concentration is short relative to the half-life of the daughter. This process is described quantitatively below, but the effect is readily visualized with the help of Fig. 9.4, which shows how the relative radioactivity of dissolved isotopes of a given element, in this case Ra, can change dramatically as a result of U and Th heterogeneities in an aquifer. In this illustration (Fig. 9.4) the rate of flow of water is such that the exposure time in the mineralized zone is about equal to the half-life of ^{228}Ra, but much less than the half-life of ^{226}Ra, and much greater than the half-lives of ^{223}Ra and ^{224}Ra. Under these conditions, the dissolved activities of the shorter-lived isotopes rise and fall much more quickly.

This aspect of the relative abundances of short and long-lived isotopes in ground water has not been appreciated by many geochemists, despite the fact that Cherdyntsev (1971) pointed it out in connection with very high ^{224}Ra/^{226}Ra values commonly observed in shallow ground waters.

The third way that half-life affects 'solubility' is also not commonly appreciated, primarily because of the convention of expressing levels of concentration in water in activity units (for example, dpm/l or pCi/l). Because the mass concentrations of two isotopes having the same radioactivity differs in proportion to the half-lives, the relative concentration, for example, of ^{238}U to ^{234}U in ground water having an activity ratio of 1.0 is almost 20 000 to 1. This means that the longer-lived ^{238}U encounters solubility product limits sooner than the shorter-lived isotope when measured in terms of specific radioactivity.

Certain of the 'insoluble' elements may in fact be insoluble only with respect to the longer-lived, more abundant isotopes. For example, ^{232}Th is seldom found in solution at a measurable level, but ^{228}Th has a level of detectability by radioactivity that is about 10^{-10} of ^{232}Th. It seems likely that there are natural

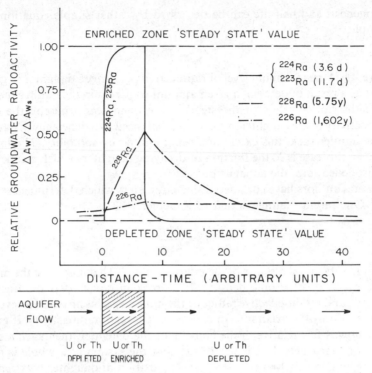

Fig. 9.4. Theoretical variations in Ra isotopes in an aquifer with a localized U and Th source. The changes depicted are relative to the theoretical maximum effects of alpha recoil mobilization for each isotope. The actual maximum values are determined in part by the relative abundance of the parent nuclides, so that the changes in ^{228}Ra and ^{226}Ra will depend on the Th to U ratio in the rock, and the ^{223}Ra maximum will be about 1/20 that of ^{226}Ra.

waters where ^{228}Th is detectably present. An example that comes to mind is that of sediment pore waters.

The common pattern of activity of Th isotopes in both sea water and terrestrial waters is ^{234}Th $\gg$ ^{228}Th $>$ ^{230}Th $>$ ^{232}Th (see chapter 2). Such an inverse relationship with half-lives is unlikely to be coincidental. Furthermore, the interfering and complicating effects of complex formation, coprecipitation, colloid formation, and adsorption on particulates, all influence solutes quite differently at high and low concentrations. Under natural conditions, it may be very difficult to decide whether a very radioactive nuclide is actually in solution, or whether its extreme dilution has delayed precipitation.

Of the many radioactivity-related factors which govern the level of abundance of U-series nuclides in natural waters, some, such as the effects of extreme dilution, are difficult to evaluate. On the other hand, the role of radiogenesis and half-life can be described by rather simple equations. For example,

$$A_{w_s}^d = F A_r^p \qquad (9.1)$$

where $A_{w_s}^d$ is the maximum level of radioactivity of a given daughter in ground water as a result of decay of a given amount of parent in the aquifer rock, A_r^p. The subscript 's' refers to the steady state condition, whereby the excess radioactivity of the daughter in solution is balanced by a deficiency of daughter in the aquifer rock, this excess not being limited by solubility. The limiting factor in this case is F, the fraction of decaying daughters which escape to, or are propelled into, the aqueous phase.

Several authors have considered the parameters which determine the value of F. One recent example is that of Kraemer (1981):

$$F = \frac{3L\rho}{4r\phi} \qquad (9.2)$$

where L is the recoil distance of the intermediate ^{234}Th daughter in the mineral grain in cm, ρ is the bulk density of the rock in g/cm^3, ϕ is the fractional porosity, and r is the effective radius of the mineral grains, in cm. However, this represents only the maximum, or steady state level of radioactivity. If ground water flow is fast relative to the half-life of the daughter, then such a steady state is not reached. This is not to say that the system as a whole is out of equilibrium, merely that the steady state partition of daughter between rock and water has not been achieved.

For a system in which the radioactivity of the daughter in the water is less than this maximum because of insufficient duration of rock/water interaction:

$$A_w^d = A_{w_0}^d + (A_{w_s}^d - A_{w_0}^d)(1 - e^{-\lambda_d t}) \qquad (9.3)$$

where A_w^d is the daughter radioactivity in the water at any given moment this value being intermediate between its initial state $A_{w_0}^d$ and the steady state value $A_{w_s}^d$; λ_d is the decay constant of the daughter nuclide, and t is the duration of water/rock interaction. Alternatively, eqn (9.3) can be rewritten as:

$$\Delta A_w^d = \Delta A_{w_s}^d (1 - e^{-\lambda_d t}) \qquad (9.4)$$

where ΔA_w^d represents the ultimate change in steady state radioactivity of the dissolved daughter as a result of movement of the water to a part of the aquifer having a different parent radioactivity. Equation (9.4) describes a system in which the water flows from a lower to a higher zone of radioactivity of the parent in the host rock. The opposite situation can also pertain, that in which the new rock radioactivity and the new steady state water radioactivity are less than before. Equation (9.4) is, however, quite general, and applies to both situations. In the case of a decreasing radioactive environment eqn (9.4) can be written in a slightly simpler form:

$$\Delta A_w^d = \Delta A_{w_s}^d e^{-\lambda_d t}. \qquad (9.5)$$

Considering the importance of A_i^p, λ_d, and t in eqns (9.1) through (9.5) it may be seen that in many cases the relative amounts of the various decay series daughters in natural waters do not seem to be controlled by solubility limits (see Fig. 9.5).

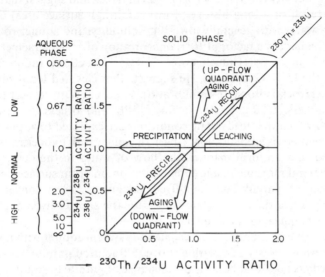

Fig. 9.5. Aquifer rock and ground water disequilibrium interactions. The right hand part of the diagram shows the variations of $^{230}Th/^{234}U$ and $^{238}U/^{234}U$ activity ratios in the solid phase (minerals and secondary accumulations). The far left-hand scale shows the corresponding variations of $^{234}U/^{238}U$ activity ratios in the ground water.

9.1.4 Radioisotope evolution and aquifer classification

The concept of changing levels of water radioactivity in response to changes in rock radioactivity as the water flows from one part of the aquifer to another is an important one when considering the geochemistry of any given radionuclide. In fact, it is a useful exercise to classify parts of an aquifer as to whether the nuclide being studied is (i) at a steady-state of radioactivity in the water, (ii) is being augmented toward a higher steady-state level with time and flow, or (iii) is diminishing by decay towards a lower steady-state level with time and flow.

This system of classification is applied most usefully to the case of ^{234}U because the level of concentration of that nuclide, along with its parent, ^{238}U, has been widely studied; also, because of its long half-life ($t_{\frac{1}{2}} = 2.48 \times 10^5$ y), the classification terms *steady-state*, *augmenting*, and *decaying* can be applied to entire aquifers.

In *steady-state* aquifers, the host rock has a uniform and low distribution of parent U, and the water moves very slowly through the system. Considering the time required to achieve the steady-state condition, there are very few studied examples. Connate waters might fall in this category; however, the U isotopic data reported for ancient water systems have yielded disparate results. Kronfeld, Gradsztajn, Muller, Radin, Yaniv, and Zach (1975) report very high $^{234}U/^{238}U$ activity ratios from a deep well in Israel, and suggest that this is the expected result of a long history of intimate (high surface area) interaction between a water with generally low ^{238}U solubility, and aquifer rock with its relatively higher (by a factor of 10^3) concentration of ^{238}U. Kraemer (1981) on the other hand, reports much lower $^{234}U/^{238}U$ activity ratios from deep waters in the U. S. Gulf Coast. He suggests that elevated temperatures may cause water/rock isotopic re-equilibration to occur. With respect to shorter-lived daughters, ^{226}Ra, ^{228}Ra, ^{224}Ra, ^{210}Pb, and others, there are many aquifers to which the steady-state conditions are expected to apply. However, this statement needs amplification. Given an aquifer with radioelements initially distributed in a uniform manner, the flow of water in time tends to cause solution, lateral movement, and reprecipitation on grain surfaces, so that the distribution of parent and long-lived daughters is no longer uniform. Necessarily, then, the steady-state values for the shorter-lived daghters vary from place to place in the aquifer.

Perhaps the best example of an *augmenting* system aquifer with respect to U is ground water slowly percolating through fractured granite. Here the rock radioactivity is high, and because of resistance of granite to solution, the ^{238}U concentration in the water is still moderate. Under these circumstances the concentration of ^{234}U and the ratio of ^{234}U to ^{238}U increases steadily with time and flow. Actually, the concentration of all of the U-series daughters is observed to increase in these waters, but the shorter-lived daughters increase the fastest. Some large isotopic ratios have been observed: $^{234}U/^{238}U$ of 10 or more, and $^{224}Ra/^{226}Ra$ of several hundred (Cherdyntsev 1971).

Decaying system aquifers are the rule wherever ground waters percolate from high to low background rocks. This in fact describes many deep aquifers, inasmuch as ground waters usually flow through enriched weathered zones, or through an enriched 'front' during recharge.

Most surface waters are decaying systems. To the extent that rivers, lakes, and the ocean are fed by ground waters, a lowered steady-state level applies where the rock/water interaction has been removed. It follows that surface waters, except near some springs, are usually low in the shorter-lived members of the U- and Th-series.

In subsequent sections the usual case of aquifers in which the rock radioactivity has been extensively redistributed as moving 'fronts' will be discussed as will the possible application of the augmenting and decaying equations to the dating of ground water.

9.1.5 Aquifer disequilibria and geochemical fronts

The study of U nuclides in ground water would be much simpler, yet still interesting, if aquifer rocks could be assumed to be homogeneous with respect to their U content and its long-lived daughters. The flow of water, however, causes a redistribution of U, Th, Pa, and Ra, from place to place in the aquifer, and also from disseminated sites in aquifer minerals to secondary coatings on solid surfaces. In order to understand ground water disequilibria, therefore, it is necessary to consider models of aquifer rock disequilibria. Simple models for confined flowing aquifer systems will be developed but it must be emphasized that such models in modified form also apply to such disparate systems as vadose (unsaturated) zones, and U ore deposits. The separation of the elements and nuclides of decay chains is primarily the result of their differing chemical character and changing environmental conditions. Of paramount importance is the property of U which allows it to exhibit two major valence states, $4+$ and $6+$. In the $4+$ state U behaves much like Th and near the earth's surface is relatively insoluble under normal conditions of Eh and pH; but in the $6+$ state U tends to be more mobile, especially because it forms soluble complex ions, such as $(UO_2)(CO_3)_2^{2-}$ (see chapter 2). As a result U is a ubiquitous constituent of the hydrosphere, especially in surface waters and the shallower more oxidized portions of ground water systems. As it moves through the hydrological cycle, U continuously generates shorter-lived daughter nuclides, some of which are less mobile and tend to be left behind to indicate its passage; and others, more mobile, which tend to move ahead and/or upward, to reveal the presence of an otherwise unsuspected accumulation (see chapter 14).

Secondary U accumulations, whether very concentrated as in sandstone type ores, or much dispersed as in normal aquifers, constitute a flow-through system which may be likened to an ion exchange column or a chromatographic analyser. The U accumulation zone itself marks a geochemical barrier, or 'front' where the character of the water is altered, such that in the up-flow direction U is mobile, but down-flow it is not. But the daughter elements are governed by different geochemical properties, and some are mobile under conditions of the down-flow water. How far down-flow such elements as Ra and Rn are found depends on the geochemical properties of the water which may vary systematically in the down gradient direction, and also the rate of flow of water relative to the half-lives of the various daughters.

At the same time, the front itself may also be moving slowly in the down-flow direction, due to the interaction of water and aquifer rocks. Such fronts often mark the point where oxidized surface water is reduced by constituents of the rock, for example, sulphides and organic debris; but the water is also oxidizing the aquifer rock, and as this happens the front moves down-flow.

The U accumulation advances with the front. Individual U atoms, precipitated on grain surfaces and in pore spaces as reducing conditions are encountered, are readily redissolved and transported forward as the oxidizing front advances. This process, in the case of long-lived ^{238}U, happens over and over again (see chapter 14).

The less mobile daughters, Th and Pa, are left behind to mark the previous position of the barrier. How far up-flow such 'ghost' accumulations are found depends upon the rate of motion of the front relative to the half-lives of the immobile constituents.

Such a flow-through system, like a chromatographic column is governed by equilibrium considerations, such that subtle differences in chemical properties can cause major disequilibria in the decay chains. In fact, even isotopic fractionation is readily effected.

This is not to say that mass fractionation of isotopes occurs. Rather, differences in the type and rate of radiogenesis can become significant (as discussed previously). Importantly, alpha decay results in nuclide recoil. In the case of ^{238}U, secondarily adsorbed on aquifer grains, there is almost a 50 per cent chance that the resulting daughter, ^{234}Th, will be expelled into the water, thus becoming mobilized despite its otherwise inert chemical behaviour. This short-lived nuclide in turn decays to ^{234}U, a long-lived, more mobile nuclide. Even if the alpha recoil event does not result in immediate expulsion into the aqueous phase, lattice disruption and ionic changes resulting from decay produces instability whereby subsequent oxidation and/or leaching of a slightly soluble ion ($^{234}U^{6+}$) is more likely to occur than that of a nearby isotope of the same element ($^{238}U^{4+}$)without its disruptive history.

The partial decoupling of ^{234}U from ^{238}U in aquifers is a commonly observed phenomenon. The same process of recoil-related mobilization must be operating for all alphagenic daughters, but the results are not so apparent because of the competing chemical fractionation effects.

Figure 9.5 shows how the various processes of leaching, precipitation, recoil mobilization, and decay produce disequilibrium of ^{230}Th and ^{234}U relative to ^{238}U at the rock–water interface.

With respect to the early longer-lived members of the ^{238}U decay series ($^{238}U \rightarrow {}^{234}U \rightarrow {}^{230}Th$), three processes control the degree of equilibrium observed in the host rock and flow-through water, especially in the neighbourhood of a redox boundary, or 'front': (i) the intermittent mobility of U and the consistent immobility of Th; (ii) mobilization of ^{234}U by recoil related processes, against the solution-only mobilization of ^{238}U; and (iii) the faster growth/decay toward equilibrium of ^{230}Th ($t_{\frac{1}{2}} = 7.52 \times 10^4$ y) against that of ^{234}U ($t_{\frac{1}{2}} = 2.48 \times 10^5$ y). The ^{238}U is essentially stable.

Simple leaching of U by oxidized and CO_2 bearing ground water causes enrichment of dissolved U at a ratio of $^{238}U/^{234}U$ near unity (left hand scale in

Fig. 9.5). At the same time it drives the ^{230}Th/^{234}U activity ratio of the solid phase to higher values (right hand arrow). Reprecipitation of the dissolved U drives the rock isotopic plot in the opposite direction, that is, leftward.

Recoil mobilization of ^{234}U, unaccompanied by leaching, causes both the ^{230}Th/^{234}U and the ^{238}U/^{234}U, values in the rock to increase (toward the upper right hand) at the same time producing an increment of ^{234}U in the water; if the dissolved ^{238}U concentration is low, a very low ^{238}U/^{234}U ratio is produced in the ground water, or in the more commonly expressed terms, a very high ^{234}U/^{238}U activity ratio (left hand scale). There is seldom much ^{230}Th dissolved in the ground water, thus the unidimensional aqueous scale.

Unequivocal examples of ^{234}U decay/growth to re-establish equilibrium are seldom seen in aquifer rocks, and probably never in water, but ^{230}Th decay/growth in rocks is a commonly observed phenomenon. The result is counter to the leaching/precipitation arrows of the diagram, and rotates the ^{234}U recoil and precipitation arrows toward the vertical.

Figures 9.6 and 9.7 illustrate the isotopic patterns of disequilibrium of the aquifer rocks as ground water sweeps past a front. In an aquifer in which the U is repeatedly mobilized, as in a 'fast moving' frontal system, considerable ^{230}Th/^{234}U disequilibrium is produced with high ratios up-flow, and low ratios down-flow. Near-equilibrium values of ^{238}U/^{234}U are the general rule, both in the solid and the aqueous phases, although high values (low ^{234}U/^{238}U activity ratio) may be found.

The arrows of Fig. 9.6a show the sequence of processes from up-flow to down-flow across a frontal system. Consideration of the sequence of processes at a given point in the aquifer rock as the frontal system passes would produce a set of arrows in the opposite direction. The slight incline of this evolution pattern is to suggest the influence of recoil related mobilization of ^{234}U.

Figure 9.6b shows the sequence of dominant nuclides (with respect to radioactive equilibrium) on a transect across such a front. The ^{230}Th is left behind, to decay away, as the two U isotopes are dissolved and move ahead. The ^{234}U is ahead of the ^{238}U because of the added mobilizing effect of recoil processes.

Figure 9.6c shows the resulting 'pulses' of these nuclides in the ground water. The ^{238}U is often more abundant up-flow from the front, with ^{234}U enriched in the down-flow direction, to a modest degree in this case. The ^{226}Ra, produced by decay of ^{230}Th, is more abundant up-flow from ^{234}U as well.

Figure 9.7 shows the rock and water disequilibrium effects in 'slow-moving' frontal systems. If U, as well as Th, is relatively immobile in a ground water system, the dominant fractionation process is likely to be the ^{234}U recoil process. This shown by an inclined evolution pattern in this diagram, with both high ^{230}Th/^{234}U and high ^{238}U/^{234}U ratios on the up-flow side of the front, and low ratios on the down-flow side. That this pattern occurs in real aquifers is demonstrated by the extremely high ^{234}U/^{238}U observed in many

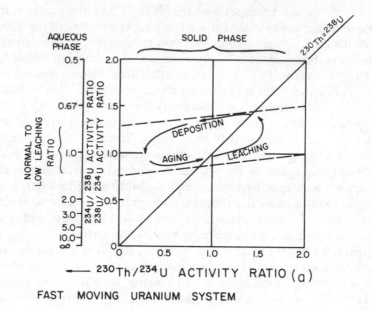

FAST MOVING URANIUM SYSTEM

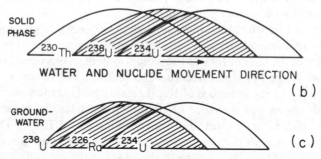

Fig. 9.6. Rock and water disequilibrium in fast moving systems. (a) Activity ratios in ground water and aquifer rock. (b) Sequence of abundances of nuclides in the solid phase on a transect across a secondary accumulation, or front. (c) Sequence of abundances of nuclides in the ground water on a transect across such a front.

deep aquifer waters (low $^{238}U/^{234}U$), and also by the commonly observed $^{234}U/^{238}U$ disequilibria in soils, weathered rocks, and ore bodies.

Because the U in the solid phase is relatively immobile, there is time for the re-equilibration of ^{230}Th with ^{234}U to occur, and thus, the apparent rotation of the evolution zone (dotted lines of Fig. 9.7a).

Figure 9.7b shows the sequence of dominant nuclides in the solid phases of such an aquifer system, and Fig. 9.7c the resulting occurrences of ^{238}U, ^{234}U, and ^{226}Ra (daughter of ^{230}Th) in the ground water. In this case, the tendency

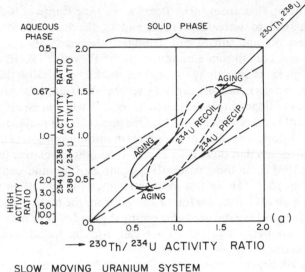

SLOW MOVING URANIUM SYSTEM

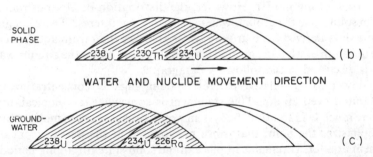

Fig. 9.7. Rock and water disequilibrium in slow moving systems. (a) Activity ratios in ground water and aquifer rock. Aging effects cause rotation of pattern relative to fast moving systems. (b) Sequence of abundances of nuclides in the solid phase on a transect across a front. (c) Sequence of abundances of nuclides in the ground water on a transect across a front.

for the ^{230}Th to be located with the ^{234}U in the rock results in a pulse of ^{226}Ra down-flow from that of ^{234}U.

Given the logic of this model, up-flow and down-flow regimes in an aquifer could be determined from the isotopic analyses of a few rock samples. Note that, as usual, these relative isotopic ratios are given in activity units. Chemically, the only easily measured distribution would be that of ^{238}U. In the ^{235}U series, ^{231}Pa ($t_{\frac{1}{2}} = 3.43 \times 10^4$ y) behaves geochemically like ^{230}Th in the ^{238}U series. However, the ^{231}Pa/^{235}U system is simpler in that there is no intervening mobile daughter. This means that immobile ^{231}Pa should always be up-flow from ^{235}U. Because the time frame of these migrations is slow relative to water flow rates and to the half-lives of other U-series daughters, Ra and subsequent daughters are nearly always associated with, and in near equilibrium amounts with, ^{230}Th and ^{231}Pa in aquifers.

The gross radioactivity of aquifer rocks, especially their gamma ray activity, is determined primarily by several short-lived daughters. Very often this gross radioactivity is measured, especially when prospecting for U, as eU, equivalent U units (see chapter 14).One ppb eU is the radioactivity of one part per 10^9 U along with all of its daughters in equilibrium. An eU/U value greater than 1.0 means that there is an excess of daughters relative to the chemically analysed ^{238}U. In most cases eU is controlled by, and can be taken to be equivalent to, ^{230}Th. In fast moving fronts, as is very often the case for ore bodies, high eU values are found *up-flow* from the main U accumulation, but from what has been deduced concerning the kind of slow moving fronts found in deep aquifers, high eU values can also be found *down-flow* from U accumulations.

The mobility models discussed above, and schematically presented in Figs. 9.6 and 9.7, are concerned primarily with the locations of the radionuclides in the host rock of an aquifer. However, the distribution of U-series radioelements in solution in the ground water is also of great interest. There is a great deal more data derived from ground water analyses than from aquifer rocks and the question of what the ground water data reveal about the aquifer/water history is discussed in the following paragraphs.

In a slow moving, relatively stable system, high U concentrations are seldom found even up-dip. This statement is somewhat tautological, for if there were much U in solution the system would obviously be in motion. But it also emphasizes the point that, once precipitated at the barrier front of a permanent aquifer, U remains in place in the solid state for a long period of time. The same is not true for ^{234}U. If conditions at the front are near chemical equilibrium, the effects of decay and recoil may be just enough to cause mobilization, and its much lesser mass/radioactivity ratio may result in ^{234}U being soluble at the same time as ^{238}U is insoluble. It may be expected then, that in the neighborhood of such stable fronts, high ^{234}U/^{238}U ratios may be found in the ground water, and possibly also for some distance downflow. Activity ratios of 5, 10, and 20 or higher are sometimes observed in large confined aquifers with steady long-term flow systems such as the Carrizo (Cowart and Osmond 1974) and Hosston Sandstones of Texas (Kronfeld and Adams 1974), and the Cambro—Ordovician aquifers in sandstone of Illinois (Gilkeson and Cowart, work in progress).

With such efficient separation of ^{234}U from ^{238}U, ^{230}Th is generated far down-flow from the primary front, and is effectively precipitated at a secondary zone. This secondary zone of ^{230}Th generates ^{226}Ra, again by alpha decay with its recoil effects. A ground water sample from this part of the aquifer may exhibit a high Ra concentration relative to U (unless there is appreciable sulphate in the water). High ^{222}Rn concentrations may also be associated with the ^{226}Ra. In accordance with this model of U series chromotography in aquifers, stable and long-standing systems are associated with high ^{234}U/^{238}U activity ratios down-flow from the primary front. The ratio of ^{226}Ra to ^{234}U near the secondary front, however, may be more strongly influenced by the water flow rate, in accordance with the augmenting system effects discussed in section 9.1.4.

Given the various combinations of aquifer flow rates, solubility and ranges of the nuclides, and mobility of the fronts, a large variety of elemental and isotopic ratios can be found in various ground waters. However, a few ratios and patterns are considered distinctly anomalous. An example of such an isotopic ratio is that of a ^{234}U/^{238}U in ground water that is less than 1.0. Under all conditions, it is believed that ^{234}U is mobilized preferentially and therefore, always in excess or equal to ^{238}U in the aqueous phases. That such low activity ratios are sometimes found, especially where high U concentrations are the rule, indicates that normal conditions do not pertain. The simplest explanation is that sudden, rather than gradual shifts in conditions have occurred, and that the normal low activity ratio of U at or near the boundary of the host rock is temporarily put in solution. This causes a simultaneous increase in concentration and a decrease in activity ratio. Such circumstances must be, in terms of a geological time frame, quite transitory. The commonest examples are waters which circulate through shallow U ore deposits and phosphorite beds.

An example of a pattern of disequilibrium in an aquifer indicative of anomalous history is the occurrence of an excess of ^{234}U in the host rock up-flow from an accumulation of ^{230}Th, or ^{235}U up-flow from ^{231}Pa. This would seem to indicate that the Th and Pa daughters were more mobile, which they are not. One interpretation of such a situation would be that for some reason the reducing barrier has moved in an up-flow direction. Such an event would have to have happened rather quickly on the ^{230}Th/^{231}Pa time scale. The evidence in the water itself would be hard to ascertain, inasmuch as at the ^{230}Th/^{231}Pa front, all of the pertinent nuclides would be relatively insoluble. (Given a low ^{238}U concentration in the water which is flowing past a previously formed accumulation, the recoil impact of ^{234}U might produce a very high ^{234}U/^{238}U activity ratio.)

The basic models of U series mobilization in aquifers have been dealt with at length. This is not only because the commonly observed U-series disequilibria in ground waters are explained by them but also because such models can be applied usefully to the study of mobilization and residual accumulations in weathered rock and soils, and in U deposits, geothermal systems, and very old ground waters. Furthermore, if these disequilibria are ever to be used in 'dating' ground water or in the determination of ground water flow rates, the initial and interval conditions will have to be defined in terms of such models (see section 9.3 below).

9.2 Case studies

9.2.1 Introduction

There have been relatively few cases of the application of disequilibrium phenomena to the solving of real hydrological problems, at least in the same sense that coral reefs can best be dated by the $^{230}Th/^{234}U$ method. In a few instances, however, the use of U-series disequilibrium in the description of aquifer processes has been successful to the extent that the potential utility of this approach is widely recognized as being very great. For example, the sources and mixing proportions of aquifer waters in limestone regimes can be identified by $^{234}U/^{238}U$ ratios. In an entirely different context, the presence of U ore bodies or the build-up of strains precursory to earthquake rupture can be recognized by the migration of Rn to the earth's surface. In the former case (mixing proportions) the usefulness of the approach depends on an early development of a disequilibrium between U-series nuclides followed by a long period in which the radionuclides behave conservatively, that is, not reacting with aquifer rocks either by precipitation or dissolution. Such applications and their potential will be discussed in section 9.2.2. In the other kind of uses (U exploration or earthquake prediction) the application depends on continuous or intermittent interaction between host rock and ground water. These systems are labelled non-conservative, and are discussed in section 9.2.3.

The dating of ground water by disequilibrium techniques appears to be a formidable if not impossible challenge. Yet the need for any kind of dating tool for ground water is so great that disequilibrium studies are being pursued vigorously in this direction. Such efforts, and some of the theoretical possibilities, are described in section 9.3.

In order to be useful as a fingerprint, a constant value for concentration and activity ratio of the subject radionuclides must emanate from the source rocks through a suitable duration of time. Supporting evidence of the temporal constancy of activity ratios of the same element can be determined by measuring the activity ratios found in materials precipitated from the water or taken up by materials in contact with the waters. Examples of the former include corals, manganese nodules, tufas or travertines, and the latter by fossil teeth and bones. In none of these cases, however, can constancy or the absolute value of concentration be verified.

Actual analyses of water demonstrating constancy of the parameters go back only a few decades, at most. Osmond and Cowart (1976a) report analytical values for U content and $^{234}U/^{238}U$ activity ratios for a small number of locations which were collected at different times. Since then, two of the sites have been recollected; the analytical values remain within the statistical uncertainty of the original values. Table 9.1 reports the values determined for these two locations. Other investigators have reported more or less constant values for shorter durations (Zielinski and Rosholt 1978, Rydell 1969).

Table 9.1. Temporal constancy of U content and $^{234}U/^{238}U$ activity ratios in ground waters

Site	Date of collection	U content (ppb)	$^{234}U/^{238}U$ activity ratio	Ref.
Wakulla Springs,	Sep 1966	0.584 ± 0.024	0.878 ± 0.020	Rydell (1969)
Wakulla County,	Apr 1967	0.608 ± 0.029	0.854 ± 0.028	Rydell (1969)
Fla	Oct 1971	0.58 ± 0.03	0.88 ± 0.05	Osmond and Cowart (1976a)
	Dec 1974	0.50 ± 0.03	0.87 ± 0.04	Osmond and Cowart (1976a)
	May 1977	0.58 ± 0.03	0.86 ± 0.04	Previously unpub.
	Oct 1979	0.61 ± 0.04	0.88 ± 0.05	Previously unpub.
City of Charlotte,	July 1972	0.006 ± 0.002	6.55 ± 1.73	Osmond and Cowart (1976a)
Atascosa Co,				
Texas	Jun 1973	0.006 ± 0.001	6.68 ± 0.85	Osmond and Cowart (1976a)
	Aug 1979	0.006 ± 0.001	6.97 ± 1.03	Previously unpub.

For the ensuing discussion of the tracing and mixing by the use of U content and $^{234}U/^{238}U$ activity ratio, it will be assumed that these parameters remain constant at a given location over a considerable time. This seems to be true for all natural sources (springs) and municipal water supply wells analysed, but in those cases where an aquifer is sufficiently stressed, changes may occur. For instance, a high production well used to dewater the U ore zone at a mine in Wyoming was analysed to have a U content of 12.5 ppb and $^{234}U/^{238}U$ activity ratio of 4.43. Seventeen months later water from this same borehole was found to have a U content of 160 ppb and a $^{234}U/^{238}U$ activity ratio of 1.53. Evidently the original water had been replaced by more oxidizing waters which were more aggressive in mobilizing U.

Michel and Moore (1980) report that the radium content and the $^{228}Ra/^{226}Ra$ activity ratio of waters from specific locations in a coastal plain aquifer were nearly constant for the duration of their investigation (2 years) even though there was considerable spatial variation of the parameters.

9.2.2 Conservative behaviour

(i) *Application requirements*. The uses of U-series disequilibrium in the study of hydrogeological phenomena can be grouped conveniently according to their geochemical behaviour as conservative and non-conservative. By conservative what is meant is that the relative and absolute proportions of the various nuclides in a ground water or surface water remain fixed as the water moves from one regime to another, or that any changes in such proportions can be interpreted as the result of mixing of two water masses. This definition applies to (i) waters which have achieved the steady-state condition with respect to partitioning of nuclides between the dissolved and solid states, and (ii) waters which move rapidly from place to place relative to the lifetimes of the dissolved radioisotopes (as for example in rivers). No drastic changes in geochemistry are involved in either case.

The uses of dissolved radioisotopes as tracers and for determining mixing proportions are closely related. To serve as tracers, it is necessary that the concentration/activity ratio fingerprints have an identifiably different field of values from other waters in the vicinity. For mixing proportions to be obtained it is necessary that each of the sourcewaters has its own distinguishing fingerprint.

(ii) *Tracing.* The single most widespread U isotope fingerprint is that of sea water. In the oceans of the world the water has an essentially constant content of 3.3 ppb and an activity ratio of 1.14 (Ku *et al.* 1974). This constant value was used by Cowart, Kaufman, and Osmond (1978) to serve as a tracer in a cavernous dolomite called the 'Boulder Zone' in south Florida. It has been hypothesized that the relatively cold saline water is actually sea water which enters the Boulder Zone in the Straits of Florida and moves landward (Kohout 1965). In sample 508 in Fig. 9.8 the U content and activity ratio are within the statistical uncertainty of the sea water value. This sample site is the closest Boulder Zone well to the hypothesized input of sea water. Further movement inland or parallel to the coast would be predicted to have the same or smaller concentrations and higher activity ratios which would change progressively away from the input location. This seems to be the case for locations both north and south of the inferred input. Thus, the Boulder Zone, which is used for waste injection, is likely to be an active flow zone.

Another, more speculative, tracing of groundwater movement was proposed by Cowart *et al.* (1978) for the Floridan Aquifer. Within this aquifer there are highly transmissive cavernous zones which have been mapped in south Florida by Puri and Winston (1974). Always within these zones and only within these zones in south Florida are found $^{234}U/^{238}U$ activity ratios at or less than secular equilibrium (see Fig. 9.8). A possible explanation for this pattern is that the U found within the transmissive zones is relict from times of lower sea level when the transmissive zones served as conduits from the recharge zones to the offshore discharge zones. Water with less than secular equilibrium is commonly found in the areas where the Floridan Aquifer rocks outcrop north of Tampa. The spring values shown in Fig. 9.8 exemplify these waters.

Zones of active water movement may be determinable if the fingerprint is

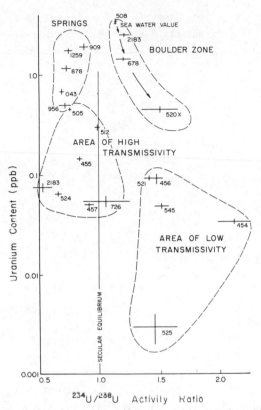

Fig. 9.8. Scattergram of South Florida ground water samples. (Modified from Cowart *et al.* 1978.)

distinct. A series of boreholes drilled by the Southwest Florida Water Management District were sampled at various depths during drilling. The samples were filtered so that the filtrate was free from visible particles. The vertical profiles of the U analyses are shown in Fig. 9.9. It may be seen that the U content and activity ratio generally vary antithetically so that high U content is associated with low activity ratios and vice versa. This is the same relationship displayed by springs in the region (Fig. 9.10), and it suggests that the sources of water issuing from the springs may be traceable by this means. Furthermore, it suggests that those zones having high U content-low activity ratio water are the zones of active water transport. It must be pointed out that the U content of these waters is generally considerably greater than for nearby springs and developed wells. This may be due to the pulverizing effect of drilling which would expose more fresh surface area of the carbonate rock to water samples, thus liberating more U than usual, but of the same activity ratio.

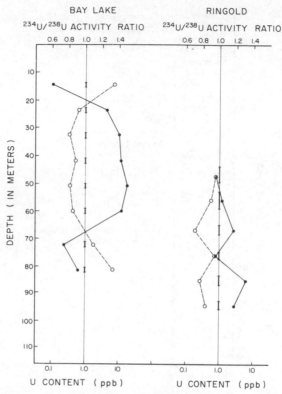

Fig. 9.9. Vertical profile of U content and $^{234}U/^{238}U$ ratio of water samples from boreholes in the Floridan Aquifer. The solid line joins the U content points and the dotted line joins the activity ratio points.

Ground waters can be characterized by their Ra content and their $^{228}Ra/^{226}Ra$ ratios in much the same way that ^{234}U and ^{238}U have been used. This is considered to be a conservative application of U-series disequilibrium even though the initial entry of the two Ra isotopes into the ground water is necessarily the result of interaction with the aquifer, and the time scale of water characterization much shorter ($^{226}Ra\,t_{\frac{1}{2}} = 1{,}602$ y, and $^{228}Ra\,t_{\frac{1}{2}} = 5.75$ y). The almost universal occurrence of Ra in ground waters attests to its generally conservative behaviour in aquifers.

The Ra content of ground waters appears generally to be less variable than that of U, though by no means constant, primarily because of its more consistent alkaline earth behaviour (2^{+} valence) as opposed to the divergent solubilities of the 4^{+} and 6^{+} U ions. On the other hand, the $^{228}Ra/^{226}Ra$ ratio is more variable than $^{234}U/^{238}U$ because their origins are in different decay series, in which the parent ^{232}Th and ^{238}U have quite dissimilar chemical behaviours in aquifers.

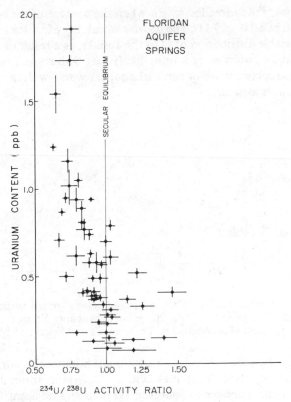

Fig. 9.10. The U isotope data for springs issuing from the Floridan Aquifer. Almost all springs with $^{234}U/^{238}U$ activity ratio greater than 1.0 are located in areas surrounded by a confining layer (Cowart 1980b).

Although ^{226}Ra content and gross Ra/U ratios have been measured in many aquifers (for example, Scott and Barker 1962) the content of ^{228}Ra has generally been ignored. This has been in part due to the greater difficulty of measurement by emanation methods (^{228}Ra is a beta emitter; its alpha emitting daughter ^{228}Th has an inconveniently long growth time ($t_{\frac{1}{2}} = 1.913$ y), and its alpha emitting granddaughter ^{220}Rn has an inconveniently short decay time ($t_{\frac{1}{2}} = 55$s). In part, however, the neglect of ^{228}Ra as a ground water component seems to be based on an assumption that it is quantitatively unimportant. That this is not the case was demonstrated by early measurements made by Soviet investigators (Cherdyntsev 1971, p. 62).

Michel and Moore (1980) observed that in a surficial aquifer in South Carolina: (i) ^{228}Ra and ^{226}Ra content is stable at a given site when recollected from season to season; (ii) a large variation occurs in both isotopes over a geographic range of a few km; (iii) there is greater variation in ^{226}Ra than

^{228}Ra; and (iv) ^{228}Ra exceeded ^{226}Ra when the content of the latter was less than about 10 pCi/l (Fig. 9.11). The greater variability of ^{226}Ra was attributed to a more variable distribution of U in the aquifer, as a result of mobilization and precipitation processes. It is quite likely that with respect to Ra isotopes this aquifer is representative of surficial aquifers world-wide, and possibly of confined aquifers as well.

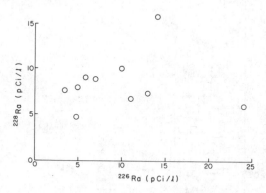

Fig. 9.11. Plot of ^{228}Ra against ^{226}Ra in ground waters from a surficial aquifer in South Carolina. Variations in ^{226}Ra are greater, probably because of the greater irregularity of U distribution relative to Th in aquifers (Michel and Moore 1980).

   (iii) *Mixing of ground waters.* The variation of both total U content and the ^{234}U/^{238}U activity ratio in springs and wells issuing from the extensive Paleozoic carbonate aquifer in southern Nevada and adjacent California is instructive in that apparently the entire system is oxidizing and the U acts conservatively. Even waters from boreholes penetrating the confined aquifer at considerable depths and tens of kilometers distant from recharge areas have measurable dissolved oxygen (I. J. Winograd, personal comm.). This being the case, mutually exclusive fields as defined by the combination of U content and excess ^{234}U can be used to calculate the relative mixing volumes of the probable recharge areas (Cowart 1979, and work in progress).
   The aquifer rocks outcrop and are recharged in the Spring Mountains and Pahranagat Valley areas; discharge occurs in the Ash Meadows area and it is likely that part of the water underflows as far west as the Death Valley–Furnace Creek area in California. The principal recharge area for the Ash Meadows discharge zone is the Spring Mountains area but the distant Pahranagat Valley (approximately 130 km distant) may contribute as much as 35 per cent (Winograd and Friedman 1975).
   When the data from the springs and boreholes are plotted in a form suitable to display mixing (see Fig. 9.12), almost all of the waters sampled from the carbonate aquifer plot linearly. It will be noted that samples from the two recharge areas fall at the end of the linear trend and that the discharging waters

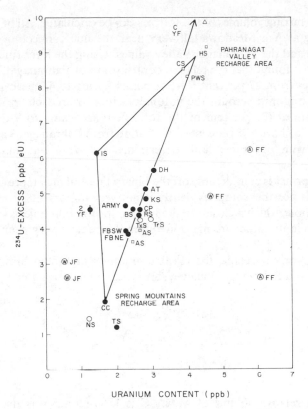

Fig. 9.12. Mixing plot showing linear relationship between Pahranagat Valley and Spring Mountain recharge areas and the Ash Meadows and Death Valley discharge areas, Nevada. Key: HS = Hi Ko Spring; CS = Crystal Spring; PWS = Pederson Warm Spring; AS = Ash Spring; DH = Devils Hole; AT = Armagosa Tracer Well; KS = King Spring; CP = Crystal Pool; BS = Big Spring; RS = Rogers Spring; FBSW = Fairbanks Spring, S.W.; FBNE = Fairbanks Spring, N.E.; IS = Indian Spring; Army = Army Well Number 1; CC = Cold Spring; TS = Trout Spring; TrS = Travertine Spring; TxS = Texas Spring; NS = Nevares Spring; WJF = Jackass Flats, welded tuff aquifer; WFF = Frenchman Flats, welded tuff aquifer; AFF = Frenchman Flats, alluvium aquifer; YF = Yucca Flats, Paleozoic carbonate aquifer.

fall between the two. A departure from the linear trend is the Indian Springs sample and, to a lesser degree, Army-1. The two Yucca Flats carbonate aquifer samples are, based on independent evidence, thought to contribute only negligibly to the Ash Meadows system (I. J. Winograd, personal comm.). The other samples in Fig. 9.12 are from alluvial or welded tuff aquifers, and they fall outside of the Paleozoic aquifer field.

If mixing equations are applied to the carbonate aquifer data, using either the form of Osmond, Kaufman, and Cowart (1974) or Briel (1976), values of the relative mixing volumes of the sources can be calculated. All of the springs discharging at Ash Meadows fall very near the line connecting the Spring Mountains and the Pahranagat Valley values. Using the lever rule for binary mixtures, it is found that the relative contribution of Pahranagat Valley type water ranges from 27 per cent for Fairbanks Spring to 42 per cent for King Spring. Taking into account the volumes issuing from each spring, the five springs sampled (77 per cent of the total flow according to Winograd and Pearson 1976), have a 35 per cent contribution from Pahranagat Valley. This is consistent with the upper limit determined by Winograd and Friedman (1975).

Indian Springs is considerably off the binary line and may represent another source. This possible source evidently is not very important in that only the Army borehole, which is closest to the Indian Springs borehole, is significantly affected while the other springs and boreholes remain very near the binary mixing line of Fig. 9.12.

For the Army borehole, the relative amounts of Indian Springs, Spring Mountain, and Pahranagat Valley water types can be calculated by use of the equation:

$$\frac{V_{SM}}{V_{AR}} = \frac{(C_{AR} - C_{IS})\left(\dfrac{X_{PV} - X_{IS}}{C_{PV} - C_{IS}}\right) - (X_{AR} - X_{IS})}{(C_{SM} - C_{IS})\left(\dfrac{X_{PV} - X_{IS}}{C_{PV} - C_{IS}}\right) - (X_{SM} - X_{IS})} \tag{9.6}$$

where V is relative volume of water, C is U content, X is the excess (or deficiency) of ^{234}U [$X = (^{234}U/^{238}U$ activity ratio -1) U content], and the subscripts IS, AR, and SM refer to Indian Springs, Army-1, and Spring Mountain sites, respectively (after Briel 1976). The respective values obtained from eqn (9.6) for Indian Springs, Spring Mountains, and Pahranagat Valley are 33, 44, and 23 per cent.

It has been suggested that the springs in the Death Valley–Furnace Creek area result from the same sources which supply Ash Meadows. The U isotope analysis tends to support this suggestion in that the analyses marked TxS and TrS (Texas Spring and Travertine Spring) are similar to the Ash Meadows water. Nevares Spring (NS) is quite different and may have a different source (unlikely) or may mix with locally derived waters.

The Ash Springs (AS) sample is anomalous in that it plots far from the other Pahranagat Valley samples and is in fact near the Ash Meadows discharge field. The spring was sampled at two different times, and each gave virtually identical results. The reason for this difference is not known.

In general, the application of U isotopes for relative mixing volumes in this carbonate aquifer system gives results consistent with those obtained from independent hydrogeological and hydrochemical studies.

(iv) *Mixing of surface and ground waters.* A study by Briel (1976) demonstrates mixing between the 'surface' and 'ground waters' in an area where the distinction between the two is not always clearcut. He found that the water in the Santa Fe River in Florida could be represented by at least two ground water regimes and by a hypothetical surface water component (see Fig. 9.13). The latter could be a function of the amount of local recharge in various parts of the basin. For samples collected progressively downstream in the Santa Fe River, the $^{234}U/^{238}U$ activity ratio approximates more and more closely the average activity ratio of the ground water.

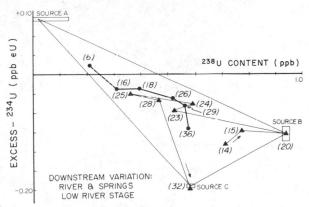

Fig. 9.13. Ternary mixing diagram showing contribution of each of the sources (two ground water, one surface water) at various locations along the Santa Fe River in north Florida (Briel 1976).

9.2.3 Non-conservative behaviour

(i) *Application conditions.* Given time, most ground water masses and some surface streams will encounter differing geochemical conditions which will cause changes in elemental concentrations. In fact, time alone will cause such changes given the varying decay rates and lifetimes of the radionuclides. Such changes in radioelement fingerprints diminishes their value in tracing and mixing studies, but leads to other kinds of investigations in which the observed isotopic evolution is used as a clue to aquifer history, and the nature of water/rock interactions. Some of these evolutionary patterns are best understood in terms of the decay rates of the dissolved species, such as the 'augmenting' and 'decaying' aquifer systems defined earlier (see section 9.1.4). Other changes are related to simple solution and precipitation processes.

(ii) *Secondary accumulation.* The accumulation of some U apparently occurs in any aquifer having a reducing environment with Eh low enough to cause U precipitation. This does not imply that an economic accumulation has or will form in every such aquifer but rather that the same mechanisms may act

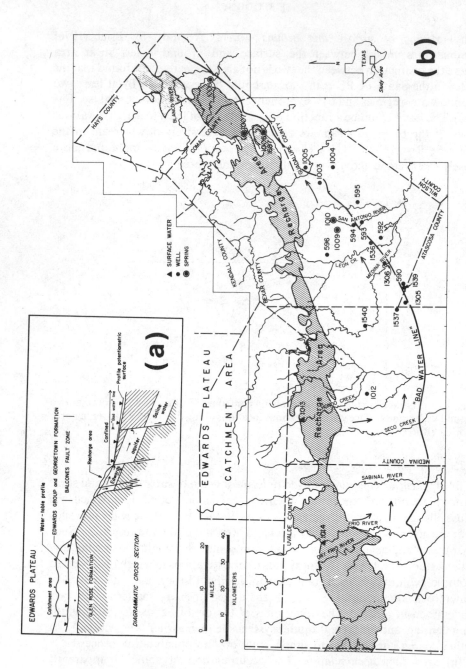

Fig. 9.14. The extent of the Edwards Aquifer, Texas, both in section (a), and in plan (b). The Bad Water Line is apparently fault controlled in some places but not in others (Cowart 1980a).

in both 'mineralized' and 'non-mineralized' aquifers although to different degrees. The fact that shallow ground waters almost always have a greater U concentration than deep waters makes this apparent.

An aquifer which has no known economic deposits but which displays a hallmark of accumulation is the carbonate Edwards Aquifer in Texas. This aquifer consists of an up-dip oxidized portion and a down-dip reduced portion. The boundary between the two is abrupt and well defined; locally it is called the Bad Water Line. Figure 9.14 shows the geographic location of the samples and Fig. 9.15 the U isotope values. The oxidized samples and the reduced samples each fall into well defined fields when plotted as U content against activity ratio. However, those samples collected in the vicinity of the

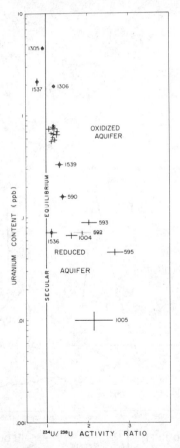

Fig. 9.15. Scattergram of U analyses of samples from the carbonate Edwards Aquifer, Texas. The oxidized aquifer is represented by samples 1013 to 1003; the boundary, by samples 1306 and 1537; and the reduced aquifer by samples 1005 to 593 (Cowart 1980a).

Bad Water Line fall into neither field and only two can be considered as transitional. Cowart (1980a) interpreted the data to be the result of long term precipitation of U at the Bad Water Line redox barrier with recent remobilization of U as the barrier moves coastward (see Fig. 9.16). Especially important to this interpretation are samples 1305 and 1537 which have relatively high concentrations and less than 1.0 activity ratios. These are thought to result from the remobilization of the U accumulation.

Secondary U accumulations are, of course, more commonly associated with sandstone. Cowart and Osmond (1976 and 1980) have described the isotopic content of ground waters in such aquifers (see chapter 14).

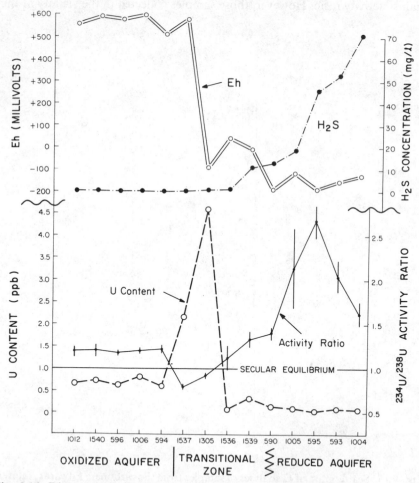

Fig. 9.16. Diagrammatic profile of redox, H_2S, and U isotopic variations in the carbonate Edwards Aquifer, Texas. The oxidized part of the aquifer is to the left, the reduced to the right (Cowart 1980a).

In the Red Desert region of Wyoming, a regional sampling of water from a sandstone aquifer reveals two distinct ground water types: high concentration and moderate activity ratio waters to the northeast, and low concentration and high activity ratio waters to the southwest. Although the water chemistry in the two areas is not markedly different, the aquifer isotopic pattern gives a clear indication of an U barrier between the two regions (see Fig. 9.17). At several sites along this boundary drilling has verified the existence of ore deposits of economic importance.

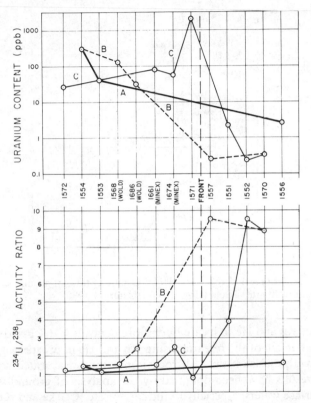

Fig. 9.17. U content and $^{234}U/^{238}U$ ratio variations across the Red Desert area, Wyoming (not to scale). (Cowart and Osmond 1980).

In the U district of South Texas, major fronts are less well defined. Often a broad zone of aquifer has several barriers, perhaps controlled by the distribution of contained organic debris; more likely by reducing gases seeping upward through faults. (Reynolds and Goldhaber 1978). Figures 9.18 and 9.19 show the possible importance of one such fault, and also the quite variable U isotopic signatures in ground waters near such a deposit.

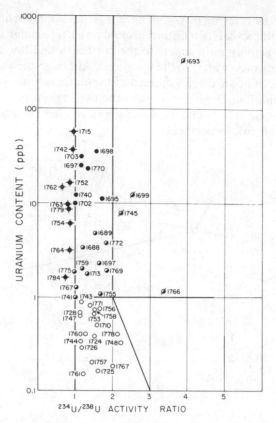

Fig. 9.18. U content plotted against $^{234}U/^{238}U$ activity ratio of ground water samples from the Lamprecht site, Texas (Cowart and Osmond 1980).

The use of Rn and He to locate buried ore deposits is discussed in chapter 14. Although much of the research using Rn and He in prospecting efforts is concerned with diffusion through rock joints and unsaturated soils, recent promising results have been obtained by the analysis of ground waters and escaping surface waters, especially in the case of He (Clarke and Kugler 1973; Dyck and Tan 1978; Severne 1978; Reimer, Denton, Friedman, and Otton 1979). Because of its mobility and stability, anomalous amounts of He in ground water may be more indicative of faults, however, than of U accumulations.

In the case of Rn, its short half-life limits its range in slowly percolating waters; also, given the disequilibrium which can exist in secondary U accumulations, an anomalously high Rn occurrence in ground water may only point to an occurrence of ^{226}Ra or ^{230}Th, rather than U itself. An interesting example of extremely high ^{222}Rn in ground water accompanied by relatively modest amounts of U and Ra is reported by Asikainen and Kahlos

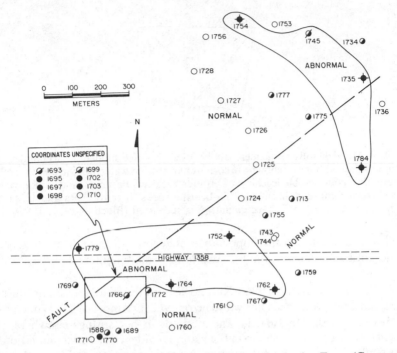

Fig. 9.19. Ground water sample locations at the Lamprecht site, Texas (Cowart and Osmond 1980).

(1979) for the Helsinki region in Finland. Whatever the cause of the disequilibrium in water, considerable Ra must be present nearby.

Bloch (1980a) has suggested that ^{226}Ra in springs, especially in regions of hydrocarbon traps, may be a clue to the presence of U deposits at depth. Such deposits might be expected to form where oxidized or CO_3^{2-} bearing ground waters are reduced by natural gases or oil. However, high Ra content alone does not constitute an infallible indicator. Certain ground waters can become enriched in Ra by the leaching of rocks of ordinary U and Th concentration; conversely, others percolating through ore deposits may do little leaching. High ^{226}Ra/^{228}Ra and high Ra/Ba ratios are diagnostic of the latter. The high ^{226}Ra in this case is entrained by recoil from the U-series daughter ^{230}Th. High ^{227}Ac would also be expected in this situation. Any Ba would have to have been leached directly and ^{228}Th could be supplied by ordinary Th-bearing rocks. Figure 9.20 shows Bloch's argument diagrammatically.

The mechanism of release of Rn and Ra by recoil processess has been studied by Tanner (1980), Torgersen (1980), and others. With respect to ground water applications, two conclusions can be derived from such studies: (i) the presence of pore water greatly promotes release and mobilization of

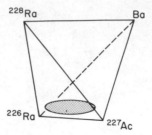

Fig. 9.20. How dissolved nuclides can be used to infer hidden U accumulations. All four of the components shown are common in ground water, but the presence of Ba suggests that considerable leaching of aquifer rock has occurred, and ^{228}Ra can be derived from recoil mobilization of Th-bearing rocks. If only ^{226}Ra and ^{227}Ac are present then recoil from a U accumulation is suggested (Bloch and Key 1981).

recoil nuclides; and (ii) the importance of recoil as a mobilization process should not be minimized even for such long-lived daughters as ^{234}U, ^{230}Th, and ^{226}Ra.

(iii) *Geothermal systems.* That Rn and Ra are frequently abundant in hot spring waters has long been known (Mazor 1962; D'Amore 1975). Wollenberg (1975) suggests that in Nevada, and perhaps in hot springs elsewhere, the presence of these radio nuclides in the water is evidence of an accumulation of U at depth. The basis for his conclusions rests on a convective geothermal model: normal regional ground waters are heated by hot rocks near the base of the system, flow upward, and are replaced by more inflow from the outer parts of the regional aquifer.

Figure 9.21 is a pictorial representation of the Wollenberg model. The two

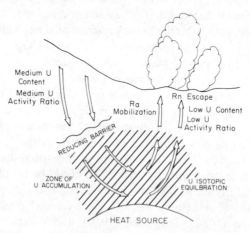

Fig. 9.21. Diagrammatic model of U-series isotopic geochemistry in geothermal systems.

unusual aspects of such a ground water system are: (i) the rapid circulation rate (driven by the elevated temperatures); and (ii) the enhanced tendency for geochemical reactions. At several sites Wollenberg determined that the U concentration in the hot springs was quite low relative to cold springs nearby. By mass balance calculations he showed that the excess Rn in thermal springs could be supplied by a few hundred kilograms of U at depth.

The radioactive springs studied by Wollenberg (1975) were depositing $CaCO_3$ tufa at the surface. Other springs depositing siliceous sinter carried only ordinary levels of Ra and Rn. The implication is that only in carbonate rich geothermal waters is Ra mobilized and carried upward from the U accumulation below. A second implication is that in silica rich waters the flow of water from accumulation to surface takes more than a few days; otherwise ^{222}Rn ($t_{\frac{1}{2}} = 3.8$ d), assumed to be mobile under all conditions, would be present at the surface.

That most geothermal waters are deficient in U was confirmed by Osmond and Cowart (1976b) who studied the $^{234}U/^{238}U$ activity ratios at Yellowstone National Park, Wyoming, Imperial Valley, and Mt Lassen, California. The expectation was that sizeable accumulations of U at depth should result in the kind of high activity ratios characteristic of ore deposits and deep aquifer barriers in general, but such was not the case. Only activity ratios close to the equilibrium ratio of 1.0 were found. In fact, on a scattergram of U concentration against activity ratio (as in Fig. 9.1) geothermal waters plot as a somewhat isolated field, which is low in U concentration and also low in $^{234}U/^{238}U$ activity ratio. This suggests that the elevated temperatures and strongly reducing conditions at depth in geothermal systems cause an acceleration of isotopic exchange processes, such that in the flow-through waters the isotopic ratio in dissolved U is essentially the same as that of the U precipitated on the host rock.

(iv) *Brines*. Brines from considerable depth in the earth are incontestably very old waters. They may or may not have been trapped at the time of sediment deposition ('connate') but they have surely resided below the zone of rapid hydrological circulation for great lengths of time, and likely for millions of years. Such waters are essentially in chemical equilibrium with the host rock, and the extent of disequilibrium observed in U-series nuclides is primarily the steady state balance between alpha decay-generated daughters mobilized by recoil and radioactive decay of the excess in solution.

The observed disequilibrium between ^{234}U and ^{238}U is of special interest because of the long half-life of the equilibration process. The environment of brines in deeply buried sedimentary rocks is reducing so that the mobility of U is small, as should be the effects of chemical leaching. Input of ^{234}U into the brine by direct alpha recoil (and to some extent by auto-oxidation) is not controlled by the reducing nature of the brine. If this is the case, the content of U isotopes in brine can be thought of as an equilibrium between the long term background input of ^{234}U into the water, and the long term loss of ^{234}U from the water by adsorption on grain surfaces and by radioactive decay.

Kronfeld *et al.* (1975) reported ^{234}U/^{238}U activity ratios of 10.1 ± 1.6 and 8.6 ± 1.2 for two brine samples taken from a deep borehole in Israel. These high ratios were attributed to an aging effect which resulted from the long term input of the recoiling precursor of ^{234}U. Analyses of brines from other locations do not indicate that high ^{234}U/^{238}U activity ratios necessarily result from interactions of long duration, in fact high ^{234}U/^{238}U activity ratios appear to be unusual in brines (see Table 9.2).

The measured value of ^{234}U/^{238}U activity is a function of both time of exposure to the ^{234}U flux (recoil, auto-oxidation) and the rate of flux (a function of the effective U content of the aquifer rock). Thus the ^{234}U/^{238}U

Table 9.2. Uranium isotope analysis of brines

Location; Age; Depth (m)	U content (ppb)	^{234}U/^{238}U activity ratio	Reference
Mulberry, Fla.; Paleocene; ~ 1525	0.60 ± 0.06	1.72 ± 0.11	(1)
Gwinville Field Ms. (Well 101-A); Cretaceous;	0.009 ± 0.002	1.23 ± 0.40	(2)
Winchester Field, La.; Tertiary; ~ 4575	0.085 ± 0.029	1.68 ± 0.60	(2)
Anse La Butte, La.; Tertiary; ~ 1550	0.073 ± 0.006	1.01 ± 0.10	(2)
Anse La Butte, La.; Tertiary ~ 3350	0.047 ± 0.003	1.31 ± 0.11	(2)
Charlotte Field, Tx.; Cretaceous; ~ 2000	0.019 ± 0.022	1.16 ± 0.14	(2)
Edna Delcambre Well No. 1, La.; Tertiary; 4092—4136	0.010 ± 0.001	1.62 ± 0.15	(3)
Edna Delcambre Well No. 1, La.; Tertiary; 4223—4236	0.005 ± 0.001	1.06 ± 0.10	(3)
Fairfax Foster Sutter Well No. 2, La.; Tertiary; 5178—5223	0.030 ± 0.002	1.21 ± 0.10	(3)
Beulah Simon Well No. 2, La.; Tertiary 4815—4846	0.008 ± 0.001	1.29 ± 0.12	(3)
Beulah Simon Well No. 2, La.; Tertiary	0.010 ± 0.001	1.14 ± 0.13	(3)
Pleasant Bayou Well No. 2, Tx.; Tertiary; 4805—4824	0.003 ± 0.001	1.42 ± 0.14	(3)
Ramallah Well No. 1, Israel 1861—1875	0.51 ± 0.09	10.1 ± 1.6	(4)
Ramallah Well No. 1, Israel 2253—2264	0.59 ± 0.09	8.6 ± 1.2	(4)
Chetopa Field, Ks.; Ordovician	0.010 ± 0.002	1.68 ± 0.32	(5)
Wackerle Field, Ks.; Ordovician	0.013 ± 0.002	1.46 ± 0.27	(5)
Schreppel Field, Ks.; Ordovician	0.010 ± 0.002	2.08 ± 0.41	(5)

(1) Cowart *et al.* (1978)
(2) Unpublished data of authors
(3) Kraemer (1981)
(4) Kronfeld *et al.* (1975)
(5) Cowart, J. B. (1981). *Jour. Hydrol.* (in press)

activity ratio itself may not be a useful measure of the age of water either between different aquifers or even within the same aquifer, if the U content of the rock varies significantly.

(v) *Earthquake precursor studies.* Anomalous changes in the ground water content of various isotopes of Rn, He, and U have been attributed, at least in part, to seismic activity. The Rn and He values for ground water in the vicinity of Tashkent, Uzbek, U.S.S.R., were collected before and after a series of earthquakes which began on 26 April 1966 (Ulomov and Mavashev 1967; Spiridonov and Tyminskii 1971). A sharp decrease in Rn and an increase in He were reported. The $^{234}U/^{238}U$ isotope analyses were only reported for samples collected after the earthquake, but the values were anomalously high and decreased with time.

Several studies of Chinese earthquake precursory phenomena have included Rn as a particularly significant parameter, and one or two accurate predictions are claimed to have been based in part on increasing levels of Rn in the ground water (Sykes and Raleigh 1975). Recent efforts to monitor earthquake-related phenomena in California have also focused on U-series isotopes in ground water, especially Rn and He, but also including U. These studies are well designed, yet so far the experiments do not have enough high energy seismic events with which to compare the monitoring data (Teng, Ku, and McIlreath 1975; Shapiro, Melvin, Tombrello, and Whitcomb 1977; King 1978).

With respect to He and Rn in ground water, both long term and short term variations have been identified (some associated with the seasons), and both positive and negative anomalies have been reported in connection with earthquake events (Moore *et al.* 1977; Tanner 1980; Smith *et al.* 1980). Available data suggest that this approach might sometimes yield precursory information, but criteria have yet to be established by which sampling sites and sampling frequencies are selected, and anomalous variations identified (Teng 1980).

There are two principal models on which Rn anomalies of the positive kind are based: (i) dilatency and fracturing, which increases the surface area of aquifer rocks and promotes diffusion and alpha recoil into the pore water; and (ii) increased lithostatic pressure in the principal zone of stress causing migration of Rn-laden waters upward and outward (Tanner 1980). The inconsistencies in field results to date may stem from the varied effects of these two quite different phenomena.

The $^{234}U/^{238}U$ data from Tashkent are interesting to the extent that if the ratio does vary in concert with earthquake precursor processes then the model of increased surface area and dilatency is supported. This is because it is believed that recoil mobilization is the principal mechanism causing changes in this activity ratio (see section 9.1.2). If this is the case, then variations among several of the U-series daughters (especially Ra isotopes) ought to be studied, so that the various half-lives of the isotopes might provide both short-term and long-term response times, as illustrated by Fig. 9.4.

9.3 Ground water dating applications

9.3.1 General requirements

One of the unsolved problems of U-series geochemistry is how to apply disequilibrium principles to the dating of ground water. Just as speleothems and coral reefs are datable (see Chapter 3) so it would seem that water masses of great age could be datable. Tritium ($t_{\frac{1}{2}} = 12.3$ y) forms an integral part of the water molecule, and ^{14}C ($t_{\frac{1}{2}} = 5.73 \times 10^4$ y) occurs as the conservative HCO_3^- ion; these two nuclides have been used to determine the ages of geologically young ground water under certain conditions. The problem with the U-series disequilibrium approach is the poor definition of the system being studied.

In section 9.1.4 it was suggested that aquifers might usefully be categorized as steady state, augmenting, or decaying systems. This classification is based on the observation that there tends to be a steady state of disequilibrium developed between water and host rock such that the water carries an excess of daughters and the rock a deficiency, this disequilibrium being maintained by a balance between recoil-caused transfer of daughter from rock to water, and decay of excess daughter in the water. Such a steady state develops more slowly for long-lived nuclides than for shorter-lived nuclides, as does a return to a lower steady state if ground water percolates to a part of the aquifer with less parent element in (or adsorbed on) the rock. In viewing aquifers in this way, it is assumed that chemical interactions between rock and water are minimal.

Radiometric age determination of any kind is possible only if three conditions are met: (i) the system being dated is of the same order of age magnitude as the life-time of the radionuclide undergoing decay; (ii) the initial state of disequilibrium is known; and (iii) the radionuclide content of the system has not changed as a result of chemical or physical reactions other than decay.

In ground water the most difficult condition is the last, the 'closed system' requirement. Yet old ground waters are often considered to be in a state of chemical equilibrium with aquifer rock such that dissolution and precipitation processes are minimal. Furthermore, 'open system' models have been developed for age determinations, for example in U/Pb dating, or in Th/U dating of soils (see chapter 10).

The estimation of initial parent-daughter contents in ground water is also very difficult. Here again, tactics employed by other geochronologists may prove useful. In determining sediment accumulation rates in the deep sea by excess ^{230}Th decay, the estimation of initial excess is impossible for any one sediment horizon. However, the plotting of the changes in excess ^{230}Th over the length of a sediment core permits the extrapolation to time-zero of this parameter, assuming long-term stability of the accumulation processes of ^{230}Th and sediment. Likewise the determination of disequilibrium states in an aquifer with well defined flow lines may permit the construction of a distance versus disequilibrium curve in which the initial condition can be deduced and the changes interpreted in terms of decay through time of the key nuclide.

The remaining requirement (i), above, that the nuclide have a mean life consistent with the flow rate, is a less challenging problem, because of the many daughters of diverse decay rates in the three naturally occurring series.

9.3.2 Radon and Helium

Inasmuch as closed system conditions are perceived to be the major problem in the dating of ground water, this discussion can begin with one of the shorter-lived daughters in the U-series, Rn.

Andrews and Wood (1972) and Marine (1976) have noted the accumulation of ^{222}Rn ($t_{\frac{1}{2}} = 3.8$ d) in ground water as a function of age and flow rate, recognizing that only short term and/or quite local systems are appropriate for this approach. An attractive aspect is the chemical inertness of the Rn atom; less satisfactory is the likelihood that dissolved ^{226}Ra in the water rather than in the rock is an important source of the Rn.

Another noble gas constituent of ground waters is He, a product of alpha decay. Although not a member, as such, of the U-series and Th-series decay chains, each of the alpha decay steps results in one He atom being produced. Marine (1979), Andrews and Lee (1979), and Torgersen (1980) have studied ground water systems in which the build-up of He is interpreted as a function of residence time of water in the aquifer. Particularly useful in this approach is the He/Rn ratio. Both are inert gases, but the Rn content is dependent primarily on the Ra (and therefore U) content in the host rock, whereas He is the product of both concentration and 'age' of the water. In the Bunter Sandstone aquifer in England this ratio increases in the down-flow direction, though faster than the known flow rate (based on ^{14}C) would require (Andrews and Lee 1979). Three possible causes are considered, each of which presents problems for any similar attempts to date ground water: (i) the differences in the escape processes of Rn and He; Rn by alpha recoil (10^{-6} cm) and He by alpha particle penetration (10^{-2} cm) and long range diffusion; (ii) the generation of He by Th-series daughters, so that the Th/U ratio in the country rock must be known; and (iii) diffusion and escape to the ground water of crustal He which has been accumulating over major geological time periods.

Torgersen (1980) observed that application of the He/^{222}Rn method to various ground water and gas fields yielded 'ages' of pore water that are too great for both young systems (springs) and older systems (early Paleozoic gas fields), although agreement was good for gas fields 10^6 to 10^8 y old.

9.3.3 Radium isotopes

As a matter of speculation, for short time periods there do appear to be possibilities for the use of Ra isotopes as indicators of ground water age. Considering Fig. 9.3, the growth and decay of ground water radioactivities due to Ra isotopes is seen to be responsive to U and Th accumulations on or in the host rock. For an augmenting system, such as water percolating through a granite host rock, the shorter lived isotope, ^{224}Ra, reaches its steady state level of radioactivity within a few weeks. If it is assumed that all of the Ra isotopes have about the same F value (fraction of alpha decay daughters escaping into the water, eqn (9.2)), then their dissolved activity ratios are time dependent as follows:

$$^{228}Ra/^{224}Ra = 1 - e^{-\lambda_{228}t}$$
$$^{226}Ra/^{224}Ra = 1 - e^{-\lambda_{226}t}. \tag{9.7}$$

Ground water ages in the range of 2 to 20 y in the first case, and 0.5 to 5×10^3 y in the second, are theoretically determinable. As far as is known there have been no attempts to apply such accumulation models to the dating of ground water. This would seem to be a promising avenue of research to follow, although as pointed out by Michel and Moore (1980), the irregular distribution of U in many aquifers would present problems.

In a decaying ground water system, where the water is flowing away from a radioactive rock or accumulation, as in geothermal systems, the decay equations of the following form may be used:

$$^{224}Ra/^{226}Ra = Ke^{-\lambda_{224}t}$$
$$^{228}Ra/^{226}Ra = Ke^{-\lambda_{228}t}. \tag{9.8}$$

In this case the problematic assumption is made that the ^{226}Ra has achieved a steady state, a process requiring several thousand years. Despite this problem and others, several investigators have used the decay relationship of ^{224}Ra to estimate in a qualitative way, water flow times of ground water (Tanner 1980; Wollenberg 1975; Cherdyntsev 1971; Michel and Moore 1980).

9.3.4 Uranium isotopes

Whereas the use of Ra isotopes to date ground waters seems possible but has not been exhaustively studied, the reverse situation applies to U isotopes: several investigators have attempted the task of using ^{234}U decay to date very old ground waters without notable success.

Although U is a conservative constituent of surface and shallow ground waters, in deeper aquifers it is less conservative. Even if a deep aquifer exhibits a fairly constant U content over a wide area, this value is usually quite low (0.001 to 0.1 ppb), and it is suspected that individual atoms are being dissolved and reprecipitated in a cyclical fashion, such that isotopic equilibration between the two phase states occurs, and an aqueous closed system does not pertain. Application of the augmenting model in the zone of greater conservatism (shallow aquifers) would seem to be a justifiable approach.

Kigoshi (1973) and Andrews and Kay (1978) have proposed models which assume uniform dissolution rates in conjunction with recoil processes. However, the addition of open system terms would make less likely uniform conditions over a long time span. This is essentially the conclusion of Andrews et al. (1981, in preparation) who have recently studied the ground waters percolating through the Stripa Granite in Sweden.

In the pure form, an augmenting model would use the equation

$$^{234}U/^{238}U = K(1 - e^{-\lambda_{234}t}), \tag{9.9}$$

but Fig. 9.4 suggests that shorter lived daughters may achieve steady state conditions quickly and thus be used to deduce the value for K, the steady state value of the dissolved activity ratio.

If it is assumed that neither the dissolved ^{234}U nor ^{226}Ra content of the ground water is limited by solubility factors, but only by the recoil-decay processes, then a modified form of eqn (9.9) is suggested

$$^{234}U/^{226}Ra = 1 - e^{-\lambda_{234}t} \tag{9.10}$$

This model would have to be tested in a system where the content of the parent ^{238}U in the host rock is relatively high and uniformly distributed, and the water is percolating slowly and uniformly through the system. In the Stripa study, reported by Andrews et al. (1981, in preparation) Ra was not routinely determined, but the Rn values reported can be substituted for Ra, recognizing the difficulties with this gas as mentioned above. Such calculations yield quite large water ages; however, independently determined data suggest that the older water at Stripa may have a different source than the younger water higher in the body (Andrews et al. 1981, in preparation).

Among reports of ground water ages based on decay of dissolved ^{234}U, and decrease in $^{234}U/^{238}U$ activity ratio, are those of Alekseev et al (1973) and Kronfeld and Adams (1974). However, Cowart and Osmond (1974) showed that application of this model to the Carrizo Sandstone aquifer of Texas yields ages that are much too old if ^{14}C-based flow rates are correct (see Fig. 9.22). Rather than water age, ages derived from ^{234}U decay in such systems probably can be interpreted as average time of migration of U atoms in the flow direction (since leaving the zone of maximum accumulation). Inasmuch as much of this time is spent out of solution adsorbed on aquifer minerals, this age is greater than that of the flow-through water.

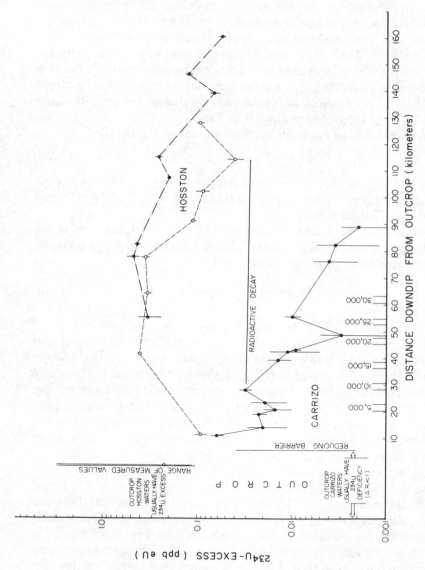

Fig. 9.22. Excess of ^{234}U as a function of down-dip flow in two Texas aquifers. The Hosston excess (from data of Kronfeld and Adams 1974) is shown along two flow lines. The Carrizo curve (from Cowart and Osmond 1974) falls more steeply than expected on the basis of decay of conservative U alone, if ^{14}C-based flow rate is assumed.

This delay factor, which amounts to the solubility product constant, might be estimated using the mixing equation

$$M_1/M_t = \frac{A_t - A_2}{A_1 - A_2} \qquad (9.11)$$

Where A_1 and A_2 are the activity ratios of U in the water and adsorbed on the aquifer grains respectively, and A_t is the activity ratio of the total system. M_1/M_t then becomes the ratio of the U dissolved in the water to the total U moving in the down-flow direction. Multiplying this factor by the age calculated for the moving U, yields the smaller age applicable to the water movement only.

The problem with this approach is that no measurements have been made on the $^{234}U/^{238}U$ activity ratio of the adsorbed U, nor is the activity ratio of the total mobile U system known. (Perhaps this could be determined by leaching a core sample.)

In the case of the Carrizo Sandstone aquifer, the activity ratio values level off at about 2.0 far down dip (Fig. 9.22), where the total activity ratio for the U in the system must be near 1.0. If 0.8 is arbitrarily selected as the activity ratio of the adsorbed U (a number suggested by the activity ratio of soluble U up-dip in oxidized waters), it can be calculated from eqn (9.11) that 1/6 is the fraction of U in solution at any one time. If this factor is multiplied by the movement rate of the total U (see Fig. 9.22) a water flow rate is obtained (about 1 m/y) which is more consistent with the value determed by ^{14}C dating for the up-dip part of the aquifer.

10   SURFICIAL CONTINENTAL SEDIMENTS
B. J. Szabo and J. N. Rosholt

10.1 Introduction

A significant part of the terrestrial surface of the earth is covered by surficial sediments which define much of the landscape occupied or used by man. The geochemical behaviour of U and Th in these loosely to well indurated surface and subsurface deposits and its application to Quaternary geochronology are discussed in this chapter. Studies have been reported recently concerned with U and Th transport in various eolian, volcanic, alluvial, lacustrine, glacial, colluvial materials as well as *in situ* weathered rocks and minerals. The carbonates in fracture infillings and interstitial lake precipitates occur frequently as integral parts of these materials. Uranium-series dating of these sediments is important because estimation of the time of deposition permits correlation of deposits between different areas and climatic regimes, provides a time frame for the study of geomorphic processes, and helps in the reconstruction of paleoclimatic and paleohydrologic conditions during the Quaternary. Furthermore, Quaternary tectonic activities may be assessed through the study of the history of movements of faults by dating carbonates which are deposited in fractures, by determining the time of deposition of surficial sediments, or by dating carbonates cementing the alluvial materials displaced by those faults. All these studies have become of increasing importance in the evaluation of landform stabilities, rates of faulting, geological and earthquake hazards, nuclear reactor site studies, and site selection for nuclear waste storage in geological media.

Two entirely different approaches have been applied successfully to dating surficial sediments. In calculating the U-series ages of carbonate deposits, the samples are assumed to have remained ideal, closed systems throughout their geological history, that is, there has been no post-depositional migration of U isotopes nor their *in situ* produced long-lived daughter products, ^{230}Th and ^{231}Pa. In U-trend dating of the time of sediment deposition, both U and Th exhibit open system behaviour, and therefore, no restrictions can be imposed on post-depositional migration of any nuclides into or out of the sedimentary units. An empirical model is used for the age calculation yielding U-trend lines that are calibrated by other radiometric methods.

10.2 Surficial carbonate deposits

10.2.1 Occurrences and dating potential of various authigenic carbonates

Inorganically precipitated carbonate materials are common components of surficial sediments in arid and semi-arid regions of the earth. These materials can be formed in a wide variety of ways, and therefore, there are many terms in the literature to describe them (Goudie 1973). They can accumulate in the zones of soil formation in different physical forms depending on their genesis and relative stages of development as well as form massive, dense deposits in alluvium, fill fractures, precipitate in lakes, and springs.

It is well known that U is readily transported in ground water mainly as carbonate complexes, in contrast, Th and Pa, the long-lived daughter elements of U, hydrolyse readily and precipitate or adsorb on the matrix material through which the ground water passes. When $CaCO_3$ is precipitated in the subaerial environment, U will be coprecipitated. Accordingly, such carbonates are potentially datable materials by the U-series methods: $^{231}Pa/^{235}U$ dating from about 1 to 180 ky, $^{230}Th/^{234}U$ dating from about 1 to 350 ky, and $^{234}U/^{238}U$ dating from about 100 to 1200 ky using standard mathematical relations discussed in chapter 3.

For the purpose of this discussion, the various precipitated secondary carbonate deposits are broadly classified into eight groups and listed in Table 10.1. With the exceptions of some travertines, tufas, and most calcite veins, these secondary carbonates are extremely impure, and contain various amounts of detrital materials mixed together with and cemented by the

Table 10.1. Descriptive classification of common surficial carbonates

Carbonate type	Description
Travertine	Hard, dense, and mostly finely crystallized carbonates precipitating from ground water (excluding all types of cave deposits).
Tufaceous travertine	Soft and porous variety of ground water precipitated travertine.
Calcite vein	Dense and often finely banded $CaCO_3$ lining or filling fractures of host rocks.
Caliche or soil-caliche	The secondary accumulations of cementing carbonates in the host materials within the zones of soil development.
Caliche-rind	The hard and dense carbonate coating on pebbles in the zones of weathering and soil formation.
Calcrete	Massive surficial conglomerates of cemented rock fragments and minerals where soil morphology is not present.
Lacustrine marl	The authigenic carbonates precipitating in lakes.
Tufa	The $CaCO_3$ deposits that display fossilized vegetable mats indicating that they are originating from spring discharges.

carbonates. Such calcareous materials are heterogeneous mixtures of at least three recognizable phases: (i) the original host or matrix material, mineral grains, and rock fragments of various sizes; (ii) the non-carbonate authigenic materials of different ages, such as, clay minerals, zeolites, and opaline silica; and (iii) the authigenic calcium carbonate component. Clearly, bulk analysis of a whole-rock sample cannot yield a meaningful age; and thus the authigenic carbonate component must be separated from the rest of the non-carbonate detritus for dating. These age determinations provide an often poorly defined average age for the process of accumulation and one must recognize that there can be different ages of authigenic carbonate cements in a single profile (Nettenberg 1978). A further complication is that the host materials may contain significant amounts of old, primary limestone or dolomite fragments (Blank and Tynes 1965) which must be separated from the secondary carbonate before chemical processing.

Because there are no simple physical or chemical means to separate the carbonate from the non-carbonate component completely, there are few published reports on dating these impure but geologically often significant materials. A relatively direct approach for dating impure carbonates is to carry out an acid leaching of the whole-rock sample, and then to analyse the acid soluble, mainly calcium carbonate fraction. To date by the acid leaching method, the authigenic carbonate must be assumed to have had initially negligible amounts of ^{230}Th and ^{232}Th, and therefore, that all ^{230}Th has been produced by the radioactive decay of the U isotopes in the carbonate phase, thus defining its age. Consequently, any ^{232}Th found in a sample must represent contribution from the detrital component via acid dissolution. Because detrital ^{230}Th is also leached by the acids together with the ^{232}Th, a correction is required for this non-carbonate-originated ^{230}Th. In addition, some U can also be leached from the detritus during the chemical processing and a correction for this effect is also necessary.

Kaufman (1971) reported successful dating of lacustrine marls of the Dead Sea Basin. He analysed the acid-leached fractions of a number of samples from each layer containing different amounts of detritus relative to the carbonate fractions. The correction for the detrital contribution of Th and U is performed using linear plots of activity ratios ^{230}Th/^{234}U and ^{234}U/^{238}U against ^{232}Th/^{234}U. Extrapolation of the lines to ^{232}Th/^{234}U $= 0$ (that is, ^{232}Th $= 0$) yields the ^{230}Th/^{234}U and ^{234}U/^{238}U activity ratios of the pure secondary carbonate fraction from which the corrected ^{230}Th age is calculated. Examples of this procedure for dating are presented in Fig. 10.1 using selected analytical data of Kaufman (1971). This procedure is suitable for dating samples from stratified deposits exposed at different localities. It is seldom useful for dating impure carbonates from single localities where all of the samples may have a similar U content and similar amounts of detrital components, and consequently, the analyses of the aliquots of a sample do not

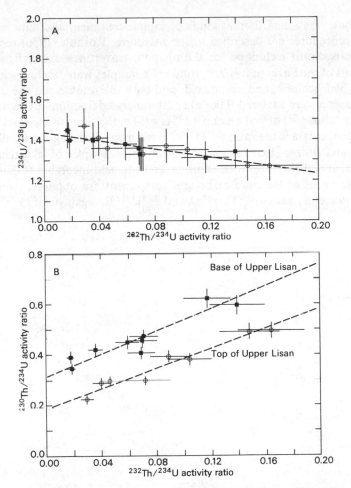

Fig. 10.1. Plots of activity ratios $^{234}U/^{238}U$ (A) and $^{230}Th/^{234}U$ (B) against $^{232}Th/^{234}U$ of Upper Lisan samples of Dead Sea Basin from sites I and IV using analytical data of Kaufman (1971). Open circles signify samples from top of formation, full squares represent samples from base of formation. Cross bars indicate 2 sigma experimental errors; lines are obtained by least squares fitting (York 1969). Calculated detritus corrected age of top of Upper Lisan is 21 ± 2 ky (1σ); calculated detritus corrected age of base of Upper Lisan is 41 ± 3 ky (1σ).

provide sufficient spread on isochron plots needed for the age correction.

Another approach to dating detritus-bearing carbonates is the multiple analyses of total rock samples. The isotopic composition of U and Th of both the acid-soluble carbonate and the acid-insoluble residue fractions are determined, and the analytical results for the residues are used to correct for the detrital contribution of ^{230}Th, ^{234}U, and ^{238}U to the acid-soluble carbonate fraction. Different kinds of sample treatment and data interpretation procedures are described in the literature. Rosholt (1976) reported a pseudo-isochron technique for dating pure travertines and caliche rinds. Aliquots of the finely pulverized, total rock samples were leached with dilute nitric, hydrochloric, and acetic acid, and both the soluble and the insoluble components were analysed. The data of the analysed fractions are plotted with activity ratios ^{230}Th/^{232}Th against ^{234}U/^{232}Th; the slope of the resulting line through the data is the ratio ^{230}Th/^{234}U from which the ages are calculated. Szabo and Butzer (1979) reported dating single samples of lacustrine marl. They analysed the total rock samples and the acid-insoluble residues. The isotopic ratios of the pure carbonate component are obtained from linear plots of activity ratios ^{230}Th/^{234}U and ^{234}U/^{238}U against ^{232}Th/^{234}U.

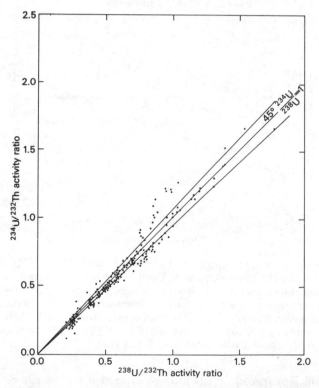

Fig. 10.2. Plots of activity ratios ^{234}U/^{232}Th against ^{238}U/^{232}Th for samples from various eolian, alluvial, colluvial, and glacial sediments mainly Quaternary in age. The 45 degree line signifies the loci of samples in which ^{238}U and ^{234}U are in secular equilibrium (or ^{234}U/^{238}U = 1); thin lines show 6 per cent experimental error envelope. Analytical data are provided by Rosholt et al. (1966); Hansen and Stout (1968); Hansen (1970); Ku et al. (1979); and Rosholt (1980b).

A slightly different approach to dating of caliches was taken by Ku *et al.* (1979). These authors leached caliche rind samples from desert carbonate soils by dilute hydrochloric acid and analysed both the leachate and the residue fractions. A mathematical model was devised for correction of detrital ^{230}Th contamination assuming that the minerals in the acid-insoluble fraction are in secular equilibrium with respect to ^{238}U, ^{234}U, and ^{230}Th. To examine the general validity of this assumption, the activity ratios of ^{234}U/^{232}Th against ^{238}U/^{232}Th (see Fig. 10.2) and ^{230}Th/^{232}Th against ^{234}U/^{232}Th (see Fig. 10.3) of surficial-sediment samples, mostly Quaternary in age, are plotted. These various eolian, alluvial, colluvial and glacial deposits are typical materials in which carbonate cementation can occur as a function of rates of both influx and redistribution. Samples grouping about the 45 degree lines in Figs 10.2 and 10.3 contain ^{238}U, ^{234}U, and ^{230}Th in secular equilibrium.

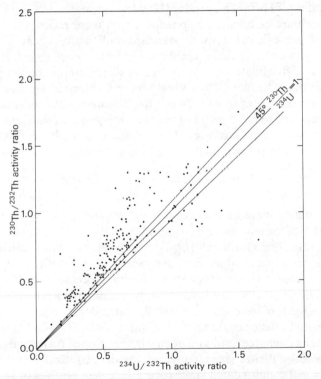

Fig. 10.3. Plots of activity ratios of ^{230}Th/^{232}Th against ^{234}U/^{232}Th for samples from various eolian, alluvial, colluvial, and glacial sediments mainly Quaternary in age. The 45 degree line shows the loci of samples in which ^{230}Th and ^{234}U are in equilibrium (or ^{230}Th/^{234}U = 1); envelope of 6 per cent experimental error is also shown. Data after Rosholt *et al.* (1966); Hansen and Stout (1968); Hansen (1970); Ku *et al.* (1979); and Rosholt (1980*b*).

Accepting a value of 6 per cent for average experimental error of these ratios, one observes that about two out of three weathered surficial samples can be expected to contain ^{238}U and ^{234}U in secular equilibrium (see Fig. 10.2), and that only about one out of four sediment samples are found to contain ^{234}U and ^{230}Th in equilibrium (see Fig. 10.3). From the published data, excess ^{230}Th with respect to its parent nuclide ^{234}U appears to be the rule rather than the exception in weathered rock fragments and minerals of different origin.

To avoid the necessity of assuming secular equilibrium between the isotopes of U and Th in detrital materials, Szabo and Sterr (1978) and Szabo, Carr, and Gottschall (1981) used simple mixing line plots of the respective acid-soluble solution and acid-insoluble residue pairs to date different kinds of detritus-bearing carbonates at the Nevada Test Site region. The slope of a plot of activity ratios ^{234}U/^{232}Th against ^{238}U/^{232}Th yields the ^{234}U/^{238}U activity ratio of the pure carbonate component, and the slope of a plot of activity ratios ^{230}Th/^{232}Th against ^{234}U/^{232}Th yields the ^{230}Th/^{234}U activity ratio of the pure carbonate component. From these ratios, the ages of the carbonates are calculated using standard radioactive decay and growth equations. Examples of these pseudo-isochron plots are shown in Fig. 10.4 using the analytical data reported by Ku et al. (1979) for the samples of the upper Pleistocene surface Q2b of Vidal Valley, California. The average age of the soil carbonate samples of Q2b soil horizon using the model of Ku et al. (1979) is reported to be 83 ± 10 ky (1σ); the average age of the same unit using the pseudo-isochron plots of Fig. 10.4 is calculated to be 75 ± 5 ky (1σ), an age which is in agreement within analytical uncertainties.

10.2.2 Examples of U-series dating of carbonates associated with surficial sediments

Results of dating more than 30 samples of various carbonate accumulations associated with tectonic movements at the Nevada Test Site area are reported by Szabo, Carr, and Gottschall (1981). Nearly all travertines, calcite crystals, and veins yield reasonable ages, but some of the tufas and tufaceous travertines appear to form an open system with respect to both U and Th, and therefore, those results may indicate only minimum ages (see Table 10.2). Other investigators have also successfully dated dense, coarsely crystallized travertines and calcite vein samples. Schwarcz et al. (1979) and Harmon et al. (1980) obtained reliable results on open-air precipitated travertines associated with archeological sites (see chapter 11 for details). Szabo and Winograd (1980, personal comm.) dated laminae of calcite vein samples in Devils Hole, Nevada. The analyses indicate that the water last flowed from the mouth of the cavern between 750 to 400 ky. This finding is of interest to ichthyologists concerned with rates of vertebrate evolution displayed by the pupfish (Cyprinodont diabolis) trapped in Devils Hole after it ceased to flow.

Szabo, Carr, and Gottschall (1981) reported dating dense, surficial

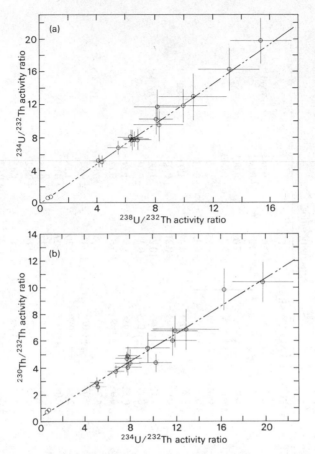

Fig. 10.4. Plots of activity ratios $^{234}U/^{232}Th$ against $^{238}U/^{232}Th$ (a) and $^{230}Th/^{232}Th$ versus $^{234}U/^{232}Th$ (b) of caliche-rind samples of the upper Pleistocene surface Q2b of Vidal Valley, California using analytical data of Ku *et al.* (1979). Samples grouping at lower left corners are the acid-insoluble residues; open circles represent acid-soluble carbonate samples; cross bars indicate 2 σ experimental errors; lines obtained by least squares fitting (York 1969). From the slopes of the lines, that is, $^{234}U/^{238}U$ and $^{230}Th/^{234}U$, the detritus corrected isochron-plot age of the geomorphic surface is calculated to be 75 ± 5 ky (1 σ).

calcretes (see Table 10.2). Some of these samples yield reliable ages for the cementation of the alluvium as confirmed by geological considerations and K/Ar ages, but others appear to reflect minimum ages only. They also reported that the analyses of samples classified as soil caliches yield questionable, young ages, probably because soil carbonates tend to evolve from less mature to more developed stages. Samples from such caliche deposits may yield better age estimates using a modified version of the U-trend

Table 10.2. Analytical data and U-series ages of carbonates at southern Nevada

Sample no.	Material	Per cent residue	Fraction	U (ppm)	^{234}U/^{238}U	^{230}Th/^{232}Th	^{230}Th/^{234}U	U-series age[10] (ky)
						(activity ratios)		
MERCURY VALLEY AREA								
30-A[1]	Trav	0	S	0.038 ±0.001	1.00 ±0.02	43.0 ±10.0	0.844 ±0.042	>700[11]
30-B[2]	Trav	0	S	0.066 ±0.001	0.987 ±0.030	77.0 ±15.0	0.992 ±0.040	>700[11]
32	Trav	0	S	0.016 ±0.002	1.00 ±0.05	4.05 ±0.41	1.12 ±0.11	>700[11]
45-A[3]	Trav	0	S	0.495 ±0.007	0.981 ±0.015	6.92 ±0.21	1.00 ±0.03	>700[11]
45-C[4]	Trav	5	S	1.10 ±0.17	0.999 ±0.015	10.2 ±0.3	1.06 ±0.03	>700[11]
			R	0.52 ±0.05	0.95 ±0.09	5.5 ±0.6	1.21 ±0.18	
47-A[5]	Trav	20	S	4.71 ±0.07	1.12 ±0.02	5.08 ±0.15	0.846 ±0.025	104±8
			R	2.84 ±0.09	0.980 ±0.015	2.20 ±0.07	1.58 ±0.05	
47-B[6]	Trav	22	S	3.16 ±0.05	1.17 ±0.02	5.37 ±0.16	0.957 ±0.029	97±8
			R	2.51 ±0.05	1.00 ±0.05	3.42 ±0.10	1.42 ±0.07	
48	Calcr	18	S	3.17 ±0.05	1.06 ±0.02	7.65 ±0.23	0.730 ±0.022	100–150
			R	21.1 ±0.3	1.01 ±0.02	16.7 ±0.5	1.08 ±0.03	

46	TTrav	23	S	5.19 ±0.08	1.04 ±0.02	38.2 ±1.5	0.999 ±0.030	≥70
			R	10.6 ±0.9	1.05 ±0.02	34.0 ±1.4	1.15 ±0.12	
31-A[7]	Calcr	70	S	19.7 ±0.3	1.45 ±0.02	13.1 ±0.4	0.750 ±0.030	102 ±8
			R	5.61 ±0.08	1.37 ±0.02	4.38 ±0.13	1.16 ±0.05	
31-B[7]	Calcr	75	S	2.21 ±0.03	1.10 ±0.02	4.66 ±0.14	0.751 ±0.030	96 ±8
			R	1.86 ±0.03	1.07 ±0.02	1.45 ±0.04	1.79 ±0.07	

ROCK VALLEY FAULT AREA

154	TTrav	55	S	2.23 ±0.03	1.17 ±0.02	5.03 ±0.13	1.72 ±0.09	—[12]
			R	0.630 ±0.010	1.03 ±0.02	2.48 ±0.07	1.98 ±0.10	
155	TTrav	80	S	4.96 ±0.07	1.27 ±0.02	9.73 ±0.29	0.650 ±0.033	70 – 110
			R	18.8 ±0.3	1.31 ±0.02	23.4 ±0.7	0.790 ±0.040	
40	Calcr	45	S	6.90 ±0.10	1.13 ±0.02	16.4 ±0.7	0.435 ±0.013	≥20
			R	12.3 ±0.2	1.13 ±0.02	11.8 ±0.4	0.824 ±0.025	
82	SC	62	S	3.79 ±0.06	1.41 ±0.02	1.81 ±0.07	0.0747 ±0.0022	>5
			R	3.02 ±0.05	1.24 ±0.02	1.10 ±0.04	1.06 ±0.04	

Table 10.2 (*contd.*)

Sample no.	Material	Per cent residue	Fraction	U (ppm)	$^{234}U/^{238}U$	$^{230}Th/^{232}Th$ (activity ratios)	$^{230}Th/^{234}U$	U-series age[10] (ky)
97	SC	48	S	1.19 ±0.02	1.35 ±0.02	1.33 ±0.05	0.119 ±0.004	>5
			R	2.91 ±0.04	1.20 ±0.02	1.22 ±0.05	1.44 ±0.06	
JACKASS FLAT								
Stop-9	SC	90	S	3.75 ±0.06	1.34 ±0.02	3.28 ±0.10	0.240 ±0.010	~24
			R	2.79 ±0.04	1.17 ±0.02	0.961 ±0.029	1.14 ±0.05	
LATHROP WELLS AREA								
60-A[8]	Calcr	35	S	5.32 ±0.08	1.23 ±0.02	6.98 ±0.21	1.05 ±0.03	345 +180 −71
			R	3.73 ±0.06	1.06 ±0.02	1.46 ±0.04	1.14 ±0.03	
60-B[9]	Calcr	46	S	5.37 ±0.08	1.20 ±0.02	5.62 ±0.17	1.09 ±0.03	345 +180 −70
			S	5.03 ±0.10	1.20 ±0.02	5.53 ±0.17	1.03 ±0.05	
			R	6.11 ±0.09	1.10 ±0.02	1.62 ±0.05	1.04 ±0.03	
59	SC	47	S	2.14 ±0.03	1.53 ±0.02	1.99 ±0.06	0.425 ±0.013	25 ±10
			R	1.94 ±0.03	1.21 ±0.02	1.13 ±0.03	1.72 ±0.05	

YUCCA MOUNTAIN AREA

Sample	Rock		S/R					
113	Calcr	75	S	2.78 ±0.04	1.03 ±0.02	0.687 ±0.027	0.146 ±0.006	>5
			R	4.17 ±0.06	0.986 ±0.015	0.620 ±0.002	0.837 ±0.025	
115	Calcr	80	S	10.6 ±0.2	1.46 ±0.03	16.7 ±1.7	0.411 ±0.017	>20
			R	9.44 ±0.14	1.51 ±0.02	22.4 ±0.9	1.19 ±0.04	
106	TTrav	70	S	9.53 ±0.14	1.26 ±0.02	4.53 ±0.14	0.660 ±0.026	78 ±5
			R	3.66 ±0.05	1.33 ±0.02	2.43 ±0.07	0.860 ±0.034	
UE-34	CV	0	S	1.37 ±0.03	1.17 ±0.02	2.37 ±0.07	1.02 ±0.04	>400
UE-611	CV	0	S	8.98 ±0.18	1.47 ±0.02	22.2 ±0.7	1.04 ±0.04	>400
UE-283	CV	0	S	6.12 ±0.12	1.29 ±0.02	72.0 ±3.0	1.19 ±0.05	>400

CRATER FLAT

Sample	Rock		S/R					
199	TTrav	30	S	1.81 ±0.03	2.16 ±0.03	2.57 ±0.08	0.290 ±0.012	~30
			R	1.58 ±0.02	1.17 ±0.02	2.81 ±0.08	1.47 ±0.06	

ELEANA RANGE

Sample	Rock		S/R					
H-1	SC	45	S	6.51 ±0.10	1.21 ±0.02	2.95 ±0.09	0.877 ±0.026	128 ±20
			R	4.09 ±0.06	1.10 ±0.02	1.37 ±0.04	1.18 ±0.04	

Table 10.2 (contd.)

Sample no.	Material	Per cent residue	Fraction	U (ppm)	$^{234}U/^{238}U$	$^{230}Th/^{232}Th$ (activity ratios)	$^{230}Th/^{234}U$	U-series age[10] (ky)
H-2	SC	40	S	16.8 ±0.3	1.34 ±0.02	2.90 ±0.09	0.0806 ±0.0024	>5
			R	11.4 ±0.2	1.35 ±0.02	1.82 ±0.06	0.331 ±0.010	
BOUNDARY FAULT AREA								
50	SC	40	S	6.21 ±0.09	1.37 ±0.02	11.2 ±0.5	0.242 ±0.007	≥24
			R	4.36 ±0.07	1.34 ±0.02	4.56 ±0.14	0.279 ±0.008	
51	SC	31	S	4.77 ±0.07	1.37 ±0.02	1.62 ±0.05	0.164 ±0.005	>8
			R	3.50 ±0.50	1.20 ±0.02	1.07 ±0.03	0.587 ±0.018	

(1) The outside, oldest part of travertine vein TSV-30
(2) The centre, youngest part of travertine vein TSV-30
(3) The outside, oldest part of travertine vein TSV-45
(4) The centre, youngest part of travertine vein TSV-45
(5) Sample represents 1/3 of full vein width of sample TSV-47
(6) Sample represents 2/3 of full vein width of sample TSV-47
(7) Different aliquots of calcrete cement sample TSV-31
(8) Inner part of dense carbonate rind growing on cobbles
(9) The softer and porous outer part of the same rind as 8
(10) Ages are calculated using pseudo-isochron plots of respective acid-soluble and acid-insoluble fractions
(11) Based on measured $^{234}U/^{238}U$ activity ratios
(12) Age cannot be calculated

Trav = travertine; TTrav = tufaceous travertine; Calcr = calcrete; SC = soil caliche; CV = calcite vein; S = acid-soluble solution; R = acid-insoluble residue. (After Szabo, Carr, and Gottschall 1981.)

method that allows for the migration of U and Th in the sedimentary column (Szabo *et al.* 1980).

Attempts to date lacustrine marl samples have been largely successful. Kaufman (1971) substantiated the age ranges of the lacustrine carbonate of the Lisan Formation of Israel. Hendy, Healy, Rayner, Shaw, and Wilson (1979) established the late-Pleistocene glacial history of Taylor Valley, Antartica, by U-series dating of lacustrine marl samples. Similarly, dating of dense caliche-rind samples appears to yield consistent results that agree with geological evidence. Ku *et al.* (1979) obtained ages for different geomorphic surfaces of California by analysing caliche-rind samples. Pierce and Rosholt (1980, personal comm.) dated caliche rinds on boulders from fan deposits near Arco, Idaho. The ^{230}Th/^{234}U ages of these rinds increase with increasing depth in the rind; the obtained dates range from 10 ky at the outer, youngest lamina of the rind to 100 ky at the innermost, oldest part of the deposit. These dates yield significant results for the evaluation of the timing and the frequency of Quaternary faulting in the region, and the rate of caliche rind accumulation for this climatic environment (see Fig. 10.5). Recent study of saline deposits of Searles Lake, California (Peng, Goddard, and Broecker 1978) indicates that bulk salt samples are datable materials by the U-series method thus revealing new ways to establish the chronology of pluvial lake sedimentary units.

In summary, U-series dating of inorganically precipitated carbonates, other than soil caliches, shows promise of yielding useful results for a number of

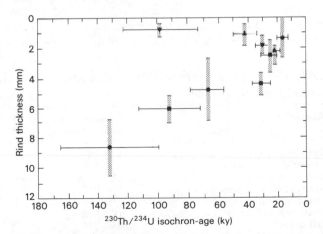

Fig. 10.5. The ^{230}Th/^{234}U age against depth within caliche rind (measured from outer layer inward) on fan gravels along Arco, Idaho, fault scarp. Vertical bar represents the interval of the rind that was sampled. Analyses indicated by(●) are from 3 concentric layers on a gravel with a large rind thickness. (■) are from 3 concentric layers on an intermediate rind thickness gravel. (▼ and ▲) are analyses from 2 concentric layers on 2 gravels with similar thin rind sequences; for each of these gravels, the outer lamellar rind did not behave as a closed system for U migration (i.e. suffered U loss).

geological applications by establishing the chronology of discrete events. However, the presence of detrital contamination always introduces uncertainty to the reliability of the dates, and therefore, it is good practice to select the densest part of the sample that is relatively free of non-carbonate contaminants. Furthermore, stratigraphic controls are still the basis against which the obtained dates must be tested. There are no absolutely certain ways to assess the correctness of the primary assumptions of this type of dating such as the assumption that the initial ^{230}Th activity is negligibly small in the samples, or that the dated carbonates remained ideally closed with respect to the isotopes of interest.

10.3 Sediments and soils

10.3.1 The behaviour of U in sediments

Significant fractionation exists between ^{234}U and ^{238}U in nearly all natural materials influenced by surface and subsurface waters (Szabo 1969; Cherdyntsev 1971; and see Table 10.3). Recent studies of U-series disequilibrium in various altered rocks indicates that U isotopes commonly exhibit open system behaviour (Rosholt 1980a). The distribution of the U-series members associated with the loosely consolidated alluvial, lacustrine, glacial, eolian, and other surficial sediments is also controlled by open system behaviour during depositional processes and during the ensuing weathering and soil-forming cycles (see Figs 10.2 and 10.3, and Rosholt 1980b).

Uranium occurs in two different major phases in the geological environment. A stationary or fixed phase (solids-dominated) exists as structurally

Table 10.3. Typical range in ^{234}U/^{238}U activity ratios for various natural materials

Material	Range in ^{234}U/^{238}U activity ratio
Open-ocean waters	1.10–1.18
Terrestrial surface waters	0.80–2.50
Underground waters	0.60–12.00
Waters of uranium mineralization	1.20–8.80
Various surficial carbonates	0.90–3.00
Fossil shells and bones	1.00–2.50
Peat deposits	0.90–2.00
Igneous rocks	0.60–2.10
Volcanic tuffs	0.50–1.60
Sandstones	0.80–2.00
Minerals and extracts of minerals	0.80–8.00
Soils	0.70–1.20

(After Szabo 1969; Cherdyntsev 1971; Osmond and Cowart 1976a; Ku et al. 1977; Rosholt 1980a.)

incorporated U in the matrix minerals. A second important part is a mobile phase (water-dominated) that constitutes the U flux that penetrates the permeable components of a sedimentary deposit. This mobile-phase U is responsible for isotopic fractionation processes in the U-series that enables the U-trend dating technique to work. Most of the deposits from semi-arid environments are dry for longer periods of time than they are wet. Nevertheless, there is a migrating component of U either in solution or on colloids which slowly works its way through the solid framework which in turn remains in place but undergoes weathering alterations. Much of this mobile-phase U spends most of its time on the surface of dry solid grains, leaving its trail of daughter products in a deposit, and only a small amount of the time in solution or suspension in actual flow through a deposit. As a deposit experiences interstratal fluids, some U and Th isotopes are leached preferentially or are etched out of the stationary phase, and join the mobile phase.

An open-system environment has a large number of geochemical variables, and therefore, the definition of the U migration within the framework of a rigorous mathematical model is not feasible. However, some observations concerning the mechanism of U isotopic fractionation can be described at this time. As dissolved ^{238}U and ^{234}U atoms decay by alpha disintegration, recoiling daughter nuclides of ^{230}Th, ^{234}Th, ^{234}Pa, and ^{234}U are adsorbed or driven into particulate matter at the solid–liquid interface. After sufficient time, this mechanism (daughter-emplacement) results in solids which have ^{234}U/^{238}U and ^{230}Th/^{238}U activity ratios significantly higher than the equilibrium ratio of unity. An alternative mechanism (^{234}U displacement) exists by which ^{234}U produced by the decay of structurally incorporated ^{238}U is selectively displaced from mineral surfaces as a leached or alpha-recoiled decay product. The ^{234}U displacement mechanism has been extensively documented in natural samples (Osmond and Cowart 1976a), and laboratory investigations of the effects of recoiling alpha-emitting nuclei have been reported (Fleischer and Raabe 1978). Kigoshi (1971) has shown that the concentration of ^{234}Th, produced by alpha decay of ^{238}U contained in fine-grained zircon crystals, increased with time in solutions in which the zircon crystals were dispersed. Fleischer (1980) reported experiments which indicate that, in addition to the direct recoil ejection into the liquid interface, a second mechanism of ^{234}U displacement can result when many of the recoiling nuclei which are ejected from mineral grains become imbedded in adjacent grains and produce alpha-recoil tracks. Subsequent track etching by natural solutions releases some of these recoiled nuclei. The recoiling nuclei accompanying alpha decay have ranges in solids of about 200 Å (Fleischer and Raabe 1978).

In contrast to ^{234}U displacement, little attention has been given to the counter process which involves preferential gain of ^{234}U relative to ^{238}U on particulate matter (daughter-emplacement). The gain results from recoiling nuclei ejected from solution at the water–particle interface, and then imbedded in adjacent grains. Most of the previously analysed alluvial units from semi-arid environments have excess ^{234}U relative to ^{238}U. This isotopic anomaly is very noticeable in altered, 0.6 to 2 my old, volcanic ashes and tuffs, from a semi-arid environment at Lake Tecopa, California (Rosholt 1980a). Zeolites formed during an early phase of alteration show pronounced effects of ^{234}U emplacement which probably are enhanced by the large surface areas of zeolites. Zeolite components formed during alteration of 10 my old tuffs of Keg Mountain, Utah, however, are 50 per cent deficient in ^{234}U relative to ^{238}U (Zielinski et $al.$ 1980). These results suggest that, given sufficient geological time, ^{234}U displacement mechanisms become predominant over ^{234}U emplacement mechanisms. In the zeolites from tuffs of Keg Mountain, the displacement processes are so effective that one half of all ^{234}U atoms, produced by the alpha decay of the structurally incorporated ^{238}U atoms, were released from these minerals.

Interpretation of radioactive disequilibrium in zeolitically altered glass at Lake Tecopa and the Keg Mountain area indicates that the predominant initial process was selective emplacement of ^{234}U and ^{230}Th. Two separate mechanisms may be responsible for emplacement of ^{234}U and other daughter products from parent atoms in solution: (i) some of the recoiling nuclei ejected at the pore water–particle interface are imbedded in adjacent grains, and (ii) some of the precursors of ^{234}U, ^{234}Th and ^{234}Pa, and ^{230}Th are adsorbed on surfaces of particulate matter at this interface. Beta decay of short-lived ^{234}Th and ^{234}Pa would leave some of their daughter ^{234}U atoms implanted on the particulate matter. These ^{234}U emplacement atoms would not be bound strongly to the solid surfaces and subsequently, over extended periods of time, significant fractions could be leached back into the water phase at rates greater than their disintegration rate ($t_{\frac{1}{2}} = 2.48 \times 10^5$ y). Gradual leaching of em-placed ^{234}U would be accompanied by the displacement mechanism where the ^{234}U produced by decay of structurally incorporated ^{238}U was selectively released from mineral surfaces by recoiling alpha-emitting nuclei and by natural etching of recoil tracks. Initial emplacement by recoil-sorption processes was probably limited by the concentrations of dissolved species and by the sorptive capacity of the solids. In contrast, displacement by recoil-leaching was controlled by the concentrations in the solid phase, and by the solubility of leached isotopes. Apparently continuous exposure of leachable ^{234}U sites during glass alteration and higher U content in the solid phases result in eventual domination of the displacement mechanism.

10.3.2 Uranium-trend dating method

Radioactive disequilibrium studies of the ^{238}U–^{234}U–^{230}Th system in soil samples have been used previously to study the migration of U and Th as a result of rock weathering (Hansen and Stout 1968), and to determine the patterns of the evolutionary trends of U and Th isotopic ratios in soils during the early stages of alterations of the transported parent materials (Rosholt *et al.* 1966). Hansen (1965) used this isotopic system to develop a model for estimating the age of soil development in San Joaquin Valley and Sierra Nevada, California. Gamma ray spectrometry, of both ^{238}U and ^{235}U decay series members, was used to measure radioactive isotope concentrations in soil on weathered granitic rock in order to estimate the weathering age of the soil samples (Megumi 1979). Determination of ^{226}Ra, as an indicator of parent ^{230}Th, and determination of ^{232}Th in soil profiles, both by gamma ray spectrometry, have been used by Hansen and Huntington (1969) to trace Th movement in morainal soils. The use of ^{234}U/^{238}U activity ratios in measuring rates of rock weathering has been proposed by Moreira-Nordemann (1980). Rate of formation of desert varnish has been studied by Knauss and Ku (1980) to establish the U-series dating potential of these weathered rock surfaces.

Rosholt (1980*b*) has investigated extensively the U–Th radioactive disequilibrium system in 15 different sedimentary deposits, some of which include multiple depositional units of different ages. The deposits are from environments which vary in climate from arid to humid. From the results of these U and Th isotopic analyses, a new concept for dating the deposition of the various surficial sediments, with or without recognizable soil development, has been developed. This model for dating, U-trend dating, does not require the standard closed-system assumptions.

Because there are a large number of variables in a system open with respect to migration of U, a rigorous mathematical model based on the equations describing the radioactive growth and decay of daughter elements cannot be constructed. Instead, an empirical model is used for the calculation of U-trend ages on several alluvial, glacial, and eolian units ranging in age from about 4 to about 720 ky. The model requires time calibration of both the isochron slope and a U-flux factor based on results from deposits of known age. The actual physical significance of the U-flux factor, $F(0)$, is not well understood. It is related to the mobile U flux through a deposit. The effect of this flux on isotope variations decreases exponentially with time. An oversimplified view of the flux in alluvium would be represented by a large volume of water accompanying deposition, followed by decreasing volumes of water available during alteration and compaction of the sediments, and subsequent soil development. Both the quantity of water affecting a deposit, and the concentration of U in this water are components of the flux; its magnitude is a function of the concentration of U in the mobile phase relative to the concentration of U in the stationary phase.

This open-system dating technique consists of determining an isochron from analyses of three to ten samples from various soil horizons in a given depositional unit. In each sample an accurate determination of the concentrations of ^{238}U, ^{234}U, ^{230}Th, and ^{232}Th is required. Whole samples are used and isotopic concentrations are determined by radioisotope dilution techniques and by alpha spectrometry measurements.

The slope of the line representing $\Delta(^{234}U-^{238}U)/\Delta(^{238}U-^{230}Th)$ yields the isochron value from which U-trend ages are calculated. A related slope used in the derivation of the model is $\Delta(^{234}U-^{238}U)/\Delta(^{234}U-^{230}Th)$. The two slopes are dependent because

$$\frac{(^{234}U-^{238}U)}{(^{238}U-^{230}Th)} = \frac{(^{234}U-^{238}U)}{[(^{234}U-^{230}Th)-(^{234}U-^{238}U)]} \tag{10.1}$$

and this form is used for computer solution of the age (derivation of equations used in the model is included in appendix E).

The results of these analyses are plotted in the form of $(^{234}U-^{238}U)/^{238}U$ against $(^{238}U-^{230}Th)/^{238}U$. Ideally, this yields a linear relationship, as shown in Fig. 10.6, in which the measured slope changes in a predictable way for deposits of increasing age. The rate of change of the slope is the function of the half-period of U-flux, $F(0)$, in the local environment. An additional parameter that can be obtained from this U-trend plot is the intercept, b', on the x-axis. The equation of a straight line having slope m and y-intercept b is $y = mx + b$, then the x-intercept, b', is $b' = b/m$. A different plot of the isotopic data can be constructed when the $^{238}U/^{232}Th$ ratios of the samples are plotted on the x-axis against the $^{230}Th/^{232}Th$ ratios plotted on the y-axis as shown in Fig. 10.7. The measured slope of the resulting line, $^{230}Th/^{238}U$, is called the Th-index, and it is another parameter used in this model. This is a plot similar to the isochron plot used by Allegre and Condomines (1976) for the dating of young volcanic rocks. The quotient of these two parameters, the x-intercept in Fig. 10.6, and the Th-index in Fig. 10.7, is used to obtain the time calibration for the empirical U-trend model where

$$\frac{x\text{-intercept}}{\text{Th-index}} = \frac{b'}{^{230}Th/^{238}U}. \tag{10.2}$$

The half period of $F(0)$ and its decay constant, λ_0 are strictly empirical values which allow selection of this variable exponential function in the equation for the U-trend model. For depositional units of unknown age, a method is required to determine the proper λ_0 for use in the equation; this is carried out with a calibration line based on λ_0 determined for units of known age. For the calibration, the quantity x-intercept/Th-index is plotted against the half-period of $F(0)$ on a log-log scale shown in Fig. 10.8. The calibration line is determined by selecting the proper λ_0 value that will yield the known age for a deposition unit using the U-trend model eqn (E.5) in appendix E. The values of x-intercept/Th-index are calculated for the deposition units used for calibration, and these values are plotted against the half-periods of $F(0)$ equivalent to their λ_0 values. The values used for the calibration line shown by index numbers in Fig. 10.8 are listed in Table 10.4. Time calibrations at 4, 12,

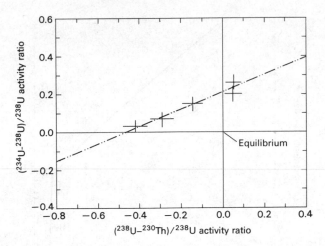

Fig. 10.6. Uranium-trend isochron of CCA section of alluvium, Shirley Mountains, Wyoming. (Reproduced from Rosholt 1980b.)

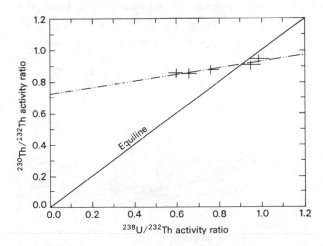

Fig. 10.7. Thorium-index plot of CCA section, Shirley Mountains, Wyoming. The slope of the line defines the 'Th-index'. (Reproduced from Rosholt 1980b.)

140, 600, and 720 ky are provided by ^{14}C and K/Ar correlations. The solution of the empirical equation using any given half period of F (0) yields a fan of U-trend slopes representing various ages. Figure 10.9 shows as an example the slopes for typical ages calculated for half-periods of 100 and 600 ky.

The ages determined for 16 depositional units using the U-trend method, are listed in Table 10.4. The procedure used to obtain the ages from isochron plots is summarized as follows: (i) determine the x-intercept (see Fig. 10.6); (ii)

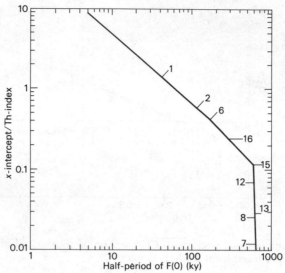

Fig. 10.8. Time-calibration line for determination of $F(0)$ from x-intercept/Th-index value. Indices on line show depositional unit number from Table 10.4.

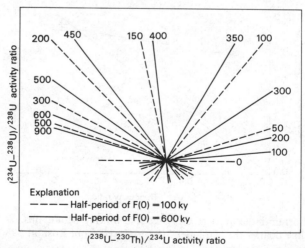

Fig. 10.9 Variation of U-trend slope with age of deposition for 100 and 600 ky half-periods of $F(0)$. Numbers on individual lines represent age (in ky).

determine the Th-index (see Fig. 10.7); (iii) using the ratio of x-intercept/Th-index, find the appropriate value for the half-period of $F(0)$ from calibration curve (see Fig. 10.8); (iv) using the half-period of $F(0)$ converted to λ_0 and the U-trend slope (see Fig. 10.6), calculate the age of the deposit by computer solution of eqn (E.5) in appendix E.

Table 10.4. Uranium-trend model parameters and ages of deposition units

Unit	Description of deposit	U-trend slope	x-intercept: Th-index	Half-period of $F(0)$ (ky)	Age (ky)
1	PC unit, Piney Creek Alluvium, Kassler quadrangle, CO.	+0.044	+1.39[1]	42	4^{+12}_{-4}
2	WF unit, upper Wisconsin loess, Wabasha County, MN.	+0.050	+0.570[1]	115	11 ± 7
3	FF unit, upper Wisconsin loess, Fillmore County, MN.	+0.080	+0.420	164	$\cdot25^{+40}_{-25}$
4	K unit, Wisconsin till, Mower County, MN.	+0.103	+0.463	145	28 ± 28
5	CCA section, alluvium, Shirley Mts area, WY.	+0.443	−2.33	23	16 ± 2
6	Parker section, lower unit Arapahoe County, CO.	+0.967	−0.431[1]	163	140 ± 23
7	P78 unit, Bull Lake end moraine, West Yellowstone, MT.	+0.114	−0.0120[1]	620	140 ± 70
8	P183 lower unit, Bull Lake loess, West Yellowstone, MT.	+0.136	−0.0258[1]	600	150 ± 25
9	P184 lower unit, Bull Lake moraine, West Yellowstone, MT.	+0.177	+0.0329	600	180 ± 80
10	GF section, upper part of soil unit, Golden fault zone, CO.	−2.52	−0.308	220	250 ± 25
11	GF section, lower part of soil unit, Golden fault zone, CO.	−0.764	−0.261	260	350 ± 40
12	Tuff A unit, Lake Tecopa, Inyo County, CA.	−0.488	+0.0700[1]	580	600 ± 60
13	Tuff B unit, Lake Tecopa, Inyo County, CA.	−0.383	+0.0269[1]	600	720 ± 120
14	Tuff C unit, Lake Tecopa, Inyo County, CA.	−0.177	−0.102	580	$\geqslant800$
15	R9 unit, upper Riverbank Formation, San Joaquin Valley, CA.	+0.130	−0.118[1]	580	140 ± 45
16	R28 unit, upper Riverbank Formation, San Joaquin Valley, CA.	+0.392	−0.239[1]	285	150 ± 20

Reproduced from Rosholt, (1980*b*).
[1] Denotes sedimentary units used for time calibration at 4, 12, 140, 600, and 720 ky. (See discussion and references in Rosholt 1980*b*.)

11 CARBONATE AND SULPHATE PRECIPITATES
M. Gascoyne and H. P. Schwarcz

11.1 Precipitation of carbonates and sulphates from ground and surface waters

Calcite, aragonite, and gypsum are commonly found in young, surficial and near-surface deposits, formed as precipitates from waters which have passed through aquifers capable of contributing Ca^{2+}, CO_3^{2-}, and SO_4^{2-} ions. In some cases, the U-series methods of analysis can be used for dating the rocks or sediments in which these phases occur.

Many natural waters are saturated or supersaturated with respect to calcite, aragonite and/or dolomite. This is especially true in areas where the bedrock is a sedimentary carbonate and ground or surface waters have had extensive opportunity to react with the rock. Typical examples of waters which have equilibrated with limestone aquifers are given by Langmuir (1971) and by Plummer, Vacher, Mackenzie, Bricker, and Land (1976). The dissolved load is controlled by, among other factors, the temperature, the amount of CO_2 available for dissolution in the water, the presence of other ions which can form stable complexes with Ca (for example, SO_4^{2-}), and the rate at which water moves through the aquifer. In general, the dissolved load increases with temperature and soil CO_2 levels; the latter is in turn controlled by density and type of vegetation.

Carbonates can precipitate from such mineralized waters as a result of various types of physical processes. The carbonate equilibrium

$$Ca^{2+} + 2HCO_3^- \rightleftharpoons CaCO_3 \downarrow + CO_2 (g) \uparrow + H_2O, \qquad (11.1)$$

can be displaced to the right by: (i) loss of CO_2 from the solution; (ii) evaporation of water; and (iii) increase of Ca^{2+} activity by breakdown of complexing ligands or dissociation of Ca complexes. However, eqn (11.1) is only a very simplified description of the chemical equilibria in the low temperature system Ca–H–C–O; temperature, pH, and concentrations of other ions in the solution can influence the degree of calcite saturation, in a manner too complex to allow a simple prediction as to when or where precipitation will occur. In addition, kinetic factors can prevent nucleation of calcite, or can inhibit calcite with respect to precipitation of aragonite or vaterite (Bischoff and Fyfe 1968). Concentrations of Mg^{2+} or SO_4^{2-} above 10^{-5} molar inhibit the growth of calcite so that increased supersaturation results in the crystallization of aragonite (Bischoff and Fyfe 1968). Certain organic substances may also have this effect. All polymorphs of calcium carbonate appear to behave similarly with respect to U-series dating, and therefore, statements in this chapter referred to calcite can be taken to be applicable to aragonite and vaterite unless specifically excepted.

Gypsum is most commonly found as an evaporite mineral resulting from the dehydration of sea water. However, minor amounts of this mineral are found as efflorescences and cements in and on continental deposits. These gypsum crusts and cements are generally produced by evaporation of ground or surface water enriched in Ca^{2+} and SO_4^{2-} ions by passage through or over bedrock containing gypsum deposits (evaporites) or passage of oxygenated water through or over rocks containing sulphide minerals. Typical of this latter process is the oxidation of pyrite as described by the reactions

$$FeS_2 + H_2O + 7/2O_2(s) = Fe^{2+} + 2H^+ + 2(SO_4^{2-}) \qquad (11.2)$$
and $\qquad Fe^{2+} + 1/4O_2 + H^+ = Fe^{3+} + 1/2H_2O.$

These reactions can be observed in weathered outcrops of sulphide-bearing sedimentary rocks. Gypsum is also locally produced by the replacement of calcite in limestone, via the reaction

$$CaCO_3 + SO_4^{2-} + 2H_2O \rightleftharpoons CaSO_4 \cdot 2H_2O + CO_3^{2-}. \qquad (11.3)$$

This reaction also describes the formation of gypsum deposits as surface alteration deposits on limestone outcrops.

Other sulphate and carbonate minerals have been reported to form in similar circumstances, for example, as crusts or speleothems in caves, or as alteration features on weathered pyritic shales. In each case the possibility exists of U-series dating of the mineral if a suitable level of U is present.

11.2 Uranium and Th geochemistry of fresh water carbonates and sulphates

The application of U-series dating to sulphate and carbonate minerals is possible because of the contrasting aqueous geochemistry of U on the one hand, and Th and Pa on the other, as described in chapter 2. In aqueous solution U is capable of being transported at concentrations up to several ppm as a carbonate, phosphate, or sulphate complex as long as the Eh is sufficiently high to keep U in the hexavalent state. If reduced to U^{4+}, U quickly precipitates as uraninite (UO_2); a process observed in the formation of 'roll' type U ore deposits.

In contrast, Th and Pa are quite insoluble in waters of near neutral pH.

Therefore, when freshly precipitated, calcite, gypsum, and other minerals deposited from surface waters contain measurable traces of U but are essentially devoid of Th or Pa. Isotopes of these two elements, [230]Th and [231]Pa subsequently grow *in situ* within the deposits, and the level of their radioactivity, compared to that of their parent, can yield the age of the deposit, as described in chapter 3. Not only is it possible to date calcite and gypsum but, in principle, any mineral formed by precipitation from such ground or surface waters is potentially datable by U-series methods. A variety of oxides, chlorides, phosphates, nitrates, and other minerals, exist as efflorescences in caves, fracture fillings in surficial deposits, evaporites in desert playas, and other geologically young settings, where U-series dating could contribute valuable temporal information. This chapter will discuss some well known examples, but the reader should consider the wider applicability of the same methods to other materials.

11.3 Uranium series dating of speleothems

The word speleothem is a term which describes any mineral precipitated in a cave environment. It is derived from the Greek, *spelaion* meaning cave, and *thema* meaning deposit, and was first coined by Moore (1952). The most common speleothems are stalactites, stalagmites, and flowstones which are usually composed of calcite, and occasionally aragonite. Carbonate speleothems are formed wherever ground water enters a cave and becomes supersaturated with respect to calcium carbonate. This may be due to outgassing of soil-derived CO_2 to the cave atmosphere and/or evaporation of the ground water itself. Generally, evaporation is only an important process in speleothem formation in cave entrances or in cave passages with strong air currents of low relative humidity. A process of speleothem deposition by slow outgassing of the seepage water with no evaporation gives rise to very slow speleothem growth, and enables the deposit to be used for paleoclimate determination through analysis of the variation in stable isotopic content of the calcite and its fluid inclusions (see chapter 13).

Typical speleothem deposits and their locations in a cave are shown in Fig. 11.1.

11.3.1 Physical properties of speleothems

Seepage water dripping from a cave roof deposits $CaCO_3$ around itself and gradually constructs a tubular or 'soda straw' stalactite that extends downwards with time. Blockage of the tube or flow over the outside will cause the stalactite to thicken near the ceiling and assume a conical shape. Stalagmites are columnar deposits which grow upwards from the floor towards the drip source. They are usually thicker than stalactites because the water splashes and flows outwards causing more lateral deposition. Flowstones are thinly

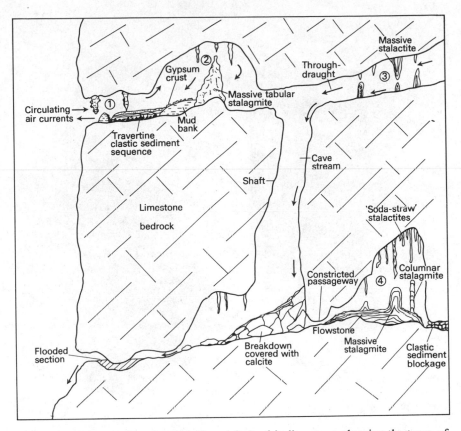

Fig. 11.1. Cross-section of a typical cave formed in limestone, showing the types of speleothem that may form and their relation to location and air circulation in the cave. Speleothem growth zones are: (1) entrance: sporadic, fast speleothem growth, mainly by evaporation; tufa-like porous deposits, often containing organic and detrital material; (2) near entrance: fast speleothem growth, aided by rapid outgassing and some evaporation, large deposits, often porous, may contain detritus; (3) interior (main passages): fairly fast growth, mainly by outgassing; large deposits, generally non-porous and free of detritus unless within reach of flood waters; and (4) deep-interior (side passages with constricted exits): slow growth by outgassing only, high relative humidity and CO_2 levels, various sizes of speleothems, generally non-porous and detritus free.

laminated deposits formed from seepage water flowing down cave walls or along floors.

In all three types of speleothem, crystal orientation is usually perpendicular to the growth surface (Kendall and Broughton 1978). Internally, they display a series of superimposed growth layers (see Fig. 11.2) which can be distinguished by changes in colour, texture, impurity content and density of fluid inclusions.

(a)

(b)

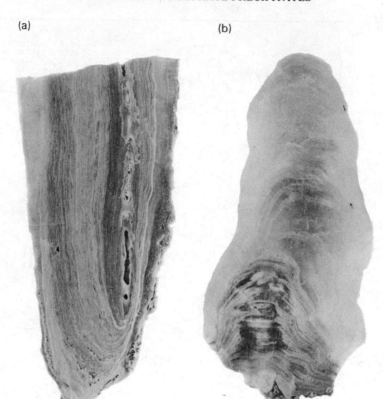

(c)

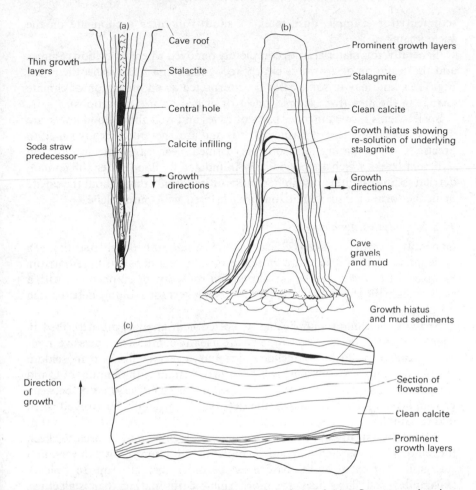

Fig. 11.2. Sections of (a) a stalactite, (b) a stalagmite, and (c) a flowstone, showing typical internal morphology.

Colours of detritus-free speleothem range from white, through orange and red, to brown or black, largely due to varying organic content (Gascoyne 1977*b*). In some cases, high concentrations of certain trace elements will also cause coloration (White and van Gundy 1974). Crystal growth is typically continuous across growth layers and sometimes, slow growing speleothems are found which consist of a few, optically-continuous crystals.

Speleothem growth is dependent on several factors but the principal requirement is continuous slow flow of water into an air-filled passageway (see Fig. 11.1). If growth stops, due to cessation of the flow, or due to submergence in ponded water, crystal continuity is ended and a thin layer of detrital sediment (for example, dust, mud, or sand) will often accumulate on the speleothem.

Generally, this material is not completely removed when deposition resumes and the cessation of growth is clearly recognized in a fossil speleothem (see Fig. 11.2), and may in some cases be interpreted as an indication of climate change in the area (for example, onset of glacial or arid conditions).

Speleothems formed in inner parts of caves and well above flood levels, are usually completely free of clastic detritus and are, therefore, ideally suited to precise U-series dating. Speleothems formed near cave entrances or on sediment banks which are periodically inundated by flood water will contain detrital sediments either distributed more or less evenly throughout the calcite or in the form of discrete horizons interlayered with pure calcite.

11.3.2 Analytical techniques

Generally, growth layers can be clearly seen in a smoothly polished section of a speleothem and it is, therefore, possible to sample a constant time horizon. Samples for analysis may be cut out with a rock saw or chipped out with a small pneumatic chisel (the latter method being best for avoiding detritus-rich horizons).

Analytical techniques applicable to speleothem analysis are described in chapters 4 and 5. Because speleothems are usually composed of massive, non-porous calcite, problems of nuclide migration or recrystallization are seldom encountered. A more common problem is in determining the amount of U and Th nuclides introduced by detritus trapped in the calcite crystals or between growth layers. With care though, the detritus can generally be avoided when taking samples from a speleothem.

The internal stratigraphy of a stalactite or stalagmite may be used to check the validity of radiometric ages determined for different growth layers: in a stalagmite, for example, the base must be older than the top. In general, stalagmites and flowstones are more suitable for dating than stalactites, because the latter usually have thin growth layers and a hole through the centre which may permit re-solution and nuclide migration after deposition (see Fig. 11.2).

11.3.3 Early studies

Rosholt and Antal (1962) were the first to apply U-series dating methods to speleothems. They analysed eight stalagmites from several European and South African caves, but found excess ^{230}Th and ^{231}Pa in all samples. This was attributed to U loss at some time following formation, and they concluded that speleothems were not suitable for U-series dating. Rosholt and Antal (1962) did not determine ^{234}U/^{238}U ratios, only U concentrations and assumed secular equilibrium existed. Subsequently, Cherdyntsev (1971) criticized the work of Rosholt and Antal on the grounds that it was difficult to understand how high excess daughter activities could occur in the calcite if almost all the U was lost, irrespective of the age of the deposit. Instead, Cherdyntsev suggested that the U determinations were low. This suggestion is supported by the fact that seven determinations of U content in various Pleistocene marine limestones (also described by Rosholt and Antal 1962) yielded values ranging between 0.2 and 1 ppm, concentrations which are low, even for partially recrystallized samples.

Cherdyntsev *et al.* (1965) were more successful in applying $^{230}Th/^{234}Th$ ($= ^{230}Th/^{238}U$) ratios to dating speleothems from Akhshtyr Cave in Krasnodar, USSR. Although three of the four speleothems contained high detrital Th levels ($^{230}Th/^{232}Th$ ratios were 0.5 to 1.0) similar ages were obtained for each (33 ky) without correction for detrital ^{230}Th. The speleothems were considered unsuitable for $^{234}U/^{238}U$ dating because of large variations in the measured ratio (1.07 to 1.38).

Komura and Sakanoue (1967) analysed a stalagmite from a cave in Japan, using the $^{230}Th/^{234}U$ dating method. Low U content (0.09 to 0.16 ppm) and the presence of some detrital Th prevented these authors from obtaining precise ages, but $^{230}Th/^{234}U$ ratios at four sites along the growth axis indicated that the stalagmite grew quickly, commencing at about 13 ky.

Fornaca-Rinaldi (1968a) also attempted to date speleothems from Italian caves by U-series methods, but erroneously assumed that the ratio $^{230}Th/^{234}Th$ was equivalent to $^{230}Th/^{234}U$ in the age calculation. This is only true if $(^{234}U/^{238}U)_0 = 1$, a relatively rare situation for most freshwater carbonates (see later). Therefore, the true speleothem ages are probably significantly less than the reported values which ranged up to 80 ky. Several pure speleothems were found to contain negligible ^{232}Th activity but a high beta activity (attributed to ^{210}Pb from rain waters). Many of the examined speleothems were considered unsuitable for dating because of the presence of Th in an insoluble detrital phase.

The first U-series dating of speleothem which was combined with stable isotopic analysis (for paleoclimate determination), was carried out by Duplessy, Labeyrie, Lalou, and Nguyen (1970). Five determinations of $^{230}Th/^{234}U$ and $^{234}U/^{238}U$ ratios were made along the growth axis of a 2.44 m long stalagmite from Aven d'Orgnac, southern France. Ages ranged from 129 ky at the base to 92 ky at the top, all in correct stratigraphic sequence, and with low error limits (± 2 to ± 6 ky). It is perhaps surprising that such good agreement, with high precision, was obtained in this work, because the speleothem contained very little U (0.06 to 0.07 ppm), a level which approaches the minimum of about 0.05 ppm required for dating.

In further study of speleothem dating, Cherdyntsev (1971) examined many archaeologically-significant travertines, tufas, stalactites, and stalagmites from several caves in the USSR. In most cases, the samples were heavily contaminated by detrital Th ($^{230}Th/^{232}Th$ ratios less than 10) and some samples, analysed without ^{228}Th tracer, showed disequilibrium between ^{232}Th and ^{228}Th. These samples did not always contain excess ^{230}Th (^{230}Th/^{234}U ratios greater than 1.0) and so would give finite, but incorrect, ages if only analysed by the ^{228}Th/^{232}U spiking method. Cherdyntsev also observed that the ^{234}U/^{238}U activity ratios for speleothems both within the same cave and between different caves in the same area, varied more than might be explained by age differences, thus demonstrating the unsuitability of this method of dating when applied to speleothem. Cherdyntsev (1971) attempted to correct the ^{230}Th/^{234}U ages of several samples for detrital Th by analysing adjacent modern deposits and assuming that the (^{230}Th/^{232}Th)$_0$ ratio is approximately constant over time and between samples. Loss of U (seen as ^{230}Th/^{234}U greater than 1.0) was observed both in porous and powdery samples such as ferruginous clays and tufas, and in more massive travertines and stalagmites, although Cherdyntsev concluded that the loss was much less extensive than that proposed by Rosholt and Antal (1962).

11.3.4. Recent work

As previously described, the problems of nuclide migration and detrital contamination are usually associated with speleothems from cave entrances or passageways subject to flooding. These problems can sometimes be avoided if samples are collected from the more inaccessible parts of a cave, such as ancient high-level tunnels which can only be reached with difficulty. In addition, these locations provide the added advantage that the speleothems may have grown under conditions of isotopic equilibrium and so might also be suitable for paleoclimate analysis (see chapter 13). Most early work on speleothem dating was carried out on deposits collected from readily accessible parts of caves, such as entrances or large open passageways, where they provided information on the chronology of archaeological or fossil-bearing deposits. However, during the last decade, an increasing emphasis has been placed on dating speleothems collected from interior areas which are less subject to flooding and calcite deposition by evaporation. Because these cave passages are generally farther below the surface, ground water seepages tend to contain much less suspended sediment and more U than shallow seepages in cave entrances. This results in the deposition of purer, more U-rich speleothems (generally greater than 0.3 ppm) which in turn results in a higher precision for isotope ratio and U abundance measurements.

Unfortunately, many Quaternary scientists are still somewhat sceptical of U-series age determinations on speleothems and travertines because of the dubiety and unreliability of the early work just cited. The sections that follow summarize successful applications of U-series dating of speleothems and demonstrate the circumstances under which reliable age determinations can be obtained.

(i) *The ^{230}Th/^{234}U dating method.* In a systematic study of the application of isotopic analysis to speleothem, Thompson (1973) developed a procedure for U and Th extraction and analysis from speleothem calcite. This author demonstrated clearly the validity of $^{230}Th/^{234}U$ dating of speleothem by obtaining eight ages along the axis of a stalagmite from West Virginia (NB 1, Thompson, Schwarcz, and Ford 1976) which were all in correct stratigraphic sequence between 137 and 35 ky. Uranium concentrations ranged from 1.5 to 4 ppm. Similar agreement was obtained for ages determined for an older stalagmite (NB 10) and a flowstone (GV 2). The $^{227}Th/^{230}Th$ dating method (see chapter 5) was used to cross-check the age of one sample from NB 10 and close agreement with the $^{230}Th/^{234}U$ age was found (115 ky and 105 ky respectively). In general, most of the speleothems analysed contained little or no detrital Th and showed no evidence of either re-solution or recrystallization following deposition.

Following the groundwork laid down by Thompson (1973), Harmon (1975) applied the $^{230}Th/^{234}U$ dating technique to paleoclimatic studies of eight areas in North America (Mexico, Texas, Bermuda, Kentucky, West Virginia, Iowa, the Canadian Rocky Mountains, and the Northwest Territories). Eighty-nine speleothems were analysed in this study, and of these, sixteen were found to have been deposited in isotopic equilibrium. Axial profiles of $^{18}O/^{16}O$ variations in the calcite of these speleothems were used to infer variations in paleoclimate over the dated period, 200 ky to present (Harmon, Thompson, Schwarcz, and Ford 1978; Harmon et al. 1978a). In addition, the method of absolute paleotemperature determination using fluid inclusions, developed by Thompson (1973) for West Virginia speleothems, was applied to six speleothems from Bermuda, Kentucky, and Mexico (Schwarcz, Harmon, Thompson and Ford 1976). This type of analysis has since been extended to the Iowan speleothems (Harmon, Schwarcz, Ford, and Koch 1979).

Many of the speleothems from the Rocky Mountains and Northwest Territories were found unsuitable for stable isotopic determination of paleoclimate because of their formation in draughty cave passages. The frequency distribution of their U-series ages, coupled with previous analyses by Thompson (1973) enabled Harmon, Ford, and Schwarcz (1977) to define periods of interglacial and glacial climate over the past 400 ky (see Fig. 13.6). The importance of U-series ages obtained for speleothem in areas glaciated during the Pleistocene is discussed in chapter 13.

On the mid-Atlantic island of Bermuda, speleothems formed in caves during low sea-stands of the Pleistocene. Uranium-series ages have been obtained on over 30 samples of speleothems recovered from depths of up to 15 m below modern sea level; 14 ages on deposits above sea level were also obtained (Harmon, Schwarcz, and Ford 1978b; Harmon, Land, Mitterer, Garrett, Schwarcz, and Larson 1981). These dates have been used to place age and elevation limits on sea stands over the last 200 ky. Most of the submerged speleothems grew in the intervals 195 to 150 ky and 120 to 10 ky. A stalactite, formed at between $+3$ and $+4$m above sea level, was overgrown with a marine aragonitic crust from 130 ± 14 to 110 ± 14 ky, indicating that sea level was higher than at present during the last interglacial. These ages were confirmed by U-series ages of corals from marine deposits which could form only when the sea was at or above its present level, at approximately 200 ky and between 125 and 120 ky. In addition, two submerged speleothems ceased growth at about 150 ky suggesting that sea level briefly rose to at least -20 m at that time. Further evidence for this event comes from comparable ages obtained by amino acid racemization analysis of gastropods from aeolianites (Harmon *et al.* 1981). Dates on corals and speleothems are concordant in indicating a high sea stand from 125 to 120 ky, in excellent agreement with ages for the last interglacial sea level maximum observed in other areas, for example, Barbados (Mesolella, Mathews, Broecker, and Thurber 1969).

Gascoyne, Benjamin, Schwarcz, and Ford (1979) also used U-series dating methods to date currently submerged speleothems in the Bahamas, as a means of determining the timing of low sea stands. They were able to place limits on a post-glacial rise of sea level using speleothems collected 45 m below present sea level from a 'blue hole' off Andros Island in the Bahamas. Initial attempts at dating the calcite core of these heavily corroded stalagmites suggested that contamination by younger, higher-U marine carbonates in worm and sponge borings was occurring. Two approaches were used to allow for this contamination: (i) careful examination of calcite chippings from the core with rejection of those showing marine alteration; and (ii) selective leaching of a crushed sample of a core to remove preferentially the more soluble aragonitic and high Mg-calcite marine deposits in earlier fractions (the extent of removal was monitored by the Mg and Sr content of the leach solutions). The greatest ages obtained by these methods were assumed to be the nearest to the true age of the speleothem calcite. The results showed that the speleothems grew during a period of significantly lower sea level (greater than or equal to 42 m, allowing for platform submergence) between 160 and 140 ky. These results were consistent with the sea level curve obtained from deep sea core isotopic data (Shackleton and Opdyke 1973) and probably correlate to the end of the Illinoian glaciation of North America (see Fig. 11.3).

Speleothems from caves in several other areas have been dated by U-series methods in recent studies. Ages of speleothems from north-west England and the Mendip area (near Bristol), England, were determined in a study by Atkinson, Harmon, Smart, and Waltham (1978). Age distributions indicated non-glacial conditions in these areas before 170 ky, 140 to 90 ky, about 60 ky, and less than 15 ky. These periods were tentatively correlated to the British Quaternary succession. Several speleothems of ages greater than 200 ky were collected from cave passages which were only a few metres above the modern water table, and these were used to infer rates of cave development and, by

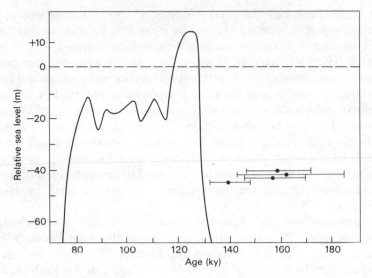

Fig. 11.3. Correlation between $^{230}Th/^{234}U$ ages of 'blue hole' speleothems from Andros Island, Bahamas, and the paleosea level curve proposed by Shackleton and Opdyke (1973). (After Gascoyne *et al.* 1979.)

extrapolation to local valley floor levels, ages of surface landforms.

In a more comprehensive study of the caves of north-west England, Gascoyne (1980) obtained 140 $^{230}Th/^{234}U$ ages·on 82 speleothems. The U-series ages were used (i) to determine minimum ages of caves in this area and rates of valley entrenchment and water table lowering; (ii) to define periods of glacial and non-glacial conditions in the area from frequency distributions (see chapter 13); and (iii) as a chronologic framework for variations in $^{18}O/^{16}O$ of speleothem calcite for paleoclimate analysis. In addition, the age data consistently demonstrated that reasonably pure, non-porous speleothems gave excellent reproducibility when dated by the $^{230}Th/^{234}U$ method, and that the ordering of age determinations for such speleothems was generally concordant with their stratigraphy. In ideal situations (i.e. high U and low Th contents, long count times, low background) the range of the $^{230}Th/^{234}U$ method could be extended from a maximum of about 400 ky to as low as 0.2 ky.

Gascoyne, Ford, and Schwarcz (1981) determined nine $^{230}Th/^{234}U$ ages on two speleothems from Cascade Cave, Vancouver Island, which were deposited in isotopic equilibrium. Ages ranged from 61 to 32 ky and were in good agreement with stratigraphy including two which contained appreciable amounts of detrital Th($^{230}Th/^{232}Th$ less than 5). Replication of these two results on less contaminated calcite gave ages similar to those with detrital contamination, suggesting that in this case, detrital ^{230}Th did not accompany

^{232}Th in the associated sediment. Analyses of ^{18}O/^{16}O variations in both speleothems were interpreted in terms of a paleotemperature record for this area of Vancouver Island for the mid-Wisconsin interstadial.

Hennig (1979) has analysed 83 stalagmites and flowstones from caves in western Europe, some of which were associated with archaeological sites. The distribution of ages showed the same periodicity as observed in the studies cited above, presumably determined by glacial/interglacial climatic change. In a few cases, Hennig was able to confirm ^{230}Th/^{234}U ages by comparison with ^{14}C, thermoluminescence (TL), or electron spin resonance dates on the same material. Excellent agreement was observed with TL dates in most cases, suggesting that this method would be well suited to extending the dating range for speleothems beyond the present limit of 350 ky defined by U-series techniques.

In addition to their paleoclimatic significance, U-series ages of speleothems may also be applied to solving problems in geomorphology. Recently, Ford, Schwarcz, Drake, Gascoyne, Harmon, and Latham (1981) have used age data for speleothems in relict cave systems in the Canadian Rocky Mountains, coupled with present altitudes of the caves, to determine minimum and maximum rates of valley deepening. Results for Crowsnest Pass and the Columbia Icefield area vary between 0.07 and 2.07 m/ky. If these rates can be extrapolated to early Quaternary and Pliocene times, then the best-estimate age of the present total relief (mean value = 1340 m) is between 1.2 and 12 my.

Speleothems from several other regions have been dated in recent studies, including Jamaica (Gascoyne 1980), Quebec (Roberge and Gascoyne 1978) and Norway (Lauritzen and Gascoyne 1980). The results were interpreted in terms of geomorphic processes and paleoclimatic significance. Speleothems from numerous archaeological sites have also been dated by U-series methods; these results are discussed in chapter 12.

(ii) *The ^{231}Pa/^{230}Th dating method.* The activity of ^{231}Pa in a speleothem at secular equilibrium is only 1/21.7 of that of ^{230}Th, and so only speleothems containing more than 1 ppm U can be dated. Thompson (1973) used ^{227}Th as a measure of ^{231}Pa activity (see chapter 5) and obtained one ^{227}Th/^{230}Th date that was concordant with the ^{230}Th/^{234}U date for the same sample of speleothem from a cave in West Virginia. Harmon and Schwarcz (1981) studied growth rates of two speleothems from Tumbling Creek Cave, Missouri, and used both methods of dating. One of the speleothems contained up to 5 ppm U and gave ^{227}Th/^{230}Th ages which were concordant with ^{230}Th/^{234}U ages over the range 86 to 53 ky, whereas another speleothem with lower U content (less than 0.6 ppm) gave apparently anomalous (negative) ^{227}Th/^{230}Th ages.

Gascoyne (1980, unpub. results) used both methods to date each of 15 speleothem samples and one coral whose U contents ranged from 0.3 to 18 ppm (see Table 11.1). Two porous samples from a flowstone that was at -15m

Table 11.1. Comparison of ages obtained by ^{230}Th/^{234}U and ^{227}Th (= ^{231}Pa)/^{230}Th dating methods for calcite speleothems (from Gascoyne 1980, unpub. results)

Speleothem number	Location	U† (ppm)	$\dfrac{^{234}U}{^{238}U}$	$\dfrac{^{230}Th}{^{232}Th}$	$\dfrac{^{230}Th}{^{234}U}$ age ($\pm 1\sigma$) (ky)	$\dfrac{^{227}Th}{^{230}Th}$ age ($\pm 1\sigma$) (ky)
76016	Bahama Blue Holes	0.26	1.028	70	128.8 ± 11.6	122.3 ± 32.9
77032	Castleguard Cave, Alta.-B.C.	2.51	1.362	274	277.8 ± 23.9	> 350
RKM-6	USIP coral standard	3.19	1.100	31	128.8 ± 5.8	107.6 ± 13.0
76001	McMaster dating standard	0.81	1.868	> 20	47.7 ± 1.8	18.3 ± 16.5
76121	Lancaster-Easegill Caverns, England	1.49	1.130	129	114.0 ± 7.3	96.4 ± 14.6
76125	„ „	2.01	1.628	105	38.1 ± 1.4	54.4 ± 11.7
76127	„ „	13.1	0.878	255	225.2 ± 22.2	203.4 ± 24.2
79005	„ „	0.43	1.429	12	43.3 ± 3.1	88.8 ± 6.6
77126	„ „	3.93	1.152	240	106.0 ± 3.6	108.7 ± 9.1
77162-1/2*	Lost John's Cave, England	6.23	0.944	127	92.0 ± 4.2	88.8 ± 6.6
77162-3/3b*	„ „	7.21	0.916	92	112.7 ± 5.1	165.7 ± 16.6
77162-10/11*	„ „	8.43	0.927	> 1000	109.4 ± 4.5	93.1 ± 7.7
79007	Peak Cavern, Derbyshire, England	1.02	1.473	73	51.2 ± 1.6	45.1 ± 8.1
78020	Winnat's Head Cave, Derbyshire England	17.6	1.182	557	$175.5 + 7.3$	285.5 ± 47.6
78010-1/8*	Warm Mineral Springs, Florida	3.47	0.992	31	12.6 ± 0.3	(-92.0 ± 5.1)
78010-2/9*	„ „	1.91	1.006	6	14.4 ± 0.6	(-69.3 ± 10.0)

* For speleothems 77162 and 78010, duplicate analyses were made at more than one site along the growth axis.
† Uranium concentrations and nuclide ratios are those for the ^{230}Th/^{234}U analyses.

in a Florida hot spring, (78010), gave grossly discordant ages, suggesting that migration of radionuclides had occurred after deposition, probably due to inundation associated with sea level rise. The discordance was thought to be due to ^{227}Th derived from ^{227}Ac adsorbed from the highly-mineralized springwater, rather than preferential loss of ^{230}Th or adsorption of ^{231}Pa. Twelve of the remaining 14 samples in Table 11.1 show agreement of ages within 2σ error limits. The ^{227}Th/^{230}Th ages showed no significant bias to younger or older values than the ^{230}Th/^{234}U ages, thus suggesting that there is no reason to reject the 32.5 ky half-life for ^{231}Pa (Kirby 1961), in favour of 34.3 ky, as advocated by Ku (1968). The main problem in dating low-U samples by the ^{227}Th/^{230}Th method lies in the numerous corrections for background activity over the wide energy range at which ^{227}Th is counted and the ingrowth and decay of several nuclides in this region (see chapter 5).

(iii) *The* $^{234}U/^{238}U$ *dating method.* In order to use this method to date speleothems it is necessary to know the value of the $^{234}U/^{238}U$ ratio at the time of formation of the deposit. If the speleothem is within the range of the $^{230}Th/^{234}U$ method then this value may be calculated from the determined age. For older speleothems, however, the initial $^{234}U/^{238}U$ ratio cannot be determined independently, and it must be estimated by other methods instead. One such method is to analyse modern cave waters and/or calcites and assume constancy over time and space. However, in a comparison of $^{234}U/^{238}U$ ratios of seepages and associate calcite deposits for several sites in three caves in West Virginia, Thompson, Ford, and Schwarcz (1975) found that: (i) there appeared to be seasonal variations in the $^{234}U/^{238}U$ ratio of seepage water; (ii) there was little relationship between the ratios for the water and its calcite, possibly due to isotopic fractionation of different U species (for example, organic against inorganic); and (iii) ratios varied both within the same cave and between different caves. In spite of this, fossil speleothems dated by the $^{230}Th/^{234}U$ method showed reasonable constancy of the calculated initial $^{234}U/^{238}U$ ratio over periods as long as 100 ky. Thompson, *et al* (1975) concluded that seepage waters were not a good measure of initial $^{234}U/^{238}U$ ratio for a fossil speleothem in the same cave because large, short-term variations in ratio can occur, which only become averaged out when a bulk sample of speleothem is analysed. Instead, these authors proposed that the dating method could be best applied to an ancient speleothem whose upper portions were within the range of the $^{230}Th/^{234}U$ method so that values of $(^{234}U/^{238}U)_0$ could be established. If these were found to be constant over the dated portion, then it could be inferred that the ratio had remained constant and of the same value, over the preceding portion and then eqn (3.8) could be used to date the remainder of the speleothem.

At about the same time, Thompson, Lumsden, Walker, and Carter (1975) analysed three stalagmites from Blanchard Springs Caverns, Arkansas, and found evidence of U loss in two speleothems, and both U and Th loss in the remaining speleothem. Ten $^{230}Th/^{234}U$ ages were determined along the axis of the latter and a generally poor relationship to the stratigraphy was found (see Fig. 11.4). Ages ranged from 400 ky at the base to 40 ky at the top, but three age inversions in the middle indicated either loss of Th or enrichment in U. The $^{234}U/^{238}U$ ratios, however, were found to decrease in a more systematic manner from the base to the top and fitted the ^{234}U decay curve defined by a constant initial $^{234}U/^{238}U$ ratio of 1.54 (see Fig. 11.5). This value was determined by analysis of an adjacent seepage water in the cave (the average of all five waters sampled was 1.56 ± 0.1). Ages calculated from eqn (3.8) showed growth over the period 786 to 135 ky (Fig. 11.4) with a break of about 125 ky in the middle of this period. The close agreement of the

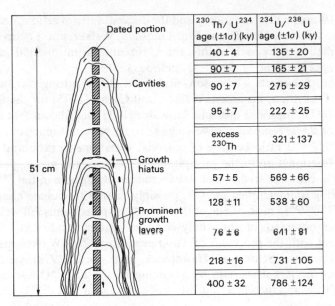

	$^{230}Th/U^{234}$ age ($\pm 1\sigma$) (ky)	$^{234}U/^{238}U$ age ($\pm 1\sigma$) (ky)
	40 ± 4	135 ± 20
	90 ± 7	165 ± 21
	90 ± 7	275 ± 29
	95 ± 7	222 ± 25
	excess ^{230}Th	321 ± 137
	57 ± 5	569 ± 66
	128 ± 11	538 ± 60
	76 ± 6	641 ± 81
	218 ± 16	731 ± 105
	400 ± 32	786 ± 124

Fig. 11.4. Comparison of $^{230}Th/^{234}U$ and $^{234}U/^{238}U$ ages for a stalagmite from Blanchard Springs Caverns, Arkansas, USA, (with permission, from Thompson, Lumsden, Walker, and Carter 1975).

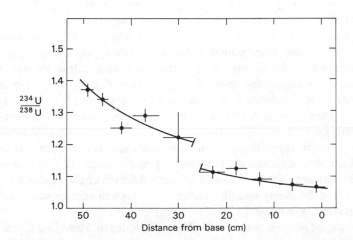

Fig. 11.5. Variation of $^{234}U/^{238}U$ activity ratio with distance from base of a stalagmite from Blanchard Springs Caverns, Arkansas, USA, (with permission, from Thompson, Lumsden, Walker, and Carter 1975). The broken curve shows the change in measured ratio for a constant initial value of 1.54.

$^{234}U/^{238}U$ ages with stratigraphy and the discontinuity of ratios on either side of the growth hiatus led Thompson, Lumsden, Walker, and Carter (1975) to suggest that in cases such as this, the U-disequilibrium method gave more realistic ages than the $^{230}Th/^{234}U$ method.

Harmon, Schwarcz, Thompson, and Ford (1978) strongly criticized the work of Thompson, Lumsden, Walker, and Carter (1975) for the following reasons: (i) Thompson, Ford, and Schwarcz (1975) had shown that results of analysis of single samples of cave seepage water were not representative of a mean $^{234}U/^{238}U$ ratio because of seasonal variations in ratio and possible isotopic fractionation in the precipitation of calcite; (ii) many speleothems analysed in previous studies had shown sudden changes in initial $^{234}U/^{238}U$ ratio during uninterrupted growth, possibly related to climate change; and (iii) the $^{234}U/^{238}U$ ratio of one of the Arkansas speleothems fell well outside the range observed in the analysed cave waters, and was, therefore, inconsistent with the proposals of Thompson, Lumsden, Walker, and Carter (1975). Harmon, Schwarcz, Thompson, and Ford (1978) went on to reinterpret the Arkansas data by assuming the $^{230}Th/^{234}U$ age data were more realistic, and therefore, indicative of a change in $^{234}U/^{238}U$ ratio of the seepage water.

Using the abundant data obtained from U-series dating of speleothems from North American caves, Harmon (1975) attempted to determine whether correlations existed between age, U content, $^{234}U/^{238}U$ ratio, and geographic location. For over 80 per cent of the samples analysed, the U concentration fell in the range 0.5 to 0.8 ppm, with highest concentrations in speleothems from the Canadian Rockies and N.W.T., and the lowest concentrations from Mexico and Bermuda. Comparison of the stratigraphy of the areas suggested that high U abundance was related to the presence of organic-rich shales interbedded with or overlying the cavernous strata. Variations of an order of magnitude or more in U concentrations were seen in analyses of an individual speleothem, possibly due to changing climate. Fifteen dated speleothems were examined for variation of initial $^{234}U/^{238}U$ ratio over their growth period, and for general relationships between initial ratios and age. For most speleothems the isotopic ratio was found to vary in a random manner along the growth direction and no clear correlation with speleothem age was found. However, a slight negative correlation was observed between initial ratios and U content which could be due to pH and Eh control in the overlying soil, similar to that seen in aquifer waters of Texas and Florida (Osmond and Cowart 1976a).

In a study of seepage waters and modern calcites in Tumbling Creek Cave, Missouri, Harmon and Schwarcz (1981) observed a significant difference in the $^{234}U/^{238}U$ ratio of U coprecipitated by the ferric hydroxide method from a sample of water (4.68), and U incorporated in recent calcite precipitated from water (4.00). They also observed large variations in the ratio for seepages at four different sites within the cave taken at two different times of the year. One seepage demonstrated a remarkable variation in $^{234}U/^{238}U$ ratio, changing from 4.68 during May 1977, to 1.50 in November 1977. Unfortunately, no longer-term measurements were made to determine the timing and extent of this variation. However, one stalagmite from this cave, deposited from 240 to 55 ky, displayed a constant initial $^{234}U/^{238}U$ ratio (as calculated from $^{230}Th/^{234}U$ ages). Apparent $^{234}U/^{238}U$ ages for this speleothem, calculated using eqn (3.8), agreed with $^{230}Th/^{234}U$ ages over its entire length. In contrast, another speleothem from the same cave passage showed variable initial $^{234}U/^{238}U$ ratios over its growth period, (140 to 25 ky), an interval which overlapped with that of the previous speleothem (see Fig. 11.6). From these results, therefore, it would appear that $^{234}U/^{238}U$ dating is a risky procedure, at best.

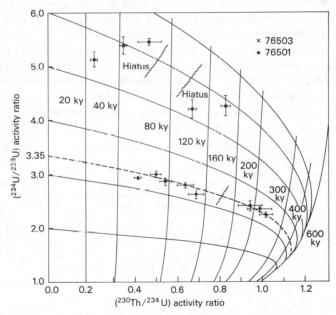

Fig. 11.6. Variation of $^{234}U/^{238}U$ and $^{230}Th/^{234}U$ ratios in two stalagmites from Tumbling Creek Cave, Ozark Mts, Missouri. Note that initial $^{234}U/^{238}U$ ratio remained constant through growth of only one of these speleothems.

In an effort to resolve the status of $^{234}U/^{238}U$ dating of speleothems, Gascoyne (1980) examined spatial and temporal variations in the ratio for 27 seepage waters and associated speleothems from eight caves in four different limestone regions. Uranium was extracted using a simpler technique than used by previous workers (U was adsorbed directly onto an anion exchange resin column attached to the drip source so that no pretreatment of the water was required, Gascoyne 1981). The spatial variation in $^{234}U/^{238}U$ ratios for 27 calcite-water pairs is shown in Fig. 11.7. Overlap of 1σ error limits (from counting statistics) can be seen for 22 pairs, and overlap within 2σ limits is found for 26 pairs. Therefore, it may be concluded that the $^{234}U/^{238}U$ ratios in ground water are faithfully reproduced in associated carbonates almost without exception. This is perhaps a surprising result because the U was collected over periods of up to three months, whereas the calcite analysed may have taken hundreds of years to form, and no apparent fractionation is found between the expected variations in amounts of different U species incorporated in the calcite. As can be seen in Fig. 11.7 the most significant aspect of this work is the potential variability of $^{234}U/^{238}U$ ratio over short distances. For instance, within 1 km of cave passage in Ingleborough Cave, the activity ratios range from 0.77 to 1.6. However, for Castleguard Cave, the three samples are spread out over about 4 km of passageway, yet all ratios overlap within 1σ limits. Similarly, sites 15 and 16, and sites 27 and 28 are 2 and 20 m apart respectively, but no significant difference in ratio is found between each pair. Temporal variation was investigated on a more limited basis by Gascoyne (1980) using two sites within White Scar Cave, England. A variation in activity ratio of up to 0.14 was found over a fourteen-month period (typical error limits were ± 0.025). The variations for both sites were synchronous, having ratios with a late-summer high and an early-spring low.

The above results indicate that sudden changes in $^{234}U/^{238}U$ ratio may be found between adjacent ground waters, but that for the most part, their ratios

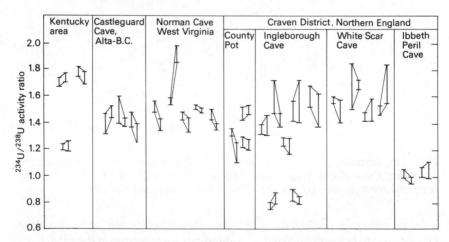

Fig. 11.7. Variations of $^{234}U/^{238}U$ ratios of cave seepage waters and associated calcite deposits for several limestone regions. Each ratio is shown as a vertical bar describing $\pm 1\,\sigma$ error limits (obtained from counting statistics); values for seepage waters form the left vertical of each figure and calcites form the right vertical.

will be very similar. Seasonal variations in ratio may be due to isotopic fractionation between different U species, the relative concentrations of which vary according to activity of vegetation, amount of precipitation, surface temperature, and other factors. In the long term, therefore, secular climate change may influence the $^{234}U/^{238}U$ ratio of a speleothem so that a constant initial ratio cannot be defined for use in disequilibrium dating.

11.4 Uranium series dating of travertines and tufas

The term travertine includes all massive occurrences of calcium carbonate formed by the flow of surface waters. Tufa is merely a very porous variety of this material, commonly containing abundant pseudomorph casts of plants around which the carbonate was deposited. The importance of the dating of travertine deposits arises from their occurrence as significant continental, Pleistocene constructional features. By obtaining absolute ages of such deposits we may date various surficial geological and biological phenomena including the following: (i) fluvial erosion of surficial deposits where these include travertines; (ii) glacial or fluvial deposition where the sediment has been deposited on pre-existing travertines; (iii) occupation of a region by a particular flora or fauna (including hominids), when fossils of these organisms are found embedded in the travertine or over- or underlain by it; (iv) periods of higher rainfall, during which travertine deposition is more active; during periods of dessication (in arid or sub-arid regions travertine deposition may cease); and (v) geologically young tectonic activity, where this has either disturbed previously horizontal travertine sheets, or where travertine deposition has been initiated as a result of flow of water from springs localized along faults.

The dating of such materials depends as usual on the low concentration of Th and Pa in spring waters and in the minerals deposited from them. This appears to be valid even for hot springs as long as they are not emitting acidic waters. Some hot spring waters associated with volcanic centres may contain significant amounts of Th leached from the volcanic rocks. The physical properties of most travertines are such that they may be somewhat less reliable for U-series dating than speleothems. In particular, they are generally porous and permeable because they have been deposited on a mat of vegetation which later decays away, leaving an open network of casts and molds. Also, because they were deposited in the open air these carbonate deposits are subject to detrital contamination by wind, streams, colluvial transport, and other effects. Furthermore, the open, porous structure provides an excellent sediment trap. Consequently many fossil travertines are partly filled with a matrix of clastic sediment. Another sort of infilling may occur as calcite-saturated waters continue to percolate through the older travertine deposits. Secondary overgrowths of calcite can gradually infill all pore spaces. In general, older travertine deposits tend to be less porous than younger ones, due to the combination of these effects. Care must be taken to select travertine samples for dating that show as little initial porosity as possible, as determined by microscopic analysis. Fortunately, most travertine-depositing springs periodically deposit less porous, more speleothem-like layers which are not subject to so much later infilling. These should be preferentially selected for dating purposes.

Another possible consequence of this relatively open structure is that radionuclides in the U-series may be mobilized by water percolating through the rock. Both addition and subtraction are possible. Cherdyntsev *et al.* (1975) believe that this has affected the travertine deposits at Ehringsdorf, DDR, causing the $^{230}Th/^{234}U$ and $^{231}Pa/^{235}U$ dates to be discordant and out of stratigraphic sequence. If possible, concordancy tests should be made on such deposits to affirm that the dates are valid. Cherdyntsev (1971) observes that travertines commonly have $^{228}Th/^{232}Th$ ratios considerably greater than unity (up to 67). This is presumably due to the mobilization and redeposition of ^{228}Ra ($t_{\frac{1}{2}} = 5.75$ y). This phenomenon is observed even in travertines in isotopic equilibrium which give concordant $^{230}Th/^{234}U$ and $^{231}Pa/^{235}U$ ages. Presumably Ra is more mobile than either U or Th, and its anomalous activity should not be taken as an indication that the sample is unsuitable for dating. However, if a $^{228}Th/^{232}U$ spike is being used in the dating procedure, obviously account must be taken of the presence of excess ^{228}Th.

Most U-series analyses of travertines reported so far in the literature have been made in order to provide dates for archaeological sites (see chapter 12). While a large variety of other purely geological or geomorphological applications can be envisaged, relatively few have so far been attempted. These will be summarized below.

Cherdyntsev (1971) summarizes his earlier studies of travertines, finding that in many cases the $^{230}Th/^{232}Th$ ratio is quite low (less than 2) with the consequence that a significant correction for an initial common ^{230}Th component must be made. In one case (the archaeological site of Tata, Hungary) he does this by analysing a historic travertine from a nearby deposit. However, this is a risky procedure, as it can be shown that the $^{230}Th/^{232}Th$ ratio is principally decreased by the presence of common Th leached from detrital sediment included in the sample, and not from Th coprecipitated with calcite from solution. Therefore, modern samples do not necessarily give meaningful estimates of the initial $^{230}Th/^{232}Th$, unless they contain equivalent detrital contaminants.

The travertines from the Leninakan region (Armenian SSR) are associated with a volcanic centre of the Aragats volcanoes; the tufas may be lacustrine. An age of 22 ± 3 ky was obtained (Cherdyntsev 1971) but the $^{230}Th/^{232}Th$ value of 0.95 makes this rather doubtful. In the same state, travertines are found at Armin and Dzhermuk. These have ages of 5 and 11 ky respectively, which Cherdyntsev takes to be the minimum estimates of the age of the associated volcanic field. There is no indication as to whether this is compatible with other dating estimates.

On Mt Mashuk (Armenian SSR) a sequence of early Pleistocene travertines are found associated with another volcanic centre. The travertines are too old to be dated by the $^{230}Th/^{234}U$ method. However, their $^{234}U/^{238}U$ ratios decrease steadily with increasing stratigraphic age. Using analyses of hot waters and young travertines to estimate the initial $^{234}U/^{238}U$ ratio, Cherdyntsev (1971) was able to estimate the age of the middle and lower Pleistocene deposits, giving ages between 280 and 1500 ky. Travertines from Mt Ararat, containing Pliocene leaf imprints, have equilibrium U isotope ratios (0.97 ± 0.04) whereas contemporary hot springs and travertines from the region have $^{234}U/^{238}U$ ratios of 2.4 to 2.6; a minimum age for the Pliocene deposits is, therefore, 2 my.

An interesting application of U-series dating to travertines associated with river terraces is presented by Pécsi (1973) in his study of the Tata river valley. A series of terraces occur above the flood plain of this river (see Fig. 11.8) which have locally been capped by warm-spring deposits of travertine. Warm springs are actively depositing travertine in the area today. The travertines were dated by J. K. Osmond; only the ages are quoted by Pécsi (1973). Terrace IIb is believed to have been formed either in the Riss/Würm interglacial or in the late Riss. It rests on the cryoturbated terrace IIa of Riss age. A travertine mound containing culture layers was deposited on IIb, at Tata. Osmond finds this layer to be 70 ± 20 ky in age. This is consistent with a Riss/Würm age for

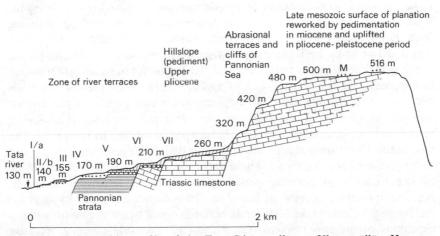

Fig. 11.8. Geomorphic profile of the Tata River valley at Vértesszöllös, Hungary (Pécsi 1973). I–VII, number of river terraces: I, recent flood plain; II/a, first flood-free terrace (Würm); II/b, second flood-free terrace (Riss/Würm interglacial); III, Riss terrace; IV, Mindel terrace; V, Günz terrace; VI–VII, Pre Günz terraces. M-Miocene terrestrial gravels.

the terrace, allowing for some intervening period of exposure and soil formation. However, other estimates of the age of this travertine suggest that it might be somewhat older, about 120 ± 6 ky (Schwarcz 1980); earlier analyses by Cherdyntsev (1971) from unspecified positions in this travertine mound gave an age of 116 ± 16 ky. The second terrace of the Danube River (IIb) at Budapest is also attributed to a last interglacial stage of fluvial erosion. A travertine deposited on it was dated as 60 ky.

Terrace No. III of the Tata and Danube rivers is covered by a travertine dated as 190 ± 45 ky. The terrace had been estimated on geomorphic grounds to be of Riss glacial age. Pécsi infers that the absolute age corresponds to the Riss I/Riss II interstadial, although present-day chronology of the Pleistocene would allow this date to correspond to the penultimate interglacial (Mindel/Riss). The archaeological site of Vertesszöllös occurs in the upper layers of a travertine deposited on terrace No. V, 60 to 70 m above the present flood plain. The travertine is dated at greater than 350 ky; this agrees with an earlier estimate by Cherdyntsev (1971). Pécsi attributes the terrace to the Günz/Mindel interglacial, partly because there is a red clay soil underlying the travertine, formed on the coarse alluvial fan deposits.

On the northeast slope of the Wind River Mountains, near Dubois, Wyoming, a series of warm- and cold-spring-deposited travertines overlies glaciofluvial outwash gravels. The latter can be related to terminal moraines formed by successive advances of the glacial ice cap of the Wind River Range through the late Pleistocene. The youngest till in this area is the Pinedale, correlated with the last advance of Wisconsin ice. Underlying it is the Bull Lake till, whose end moraines prograde into the Circle Terrace, along the Wind River. Travertines deposited on this terrace have yielded ages ranging from 95 ± 9 to 28 ± 4 ky. In Yellowstone Park, a supposedly correlative till was dated at 85 ky by K/Ar dating of an overlying ash bed. The dates are puzzling in that this was a period of ice advance and not of significant outwash development, in the central U.S. Possibly, these dates are on materials deposited considerably after the deglaciation which produced the terrace deposits, at the end of the Illinoian. An older till, the Sacajawea Ridge, progrades to a higher terrace which is in turn partly covered by travertine. The base of this travertine yields a date of greater than 350 ky, while the top (30 m higher) is dated at 233 ± 49 ky. These dates suggest that the Sacajawea Ridge till is not correlative with the pre-Illinoian till, but is somewhat older yet. Another travertine, somewhat lower in elevation than the 233 ky deposit, yields an age of greater than 350 ky as well; this could be attributed to crossing elevations of the two terraces (G. Richmond, personal comm.). Because $^{230}Th/^{232}Th$ ratios of these travertines are in all cases greater than 6, and mostly greater than 10, age correction for initial ^{230}Th is not large. These dates provide a chronology for the late Pleistocene glacial history of the Wind River–Yellowstone region, which in turn has been used extensively as a basis of correlation with other areas of the north-western U.S. (Schwarcz and Richmond, current research.)

While many other occurrences of travertine in association with critical geomorphic or paleoclimatic features are known, these authors know of no other efforts to date them. Clearly this is a potentially rich field for investigation. For example, many travertine deposits have been mapped in the Western U.S. (Feth and Barnes 1979) and some of these can be shown to be interstratified with alluvial and glacial deposits. In as much as they are readily datable, they should be considered to be as important chronological markers as volcanic deposits.

11.5 Lacustrine sediments

The lacustrine sediments of interest in Quaternary research are generally those which are formed during pluvial periods—defined as 'episodes of widespread long-term increase in effective precipitation of sufficient duration and intensity to be of geological significance' (Butzer 1961). Pluvial lakes characteristically have experienced considerable fluctuation in volume and level in the past, usually in response to climate changes (Morrison 1965). Interpluvials, or non-pluvials, define the drier periods between pluvials.

Pluvial lakes are generally restricted to relatively arid regions and are often found in closed basins formed by tectonism or volcanic activity. Pluvial lakes occur in most desert and semi-arid areas of the world including parts of S. America, Africa, Australia, and Asia, but some of the best known are found in the United States. Pluvial lake sediments are often well-preserved long after dessication of the lake, and even during interpluvial periods, sediments may still be accumulated by sub-aerial and shallow lake sedimentation (Morrison 1965). The synchroneity of pluvial periods and northern hemisphere glaciations during the Wisconsin stage has been fairly well established for North American pluvial lakes by ^{14}C dating of lake sediments and stratigraphic relationships of sediments to glacial deposits (see Flint 1971). This coincidence is regarded as due to the displacement of Arctic and Antarctic convergences towards the equator during glacial times, such that areas which were normally arid are traversed by more humid airmasses, while evaporation from the lake surface was diminished by lower air temperatures and higher humidities.

Although ancient pluvial lakes are widespread features in most continental areas (110 are known in North America alone, Flint 1971) the detailed stratigraphy and ^{14}C ages of sediment sequences are known for relatively few of these. Dating by U-series techniques has been applied to even fewer sequences, and usually only to give a comparison with ^{14}C ages. These studies are described below.

11.5.1 Physical character of the deposits

As is typical of lacustrine environments, an enormous variety of sedimentary deposits are found at the sites of dessicated pluvial lakes. Amongst these are chemical precipitates and organogenic carbonates which are datable by U-series methods.

Many pluvial lakes are formed in basins which are sufficiently voluminous that they remain closed even during pluvial episodes. Therefore, as at present, evaporation of water from the lake results in concentration of dissolved salts and, ultimately, in precipitation of the least soluble species, as sedimentary laminae on the lake bottom. Most commonly, $CaCO_3$ is precipitated either in the form of aragonite or calcite; for example $CaCO_3$ can be seen precipitating today from the Dead Sea at times of high photosynthetic activity and consequent depletion of HCO_3^- ions. At higher degrees of salt saturation, sulphates and chlorides may also form, though usually these are limited to the last stages of dessication of the lake. Evaporitic deposits of carbonates and sulphates can also form along the shores of the lake and become interstratified with, or encrusted upon, beach deposits, marking former stands of the lake level. Detrital sedimentary material is invariably interstratified with chemical sediments on the lake bottom. Where the surrounding bedrock includes carbonate sedimentary rocks, the detritus will include fine particles of carbonate which are chemically and physically inseparable from the precipitated carbonate, except where the latter is still preserved as aragonite. Organogenic carbonate of the same age as the chemical precipitate may also be admixed, especially fragments of algal carbonate.

Pluvial lakes also contain a flora and fauna that precipitates $CaCO_3$. Various gastropods and pelecypods inhabit the shore and lake bottom. Ostracods can flourish in the lake water and their calcite tests will contribute to the fine carbonate sediments of the lake. Algae such as *Chara* live in shallow water near the shore and precipitate aragonite which is broken up by storms and currents, and settles on the lake bottom.

Although biologically and chemically precipitated carbonate is a common constituent of lacustrine sediments, it is only rarely that these two components are sufficiently dominant that they can be isolated for dating. The typical lacustrine sediment is a *marl*, a heterogeneous mixture of detrital, authigenic, and organogenic components (including organic matter), and U-series analysis of this material generally does not provide an accurate age of deposition. Rare instances are known, however, of rich concentrations of chemically precipitated carbonates, such as the alternating layers of pure white carbonate and grey detritus of the Lisan Formation, exposed along the shores of the Dead Sea (Kaufman 1971).

11.5.2 Analytical techniques

Because carbonate lake sediments are predominantly composed of silt, marl, and clay (usually greater than 95 per cent by volume), in order to analyse shells and calcareous plant material, it is first necessary to sort and clean the specimens. Kaufman (1964) described a procedure for isolating specimens of ostracods, gastropods, *Chara*, and marl from a bulk sample of lacustrine sediment (shown as a flow chart in Fig. 11.9). After cleaning, the samples were dissolved in acid, spiked, and the liberated CO_2 removed for ^{14}C dating. The resulting solution was analysed for ^{226}Ra (by ^{222}Rn counting) and then analysed for U and Th nuclides by methods described in chapter 4.

Acid-insoluble detritus constituted about 20 per cent by weight for marl, 7 to 15 per cent for ostracods, and 1 to 5 per cent for *Chara* and gastropods. This detritus was dissolved by HF and $HClO_4$ and added to the soluble portion. Kaufman discusses at length the problem of deciding whether to perform the latter operation. He used the results of analysis of several spiked leachates and

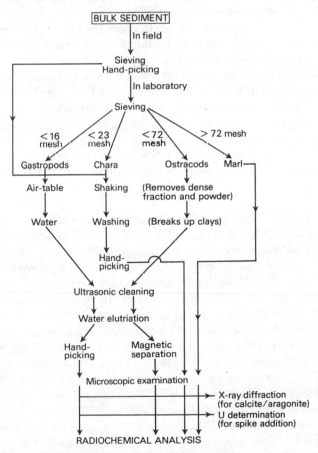

Fig. 11.9. Method of sorting and cleaning lacustrine sediments for use in U-series age determination (after Kaufman 1964).

complete dissolutions to show that a significant fraction of the carbonate-bound ^{230}Th could not be separated from the insoluble residue if the sample was only leached with acid, possibly because of adsorption of ^{230}Th on silicate grains. Complete dissolution was preferable even though all detrital Th nuclides were incorporated into the analysis. Because the ^{230}Th/^{232}Th ratio of the silicates was likely to be constant, it could, therefore, be estimated from multiple analyses of contemporaneous deposits, and the detrital ^{230}Th corrected for. This technique is described in the following section.

Datable non-carbonate lacustrine sediments such as halite and gypsum are less soluble in acid than in pure water, and different procedures are sometimes used in their analysis. Peng et al. (1978) dissolved powdered inner parts of core samples consisting of trona and halite, in water, and then acidified the solution to between pH 1 and 2 before filtering and spiking. In an experiment to determine the effect of acidification, they found that more than ten times the amount of U in solution was obtained from an acidified solution. Although they observed no differential loss of Th between the pure water and acidified solutions they concluded that U-rich material was present in the sediment which could only be dissolved by acid.

Hendy et al. (1979), in the analysis of gypsum deposits from Antarctica, also dissolved the sample in water before acidifying and spiking. However, before the coprecipitation step they passed the solution through an anion exchange column in the chloride form, to remove sulphate ions. They assumed SO_4^{2-} would interfere with the coprecipitation step, by complexing U. It is doubtful, however, that this problem would arise, because at the pH of coprecipitation (about 4), uranyl sulphate complexes are relatively unstable (see Fig. 2.3).

11.5.3 Previous work

In a comprehensive study of the Pleistocene lakes Lahontan and Bonneville in Nevada and Utah, USA, Kaufman (1964) obtained 53 samples from surface exposures of lacustrine sediments forming strandlines and beaches. These were sorted, cleaned, and dated by ^{14}C and ^{230}Th/^{234}U methods as described in section 11.5.2. The ^{14}C ages were corrected for the slightly lower ^{14}C/^{12}C ratios found in modern fish, molluscs, and plants by subtracting 500 y from the calculated ages (Broecker and Kaufman 1965). In all cases, ^{230}Th/^{234}U ratios were corrected for detrital ^{230}Th by assuming a constant value for the initial ^{230}Th/^{232}Th ratio (Kaufman and Broecker 1965). This value was determined by analysis of 12 samples containing the highest detrital Th concentrations, and assuming the ^{14}C age to be correct, and the system to have remained closed, the amount of ^{230}Th produced by the decay of ^{234}U over this period was calculated. This authigenic ^{230}Th component was then subtracted from the total ^{230}Th, and a correction made for unsupported decay since formation. A mean value of 1.70 ± 0.17 was determined for the initial ^{230}Th/^{232}Th ratio, from results which showed remarkable constancy in view of the widely varying age, geographic location and material type of the 12 samples used.

The ^{226}Ra/^{230}Th ratios, were determined for all analysed samples. Several showed ratios between 1.1 and 1.5 indicating either Ra addition or Th loss. These samples often, but not always, showed lack of concordance between ^{14}C and ^{230}Th/^{234}U ages. In some cases, nuclide migration could be blamed, and in others (principally the samples older than 22 ky), it was concluded that the ^{14}C ages were incorrect, due to contamination by young carbon. Kaufman and Broecker (1965) also observed that calculated initial ^{234}U/^{238}U ratios were not always constant, either with location or age. Therefore, these authors concluded that the ^{234}U/^{238}U dating method could not be used reliably on older sediments from the lakes. However, generally good agreement was found between ^{230}Th/^{234}U and ^{14}C ages greater than 10 ky old, for samples which did not show Ra anomalies.

In a subsequent study, Kaufman (1971) dated 22 samples of three Pleistocene chalk deposits (the Upper and Lower Lisan, and Amora Formations) in the Dead Sea Basin. The formations were composed of lithologically similar units, consisting of alternating light and dark laminae of aragonite, and silicates plus calcite respectively. The former were thought to be the result of sudden precipitation from supersaturated lake water, whilst the darker layers contained more clastic detritus and were presumably the result of normal sedimentation processes. To date the formations and study the effect of the detrital-rich layers, Kaufman separated the laminae into four different portions for each collected sample, depending on detrital content. He labelled these very clean, fairly clean, fairly dark, and very dark. A linear relationship between ^{232}Th content and per cent acid-insoluble residue of each aliquot showed that ^{232}Th was associated only with the dark (detrital) component. By plotting ^{230}Th/^{234}U against ^{232}Th/^{234}U for all samples from the same geological formation, he obtained the ^{230}Th/^{234}U ratio (and hence age) for each formation when ^{232}Th/^{234}U was extrapolated to zero (i.e. when no detrital Th was present). Kaufman's results showed that the Upper and Lower Lisan Formations were respectively 40 to 18 ky and 60 to 40 ky old, and that the underlying Amora Formation was about 200 ky old. The slope of each line in the above plot is equal to the ^{230}Th/^{232}Th ratio of the detrital component and, using the calculated age, the initial value of this ratio may be determined. Initial ratios for 14 samples from the Upper and Lower Lisan Formations are plotted against age in Fig. 11.10. Variations in initial ^{230}Th/^{232}Th ratio ranging from 1.3 to 4.5 (mean = 2.84 ± 0.86) can be seen, with no apparent relationship to the sample age. These results, unlike those determined for Lakes Lahontan and Bonneville (Kaufman and Broecker 1965), show greater variation of initial ^{230}Th/^{232}Th of the detrital component of the lacustrine samples, although this may be due to the greater errors incurred with older samples in this type of calculation. The apparent

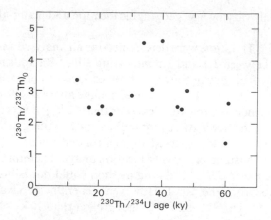

Fig. 11.10. Variation of initial $^{230}Th/^{232}Th$ ratios with $^{230}Th/^{234}U$ age for the Upper and Lower Lisan Formations, Israel (data from Kaufman 1971).

variation in initial ratio does not appear to influence significantly the age because good agreement was found between ^{14}C and $^{230}Th/^{234}U$ ages for three out of four samples.

The $^{234}U/^{238}U$ ratios for the three formations fitted closely a single decay curve defined by $(^{234}U/^{238}U)_0 = 1.54$ indicating that the initial ratio had remained constant for more than 200 ky. Although not discussed by Kaufman (1971), the $^{234}U/^{238}U$ ratios for the fractions of each sample can be seen clearly to decrease with increasing content of insoluble residue (see Fig. 11.11). This is probably caused by leaching, from the detritus, of U whose isotopic ratio is less than that of U in the pure carbonate. The apparently linear relationship between isotopic ratio and detrital content seen in Fig. 11.11 indicates that the U concentration leached from the detritus is similar to that in the carbonate, and that its $^{234}U/^{238}U$ ratio is about 1.0.

Cherdyntsev (1971) has dated several lake deposits in the USSR. One sample of tufa from Karatau contained very little detrital Th ($^{230}Th/^{232}Th$ = 30), and was found to be in secular radioactive equilibrium. It was presumed to be Upper Pliocene in age. A tufa from the Lenikan region was dated at 22 ky but the low $^{230}Th/^{232}Th$ ratio (0.95) indicated significant detrital contamination. Lake tufa and travertines from Pleshcheevo Lake, Vas'kovo, were also analysed by Cherdyntsev (1971). Two $^{230}Th/^{234}U$ ages for the deposits were in good agreement with a ^{14}C age (about 7 ky) in spite of a large ^{228}Th excess in one of the travertines ($^{228}Th/^{232}Th$ greater than 30). The $^{234}U/^{238}U$ ratios of four samples of these deposits ranged from 1.10 to 1.22, whereas the lake water ratio is currently 1.53. No reason for this difference was proposed.

Szabo and Butzer (1979) attempted to date two lacustrine sediment samples

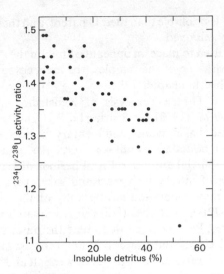

Fig. 11.11. Variation of $^{234}U/^{238}U$ ratio with per cent insoluble detritus in analyses of fractions of samples from the Upper and Lower Lisan Formations, Israel (data from Kaufman 1971).

formed by an ancient seasonal lake at Rooidam, near Kimberly, South Africa. Both samples were marl which had originally been precipitated as aragonite, but was now partially recrystallized to calcite. They dated each sample by $^{230}Th/^{234}U$ and $^{231}Pa/^{235}U$ methods, using an isochron technique similar to that of Kaufman (1971) to allow for nuclides leached from the detrital phase. However, because only one specimen was available from each deposit and the detrital contaminants were evenly distributed throughout it, an isochron line defined by several points could not be obtained. Szabo and Butzer (1979) used only two points, determined by analysis of the whole rock and the detrital fraction (obtained by prior removal of the carbonate with acetic acid). To determine the $^{230}Th/^{234}U$ and $^{234}U/^{238}U$ ratios of the pure carbonate phase, two plots of $^{230}Th/^{234}U$ against $^{232}Th/^{234}U$ and $^{234}U/^{238}U$ against $^{232}Th/^{238}U$ for each sample were extrapolated to zero ^{232}Th content. The $^{231}Pa/^{235}U$ ages were corrected for initial ^{231}Pa content by assuming that initial $^{231}Pa/^{232}Th$ was the same as initial $^{230}Th/^{232}Th$, determined as 0.6 from the slopes of the relationship between $^{230}Th/^{234}U$ and $^{232}Th/^{234}U$ ratios. Although ages determined by both methods were concordant, they did not agree with the stratigraphy of the samples. This was tentatively explained in terms of the varying degree of recrystallization of each sample and associated nuclide migration. It was argued that concordance between the ages must, therefore, be due to the complete resetting of both radiometric clocks but this is difficult to envisage since the radiogenic daughters, Th and

Pa, are appreciably less soluble than their U parent, and therefore, are unlikely to be preferentially removed.

The results were used to place an upper age limit on the Acheulian artefacts found associated with one of the samples (the archaeological importance of this work is described in chapter 12).

A recent detailed U-series analysis of lacustrine sediments has been reported by Peng *et al.* (1978) for Searles Lake, California. This lake is at present a dessicated basin whose sedimentary sequence extends beyond Brunhes/Matuyama boundary (Liddicoat, Opdyke, and Smith 1980). Its upper 50 m records pluvial and interpluvial periods of Late Quaternary age mainly in the form of interbedded marl and salt layers. A comprehensive description of the mineralogy and stratigraphy of the sediments has been given by Smith (1979). Unlike the studies of Kaufman and Broecker (1965) and Kaufman (1971), Peng *et al.* chose to date the purer salt deposits rather than the marls. The Upper and Lower Salt units consist mainly of trona and halite (see Fig.13.2). Twelve samples of the Lower Salt and two samples of the Upper Salt were analysed, many in duplicate. Corrections for detrital Th content were made in all cases by assuming tentatively the same initial $^{230}Th/^{232}Th$ ratio as that observed for detritus for Lakes Lahontan and Bonneville (1.7, Kaufman and Broecker 1965). Corrected ages of the Lower Salt unit ranged from 33 to 23 ky, most showing agreement with stratigraphy. Measured $^{230}Th/^{232}Th$ ratios were generally quite high (from 8 to 50) so that the age correction was seldom greater than 5 ky. Although $^{234}U/^{238}U$ ratios were fairly constant (about 1.24), a slight *increase* with increasing depth was observed, suggesting that the $^{234}U/^{238}U$ ratio of the lake water may have changed with time.

Comparison of corrected $^{230}Th/^{234}U$ ages with ^{14}C ages determined from disseminated carbon in the marl layers of the Lower Salt unit, showed excellent agreement provided that the ^{14}C ages were first corrected for the low $^{14}C/^{12}C$ ratio of lake water during pluvial times. This correction was determined by calculation of gas exchange rates, lake volumes and surface areas, and likely carbonate content of the lake water. Using this method, an apparent age of 0.9 ky was obtained for all carbonates at the time of formation, and this was subtracted from all ^{14}C ages to obtain a corrected age. The difference between corrected $^{230}Th/^{234}U$ and ^{14}C ages varied between + 2.4 to − 4.4 ky and showed no apparent relationship to the magnitude of the $^{230}Th/^{232}Th$ ratio (see Fig. 11.12). The even scatter about the zero difference line in Fig.11.12 suggests that the initial $^{230}Th/^{232}Th$ ratio of the detritus was close to 1.7 as originally assumed.

As the foregoing summaries demonstrate, there have been relatively few applications of U-series dating methods to lacustrine sediments, in spite of the abundance of known pluvial lake basins. One recent application, in an environment different from the warm deserts described above, has been the

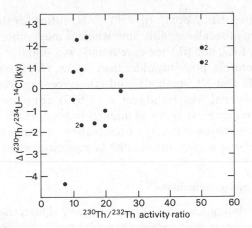

Fig. 11.12. Difference between corrected ^{230}Th/^{234}U and ^{14}C ages as a function of ^{230}Th/^{232}Th ratio of the sample for lacustrine sediments in Searles Lake, California. (With permission, from Peng *et al.* 1978.)

dating of carbonates and evaporites found in lacustrine sediments in the Taylor Valley, Antarctica (Hendy *et al.* 1979). These sediments were deposited during periods of high lake level, contemporaneous with local (and presumably global) glacial advances. The carbonates are algal limestones composed of micritic aragonite. They are relatively rich in U and poor in Th, and therefore, suitable for dating. Several types of sediments in the Taylor Valley area were analysed: (i) aragonite and gypsum laminae within halite layers from sediment cores; (ii) shallow lacustrine micrites from dissected sediment sequences; (iii) carbonate cements, nodules and clasts in exposed deltaic sequences and from boreholes; and (iv) carbonates occuring as desert lag, found on the surface. Detrital Th was present in some samples, particularly of gypsum, and ages were corrected by assuming that the ^{230}Th/^{232}Th ratio of the detritus was unity. Although this value was not determined independently, Hendy *et al.* (1979) considered it to be no greater than 1.0, otherwise corrected ages for young gypsum samples would have become 'negative'.

The ^{230}Th/^{234}U ages determined were found to agree closely with the geomorphological setting of the samples and they described the following intervals: (i) less than 5 ky—a period of lake evaporation, and gypsum and aragonite deposition; (ii) 90 to 105 ky and 120 to 130 ky corresponding to periods of higher lake level than present due to melting of the main Taylor Valley glaciers; (iii) periods 200 to 210 ky and older, possibly also related to glacier melting and lake formation. Hendy *et al.* (1979) distinguished ages of 300 ky from those of greater than 400 ky—a practice which is normally of dubious merit, but in this case could be justified because: (i) U contents were fairly high (greater than 3 ppm); (ii) $^{234}U/^{238}U$ ratios for the samples older than 400 ky were at secular equilibrium, whereas most other carbonates had ratios in excess of 1.2; and (iii) one core sample was found to be magnetically reversed, and therefore, probably older than 700 ky. One sample was dated by both $^{230}Th/^{234}U$ and ^{14}C methods, and fair agreement was obtained after correction for detrital Th. Hendy *et al.* (1979) considered that lack of concordance of radiometric ages was unlikely to occur in the samples studied because of their almost continuous preservation at temperatures less than $0\,°C$, thus preventing nuclide migration via aqueous media.

11.6 Summary and conclusions

Uranium-series dating can be applied to any Quaternary deposits of sufficiently young age in which aqueous transport of U has occurred followed by a chemical or biogenic precipitation process. Chapter 11 has described a variety of continental settings in which such deposits are found. Of these various environments, only two, lake sediments and speleothems, have been investigated extensively thus far. Only speleothems furnish chemically precipitated carbonate which can be sufficiently pure and rich in U to allow accurate dating. Furthermore, progress in the dating of speleothems has been spurred on by the light that it sheds on fundamental questions of paleoclimatology and archaeology. Nevertheless, other types of continental deposits are adequately datable and should be further investigated. Unfortunately, speleothem analysis is limited by problems of access to the deeper parts of caves and the requirement for conservation of these often beautiful deposits. Furthermore, it is impossible to determine a speleothem's age and growth duration prior to removal, or whether it is suitable for stable isotopic analysis. Continuous sedimentary sequences have no such limitations.

As will be seen in chapter 13, all continental deposits occur in a geological context which can provide paleoenvironmental information. Therefore, the dating of such precipitates should be looked upon as equal in significance to the dating of lava flows or ash fall tuffs so commonly cited as chronostratigraphic markers in detrital sediment sequences and soils.

In addition to the chemical precipitates mentioned in this chapter, marine evaporitic and chemically precipitated sediments are also potential targets for U-series dating, although so far no attempt has been made to do so. Deposits of gypsum and halite are formed in tidal flats such as the sabkhas along the coast of the Persian Gulf or the Mediterrannean on the north coast of the Sinai Peninsula. Only the younger parts of these sequences have been studied so far, and they are all in the range of ^{14}C dating. Deeper layers might be treated as the Searles Lake sediments studied by Peng *et al.* (1978).

Not discussed in this section, are phosphate precipitates, which can be found locally as alteration products of organic deposits (guano and bone) both in caves and in exposed subaerial deposits. Generally such materials are too highly contaminated to be used for dating; concretions of phosphate entrap detrital particles as do carbonate concretions. However, the U content of phosphate may be sufficiently high in such concretions, that it outweighs any contribution from the detritus. Also, due to the great insolubility of Th in the presence of phosphate ions, little ^{230}Th may be mobilized to become included in these concretions. Therefore, they may prove to be ideal for U-series dating although they are extremely rare occurrences in Quaternary deposits.

12 APPLICATIONS OF U-SERIES DATING TO ARCHAEOMETRY
H. P. Schwarcz

12.1 Introduction: the problem of absolute calibration of human evolution

The present human species, *Homo sapiens sapiens* is the product of a long process of organic evolution. The earliest fossil hominids, found buried in detrital and volcanic sediments of East Africa, are at least 3000 ky old. One of the most persistent problems in the study of early man, has been the establishment of an absolute time scale for human evolution. The early evolutionary stages leading to modern man, cover a sufficiently long period of time that traditional methods of geochronology (K/Ar, fission track) have been capable of yielding an adequate chronology. However, for the more recent stages immediately preceding the appearance of our own species, the middle and upper Pleistocene, successive bench-marks in organic and cultural evolution appear with such close temporal spacing and over such a short time range, that new methods have had to be developed to deal with this interval. Of course, the same problems of absolute dating have been encountered in the geological study of this time interval, especially in the construction of a time scale for the glacial cycles which characterize the Pleistocene. The purpose of this chapter is to show how U-series dating has been applied to the problems of archaeometry in the late Pleistocene.

There are several facets to this problem. First, in specific archaeological sites which have been systematically excavated, archaeologists have recorded stratigraphic sequences through which cultural progressions can be traced. The first archaeometric requirement is to find material in such sites that can be dated, so as to place an absolute time scale on the strata. The second requirement, given the few sites from which human skeletal material has been recovered, is to obtain dates as closely related as possible to the stratigraphic position of these remains, so as to be able to establish both an absolute and relative evolutionary ranking. Finally, where artifacts or skeletons cannot be dated *in situ*, it may be possible to relate them to vertebrate or other organic fossil remains, which may, in turn, be absolutely dated in other, non-archaeological contexts.

Most of human evolution occurred during periods of alternating cool and warm climate, the glacial cycles of the Pleistocene. Consequently, the chronology of successive glacial periods has been used extensively as a time frame for human evolution. An absolute chronology for these cycles has been provided recently with the advent of studies on deep-sea oxygen isotope variations (Emiliani 1955) which can be dated by radiometric and paleomagnetic methods (Shackleton and Opdyke 1973; Kominz *et al.* 1979).

The absolute dating techniques available to archaeometrists are summarized in Fig. 12.1. Although the time range of these various methods spans the period of rapid human evolution, the applicability of some of the methods to archaeological materials is limited. The K/Ar method has been the most fruitful method in establishing a time-scale for the early stages of hominid evolution in Africa. This is due to the fortuitous occurrence of numerous pyroclastic deposits which are interstratified with the hominid bearing sedimentary beds (McDougall, Maier, and Sutherland-Hawkes 1980). Fission track dating has also proved very useful in the analysis of these deposits (Gleadow 1980). However, K/Ar dating is applicable only where contemporaneously erupted volcanic debris is associated with the archaeological site, a situation only rarely encountered in the vast majority of sites too old for ^{14}C

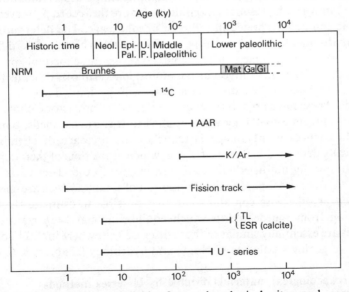

Fig. 12.1. Dating methods available for archaeological sites and corresponding durations of cultural stages in Europe and Asia. U.P. = Upper Paleolithic; Epi-Pal = Epipaleolithic (= Mesolithic); Neol = Neolithic; AAR = amino acid racemization; TL = thermoluminescence; ESR = electron spin resonance; NRM = natural remanent magnetism (shaded = reversed); MAT = Matuyama; Ga = Gauss; Gi = Gilbert; older chrons not shown.

dating. Likewise, fission track dating requires the presence either of fresh volcanic debris or of materials which have been heated sufficiently to have annealed all pre-existing (geological) tracks from the material being analysed. This situation is encountered only rarely in fireplaces predating the advent of ceramic technology. Therefore, these two most promising radiometric methods are not applicable to many of the classic prehistoric sites of Europe, Asia, and Africa.

Radiocarbon dating is, of course, the basic method of the archaeometrist, but it is only applicable to the last 50 to 60 ky. Isotope enrichment methods (Stuiver, Heusser, and Yang 1978) and mass spectrometric detection of ^{14}C will increase the sensitivity for ^{14}C and increase the range of the method, but will not solve the more serious problem of contamination which affects the dating of samples from subsoil environments. Furthermore, the dendrochronological adjustment of ^{14}C dates is at present applicable only back to about 8 ky (Clarke 1975). Beyond this limit '^{14}C years' may deviate by an unknown amount from chronological dates.

At present, other methods of dating listed in Fig. 12.1, thermoluminescence, amino acid racemization, and electron spin resonance dating are all still at a developmental stage of application to prehistoric sites. For each of them, certain environmental parameters (for example, temperature, radiation levels) must be known with reasonable confidence before the record preserved within a sample can be interpreted in terms of absolute age. The determination of these environmental variables has generally proved to be at least as difficult as the analysis of the samples. In principle, however, these methods promise to provide dates for a time-range which is almost inaccesible by other techniques.

The remainder of the discussion here will be limited to the U-series methods. These have proved to be quite effective in prehistoric contexts over the interval from 5 to 400 ky when applied to travertines, shells, bones, and teeth. In each case, advantage is taken of the presence of chemically or biologically deposited materials at a site, which at the time of their deposition were isotopically out of equilibrium with respect to the U-series daughters. When due attention is paid to the elimination of samples which are chemically disturbed or otherwise contaminated, it is possible to estimate the time of deposition from the degree to which equilibrium has been restored. The methods are essentially similar to those discussed elsewhere in this book in the context of geology, paleoclimatology, and hydrology (chapters 3, 6, 11, 18).

12.2 Archaeological materials datable by U-series methods

12.2.1 Inorganic deposits of calcium carbonate

(i) *Mineralogy and chemistry.* Calcium carbonate is found as an inorganic precipitate in a variety of natural environments. Both sea water and fresh water in limestone terrains are commonly close to saturation with respect to calcite. As discussed in chapter 11, such waters may contain a few ppb of U, generally as a carbonate or phosphate complex of the uranyl ion. Thorium is generally absent from such waters due to its low solubility. Therefore, $CaCO_3$ precipitates are ideally suited for U-series dating, containing up to a few ppm of U and virtually no Th at the time of deposition. While the most common crystalline form to precipitate is calcite, natural precipitates of the polymorphs, aragonite and vaterite are also known. Nothing is known about the U chemistry of the latter, but aragonite behaves identically to calcite, possibly having a somewhat higher partition coefficient for U than calcite. Subsequent to deposition, aragonite may invert to calcite, resulting in an opening of the chemical system, to allow redistribution of U and Th isotopes especially if a fluid phase was involved. This would cause at least a partial resetting of radiometric dates; therefore, the archaeometrist must always be on guard for petrographic or geochemical evidence that inversion has occurred (Folk and Assereto 1976). Szabo and Butzer (1979) account for an apparent age inversion in a sequence of lacustrine limestones, by invoking resetting of the U/Th systems during inversion of aragonite to calcite.

(ii) *Cave deposited travertine (speleothem)*. Caves and rock shelters ('abris') have been used by man as temporary shelters or permanent homes for at least 250 ky. The cave of Zhou Kou Dien near Peking was inhabited by *Homo erectus*, possibly as long ago as 500 ky. Prior to the human occupation of caves, they were inhabited by a wide variety of larger vertebrates, principally carnivores. Subsequently, they may have been driven out by man when he aquired the use of fire and weapons adequate to deal with such formidable opponents. Thus, older cave deposits contain human remains which were evidently taken into the caves as the prey of carnivores (Brain 1970).

Most of these caves have been formed by the dissolution of limestone by ground water (see chapter 11) and were subsequently partly filled in with detrital sediments washed, blown, or carried into the caves. These layers of debris are deposited on cave floors interstratified with fragments fallen or broken off the cave roof and walls. Intercalated with these variegated deposits at the mouths of the caves, are layers of speleothem, formed either by water dripping from the roof of the cave, or by slow ground water seepage from the walls of the cave across its floor. Even relatively modestly sized rock shelters can contain such carbonate encrustations. Often these stratiform carbonate deposits vary greatly in thickness as they are traced across a particular stratigraphic horizon; they may be relatively thick near the cave wall at a former seepage site, and diminish to zero thickness near the cave mouth (see Fig. 12.2a). Such speleothem layers can be considered to be chronological markers in the succession of cave deposits, and can be dated by U-series methods to place absolute age limits on under-and overlaying artefacts or skeletal remains.

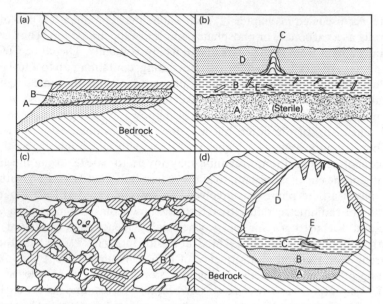

Fig. 12.2. Typical occurrence of speleothem in archaeological contexts. (a) Cultural layer B (containing artefacts, skeletal material, etc.) interstratified between flowstone layers A and C; $t(C) < t(B) < t(A)$ (t = age); note that flowstone A is discontinuous. (b) Stalagmite C deposited on top of cultural layer B: $t(C) < t(B)$; 'soda straw' stalactites (E) embedded in layer B: $t(E) \simeq t(B)$. (c) Culture-bearing breccia of fallen blocks (A) cemented by calcite (B); calcite also infills marrow cavity in bone fragments (C): $t(C) = t(B) < t(A)$. (d) Cross-section of cave passage partly filled with cultural layers A, B, D; wall deposits (stalactites, curtains, etc.) are partly younger than cultural layers; stalactite (E) has fallen into cultural layer C; its outermost layers are closest in age to layer C.

Several other types of related carbonate deposits may also be useful in a similar context. Stalagmites deposited on the floor of the cave, may provide a minimum age for the strata upon which they are formed. In some cases they can also be used to date younger deposits which partially engulf them , if surface features of the stalagmite show that it definitely ceased growth prior to burial (see Fig. 12. 2b). Stalactites are not generally very useful for dating, nor are sheets of speleothem (curtains) that formed initially on the cave roof and were subsequently spalled off onto the floor. At best, the youngest layers of such deposits can be used to place an upper limit to the age of the stratum in which they have fallen. However, small 'soda straw' stalactites are known to grow quite rapidly and can fall or be knocked off onto the cave floor at a more or less constant rate. This component of the cave sediment may prove as useful as a speleothem layer, if it is sorted out during the processing of the sediment by the archaeologist. For example, in Combe Grenal, south-western France, soda straw stalactites constitute over 15 weight per cent of some layers of the sedimentary fill of this rock shelter (Bordes 1972). As carbonate-saturated water percolates through the cave fill, it can precipitate calcite in open spaces such as between large fallen blocks ('eboulis') or within the marrow cavities of bones. Such secondary infillings can provide a minimum age for the stratum in which they are found (see Fig. 12.2c). Deposits formed on the wall of the cave are not generally useful for dating of the cave fill. However, detailed study of stratigraphic relationships at the junction between the cave fill and encrusting layers may reveal a relative chronology which makes possible dating of the occupation of site from absolute ages of the wall deposits (see Fig. 12.2d). Wall encrustations may also cover or underlie art works painted or inscribed on the wall, as is common in south-western France and northern Spain.

Whereas caves are in one sense a very fruitful site for the discovery of datable carbonate deposits associated with cultural remains, in another sense they are far from ideal. Generally hominids only used or inhabited the outermost parts of what may have been quite extensive cave systems. These cave mouths were, as already noted, the site of much detrital deposition, and consequently the carbonate precipitates formed there tend to be contaminated by such detritus. Much of the detritus consists of fine particles of the local limestone rock which is indistinguishable during normal isotopic analysis from the speleothem in which it is embedded. However, being generally many millions of years old, it contributes U and Th which are in secular equilibrium and this contribution leads to erroneously high ages for the deposit. Even non-carbonate detritus can have this effect, if U and Th are leached from it during sample analysis. Therefore, the samples selected for dating must be as free as possible of detrital material. Detrital contaminants can be detected in various ways.

(a) Microscopic analysis. Study of thin sections of the speleothem can reveal the prescence of particles of limestone or dolomite which are texturally different from the coarsely crystalline speleothem, and often will contain relict organic structures. Quartz and feldspar grains or clay particles are generally a clear indication that a detrital component is present, the carbonate fraction of which may not be so readily discernable although it is quite abundant. It may be useful to compare the mineral composition of layers of pure cave sediment, to estimate what fraction of carbonate might have accompanied a given per cent of quartz, and so on. (b) Isotopic analysis. The presence of the 4.0 MeV peak of alpha particles from ^{232}Th is an indication that some contaminant was present, as this isotope is generally absent in freshly precipitated calcite (see chapter 11). Various methods for quantitative correction of contaminated samples based on the measurement of ^{232}Th activity are discussed below. (c) Thermoluminescence, cathodoluminescence. Particles of rock from the wall of the cave or windblown sediment are likely to have retained a significant TL signal due to their age and exposure to radiation. If a slice of the sample is heated in an appropriate television or camera optical system, a record of the distribution of detrital particles can be obtained (Walton and Debenham 1980). Similarly, such particles are likely to have different trace element content from the chemically precipitated calcite. Therefore, under cathodoluminescent illumination, the detrital particles may stand out from the matrix of speleothem.

Ideally, when such detrital material is detected the sample should be discarded in favour of a pure sample. However, if none is available, the fraction of detritus in the sample can be used to estimate a correction factor for the true age of the sample, if some estimate of the U and Th contribution due to the detritus is available. It is, therefore, useful to collect samples of 'pure' contaminant as well when sampling a contaminated speleothem layer.

It is perhaps remarkable that the speleothems found in cave entrances are as free of detritus as they are. The growth surface of the speleothem is wet, and therefore, capable of trapping wind-blown particles or material tracked into the cave by animals and man. However, in some cases the flow of water from a source in the cave roof or wall may be sufficiently strong to wash clean the growing surface, and therefore, leave it uncontaminated. Spring deposits may be washed continually clean in the same fashion (see below).

Another difficulty with carbonate deposits at cave entrances is their high porosity and permeability, because they are commonly formed by relatively rapid evaporation of carbonate-bearing water. After deposition, the vacant pore space may be filled in with later deposits of calcite, or may trap detrital particles which are very difficult to separate during preparation of the sample for analysis.

Although the principal application of speleothem dating in archaeology has been to those sites where artefacts or human skeletons have been found, the same techniques are being applied to the sites containing only non-hominid vertebrates, for example at the Joint-Mitnor and Victoria cave sites, England (Schwarcz and Suttcliffe 1981; Gascoyne 1980). These vertebrate assemblages are the most important index fossils for the Pleistocene, and in other sites they are closely associated with cultural remains, either as prey or as presumed predators. Therefore, even archaeologically 'sterile' caves ought to be considered important targets for U-series dating by archaeometrists.

(iii) *Spring deposited travertine, tufa, and marl.* Calcium carbonate is precipitated by spring waters as they become supersaturated in $CaCO_3$ through loss of CO_2 to the atmosphere. This process may be abetted by the action of plants extracting dissolved CO_2 from the water. Such deposits can take many shapes: a series of stepped terraces over which the spring waters descend, passing through pools rimmed with natural dams of calcite; broad, featureless sheets; hemispherical mounds centered around an emerging spring. Such carbonate deposits are commonly very porous and permeable, because they tend to be deposited around plants present at the site. As this organic matter rots away, a system of tubes and open spaces is left which may subsequently be partly or wholly filled in with a later generation of calcite. Interstratified between those more porous layers, one commonly finds discontinuous layers of dense, impermeable travertine, which is of course much more suitable for dating.

In arid regions, springs were commonly favoured centres of human activity. Not only did hominids seek water there, but they preyed on other animals that came to these watering holes. Therefore, it is not surprising that several important prehistoric sites are found interstratified with the calcareous deposits of fossil springs, for example, at Vértesszöllös, Tata, Bilzingsleben and Ehringsdorf. The dating of such carbonate deposits would clearly allow a precise date to be attributed to the artefacts and human remains at these sites. Unfortunately, as with cave deposits, there are certain problems.

The most serious difficulty is that, when first precipitated these travertines are invariably porous. However, when collected today, much of the original porosity has been reduced by overgrowths of calcite, which are clearly discernable in thin sections (see Fig. 12.3). At least three consequences of this change in porosity must be considered: (i) Post-depositional recrystallization and overgrowths. As calcite is continually deposited in the pores of these strata, the apparent U-series age will be made correspondingly younger, finally achieving an average value which relates to the time taken to reduce the permeability fully in the travertine. For very old deposits (for example, Ehringsdorf) this is probably a small fraction of the total age of the deposit. (ii) Detrital trapping. Any detritus which is being carried in suspension by spring water permeating through the travertine can be trapped in these pores, especially as the flow of water is slowed by the small channels in the travertine. Therefore, it is observed that the pores are sometimes clogged with clay or silt, which must be removed before an accurate date can be obtained.

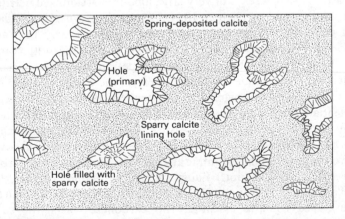

Fig. 12.3. Spring-deposited travertine at Ehringsdorf, D.D.R., showing overgrowth of secondary calcite partly filling primary porosity of travertine.

(iii) Leaching. The porous and permeable calcite presents a large surface area to the action of percolating water. To some extent this water may be able to remove selectively readily soluble ions or exchange them for more reactive ions. In particular, U which is readily soluble in carbonate-bearing waters, may be leached if the U content of the water periodically drops well below that of the water from which the calcite was first precipitated. Cherdyntsev *et al.* (1975) found some evidence at the Ehringsdorf site for leaching of U. However, it is likely that such leaching can only occur if there is also recrystallization of the calcite. Spring-deposited carbonates showing textural evidence of gross recrystallization should thus be looked upon with scepticism.

Tufas can also be formed where streams fed by karstic springs are aerated by passing over waterfalls or rapids. Large mounds of tufa built up in this way along the lip and sides of the waterfall, may enclose artefacts or vertebrate skeletal remains.

Marl, the product of diffuse spring activity in a marshy area, can constitute an abundant surficial deposit, sometimes associated with important archaeological materials. This is evidently the case at Torralba-Ambrona, Spain (Butzer 1965). Such marls are, however, likely to be seriously contaminated with detritus washed off adjacent slopes and trapped in the marshes in which the marls are deposited. Hopefully, such contaminants can be recognized by one of the techniques described above.

(iv) *Concretions and calcrete.* In arid or sub-arid regions, $CaCO_3$ can accumulate in the soil due to the rise of carbonate-bearing waters from the capillary fringe of the water table. Such carbonate, deposits formed either as discrete nodules or as discontinuous sheets of heavily calcified soil are called calcrete or caliche (Reeves 1976). The dating of such deposits is greatly complicated by the presence of very large amounts of admixed detritus. This problem is discussed in chapter 10. In some areas, where aggradation of surficial deposits is fairly rapid, such calcrete layers may be stacked in successive sheets, corresponding to cycles of alternating wetter and drier climate. In a few localities hominid occupation sites have been found associated with such sequences, and therefore, the dating of these deposits becomes archaeologically significant (Nettenberg 1978).

A potentially important application of U-series dating to calcrete soils is found in certain regions of East Africa where volcanic or volcaniclastic deposits have been successively subjected to pedogenesis and then covered by further volcanic material. Paleolithic artefacts have been excavated from some of these sequences (Wendorf, Laury, Albritton, Schild, Haynes, Damon, Shafiqullah, and Scarborough 1975). Because the detritus from which the soils were developed contains essentially no carbonate clasts, the carbonate content of these soils can be assumed to be entirely authigenic. This is the ideal situation for the dating of calcretes (Ku *et al.* 1979), and could be used to confirm K/Ar dates younger than 300 ky, where the latter are subject to great uncertainty due to the possibility of contamination by minute amounts of ancient detritus.

(v) *Inorganic lacustrine limestones.* Inland, closed basins in arid regions commonly contain hypersaline lakes from which calcite or aragonite is periodically precipitated. These lakes may become dessicated during episodes of relatively dry climate (such as the present). Although it is rare to find archaeological materials associated with such deposits, a few are known from South Africa (Szabo and Butzer 1979). The dating of such deposits has been useful in defining the cycle of pluvial/interpluvial climate in the regions in which these ephemeral lakes are found. This, in turn, can shed some light on the cycle of human occupation. Dating of deposits of this type has been discussed in chapter 11; note that, as shown by Kaufman (1971), the problem of detrital contamination is very significant for such deposits.

12.2.2 Biogenic materials

(i) *Calcareous tests of invertebrates.* Many prehistoric peoples made use of marine or lacustrine invertebrates as a part of their food supply; middens of shell refuse are common in archaeological sites ranging back to the lower Paleolithic. Uranium series dating of such shell material has been attempted by many workers. However, Kaufman *et al.* (1971) showed that there are significant differences between ^{14}C and U-series dates on shells over the last 40 ky (see Fig. 12.4). Also, pre-Pleistocene shells sometimes give finite U-series dates. Evidently the shells do not behave as a chemically closed system after the death of the organism. This is easily shown by the almost ten-fold increase in U content which occurs shortly after the burial of the shell. Much of this post-mortem uptake of U may be onto the organic matrix between crystallites of carbonate, although attempts to remove the excess U component by destruction of the conchiolin matrix have so far been unsuccessful (Schwarcz, unpublished studies). In spite of this apparent uncertainty in the significance of U-series dates on moluscan shells, many workers continue to generate or to quote such results with apparent confidence, especially where the dates agree with preconceived notions of the age of deposits. For example, Butzer (1975) bases his analysis of Mediterranean sea-levels on the U-series ages of molluscs obtained by Stearns and Thurber (1965), while admitting that the dates might be uncertain. In an attempt to resolve the problem, Szabo and Rosholt (1969) devised an open system model to correct for the migration of U after death of the organism. In some cases this has yielded apparently satisfactory ages (Szabo and Vedder 1971; Szabo 1979*a*). As with travertine, a good test of the validity of a closed-system ages is concordance between the $^{230}Th/^{234}U$ and $^{231}Pa/^{235}U$ ages of the sample.

Other types of calcareous tests have proved more reliable than molluscs. Coral has generally been found to yield excellent dates by the $^{230}Th/^{234}U$

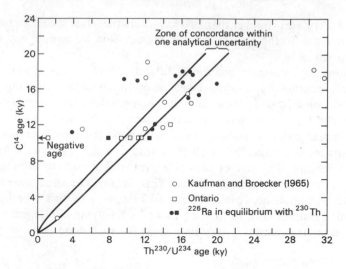

Fig. 12.4. Comparison of ^{230}Th/^{234}U ages and ^{14}C ages for mollusc shells. Data were selected from two regions where the initial ^{230}Th/^{232}Th ratio is sufficiently well known to allow correction for initial ^{230}Th. Note tendency of ^{230}Th/^{234}U ages to be too young, probably due to postmortem uptake of U. Kaufman and Broecker's data are all from the deposits of Lakes Lahontan and Bonneville (from Kaufman *et al.* 1971).

(Bender *et al.* 1979) and ^{231}Pa/^{235}U (Ku 1968) methods. However, Bender *et al.* (1979) find evidence for some addition of U and Th to corals older than 125 ky on Barbados. Unfortunately, coral is rarely found associated with Paleolithic occupation sites.

In the absence of any other datable material, mollusc shells are probably capable of yielding at least an approximate estimate of the age of a site, bearing in mind that the ages will probably be underestimates. All invertebrate tests used for U-series dating should be analysed by X-ray diffraction to ensure that tests which were initially aragonitic have not inverted to calcite.

(ii) *Bones and teeth.* Uranium is present in relatively high concentrations in fossil bones and teeth, from 1 to over 1000 ppm. The concentration in the bones of freshly killed animals is, on the other hand, less than 0.1 ppm. Uranium is taken up from ground water by buried bones, and apparently enters the crystal lattice of apatite which forms during the recrystallization of fossil bones. Szabo (1979b) suggests that the decay of organic matter in the bone causes the local reduction of U and some form of precipitation. The U/Th ratio of bones is generally quite high, and the ^{230}Th/^{232}Th ratio of fossil bones is normally above 20. Therefore, if the uptake of U by buried bones is relatively rapid, and reaches some saturation level after a time which is short compared with the age of the bone, it should be possible to obtain ^{230}Th/^{234}U ages for fossil bones. This was first attempted by Cherdyntsev (1971) who obtained results of variable quality. At the site of Molodovo, he found that younger strata contained bones that gave ages consistently younger than the ^{14}C ages for the site, over the time interval 10 to 25 ky. However, a Mousterian cultural layer in the same site gave an apparent ^{230}Th/^{234}U age of 126 ± 2 and 129 ± 11 ky, in good agreement with the expected value. Cherdyntsev noted that the age ratio $t(^{14}C)/t(^{230}Th/^{234}U)$ increased steadily downward in the upper layers, suggesting recent, downward migration of U into these layers. He further proposed that this U diffusion had not yet reached the lower Mousterian layers, accounting for their more acceptable age. Of course, one cannot be certain of that older age; it too could be somewhat reduced relative to the true age of deposition. The high ^{230}Th/^{232}Th ratio of these older bones (greater than 120 ky) does not prove lack of U migration into them; the U concentrations are unfortunately not reported. Szabo (1980), summarizing several studies of this problem, notes that the $t(^{14}C) - t$ (^{230}Th/^{234}U) difference is fairly consistent for deposits within the range of ^{14}C dating (see Table 12.1) and suggests that U migration ceases after about 2.7 ky (for bones buried in alluvium). The dates on older samples (about 100 to 200 ky) are consistent with other geomorphological and archaeological data, for several sites which he has studied (Szabo and Collins 1975; Szabo, Stalker, and Churcher 1973; Howell, Cole, Kleindienst, and Szabo 1972), lending

Table 12.1. Comparison of U-series and ^{14}C ages on bones from different archaeological sites[†]

Location	U-series age (y)	^{14}C age (y)	^{14}C age − U-series age (y)
Lindenmeier site, Colorado	4250 ± 500[a]	$10\,780 \pm 375$[b]	6530
Lindenmeier site, Colorado	5500 ± 500	$10\,780 \pm 375$[b]	5280
Dent site, Colorado	7700 ± 500	$11\,200 \pm 500$[c]	3500
Lehner site, Arizona	7700 ± 1500[d]	$11\,115 \pm 500$	3415
Domebo Mammoth site, Oklahoma	$11\,500 \pm 2000$[d]	$11\,045 \pm 647$[c]	−455
Murray Springs site, Arizona	$10\,250 \pm 2000$[a,d]	$11\,230 \pm 340$[e]	980
Caulapan, Mexico	$18\,500 \pm 1500$[a,f,j]	$30\,600 \pm 1000$[g]	12 100
Caulapan, Mexico	$21\,000 \pm 1500$[a,f]	$21\,850 \pm 850$[g]	850
Medicine Hat, Canada	9500 ± 1000[a,h]	$11\,200 \pm 200$[i]	1700

(a) Average concordant ^{230}Th/^{234}U and ^{231}Pa/^{235}U age; (b) Haynes and Agogino 1960, Folsom level; (c) Haynes 1967, Clovis levels; (d) Unpublished results of B. J. Szabo; (e) Haynes 1968, Clovis level; (f) Szabo *et al.* 1969; (g) Kelley, Spiker, and Rubin 1978; (h) Szabo *et al.* 1973; (i) Lowdon and Blake 1968; (j) date obtained on dentine, now considered unreliable for dating.
† From Szabo (1980) with permission.

some credence to the method. If the method were generally applicable, it would be of inestimable value in archaeological studies since bone material is frequently associated with prehistoric sites.

Nevertheless, certain points of caution must be raised with respect to the application of this method. The uptake of U by fossil bones is well known in archaeological studies and has been itself used to estimate the age of the bones, for times of the order of 300 ky (Oakley 1969, 1980). If U uptake occurred over such a long time-scale, it would be difficult to estimate the age of a bone as the factors controlling that process would likely have varied throughout the burial history of the bone. There is a curious caveat, however, to the long-standing claims of archaeometrists that U is progressively taken up: almost all the measurements to demonstrate this phenomenon were made by measurement of the total radioactivity of the bone, usually by beta particle counting (for example, Oakley 1969). This method of analysis would in fact be a measure not only of the concentration of U but also of the extent of ingrowth of the daughter nuclides, many of which are beta emitters (^{214}Bi, ^{210}Bi, and so on). Therefore, even a bone whose U content is constant through time would appear to increase in 'eU' (equivalent U) with time. Unfortunately, only a very few analyses of U concentration by fluorimetric or other direct methods have been made on stratigraphic sequences of bones, so the claim of increase with age is not as yet substantiated (see, for example, Haynes, Poberenz, and Allen 1966).

Not only is U taken up by bones *post mortem* but the distribution of this U in the bones is moderately inhomogeneous, as shown by fission track mapping (Szabo, Dooley, Taylor, and Rosholt 1970) and neutron activation analysis of increments drilled from bones (Farquhar, Bregman, Badone, and Beebe 1978). While Szabo *et al.* 1970 only observed a general deficiency near the Haversian canal and enrichment in organic-rich regions, Farquhar *et al.* observed a regular gradient in U concentration, reaching a minimum at the centre of the bone wall and rising toward the outer surface and the hollow interior of the bone (see Fig. 12.5). Other trace elements (for example, Mn, F, Ba) showed very similar parabolic patterns of concentration, with higher values at the outside than inside. The U concentration changed by almost 10 times over this distance. These data, obtained on bones whose age is estimated to be greater than 70 ky, show that a steady-state concentration of U has not been reached yet, and that diffusion was probably still in progress when the bones were recovered. Thus, the conclusion must be drawn that radiometric dating of bones should be accompanied by fission track mapping or other analysis of U distribution. The bones analysed by Szabo *et al.* (1970) may well have reached steady state, and the residual variations in concentration could be attributable to variations in abundance of organic matter, capable of taking up U from the solution, or reducing uranyl ions to the less soluble 4+ state.

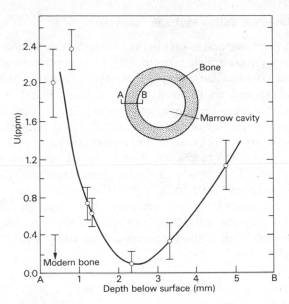

Fig. 12.5. Uranium concentration variations with depth in the humerus (leg bone) of a fossil bison (age greater than 70 ky) from sediments of the Old Crow River, Northern Yukon Territory, Canada. Typical concentration in modern bone is less than 0.4 ppm. Other trace elements (F, Ba, Mn, V) show similar parabolic trends. Analysis by neutron activation (from Farquhar *et al.* 1978).

(iii) *Other biogenic materials.* The shells of avian eggs are composed of calcite crystals embedded in a small proportion of organic matrix. Ostrich shells are up to 2 mm thick and occur as a component of some occupation sites in South Africa and Israel (Marks 1977). These would be potentially datable although some of the problems encountered with molluscan shells may also occur with these materials.

Peat deposits are known to take up U by adsorption from ground water (see chapter 2) and are reported in some deposits to have high $^{230}Th/^{232}Th$ ratios (Cherdyntsev 1971) and to give ages concordant with ^{14}C dates. However, other deposits give low $^{230}Th/^{232}Th$ ratios and, surprisingly, $^{230}Th/^{234}U$ ages which are too low. Uranium is apparently continuously adsorbed by the peat; however, the $^{234}U/^{238}U$ ratio of the peat is lower than that of the present ground water (Cherdyntsev 1971, Table 46). Peat is composed almost entirely of decayed organic material and, therefore, contains very little common Th. Megumi (1979) has attempted to estimate the age of a brown, loamy soil with little organic matter using an open system model; he concluded that the soil had been weathered for more than 1 ky, largely based on the $^{230}Th/^{226}Ra$ ratio. Soils containing more humic material would presumably behave in a manner transitional towards that of peat.

12.3 Procedures for dating and age ranges

The materials described above vary in their chemical characteristics from well behaved, pure systems which take up only U at deposition and thereafter remain as closed systems (for example, speleothem) to materials which take up both Th and U and continuously gain or lose U and possibly other radionuclides after deposition (for example, soils). Thus, procedures for dating must be adapted to suit the material under study. In general, the radiochemistry of the separation of U and Th for measurements of specific activity is carried out as described in chapter 4 with the added precaution that the concentration of U in many archaeological materials (for example, travertines) is very low, 0.05 to 0.1 ppm. The preparation of samples for dating must be adapted to the presence of various possible types of contaminants in the sample and to its chemical and physical composition. The following sections describe possible procedures for some of the more common archaeological materials.

12.3.1 Travertine (including speleothem, tufa, spring-deposits)

The preparation depends on the purity of the material, which can generally be estimated by hand specimen or microscope inspection. The most common contaminants are detrital particles of silicates and carbonates. If the sample is dissolved in 2N HNO_3, as is the common practice, the former contaminant will be left as an insoluble residue from which some fraction of U and Th may have been leached. The carbonate component will be completely dissolved; generally, this component is derived from an ancient limestone bedrock. Thus, both of these detrital components will contribute excess U and Th, tending to give the sample an older apparent age. Various methods of correcting for this effect of contamination are described in Schwarcz (1980) and Ku et al. (1979). All such techniques make use of the fact that detritus contributes ^{232}Th to the sample, whereas biogenic or chemically precipitated carbonate does not (as observed in modern samples). Therefore, the ^{232}Th content of the Th extract can be used as a monitor of the amount of contamination of the sample. In the simplest approach, the initial $^{230}Th/^{232}Th$ ratio is assumed, based either on the value in pure, modern detritus from the site or as inferred from observations in other deposits. Then, the ^{230}Th activity of the ancient sample is corrected for its non-radiogenic component by setting the total activity equal to the sum of the radiogenic and age-corrected common ^{230}Th, and solving the usual age equation for $^{230}Th/^{234}U$ and $^{234}U/^{238}U$ ratios (see chapters 1 and 3).

This method is not very satisfactory if the major contaminant is carbonate detritus, for the latter contains very little ^{232}Th (Th/U ratios are generally less than 1 as compared with values of 3 to 5 for clastic sedimentary material). Fortunately, this component will go completely into solution so that a correction for it can be made if its isotopic composition is known, and if the weight fraction of this component is determined separately. In some cases this can be done for stalagmitic deposits by microscopic analysis; also, cathodoluminescence is capable of revealing the presence of detrital calcite particles in a calcite matrix due to the difference in trace element content of the two calcite phases.

Where an insoluble detrital residue remains after carbonate dissolution by 2N HNO_3, the residue may itself be brought into solution, and analysed to determine the ^{232}Th, ^{230}Th, and U contents, for use in correction procedures (Ku *et al.* 1979). Some laboratories prepare samples of travertine, bone, and so on, for dating by igniting them in air at 800 to 1000 °C. This presumably has the effect of oxidizing organic constituents which may act to complex U or Th during hydroxide coprecipitation. Ignition of calcite results in the formation of CaO, so that water should be added before adding acid, to prevent an excessively exothermic reaction during dissolution. Samples with detrital contamination may sometimes be purified by various sorts of physical separation procedures (grinding, sieving, magnetic, or electrostatic separation). Unfortunately, the most common contaminant in archaeological sites is clay or clay-sized particles of silicates and oxides which are extremely difficult to separate from the carbonate component. Advantage may be taken of the coarser crystal size of the calcite which allows it in some cases to be enriched by sieving or even hand picking.

12.3.2 Biogenic calcium carbonate

As with travertines, some fossil molluscs, especially those buried in detrital sediment, display significant ^{232}Th contents, and some corrections for presumed common ^{230}Th must be made. In addition, as already noted, the closed system ages of many molluscs disagree with independent age estimates. At least two open-system models for mollusc dating have been derived (Szabo and Rosholt 1969; Hille 1979). However, these models make specific assumptions about the interpretation of discordance between the $^{231}Pa/^{235}U$ and $^{230}Th/^{234}U$ dates of the molluscs. In some cases the model ages appear to be in good agreement with independent criteria. The applicability of the models has been questioned by Kaufman *et al.* (1971) and Kaufman (1972). The surest test of reliability of the U-series age is concordance between the $^{231}Pa/^{235}U$ and $^{230}Th/^{234}U$ dates. Because fossil molluscs typically contain 1 to 2 ppm U it is generally possible to measure $^{231}Pa/^{235}U$ activity ratios (^{231}Pa activity is difficult to measure in samples containing less than 1 ppm U.) When dating marine molluscs, attention should be paid to the $^{234}U/^{238}U$ ratio; commonly, this is significantly different (higher, in general) from that of modern molluscs (see chapter 17) showing that U has been taken up *post mortem* from ground water. The uniformity of this ratio between separate aliquots of a mollusc sample may be a guide to the degree of disturbance of the sample. Another approach which has been attempted by this worker is to attack selectively the organic matrix (conchiolin) of the fossil shell with various oxidizing agents (for example Chlorox, hydrogen peroxide) because others have shown that organic coatings on the mollusc shell are associated with high U contents (Lahoud, Miller, and Friedman 1966; Komura and Sakanoue 1967). Destruction of these coatings should eliminate much of the adsorbed U; however, care must be taken not to allow the ^{230}Th which was produced by it to be redeposited on the shell material. Many molluscs deposit aragonitic shells; shells which have inverted from aragonite to calcite are unlikely to give reliable dates.

12.3.3 Bones and teeth

As with molluscs, U is taken up *post mortem* and, therefore, some correction for this open-system behaviour should be made. Szabo (1980) has found that the period over which U is taken up is generally short enough that bone dates ought to be reliable for samples older than about 20 ky. As with molluscs, the U concentrations are high enough to permit testing ^{231}Pa/^{235}U ages for concordancy with ^{230}Th/^{234}U ages. Where excess ^{231}Pa is observed, the open-system model of Szabo and Rosholt can be applied (Szabo 1980). Since phosphate interferes with the extraction of U and Th, special care must be taken to assure high yields (Kolodny and Kaplan 1970), although the U concentrations are sufficiently high in many fossil bones that low yields are acceptable.

12.3.4 Other criteria

In addition to measuring the ^{230}Th/^{234}U, ^{231}Pa/^{235}U, and ^{234}U/^{238}U ratios, other daughter nuclides have been studied to test for the reliability of the samples. The ^{226}Ra/^{230}Th ratio for samples older than 10 ky ought to be unity. If it is not, the sample has been chemically disturbed within the relatively recent past. Similar tests can be made for the other short-lived daughters (see chapter 2). In some cases it may be preferable to assay the ^{230}Th content of a travertine by use of the more readily soluble ^{226}Ra, thus allowing one to use a weaker acid to dissolve the sample and avoid leaching of ^{230}Th from a detrital component (Turekian (1979) personal comm.).

The upper time limit for the age of archaeological samples datable by U-series methods appears to be about 350 ky, that is, the later part of the Lower Paleolithic. This limit is largely determined by the concentrations of U and size of sample available as these factors determine the total number of alpha counts available, and thus the precision of the isotope ratios. As this ratio reaches a limit near unity for old samples, the uncertainty in the age estimate increases drastically for a given precision of the ^{230}Th/^{234}U ratio (see chapter 3). The lower limit of dating also depends on the U content; high-U samples can be dated down to about 3 ky, with an error of about 10 per cent. Obviously over the low end of the time range, ^{14}C is generally a preferable method of dating. Reported ages for travertines and bones generally have quoted precisions ranging between 4 and 10 per cent.

12.4 Applications

U-series dating has not yet been extensively exploited in archaeometry. The principal published examples have been in the work of Szabo and his associates on bones and molluscs, that of Cherdyntsev on bones and travertines, and the work of the present author. Some recent examples are given below.

12.4.1 Travertines

(i) *Nahal Zin, Israel* (Schwarcz *et al.* 1979). A series of spring-deposited travertines have been found in this valley of the northern Negev desert. Artefacts have been found embedded in two of these spring deposits, apparently left there by hominids who were occupying nearby sites found in the same valley (Marks 1977). At 'En Mor, near the site of Avdat, artefacts were found embedded in a block of travertine that had fallen from a dissected spring deposit high on the wall of the canyon; the artefacts are attributed to the transition from Middle to Upper Paleolithic culture in this region. Layers of travertine above and below the tool-bearing layer give a concordant age of 46.5 ± 2.9 ky. At a nearby site in alluvial sediments, similar artefacts were found, associated with charcoal which gave ^{14}C dates of 45 ± 2.4 ky.

At another site, about 5 km distant, a mound of travertine was found, inter-stratified with artefact-bearing colluvium. The artefacts are of Levalloiso–Mousterian affinity; the travertine layers immediately underlying them give a date of 80 ± 10 ky, which is considered to be quite consistent with the technological level of the artefacts. The same spring was apparently active from about 260 ky to the time of occupation. Such springs are commonly found to be active intermittently during periods of higher rainfall, and the dating of such deposits may eventually allow us to work out a detailed chronology for the Pluvial/Interpluvial cycle in the Levant.

(ii) *La Chaise, France* (Schwarcz and Debenath 1979; Blackwell 1980). This cave in the Charente district of central France, contains a thick series of detrital sediments in which three distinct travertine horizons are interstratified (see Fig. 12.6). Artefacts in the sediments range from Lower Paleolithic to Middle Paleolithic industries. The lowermost travertine, layer 53', found exposed in Abri Suard, yielded an age of 185 ± 30 ky, placing its time of deposition within the penultimate glaciation in agreement with pollen evidence from the overlying sediments. Above this level are found the remains of a pre-Neanderthal hominid. The age of the middle layer (bed 11 in Abris Bourgeois-Delaunay) is 151 ± 16 ky. Immediately below this layer or embed-

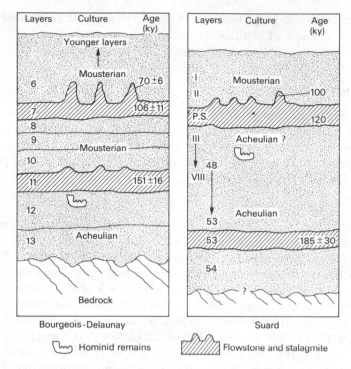

Fig. 12.6 The ^{230}Th/^{234}U dates on speleothems from the cave of La Chaise, Charente district, France. Bourgeois-Delaunay and Suard are two adjoining chambers of this cave. The stratigraphy and cultural assignments were made by Debenath (1976); U-series dates are from Blackwell (1980) and Schwarcz et al. (1979).

ded in it was found an early Neanderthal mandible. The upper travertine, dated at 106 ± 10 ky, caps beds containing Mousterian artefacts. Pollen and sedimentological data suggest that bed 11 should have been formed during the last interglacial; the age given here is distinctly too old for isotopic stage 5e, and may represent an interstadial in the preceding glacial ('Riss II/III' of the French terminology).

(iii) *Pontnewydd cave, Wales* (Green, Stringer, Colcutt, Currant, Huxtable, Schwarcz, and Bevins 1981). A cave opens off the steep side of a glaciated valley near the Irish Sea coast. The cave is partly filled with a sequence of detrital sediments including at least two poorly sorted, very coarse deposits interpreted to be the products of mud flows. These beds contain stone artefacts attributable to an Acheulian industry. One hominid tooth has been found near the top of the lower mud flow. Both mud flows contain fragments of stalagmite, and stalagmitic crusts cap the lower mud flow. The ^{230}Th/^{234}U dates on the stalagmitic cap of the lower flow range from 177 ky to 90 ky, indicating that there was no deposition of detrital matter during the Riss glacial stage.

The older date places a lower limit on the age of the hominid, and is in good agreement with a thermoluminescence date on a burnt flint core from within the lower mud flow, and close to the site of the tooth. Ages on stalagmitic fragments within the flows range from 220 to 130 ky and suggest that the fragments are reworked from cave-mouth deposits formed in an outer part of the cave that has been obliterated by glacial erosion. The artefacts were presumably also originally laid down in this outer section.

(iv) *Bilzingsleben, DDR* (Harmon, Glazek, and Nowak 1980). Within layers of spring deposited travertine were found various fossil fauna and flora as well as parts of the skull of a hominid identified as *Homo erectus* by Vlcek and Mania (1977). The travertine has been dated by ^{230}Th/^{234}U ratio measurement and yields an age of 228 ± 15 ky. Various size fractions of the travertine all yield the same age within the precision of the analysis. This result is especially interesting because the nearby site of Ehringsdorf, in which a Neanderthal skull fragment was recovered appears to date from about the same time (Schwarcz 1980). This possible coexistence of two species of the genus *Homo* in Europe in the middle Pleistocene may require some modification of theories of the evolution of man.

Other sites at which travertine associated with archaeological materials have been dated are described in Schwarcz (1980); Schwarcz, Goldberg, and Blackwell (1980); Pecsi (1973); Cherdyntsev (1971); Turekian and Nelson (1976), and Cherdyntsev *et al.* (1975). Szabo and Butzer (1979) have dated a series of lacustrine deposits which contain an archaeological site. Fornaca-Rinaldi (1968*b*) measured ^{230}Th/^{238}U ratios of travertines associated with Middle and Upper Paleolithic sites but failed to correct for the departure of ^{234}U/^{238}U ratios from unity; her age determination are presumably erroneously large, by an indeterminable amount.

12.4.2 Molluscs

Although many archaeological sites contain mollusc shells, few studies based on U-series dating of such sites have been undertaken. Stearns and Thurber (1965) dated a number of raised beaches around the Mediterranean and North Africa. These data were later used by Butzer (1975) to establish a chronology for the sea-level history of Mallorca. Butzer, Brown and Thurber (1969) dated an oyster reef interbedded with fluviatile sediments in the Kibish formation, southern Ethiopia, at 130 ± 5 ky. This reef underlies a poorly defined artefact assemblage and a *Homo sapiens* site. An overlying bed was dated at 30 ± 0.5 ky, while ^{14}C dates on the same molluscs gave ages of at least 37 ky.

Although corals are believed to be intrinsically better for dating than molluscs, very few archaeological sites contain coeval fossil corals. Such an occurrence might be expected in areas of pronounced tectonism or where occupation had occurred during an interglacial; in either case the regional climate must be subtropical or warmer. One such site, in the Afar district of

Ethiopia, was studied by Roubet (1969). She found two artefacts which could be approximately associated with a coral fauna that had been dated at about 200 ky. However, there was some uncertainty in the stratigraphic origin of the artefacts, and their cultural significance. One, a hand axe, could have been of late Acheulian origin.

12.4.3 Bone

(i) *Isimila, Tanzania* (Howell *et al.* 1972). An 18 m thick section of detrital sediments at this site contains Acheulian artefacts and vertebrate bones. Szabo analysed a bone sample from the lowest bed of this sequence (Sand 4) and obtained the data shown in Table 12.2. The ^{230}Th/^{232}Th ratio was so high in this sample (21 000) that no correction for inherited ^{230}Th needed to be made. This is characteristic of fossil bone. The ^{230}Th/^{234}U age of $260 \pm ^{40}_{70}$ ky is described by Howell *et al.* (1972) as being compatible with the archaelogical evidence; perhaps this is a reflection, in part, of the great uncertainty in the chronology of Lower Paleolithic cultural evolution.

(ii) *Valsequillo, Mexico* (Szabo *et al.* 1969). This site contains a series of detrital, fluviatile sediments, the upper parts of which have been dated by ^{14}C to between 9 ± 0.5 and 22 ± 0.9 ky. A proboscidean vertebra from the stratigraphic level of the 22 ky ^{14}C age determination yielded dates in excellent agreement with the ^{14}C date (see Table 12.2, sample M-B-6). A bone from lower down in the section (where molluscs had yielded ^{14}C ages of 30 ky or older) gave concordant ^{230}Th/^{234}U and ^{231}Pa/^{235}U ages of 18 and 19 ky respectively (sample M-B-5). There is some question about the provenance of the molluscs used for the ^{14}C date. The greatest enigma at this site arose, however, from dates on bones associated with artefacts, that ranged from 248 to 340 ± 100 ky by ^{230}Th/^{234}U and greater than 165 ky by ^{231}Pa/^{235}U (samples M-B-3, 8, 4). Open system dates on two of these samples were 240 ± 40 and 260 ± 60 ky. One of the samples (M-B-8) was a tooth of a mastodon that had been slaughtered by the occupants of the site. It cannot, therefore, be accounted for as a reworked older bone. This sample had the highest U content (150 ppm) and gave only lower age limits of 280 and 165 ky. It is possible that the sample was radiochemically disturbed. All the bone samples had ^{230}Th/^{232}Th ratios greater than 33, so it is unlikely that contamination by common ^{230}Th can account for these high ages which are much greater than have been generally assumed for hominid occupation of the New World. The bone samples M-B-3 and M-B-8 could conceivably represent transported bone from earlier vertebrate remains, or may have been collected from such deposits by the occupants of the site. Typologically the tool assemblage at this site (bifacially worked knives, scrapers, burins, and tanged projectile points) is known from other sites in North America where it is dated at about 11 ky. Curiously, fission track dates on zircon extracted from tephra units at this site also suggest high

Table 12.2. Uranium series dates on fossil bones

Sample	$^{234}U/^{238}U$	$^{230}Th/^{234}U$	$^{231}Pa/^{235}U$	$^{230}Th/^{234}U$ date (ky)	$^{231}Pa/^{235}U$ date (ky)	Open system date (ky)	^{14}C date (ky)	U (ppm)
1. Isimila prehistoric site, Tanzania (Howell et al. 1972)								
—	1.23±.012	0.96±.05	1.00[a]	260^{+70}_{-40}	g.t.170	—	—	1457±20
2. Valsequillo reservoir site, Puebla, Mexico (Szabo et al. 1969)								
M-B-6	1.30±.02	0.172±.008	0.38±.02	20±1.5	22±2	—	21.85±.85	77
M-B-5	1.30±.02	0.161=.006	0.31±.02	19±1.5	18±1.5	—	30.6±1.0[a]	79
M-B-3	1.22±.02	0.939=.038	1.03±.05	245±40	g.t.180	245±40	—	86
M-B-8	1.26±.02	1.03±.05	1.02±.05	g.t.280	g.t.165	g.t.280	—	150
M-B-4	1.40±.02	1.04±.04	1.11±.06	340±100	g.t.180	260±60	g.t.35	58

[14]C dates were determined on molluscan fossils associated with vertebrate remains, by M. Rubin, U.S. Geological Survey.
[a] Association of shell and bone is not clear.

ages (370 ± 20 ky and 600 ± 340 ky; Steen-McIntyre, Fryxell, and Malde 1975).

Other studies of bone in archaeological sites are those of Szabo and Collins (1975); Cherdyntsev (1971); Szabo (1979a,b), McKinney (1980); and Sakanoue and Yoshioka (1974).

12.5 Future work

As noted above, the application of U-series dating to archaeological sites is still in its infancy, in spite of its long history. The potential of the method has not been generally appreciated by most archaeologists. There is a need to inform prehistorians involved in field studies of the possibilities of the method, and of the various types of materials that are suitable for dating. For example, in many cave sediments, fine 'soda straw' stalactites are obtained during the sieving of the sediments. This component is probably suitable for dating and should be preserved for this purpose.

Besides the extension of the methods described above to more sites, there are some related dating methods which should be investigated. Thermo-luminescence dating of travertines has been attempted by Wintle (1978) with partial success. However, she notes that for samples younger than 400 ky, correction must be made for the growth of the daughters of the U isotopes, as these contribute to the radiation field which induces the trapped electrons detected by this method. Electron spin resonance dating (Ikeya 1978) also depends on the presence of trapped free electrons. Therefore, estimates of radiation dose must also include a term for approach to secular equilibrium, as computed by Wintle.

The age limits of the U-series methods at present lie between about 5 and 350 ky. One may consider how these methods may be extended to both older and younger dates. The $^{234}U/^{238}U$ ratio method has been proposed as a method for determining ages up to 1000 ky (Thompson, Lumsden, Walker, and Carter 1975). However, the method (described in chapter 3) depends on the constancy of the $^{234}U/^{238}U$ ratio in the water from which the calcite was deposited. This is clearly not true in all cases, as shown by Harmon, Schwarcz, Thompson, and Ford (1978) and Cherdyntsev (1971). One alternate approach to this problem is to measure the isotopic ratio of U that is leached from travertine by a weak acid. This U will be isotopically enriched in ^{234}U with respect to the bulk U in the sample, due to the recoil of daughter atoms into the interstitial sites in the calcite lattice. As the calcite ages, a larger fraction of the total ^{234}U atoms resides in such 'hot atom' sites, whereas in a freshly deposited calcite, all ^{234}U atoms will be in lattice sites appropriate to U (presumably, Ca^{2+} sites). It can be shown that the age t is given by the relation

$$\frac{B_s - B_x}{f - 1} = 1 - e^{-\lambda_{234}t} \tag{12.1}$$

where B_s and B_x are the $^{234}U/^{238}U$ ratios in the leach solution and total calcite respectively, and f is the $^{234}U/^{238}U$ ratio in a solution obtained by leaching an infinite-age calcite sample; λ_{234} is the decay constant of ^{234}U. Andrews (personal comm.) has obtained a value of $f = 1.7$ for leaching of a limestone with carbonic acid.

To extend the age range to younger dates, one may consider the measurement of the ^{226}Ra activity of travertines. Because of its greater chemical mobility, Ra is found in young chemically precipitated calcite in significant excess over the amount of supporting ^{230}Th. If the initial excess is known or can be estimated, then the residual ^{226}Ra activity may be used as a crude estimate of time since deposition. It would be interesting to test this method on young spring deposits where ^{14}C-dated organic matter is also present, embedded in the deposits. Hard water deposits in ancient aquaducts might be datable by this technique.

13 APPLICATIONS OF U-SERIES DATING TO PROBLEMS OF QUATERNARY CLIMATE
H. P. Schwarcz, M. Gascoyne, and R. S. Harmon

13.1 Introduction

The Quaternary period of the geological record has been characterized by intense fluctuations of global climate, and consequent changes in landforms, sedimentation patterns, biota, oceanic and atmospheric circulation. Large masses of ice have periodically advanced over and receded from the continents, while sea level has risen and fallen synchronously. These processes have left a vast variety of geological and geomorphological evidence, from which it has been possible to reconstruct the paleoclimatic history of this period. However, the time-scale of these events is such that the more traditional methods of radiometric dating are not adequate to provide an absolute chronology for them. Fortunately, U-series dating provides a temporal bridge between the range of radiocarbon (conventionally applicable to deposits younger than 50 ky) and $^{39}Ar/^{40}Ar$ or fission track dating, which are generally applicable to volcanic deposits older than 500 ky (see chapter 3). The applicability of U-series dating to establishing a late Pleistocene chronology depends, however, on the occurrence of appropriate, datable materials which are syngenetic with the climatic event being dated. The geological settings in which such datable materials can be found, and some of the evidence which has been obtained from the study of these materials are reviewed below. It should be noted that several of the other chapters have, in one way or another, provided evidence about Quaternary paleoclimate, since it is hardly possible to describe the dating of speleothems, pluvial lakes, or raised coral reefs without noting that their existence, or frequency of occurrence, is largely controlled by the alternation of glacial and interglacial climates of the Pleistocene epoch.

Although studies of Quaternary geology have largely focussed on continental or shoreline processes, the problem of establishing a chronology for the climatic cycles has so far been most effectively dealt with through the study of marine deposits, primarily the record of foraminifera in deep sea sediments. The dating of the upper layers of these deposits has been possible by use of

[14]C, but for the greater part of the sediments, a time-scale has been obtained by study of U-series nuclides, coupled with the measurement of paleomagnetic orientation. The ^{18}O-content of foraminifera in these sediment cores has been shown to be largely controlled by the volume of glacial ice on the continents, and therefore, this isotopic record does provide, at least in principle, a chronology for global climatic cycles, which should ultimately be correlative with the glacial stratigraphic chronology of continents. While some attempts to make this correlation have been made (for example, Kukla 1977) there is as yet no general agreement as to the details of this correlation, so that it is not yet possible to replace the conventional continental glacial stage names with corresponding isotope stage numbers as given by Emiliani (1955). Nor would it be desirable at this time to do so, for it is likely that many of the less intense glacial advances and retreats may not be faithfully preserved in the marine foraminiferal record, due to bioturbation and other types of smoothing of this record.

13.2 The nature of the paleoclimate record

This section describes the natural settings in which climatic fluctuations have left features which are, or might possibly be, datable by U-series techniques. The deep sea record is only dealt with briefly here, although ultimately it is the most complete and readily datable record of global ice volume, and hence global climatic state. Uranium-series methods of dating deep sea sediments are described in chapter 16. Similarly, the coastal record of sea level fluctuations, which is also a record of ice volume, is discussed in chapter 18, and will be only briefly considered here except as it relates to cave deposits or other continental sediments.

13.2.1 The marine record

(i) *Deep sea sediments.* In deep sea sediment cores it has been demonstrated by Shackleton (1967) and Shackleton and Opdyke (1973) that ^{18}O/^{16}O variations in carbonate foraminifera are a direct measure of the volume of water stored on the continents as glacial ice, and thus, also a sensitive indicator of global climate change. However, in order to have more than just a qualitative record of climatic change from a deep sea sediment core, it is necessary to relate observed oxygen isotope variations to an absolute time-scale (see Fig. 13.1). Although paleomagnetic calibration has recently become the standard means of establishing time scales (see for example, Shackleton and Opdyke 1973), U-series techniques were used by Rosholt *et al.* (1961, 1962), Ku and Broecker (1966), Rona and Emiliani (1969), and Broecker and Ku (1969) for this purpose. Here are shown the data of Kominz *et al.* (1979) for Pacific core V28-238 only to illustrate the characteristic decrease of [230]Th activity with depth in the core and the precision with which

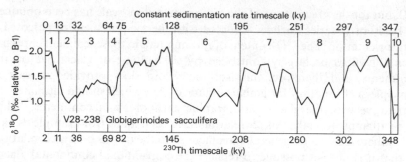

Fig. 13.1. Oxygen-isotopic composition of *Globigerinoides sacculifera* in deep-sea sediment core V28-238 expressed as per mil (‰) deviation from the Emiliani 'B-1' standard. Two time-scales are shown; the top, that of Shackleton and Opdyke (1973),is based upon a 700 ky age for the Brunhes–Matuyama magnetic reversal in the core and the assumption of a constant sedimentation rate throughout, and the bottom that of Kominz *et al.* (1979), derives from the actual ^{230}Th dating of the core which is in agreement with the time-scale derived based upon an age of 730 ky for the Brunhes–Matuyama boundary advocated by Mankinen and Dalrymple (1979). Also shown are the isotope stage numbers and boundaries of Emiliani (1955). Periods of ^{18}O enrichment in the ocean represent times of northern hemisphere glaciation when ^{18}O-depleted ice accumulated on the continents. It should be noted that the TWEAQ (TUNE-UP with exact astronomical quantification) advocated by Kominz *et al.* (1979) places the stage 1/2 boundary at 11 ky, the stage 2/3 boundary at 29 ky, the stage 3/4 boundary at 61 ky, the stage 4/5 boundary at 73 ky, the stage 5/6 boundary at 127 ky, the stage 6/7 boundary at 190 ky, the stage 7/8 boundary at 247 ky, the stage 8/9 boundary at 276 ky, and the stage 9/10 boundary at 336 ky.

Table 13.1 Excess ^{230}Th and age against depth in Pacific sediment core V28-238 (after Kominz *et al.* 1979).

Isotopic* stage	Depth (m)	Excess ^{230}Th (dpm/cm^3)	Age ($\pm 1\sigma$)[†] (ky)
1	0.0–0.2	5.30 ± 0.23	11.2 ± 0.45
2	0.2–0.5	8.01 ± 0.37	36.3 ± 1.8
3	0.5–1.1	4.81 ± 0.24	69.0 ± 3.7
4	1.2–1.3	4.59 ± 0.19	81.6 ± 4.1
5	1.3–2.1	3.21 ± 0.14	144.7 ± 9.5
6	2.1–3.2	1.44 ± 0.08	208.0 ± 18
7	3.2–4.4	0.83 ± 0.07	260.0 ± 30
8	4.4–5.1	0.52 ± 0.04	302.0 ± 41
9	5.1–5.9	0.36 ± 0.03	348.0 ± 58

* Isotopic stages are numbered according to Emiliani (1955).
† Ages are determined from the equation $t_x = (1/\lambda_{230}) \cdot \ln (1 - N_x/N_t)$ where t_x is the age of the bottom of the X th section, N_x is the cumulative ^{230}Th above the bottom of the X th section, N_t is the total unsupported ^{230}Th content of all sections in the core, and λ_{230} is the decay constant of ^{230}Th.

[230]Th ages can be determined (see Table 13.1). The $^{18}O/^{16}O$ paleoclimatic record of core V28-238 based upon this absolute dating is compared in Fig. 13.1 with that derived from paleomagnetic stratigraphy assuming a constant sedimentation rate. Times of Pleistocene glaciation (the even numbered stages) are indicated by ^{18}O-enrichment in marine foraminifera as a result of accumulation of ^{18}O-depleted ice on the continents. Ages for the last four glacial–interglacial cycles as derived from $^{18}O/^{16}O$ variations in U-series dated, deep-sea sediment cores, are given in Table 13.2. The good agreement between the [230]Th time scale for core V28-238 of Kominz *et al.* (1979) and the 'best estimate' TUNE-UP time-scale of Hays, Imbrie, and Shackleton (1976) is a strong argument that U-series methods can provide a reliable means of dating deep sea sediments.

Table 13.2. The [230]Th chronologies for deep-sea sediment cores

Isotopic stage	V28-238[1]	'TUNE-UP'[2] age (ky)	V12-122[3]
1	0–11	0–10	0–11
2 3 4	12–81	11–73	12–74
5	82–145	74–127	75–127
6	146–208	128–190	128–186
7	209–260	191–247	187–225
8	261–302	248–276	226–256
9	303–348	277–336	257–283
10	349–362	337–356	284–350

[1] Kominz *et al.* (1979)
[2] Hays *et al.* (1976)
[3] Broecker and van Donk (1970)

(ii) *Sea level variations.* Eustatic variations in sea level of up to 150 m occurred during the Quaternary as a result of episodic continental ice formation. Coupled with these shifts was a change in both the isotopic composition of sea water (see above) and temperature of surface ocean water. Marine deposits formed on the sea coasts are found to contain a record of these variations. The most striking examples are raised coral reefs which are found along many tectonically-active coastlines, such as New Guinea and Barbados. In areas of active uplift, reefs can only flourish at times of high sea stands, when the rise of sea level keeps up with the rise of the substrate. Numerous U-series dates have been obtained on aragonitic corals from these raised reefs, and coincide in general with estimated dates for low ice volume as obtained in deep sea sediment cores. Oxygen isotopic studies of both the corals themselves and molluscs associated with them have suggested that, as expected, the isotopic composition of sea water at the time of high sea stands was close to its present value (Fairbanks and Matthews 1978).

Unfortunately only *high* sea stands can generally be recognized by studies of reef deposits. However, in many areas of coastal carbonate deposition, karstic caves and speleothems are present, and the deposition of speleothems is initiated at elevations below present sea level during glacial periods. Subsequently, with rise in sea level, the caves are filled with sea water, and deposition of speleothem ceases. It is thus possible to establish points of minimum depth of sea level at a given point in time, by dating speleothem recovered from such 'drowned' caves. This has been done in Bermuda (Harmon, Schwarcz, and Ford 1978) and the Bahamas (Spalding and Mathews 1972; Gascoyne *et al.* 1979). The Bermudan work noted the general coincidence of the timing of high and low sea stands with interglacial and glacial stages, respectively, as inferred from the isotopic record of deep sea sediments, while Gascoyne *et al.* were able to establish a point in the curve of relative sea level, at −42m, which agreed well with the estimate of lowering given by Shackleton and Opdyke (1973), based on foraminiferal isotopic analyses.

Other records of high and low sea stands have been obtained indirectly by the dating of sediments on adjoining land masses. For example, on Bermuda, aeolianite dunes are the principal constructional landforms on the island and have been shown by Harmon *et al.* (1981) to have been accreted during inter-glacial periods when sea level was below its present level but above the −20m level of the Bermuda platform. The dating of these events is made possible, in part, by U-series analysis of corals found in beach deposits correlative with the aeolianites, or overlying them. Another setting for the dating of sea level variation is in sediment fill of coastal caves (just above sea level) where periods of high sea stand may cause sand to accumulate in the caves due to approach of the strand line to the cave. This has been observed in the cave of Tabun on Mt Carmel, Israel (Goldberg, in Jelinek, Farrand, Haas, Horowitz, and Goldberg 1973).

13.2.2 The continental record

Records of climatic variation during the Quaternary are preserved in sediments which have been laid down in a variety of terrestrial settings. These are summarized in Table 13.3. For the most part we are concerned with sedimentary records and we should consider how changes in climate affect sedimentation patterns. Physical changes in rates and types of sedimentation can occur in the following settings.

(i) *Lakes*. In semi-arid regions, an increase in rainfall with respect to evaporation and stream runoff will lead to an increase in lake volume, which may locally lead to freshening of otherwise saline (closed-basin) bodies of water, and also to a rise in shore lines. A record of such an event can be preserved in lacustrine sediments, shoreline deposits including beaches, and fresh water limestone encrustations, and wave-cut terraces. Material datable by U-series methods can be found in any of these types of deposits.

(ii) *Streams.* Rates of aggradation and downcutting of stream valleys also vary as a result of climatic change due to changes in stream runoff and, in some instances, melt-water discharge from icesheets. The fluvial sediments are not themselves datable by U-series techniques, but locally may be capped by spring deposits (tufa) that are datable.

(iii) *Glacial deposits.* Deposits that are clearly of glacial origin (for example, drift, loess, varved lake sediments) are a direct record of climatic fluctuations. Unfortunately there is virtually no portion of these deposits which is directly datable. Subglacial calcite precipitates have been described (Hilaire-Marcel, Soucy, and Cailleux 1979; Hanshaw and Hallett 1978) which could be dated although they tend to contain significant amounts of rock dust. Carbonate-rich loess develops concretions, which also contain some rock particles, but minimum estimates of age may still be determined.

(iv) *Spring deposits.* The flow of spring waters is often modulated by changes in rainfall, which are related to changes in atmospheric circulation. In arid regions, spring flow may be possible only during periods of water recharge into local aquifers, that is, during pluvial periods. In intermittently glaciated regions, recharge may be interrupted during glacial periods due to ice cover or permafrost formation. Tufa and travertine deposits are formed where these springs are emerging from carbonate aquifers. If these deposits are not too porous or contaminated with detritus, they can be dated by U-series methods.

Not only recharge and spring flow is necessary to assure carbonate deposition. The emerging water must be super-saturated in $CaCO_3$. This will not occur if the bicarbonate ($HCO_3{}^-$) content of the water was in equilibrium with atmospheric CO_2 levels, because such waters only dissolve a small amount of $CaCO_3$. Therefore, the presence of tufa deposits also indicates that the emergent waters have acquired excess $HCO_3{}^-$ at the time of their recharge through the soil zone, due to root respiration and the presence of decomposing organic material in the soil. Ancient tufas can thus serve as indicators of the past existence of relatively dense vegetation in areas which might be only sparsely vegetated today.

Spring deposits and other forms of tufa form datable cappings deposited on top of other surficial materials. Frequently, these materials, such as glacial till, glacial outwash deposits, loess, paleosols, pollen-bearing alluvium or colluvium, and cryoturbated soil, may be indicators of past climatic conditions. In exceptional cases, such deposits may be interstratified between successive sheets of travertine, thus giving an excellent setting for dating of the climatic events. Distinctive land forms, especially erosional features such as terraces, gullies and badlands, can locally be found to have been either eroded

Table 13.3. Summary of continental Quaternary deposits and features used to determine chronology and paleoclimate

Type	Feature or deposit	Dating methods	References
1. Climate charac-teristic deposits	till, fluvioglacial deposits, erratics, glacial striae, periglacial features U-shaped valleys, etc.	—	Flint (1971)
	pluvial lakes (in presently-arid regions)	—	Flint (1971)
	river terraces		Zeuner (1959)
2. Datable climate-characteristic deposits	any of type 1 when interbedded with datable strata (e.g., till with volcanic deposits)	—	Porter (1979)
	pollen assemblages in bog and sediment cores	^{14}C	van der Hammen, Wijmstra, and Zagwijn (1971); Woillard (1978)
	well-preserved wood, peat, etc.	^{14}C, $^{230}Th/^{234}U$, A.A.R.	Redfield (1967); Cherdyntsev (1971); Lee, Bada, and Peterson (1976)
	faunal remains (bones, teeth, beetles, etc.)	^{14}C, $^{230}Th/^{234}U$, $^{231}Pa/^{235}U$, A.A.R.	Coope (1977); Cherdyntsev (1971); Szabo et al. (1973); Bada (1972)
	lacustrine sediments	^{14}C, $^{230}Th/^{234}U$	Broecker and Kaufman (1965); Peng et al. (1978)
	molluscs (mainly marine) and gastropods	^{14}C, $^{230}Th/^{234}U$, $^{231}Pa/^{235}U$, ^{226}Ra, A.A.R.	Keith and Anderson (1963); Szabo and Rosholt (1969); Kaufman et al. (1971); Mitterer (1975)

soils, paleosols, and calcrete	^{14}C, ^{230}Th/^{234}U	Williams and Polach (1971); Valentine and Dalrymple (1976); Ku et al. (1979)
loess	^{14}C, P.M.	Kukla (1977)
varves	^{14}C, layer counting	Tauber (1970)
travertine	^{14}C, ^{230}Th/^{234}U, ^{231}Pa/^{235}U	Cherdyntsev (1971); Schwarcz et al. (1979)
archaeological sites (burnt flint, fireplaces etc.)	T.L.	Wintle and Aitken (1977)
3. Datable deposits containing climate- or temperature-sensitive component (e.g. stable isotopic content)		
ground water	^{14}C, ^{234}U/^{238}U	Wigley (1975); Chalov et al. (1966b)
tree rings	^{14}C, D.C.	Suess (1965); Yapp and Epstein (1977)
ice cores	^{14}C, ^{210}Pb, ^{32}Si, ^{39}Ar	Hammer, Clausen, Dangsgaard, Gundestup, Johnson, and Reeh (1978)
speleothems	^{14}C, ^{230}Th/^{234}U, ^{231}Pa/^{235}U, ^{234}U/^{238}U, P.M., E.S.R., T.L.	Rosholt and Antal (1962); Hendy and Wilson (1968); Thompson, Lumsden, Walker, and Carter (1975); Harmon et al. (1978b); Latham, Schwarcz, Ford, and Pearce (1979); Ikeya (1978); Wintle (1978)

A.A.R = Amino acid racemization
P.M. = Paleomagnetism
T.L. = Thermoluminescence
D.C. = Dendrochronology
E.S.R. = Electron spin resonance

into travertines, or capped by them. Thus travertine can be considered as a generally datable, absolute chronometric deposit, comparable to volcanic ash or a lava flow.

(v) *Cave deposits*. Karstic and other types of caves serve as excellent repositories for undisturbed sequences of detrital sediments, the nature of which can be interpreted in terms of changes in climate on the surface (Kukla and Lozek 1958; Bordes 1972). Speleothems are commonly found overlying and interstratified with such deposits and can be used to date the deposits (see chapter 11). Loess, congelifracted wall rock, gravel washed in by melt-waters, and characteristic fauna and pollen are some of the climatic indicators found in these deposits. The speleothems themselves also contain valuable paleoclimatic information in their isotopic and chemical composition. If formed in oxygen isotopic equilibrium with the water from which they were deposited, their $^{18}O/^{16}O$ ratios are determined by the temperature of formation and the isotopic composition of the water.

Speleothems deposited at isotopic equilibrium also trap inclusions of the water from which they grew. The isotopic composition of this water (D/H and $^{18}O/^{16}O$ ratios) is found to vary today as a function of local temperature and climatic conditions. It is inferred that paleoclimatic variations at a given site should also have brought about similar shifts in isotopic ratios, which can be observed in these fluid inclusions.

The carbon isotopic composition of speleothems can in principle also be used as a record of climatic variation, but in practice this has proved to be quite difficult. Variations in the ^{13}C content of the calcite are due to local kinetic and equilibrium isotope effects at the site of precipitation which depend on the rate of calcite growth, the extent to which the water has exchanged with the cave atmosphere or the overlying limestone through which it passed (Wigley, Plummer, and Pearson 1978; Deines, Langmuir, and Harmon 1974), and to variations in the carbon isotopic composition of organic matter in the soil zone above the cave. The latter is partly determined by the type of vegetation (Smith and Epstein 1971) and may vary as a result of changes in climate. For instance C_4 plants are mainly found in semi-arid regions and fix carbon which is 9 permil enriched in ^{13}C relative to C_3 plants, which predominate in humid temperate and tropical regions.

The frequency of deposition of speleothems in a given cave is also modulated by climate due to the effect of changes in rate of recharge of water in the overlying soil. In arid regions it is expected that speleothems will only form during pluvial periods, while in glaciated regions no, or very little, deposition can occur when the surface is ice covered. Permafrost or periglacial conditions may also diminish speleothem growth.

(vi) *Swamp and marsh deposits*. Marshes and swamps act as sediment traps from which records of changes in type of sediment can be obtained through analysis of cores. Furthermore, such cores generally contain abundant pollen, from which much paleoclimatic information can be obtained. Some attempts at U-series dating of organic-rich deposits from bogs have been made. Carbonate-precipitating organisms such as *Chara* spp. and snails may also form datable deposits.

(vii) *Soil.* The rate of pedogenesis and the type of soil generated are strongly controlled by climate, both through the direct action of rainfall as well as freeze/thaw effects, and indirectly through the influence of vegetation. In semi-arid zones, soils may contain carbonate-rich *ca* horizons of caliche (calcrete) which are datable by U-series methods. The age of the caliche becomes climatically significant when post-formational erosion can be demonstrated (for example, terrace cutting, shoreline erosion by pluvial lakes). Colluvium may contain fragments of caliche, whose age gives a lower limit to the movement of the sediment.

(viii) *Ground water.* While not a sedimentary deposit, ground water is a product of recharge that is accelerated or retarded during variations in rain or snowfall. Also, the isotopic composition of paleo-ground waters in regions of slow movement, reflects the composition of recharge, and should be an index of climatic variation. Dating of older water by U-series methods has been attempted with varying degrees of success (see chapter 9).

(ix) *Flora and fauna.* Floral and faunal assemblages are also very sensitive climatic indicators. Pollen is commonly preserved in non-oxidized sediments, and variations in ratios of arboreal and non-arboreal plants together with abundances of individual species can be interpreted in terms of changes in temperature, soil moisture, and other climatic variables (Webb and Bryson 1972). Pollen is also found trapped within speleothems (Bastin 1978) and U-series dates on the speleothem can thus be used to provide an absolute time-scale for the pollen record. Vertebrate migration patterns have been related to changes in continental climate as have distributions of invertebrates (principally molluscs and insects). Dating of these organic remains is contingent on their presence in datable sedimentary matrices. However, direct U-series dating of some fossils is possible, molluscan tests made of calcium carbonate (see chapter 18) and vertebrate bones, made of calcium phosphate (see chapter 12) offering the most potential. If such remains could be dated reliably, then they could provide a very important method of assigning dates not only to the faunal distributions, but to a wide variety of continental clastic sediments of diverse origin, for example, fluvial terrace deposits, glacial outwash and lake deposits, pluvial lake beds, sandy beach deposits, and so on. Unfortunately both molluscs and bones have proved to have serious limitations as subjects for U-series dating. Improvements in the analyses of these materials are badly needed.

Archaeological sites represent a particular form of organic deposit in which vertebrate remains, artefacts, and other sedimentary deposits are found, commonly in a stratigraphic sequence. Artefacts can be treated somewhat like fossils, to define a relative chronology. As with organic evolution, it is assumed that lateral dispersal of cultural patterns is rapid compared with evolution of artefact styles, so that, for instance, the transition from middle to upper Paleolithic technology can be treated as a time marker. (U-series dating of archaeological sites has been discussed in chapter 12.) The paleoclimatic significance of human migration has been discussed by various writers, for example, Bordes and Thibault (1977).

13.3 Examples of the dating of continental deposits of climatic significance

13.3.1 Lake sediments

Calcareous sediments which have formed in pluvial lakes are known from arid regions of North America, Africa, and Asia. Both calcareous bottom sediments, formed in part by chemical precipitation from lake water, and tufa deposits, along the extinct lake margins, have been described from such sites, and have in some instances, been directly dated by U-series methods.

In the Pleistocene lakes Lahontan and Bonneville (USA), a variety of lacustrine carbonates (gastropods, ostracods, tufa, marl, and so on) were dated by both ^{14}C and U-series methods (Broecker and Kaufman 1965). The ages were in general agreement for those samples showing no ^{226}Ra anomaly, and lay in the region 10 to 25 ky, a period corresponding to the Late Wisconsin glaciation in North America. Two older gastropod samples were found correlating to earlier pluvial events at about 120 and 250 ky, but these ages must be accepted with caution because ^{14}C and ^{226}Ra data only indicated that the samples had to be older than about 40 ky. Subsequently, Eardley, Shuey, Gvosdetsky, Nash, Dane Pritchard, Gray, and Kukla (1973) examined a 650 m long core from the Bonneville basin and were able to date the −99 m level by the paleomagnetic reversal at the 710 ky Bruhnes–Matuyama boundary. They observed many pluvial and interpluvial cycles, recognizable from alternating gravel/sand/clay/marl and soil horizons in the core, and correlated these to the deep sea core isotopic record and the Central European loess record, but without the aid of ^{14}C or U-series ages. In a study of the evaporitic deposits of Searles Valley, California, Peng et al. (1978) obtained ^{230}Th/^{234}U ages between 24 and 33 ky for a salt horizon at about 30 m depth and 5 to 9 ky for salt at about 15 m depth (Fig. 13.2). Each period is indicative of shallow lacustrine conditions with high evaporation rates and little input of detrital material. An intervening mud layer at about −20 m and a thick mud layer underlying the entire formation were probably formed under deeper water conditions corresponding to full pluvial conditions, between about 10 to 24 ky and greater than 33 ky respectively. The record of pluvial cycles for Searles Valley has recently been extended back to 3.2 my by Liddicoat et al. (1980) using known ages of paleomagnetic reversals. Correlation to records of local and global climate change is found to be

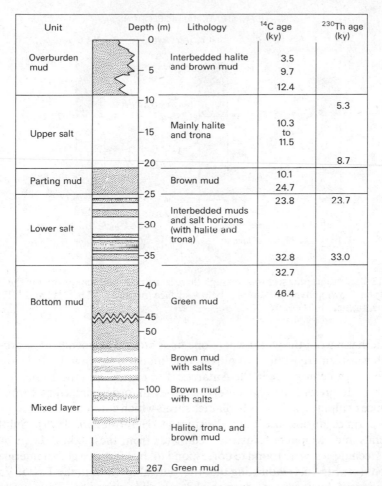

Unit	Depth (m)	Lithology	^{14}C age (ky)	^{230}Th age (ky)
Overburden mud	0–5	Interbedded halite and brown mud	3.5 / 9.7 / 12.4	
Upper salt	10–20	Mainly halite and trona	10.3 to 11.5	5.3 / 8.7
Parting mud	20–25	Brown mud	10.1 / 24.7	
Lower salt	25–35	Interbedded muds and salt horizons (with halite and trona)	23.8 / 32.8	23.7 / 33.0
Bottom mud	35–45	Green mud	32.7 / 46.4	
Mixed layer	100 / 267	Brown mud with salts / Brown mud with salts / Halite, trona, and brown mud / Green mud		

Fig. 13.2. The subsurface stratigraphy, ^{14}C and ^{230}Th/^{234}U ages, and levels for the Searles Valley lacustrine sediments, California. (After Peng et al. 1978; Smith 1979.)

generally good to 130 ky, but prior to this, the record of lake level fluctuations bears little resemblance to global records. Use of U-series dating methods in the upper part of this core would allow the synchroneity of pluvial events with climate to be better established.

In a similar study of lacustrine marls from the Jordan River–Dead Sea area of Israel, Kaufman (1971) was able to date three carbonate-rich horizons believed to be of Middle to Late Pleistocene age. Both vertical and lateral consistency was observed for the ^{230}Th/^{234}U ages and there was good agreement for younger samples dated by both the U-series and ^{14}C methods (see Fig. 13.3).

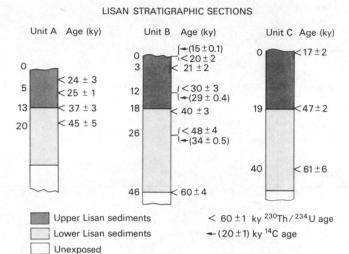

LISAN STRATIGRAPHIC SECTIONS

Fig. 13.3. Stratigraphic sections through three lacustrine sediment profiles in the Dead Sea Basin showing depth of samples below surface (m) and $^{230}Th/^{234}U$ and ^{14}C ages (after Kaufman 1971).

Another part of the world where one finds an arid environment at present, which was formerly subject to pluvial conditions, is the high latitude area of the southern hemisphere in the Antarctic. For example, in the Taylor Valley, which is at present a polar desert, a series of proglacial lakes formed in advance of alpine glaciers and polar ice sheets which entered the Taylor Valley several times during the Late Pleistocene (Hendy *et al.* 1979). Sulphate minerals and carbonate lacustrine sediments from these lakes, dated by U-series techniques, were found to correspond to the last three global interglacial periods (see Fig.13.4) indicating that the 'pluvial' periods in the Taylor Valley were caused by local ice sheet thickening and expansion, with increased ablation and lake formation. These observations have been interpreted by Hollin (1980) as evidence supporting the possibility of a major ice surge from the East Antarctic ice sheet towards the end of an interglacial period, thus triggering a rapid global cooling and onset of full glacial conditions.

13.3.2 Spring deposits

The various attempts at dating spring-deposited travertines by U-series methods have been described in detail in chapters 11 and 12. Relatively few of the published results have indicated success in dating changes in paleoclimate, mainly because most deposits suffer from contamination by detrital ^{230}Th or from nuclide migration under open system conditions following deposition (Rosholt and Antal 1962; Cherdyntsev 1971; Cherdyntsev *et al.* 1975).

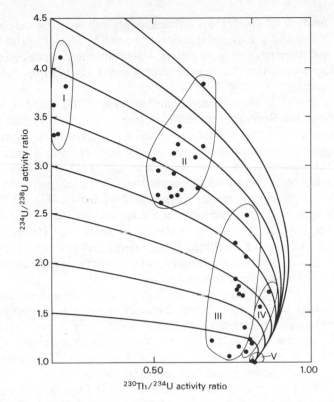

Fig. 13.4. The $^{230}Th/^{234}U-^{234}U/^{238}U$ isotope evolution diagram for sulphates and carbonate lacustrine sediments from Taylor Valley, Antarctica (after Hendy *et al.* 1979). Sulphates from Lake Bonny (Group I) give post-glacial ages whereas lacustrine carbonate and carbonate cements for older, stratigraphically distinct sample populations give ages of 70 to 127 ky (Group II), 160 to 255 ky (Group III), 260 to 350 ky (Group IV), and > 400 ky (Group V), suggesting an advance of the Taylor glacier during each of the last three interglacial periods.

However, four studies have produced reliable results which provide some information of paleoclimatic significance.

Cherdynstev (1971) reported age determinations by the $^{230}Th/^{234}U$ and $^{231}Pa/^{235}U$ methods for several spring deposits in Hungary and the USSR. This author correlated ages from the Tata travertine deposit to the Early Würm glaciation (about 95 ky) and the Vertesszöllös travertines to the warm Clactonian period (about 250 ky), and the glacial Mindel period (greater than 400 ky). Spring deposited tufas from several regions of the USSR, were also dated, but only the results from the extensive travertines of Mashuk Mountain in the Caucasus were considered reliable. The $^{234}U/^{238}U$ dating method was used here to determine the age of a Middle Pleistocene fauna (mainly *Elephas antiquus*) because ^{230}Th/^{234}U ages were strongly influenced by the presence of detrital ^{230}Th, and a reasonable estimate of the ^{234}U/^{238}U ratio could be made from contemporary spring waters. Using this method, an age of no less than 280 ky was determined, representing a lower age limit for the Mindel-Riss interglacial.

More recently Harmon, Glazek, and Nowak (1980) dated a travertine horizon from the Bilzingsleben archaeological site in East Germany which contains, in addition to hominid remains, an interglacial flora and fauna which suggest a depositional climate slightly warmer than at present. Four ^{230}Th/^{234}U ages were in close agreement and gave an average age of 228^{+17}_{-12} ky for the deposits which placed the Bilzingsleben sequence correlative with stage 7 in the marine oxygen isotope record (Fig. 13.1), that is, the penultimate interglacial period. The Bilzingsleben deposit had, at various times in the past, been assigned to the Elster-Saale (i.e., Holstein), Saale-Warthe, or Warthe-Weichsel (i.e. Eem) interglacial periods of the North European glacial stratigraphic record. The U-series age of about 228 ky indicates a Saale-Warthe stratigraphic age.

Schwarcz *et al.* (1979) have dated dissected travertines located high on the wall of a valley in the Negev Desert, Israel. The valley is currently dry, but ^{230}Th/^{234}U ages averaging 46 ky indicate that the area was considerably more humid during the interstadial between the Würm glaciations. They also demonstrated that a spring marked by a fossil travertine, 5 km from this site, was intermittently active in the interval 80 to 260 ky.

In the Wind River Mountains of northwestern Wyoming, travertines have been deposited on top of glacial outwash deposits which can be correlated with the recession of alpine glacial caps on the mountain range. These travertines, dated by U-series methods, show that the younger series of outwash deposits was formed early in the Wisconsin glacial stage, while an older series, deposited on a more highly dissected terrace, was deposited at 207 ± 30 ky, presumably close to the beginning of the Illinoian glacial stage (Schwarcz and Richmond, unpub. data).

13.3.3 Speleothem deposits

(i) *Isotopic. records.* The paleoclimatic significance of oxygen isotope variations of calcite in speleothem was first investigated by Hendy and Wilson (1968) and Hendy (1971), who demonstrated that isotopic equilibrium deposition does occur in the deep interiors of caves with 100 per cent relative humidity. In these circumstances isotopic equilibrium is established between the speleothem calcite and seepage water, such that the distribution of ^{18}O between the two phases is then a function of temperature alone. Variations of ^{18}O content along the growth axis of a speleothem deposited in isotopic equilibrium may, therefore, be correlated to temperature changes in the cave, which in turn reflect changes in mean annual temperature on the surface above the cave.

In practice, however, the isotopic composition of a speleothem also varies with changes in the isotopic composition of rainfall above the cave, which in turn is affected by factors such as temperature change at the cave site, variation in isotopic composition of the source (sea water) and so on. The paleoclimatic significance of variations of ^{18}O content of fossil speleothems will not be considered here, but has been described in detail by Harmon, Thompson, Schwarcz, and Ford (1978) and Gascoyne et al. (1978). Paleo-climate records have been obtained from speleothem and calibrated by U-series dating methods by Duplessy et al. (1970); Thompson et al. (1976), Harmon, Thompson, Schwarcz, and Ford (1978); Harmon, Schwarcz, and Ford (1978a); and Gascoyne, Schwarcz, and Ford (1980). An example of the potential of speleothems as paleoclimatic indicators is seen in the close correlation of the stable isotope record of a dated flowstone sequence from a cave in north-west England, with the marine isotope record (see Fig.13.5).

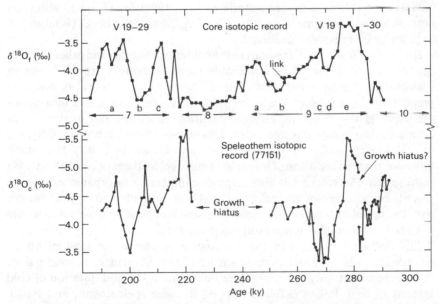

Fig. 13.5. Correlation of the O-isotopic record of speleothem 77151 from Victoria Cave, N. England (lower curve) with composite deep sea core isotopic record (upper curve) for isotope stages 7 to 9 inclusive (after Gascoyne 1980). The core record is a composite of V19-29 (Ninkovitch and Shackleton 1975) and V19-30 (Shackleton 1979, unpub. data), linked as shown. The time scale is determined from eight $^{230}Th/^{234}U$ ages for speleothem 77151 assuming constant growth rate between dated horizons. The core record is fitted to the speleothem time scale assuming constant sedimentation rate and coincidence of 9e isotope peaks.

Carbon isotopic records have been presented in most of the above isotopic studies, but generally these cannot be interpreted simply in terms of paleoclimate variation. However Goede, Green, Harmon, and Hitchman (1981), in a study of a $^{230}Th/^{234}U$ dated stalagmite from Tasmania, found pronounced variations in ^{13}C content of the calcite which were correlated with variations in U content, and which were slightly out of phase and leading variations in ^{18}O content. This was interpreted as reflecting a shift from interglacial climate marked by a forest, rich in C_3-type vegetation, to a colder regime characterized by grassy C_4-type vegetation during the glacial periods.

Variations in the D/H ratio of fluid inclusions in speleothem were first related to climate shifts by Schwarcz et al. (1976) and later studied further by Harmon, Schwarcz, and O'Neil (1979) who found that glacial-age precipitation was somewhat depleted in deuterium relative to modern precipitation. Harmon and Schwarcz (1981) inferred that the relationship between the D/H and $^{18}O/^{16}O$ ratio of meteoric water during the Pleistocene glacial stages was slightly different from the modern and interglacial relationship. Oxygen isotope studies of fluid inclusions in speleothems by Yamamoto and Schwarcz (1981) suggest that the oxygen isotopic composition of the trapped water has been disturbed in some way after trapping, although the D/H ratio still appears to be preserved unaltered.

(ii) *Trace Elements.* Variations in the content of Mg, Sr, and other divalent cations in calcite should be observable along individual growth layers in speleothem, due to enrichment of these elements in solution as calcite is precipitated. The ratio of Mg to Sr could, therefore, be used as a measure of the value of the respective element partition functions and in turn as an estimate of depositional temperature. This phenomenon has been investigated in preliminary studies by Johnson (1979) and Gascoyne (1981). It is likely, however, that the inclusion of trace elements in speleothem occurs not only via solid solution in calcite but also through complexing in organic substances present on surfaces of the calcite crystals, or by inclusion in microscopic particles or colloids on crystal surfaces. These modes of incorporation would not show such regular temperature dependence.

(iii) *Pollen.* Variations in the abundance of species of land plants as recorded in pollen are an especially sensitive index of change in climate, since they respond not only to temperature but to humidity and duration of cold seasons as well. Pollen is found trapped in some speleothems, and Bastin (1978) has shown significant variations in species abundance during the growth history of five stalagmites from Belgian caves. The cave environment is excellent for the preservation of pollen; however, the transport of pollen into the cave is limited because there is generally little air movement in the deeper interiors of caves to permit transport of air-borne grains, while water seeping into the cave from the soil zone has been effectively filtered free of particulate matter during its passage downwards.

(iv) *Variation in frequency of occurrence of speleothems.* In addition to the basic application of U-series techniques to speleothems in order to obtain a chronological framework for other paleoclimate information, speleothem age data can themselves be used directly as a proxy record of past climatic variation. Ground water seepage into a cave usually occurs at discrete sites, forming individual speleothems, most of which in any single cave are suitable for U-series dating. The very presence of a speleothem in a cave means that the cave temperature was not below freezing point and that the prevailing surface climate was not glacial or periglacial, otherwise all ground water movement would be prevented by ice formation. Additionally, high ground water bicarbonate levels are required to dissolve sufficient bedrock to permit subsequent calcite precipitation, once ground water seepage encounters a cave. Typically ground water in glacial areas has very low bicarbonate levels due to the absence of vegetation which is essential for CO_2 production in the soil. Thus, these two climate-related factors mitigate against speleothem deposition during glacial periods in areas of marked glacial–interglacial climatic contrast (for example, alpine and high latitude areas where speleothem deposition is only marginal at present). Therefore, it is possible to obtain a direct paleoclimatic record by dating large numbers of speleothems from an individual cave or, preferably, within a discrete physiographic province, and subsequently using the frequency distribution of the U-series ages as a guide to the absolute timing of past warm and cold periods. Earlier work by Franke and Geyh (1970) has shown that the frequency of ^{14}C dates on speleothems from caves in Western Europe fell to a minimum around 16 ky and rose again to a high value after 12 ky. This was interpreted as indicating a transition from non-glacial to glacial conditions at 16 ky, changing to warm interglacial conditions by 12 ky.

Two examples of paleoclimatic records obtained from age frequency distributions are shown in Fig. 13.6 for caves in the Rocky Mountains of North America, and the Yorkshire Dales and Mendip Hills in England. The periods of most abundant growth are clearly correlated to interglacials in the Middle to Late Pleistocene, by the coincidence of histogram peaks with periods of [18]O depletion in the marine O-isotope record. Also clearly visible are periods of almost no speleothem deposition which correspond to major Northern Hemisphere glaciations. Each area is known to have been fully glaciated or subject to substantial periglacial climate during these times, and therefore, speleothem growth is likely to have been reduced markedly, if not completely terminated. The biasing effects of erosional processes and burial of older deposits can be seen by the greater frequency of ages from the Holocene relative to the last interglacial, even though the two periods are thought to have been of comparable climate. An additional 'operator effect' also influences the age distributions in Fig. 13.6, because these histograms were assembled from data obtained as part of a general search for ancient and

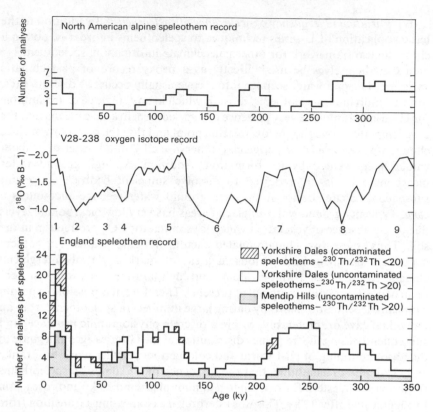

Fig. 13.6. Age frequency distributions for speleothems from (i) the Rocky and Mackenzie Mountains of Western North America (from Harmon *et al.* 1977, and Gascoyne 1980), and (ii) north-west England and the Mendip Hills of south-west England (from Gascoyne 1980 and Atkinson *et al.* 1978), compared to the oxygen isotopic record of core V28-238 (from Kominz *et al.* 1979).

morphologically-significant speleothems in an area. Therefore, obviously young, fresh-faced speleothems were sometimes rejected for analysis in favour of apparently older specimens.

13.3.4 Pedogenic carbonates

In the same way that calcareous or evaporitic sediments may accumulate as lake sediments during pluvial events, carbonates may also form as soil and sub-soil concretions in semi-arid environments during times of raised water table and high evaporation rate. Radiocarbon ages of calcareous horizons in soils have been used by several workers to date pluvial events in the Late Pleistocene (see for instance, Williams and Polach 1971). However, relatively little work has been done to extend the chronology beyond the limit of ^{14}C

dating, using U-series or other methods, mainly because of the problem of contamination of authigenic nuclides with those derived from the detrital host mineral phases. This problem has been recently treated by Ku *et al.* (1979) who measured nuclide activities in both the leachable (carbonate) and detrital fractions of soil carbonate deposits of the Mojave Desert, California. After correction for detrital Th, good age agreement was found for 14 samples of pebble coatings from one geomorphic surface (83 ± 10 ky), indicating a period of less arid conditions than present, probably corresponding to the beginning of the Early Wisconsin glaciation at higher latitudes in North America. The influence of climatic conditions on rates of accretion of pedogenic carbonate has been determined by several authors by ^{14}C dating, and more recently, U-series dating. In all situations, the rates were found to be low in comparison to travertine and speleothem growth rates, ranging from 2 to 10 ky per mm thickness of carbonate. Semi-arid climates (associated with northern hemisphere glaciations) are thought to give much higher accumulation rates than those of presently-arid climates (Ku *et al.* 1979).

13.3.5 Organic deposits

(i) *Faunal records.* Uranium-series dating methods may be applied to many types of faunal remains, particularly bones, teeth, and shells of freshwater molluscs, and gastropods. Climatic information can be obtained directly by taxonomic analysis of a single bone or shell but the study of faunal assemblages enables a more complete reconstruction of environmental conditions (for example, temperature range and amount of precipitation). The major problem in dating these types of deposits is the general lack of closed system conditions with respect to radionuclide mobility, following death and burial. Molluscs and bones are well known for their ability to concentrate U in the absence of Th and Pa daughters for a prolonged period after death (Kaufman *et al.* 1971). Open system models have been proposed to describe this process and so obtain a better estimate of the age (Szabo and Rosholt 1969, and see chapter 3) but inconsistencies are still found. This topic is discussed further in chapter 12.

Szabo and Collins (1975) used U-series ages of fossil bones from archaeological sites in southern England to determine age limits for interglacial periods in the British and North European Quaternary records. They concluded that the ages of Clacton and Swanscombe deposits, previously correlated to the Hoxnian interglacial, date this period at about 250 ky. However, this result is based on open system ages determined for some of the analysed samples ($^{230}Th/^{234}U$ and $^{231}Pa/^{235}U$ ages were discordant), and on large error limits for the ages of remaining samples, and therefore, must be regarded as tentative. Similar problems were experienced in the correlation of ages of mammoth and *Equus* remains with climatic events determined from till stratigraphy at Medicine Hat, Alberta (Szabo *et al.* 1973).

Table 13.4. The ^{230}Th/^{234}U age data for calcite flowstones overlying and enclosing Ipswichian mammal remains in Victoria Cave, northern England. (After Gascoyne, Currant, and Lord, in preparation.)

Speleothem number	Description	Analysis number	location	U (ppm)	$\frac{^{234}U}{^{238}U}$	$\frac{^{230}Th}{^{232}Th}$	$\frac{^{230}Th}{^{234}U}$	Age ($\pm 1\sigma$) (ky)
79 000	brown laminated fs enclosing *rhino* teeth	−1	adjacent to teeth	0.62	1.000	44	0.610	$102.1 \begin{array}{c}+12.0* \\ -10.8\end{array}$
		−2	∼ 1 cm away from teeth	0.63	1.019	20	0.672	$120.3 \begin{array}{c}+7.0 \\ -6.6\end{array}$
79 001	yellow-brown fs enclosing *rhino* jaw	−1	adjacent to bone	0.50	1.100	20	0.623	$103.8 \begin{array}{c}+6.2 \\ -5.9\end{array}$
		−2	∼ 2 cm away from bone	0.40	1.033	152	0.691	$126.3 \begin{array}{c}+9.7 \\ -8.9\end{array}$
		−3	∼ 2 cm away from bone	0.43	1.022	34	0.714	$134.7 \begin{array}{c}+8.7 \\ -8.1\end{array}$
79 002	calcite enclosing red deer antler	−1	adjacent to bone	0.50	1.012	16	0.702	$130.7 \begin{array}{c}+8.8 \\ -8.2\end{array}$

79021	fs enclosing *rhino* teeth	−1	2 cm adjacent to teeth	0.43	1.057	31	0.690	124.9 +6.6 −6.3
79023	fs and pool calcite enclosing giant deer teeth	−1	1 cm of clean side of fs	0.50	0.994	71	0.678	123.1 +7.0 −6.5
79025	9 cm deep section of fs block which contains large mammal bones at its base	−1	0.5 cm above base (at bone level)	0.88	1.120	111	0.484	70.8 ±2.8
		−2	top 1.5 cm	0.43	1.048	84	0.654	113.9 +5.1 −4.8
		−3	0.5 to 1.5 cm above base	0.58	1.037	47	0.672	119.7 +6.3 −6.0
79026	fs enclosing *rhino* teeth	−1	top (?) 2 cm of fs	0.46	1.022	>1000	0.668	119.1 +5.1 −4.9

* U yield = 6%, all other yields are > 20%
fs = flowstone

More reliable U-series ages for bones can be obtained by dating material associated with the bone, which is less susceptible to nuclide migration, rather than the bone itself. A recent example of the paleoclimatic significance of such studies is described by Gascoyne, Currant, and Lord (1981) where large mammal bones (including rhinoceros and red deer) were found to be associated with calcite flowstone in Victoria Cave, northwest England. (Recent analysis (A. P. Currant, personal comm.) has also identified the presence of hippopotamus in direct association with flowstone.) Following excavation over a century ago, the remains were identified as a classical, warm-climate assemblage, characteristic of the last interglacial period (Ipswichian) in the British Quaternary record. Analysis of the calcite surrounding the bones gave ^{230}Th/^{234}U ages within the range 114 to 135 ky (Table 13.4). This period correlates well with the duration of O-isotope stage 5e (Fig. 13.1), the last interglacial period in the marine record. These results not only provide absolute ages for the Ipswichian, but are able to link directly the continental climatic record to that determined from deep sea core data. Although the bones themselves were not analysed, evidence of open system conditions was found in that ages of the flowstone immediately adjacent to the bone (samples 79 000–1, 79 001–1, and 79 025–1 in Table 13.4) were significantly less than the stratigraphically-younger flowstone more distant from the bone (samples 79 000–2, 79 001–2 and –3, and 79 025–3), thus indicating migration of daughter-deficient U from the bone into the adjacent calcite.

Considerably more paleoclimatic information has been determined from U-series analyses of marine molluscs than of bone, principally because of their greater abundance and the fact that their elevation with respect to modern sea level is an indicator of past climate as well as their species relationship. Unfortunately, the same problems of post depositional nuclide migration have been documented for molluscs. Their greater abundance, however, enables clearer recognition of anomalous results by tests of reproducibility of age determinations as well as comparison with ages by other dating methods.

For example, in a study of littoral molluscs from the Mediterranean and Morocco, Stearns and Thurber (1967) used ^{226}Ra/^{230}Th ratios to test for recent closed system conditions. They then obtained several ^{230}Th/^{234}U ages on different specimens from equivalent horizons in ancient strandlines and correlated age groupings to periods of high sea level at greater than 250, 200 to 120, and about 80 ky. In a study of molluscs from marine terraces in southern California, Szabo and Rosholt (1969) found several instances of discordance between ^{230}Th/^{234}U and ^{231}Pa/^{235}U ages. However, they adjusted the ages using a model describing the open system migration of U in the shell and noted that the resulting age groupings closely corresponded to the high sea stands of isotope stage 5.

The status of mollusc dating was further considered by Kaufman *et al.*

(1971) who strongly criticized most previous mollusc results on the grounds that migration of nuclides after deposition in sediments is an unpredictable process which may or may not continue for tens of thousands of years. These authors went on to show that the Szabo–Rosholt 'Open System Model' was invalid for most cases of nuclide migration and concluded that even concordance between $^{230}Th/^{234}U$ and $^{231}Pa/^{235}U$ ages did not guarantee accuracy. It is, therefore, unlikely that molluscs will provide much paleoclimatic information in the future, unless open system migration is better understood or perhaps, concordance is found between ^{14}C, $^{230}Th/^{234}U$, $^{231}Pa/^{235}U$, and $^{226}Ra/^{230}Th$ ages—an unlikely situation.

(ii) *Floral records*. Little use has been made so far of U-series dating of organic sequences. Early analyses by Cherdynstev (1971) of peats from various regions in the USSR gave conflicting results. In peat of high detrital Th content, $^{230}Th/^{234}U$ ages bore little relation to the stratigraphically-determined age of the peat, whereas peats containing low detrital Th and high U content gave ages comparable to the ^{14}C age. Certainly further investigation of the possibility of dating peat and other similar organic deposits by U-series techniques is warranted.

As noted above, pollen can be found embedded in speleothem, and U-series dating may eventually be used to establish ages for pollen spectra as is done at present by ^{14}C dating. Similarly, some travertines deposited by springs contain well-preserved pollen, and can also be dated. In addition, casts and molds of plants which lived in the spring can be identified as to genus and species. The spring deposit at Tata, Hungary, for example, which was dated by Cherdyntsev (1971), contains a sequence of characteristic flora that indicate a transition from interglacial to glacial climate (Budo and Skoflek 1964), as do molluscs formerly living in the same deposit (Krolopp 1964).

13.4 Summary and conclusions

Wherever continental deposits of the Quaternary period have been studied, they have been interpreted in terms of a climatic variation. Quaternary and glacial (or pluvial) stratigraphy are almost synonymous. However, due to the cyclical nature of the climatic variations of this period, it is generally difficult to determine which glacial or stadial, interglacial or interstadial is represented in any particular paleoclimatic sequence, whether it be a pollen record from a sediment core, or a series of terminal moraines. Therefore, it is essential that an absolute chronology be available to place at least approximate constraints on the timing of the geological deposits. While ^{14}C and K/Ar dating have played the largest role in establishing the existing absolute chronology (with paleomagnetic records forming a bridge to otherwise non-datable deposits), it has been seen that U-series dating can often provide a crucial input into the dating of the Middle and Late Pleistocene.

Once obtained, dated continental paleoclimate records are often correlated with well-established stratigraphic sequences and stages, and it is then tempting to define radiometric ages for these hitherto undated periods. Several examples of this type of correlation have been given in this chapter. Conversely, long-standing correlations of isolated Quaternary deposits to climatic events older than the last glaciation are becoming increasingly questioned as more and more absolute age data are accumulated from datable horizons within these deposits. In the same way that the classical Alpine glacial chronology of Penck and Bruckner (1909) has been gradually rejected in favour of the multiple glacial sequence seen in the deep sea core record (Kukla 1977) so might the assigned age, duration, and ordering of climatic stages have to be revised if the type-sections and related deposits are ever dated.

In principle, the dating of bones and molluscan calcite would be a valuable adjunct to Quaternary studies, but in the past both of these materials have demonstrated serious problems for U-series dating. Nevertheless, they should be studied further in the hope that a satisfactory method can be developed for the dating of each of them. The very high U concentrations encountered in some bones makes them especially appealing as targets for dating, since it would allow the dating of very small samples, and permit replicate dating of valuable specimens. Furthermore, it should be remembered that vertebrate fossils themselves form one of the richest sources of information for Quaternary chronostratigraphers. Therefore, if an absolute time scale can be established for the evolution of land mammals, this would permit direct correlation with other deposits containing the same vertebrate assemblages.

Uranium-series dating methods will also become indispensable in establishing the chronology of short-period fluctuations in the earth's climate which have recently been observed as a fine structure of ^{18}O-variations in speleothem calcite (Wilson, Hendy, and Reynolds 1979; Gascoyne 1980, unpub. results). The prospect of obtaining a very detailed record (perhaps to within 20 y intervals) from speleothem analysis, for periods as old as 300 ky, is becoming very real. From this work it may be possible to identify climate-controlling factors other than insolation variations caused by changes in the earth's orbit; factors which, because of their rapidity and feedback effects, are more critical to climate prediction for the immediate future.

14 EXPLORATION FOR U ORE DEPOSITS
A. A. Levinson, C. J. Bland, and R. S. Lively

14.1 Introduction

Disequilibrium within the U-decay series was known to the earliest physicists and chemists concerned with radioactivity as the writings of Mme Curie, Rutherford, and others clearly show. However, most geologists and geo-chemists involved with radioactive ores have, until relatively recently, been effectively unaware of the general concepts of U-series disequilibrium, as well as the significance and interpretation of these concepts in the context of exploration. This lack of appreciation and understanding became increasingly apparent in the past decade with: (i) the development of new field and laboratory instruments designed to detect specific nuclides in the U-series (for example, Rn detectors); and (ii) the discovery of ores in certain environments (for example, the arid regions of southern Africa and Australia) which are often not in equilibrium. At present, however, there is an awakening to the importance of disequilibrium on the part of many exploration geologists involved in the search for radioactive ores, although the interpretative aspects of U-series disequilibrium in natural environments still require further elucidation.

From an historical point of view, it is interesting to note that radioactive nuclides produced by the decay of U, specifically Ra and Rn, and stable He had been recognized in spring waters and municipal water supplies from such widely separated places as England, Germany, and Connecticut at the turn of the century. Bumstead and Wheeler (1903, 1904), for example, observed (1904) that the radioactive gas found in the ground and in the surface water near New Haven was apparently identical with the emanation from Ra. With respect to this emanation they had determined (1903) that the activity fell to half its value in a time very close to four days. The emanation, of course, now we know to be Rn. Emanations of Rn from Ra in both the laboratory and in natural environments (for example, waters) had been detected about the same time, or earlier, by Mme Curie, Thomson, Rutherford, Dorn, and others (see Bumstead and Wheeler 1904; Boltwood 1904b).

Lord Rayleigh (1896) observed that gases which arise from springs at Bath contain He, although at that time the ultimate sourse of the He was not known. Strutt (1904) determined the presence of Ra in waters from the same springs at Bath. This led him to observe that He owed its origin to the same source that supplied Ra to the waters. In the same study, Strutt (1904) reported the occurrence of a red deposit, coloured by iron, with a high activity which gave off an emanation freely, even without heat. In modern terminology Ra, from which Rn emanates, is adsorbed on iron oxides which have precipitated at the spring orifices. Helium in the gases of U minerals was detected by Ramsay and Travers (1898) and Travers (1899). In the same era, Boltwood (1904a) observed that the quantity now referred to as the emanation factor for minerals can be very low. Specifically, he reported that when a certain specimen of the mineral samarskite was heated to a low redness only 10 per cent of the radioactive gas was released (20 per cent at bright redness) compared to the amount released (emanated) when the mineral was completely decomposed with hot sulphuric acid.

Boltwood (1904b) was probably the first to imply that the products of U-series decay in waters, specifically Ra and Rn, could be used for purposes of U exploration. Consider the following statements (Boltwood, 1904b, pp. 386–7):

> Even brief contact with uranium minerals can impart to water very marked radioactive properties due to dissolved radium emanation . . . An extremely minute trace of uranium minerals in the rocks and soil through which the waters percolate in their underground passage would be sufficient to impart to them radio-active properties, which could be readily detected by the sensitive method at command [an electroscope]. It can be anticipated that waters which rise through strata containing appreciable quantities of uranium minerals will be found more highly radio-active than any which have thus far been described. The results obtained from the examination of waters from springs in well-known uranium localities can be looked forward to with interest.

Lester (1918) was possibly the first to apply Boltwood's (1904b) suggestion during his study of the radioactive springs of Colorado for which he used a truck-mounted, gold leaf electroscope. Some of his results and observations are particularly pertinent as present day hydrogeochemical surveys based on the determination of U and Rn encounter similar frustrations. For example, Lester (1918) noted that the most active waters showed the highest radioactivity found in the United States. However, a careful comparison of the radioactive measurements with the data obtained from the chemical analyses showed that there was no connection between radioactivity and any chemical property. Further, he anticipated that some springs of extraordinarily high radioactivity would be found since Colorado contains quite extensive deposits of radioactive ores. This expectation, however, was not fulfilled. A number of springs, often highly gaseous, situated not far outside such regions showed, in general, the least activity of any examined. With respect to springs which did show high radioactivity, Lester (1918) observed that even though such a spring did show high activity, it did not mean necessarily that there was highly radioactive material near by.

Radioactivity in rocks was also studied in the same era. Joly (1909), for example, found a large excess of Ra over equilibrium amounts in lavas from Vesuvius. Numerous subsequent studies (summarized by Adams and Gasparini 1970, p. 216) have shown that in Vesuvius lavas there is a fairly regular decrease in the ^{226}Ra content with age, but also that disequilibrium exists in lavas erupted in historical times as the ^{226}Ra/^{238}U activity ratio exceeds the equilibrium value. Oversby and Gast (1968) have interpreted the disequilibrium found in recent volcanic rocks (the ^{226}Ra/^{238}U activity ratio generally ranges from 1.3 to 2.0) from various parts of the world in terms of chemical fractionations which occur during magmatic processes, i.e. Ra follows Ca, Sr, and other divalent elements which crystallize in mafic rocks, whereas U remains in the magma.

Rosholt (1959) pointed out that many radioactive ore samples show radioactive disequilibrium because of the numerous geochemical processes affecting ore deposits. As a result, it is often difficult to interpret disequilibrium by simply comparing radiometric and chemical assays for U and, therefore, analyses should be made for ^{231}Pa, ^{230}Th, ^{226}Ra, ^{222}Rn, and ^{210}Pb (analyses for ^{231}Pa, a nuclide in the ^{235}U series, are not normally made at this time for exploration purposes). Rosholt (1959) also found that disequilibrium in most present day ore deposits can be classified according to six basic types, the first three of which have U values in excess of ^{226}Ra, and the last three of which have ^{226}Ra values in excess of U. These are briefly summarized in Table 14.1 (from Ostrihansky 1976).

In recent years the importance of ^{234}U in disequilibrium studies has become apparent and, consequently, the following relationships are now recognized to be important (Hambleton-Jones 1979):

(1) ^{234}U $\geqslant$ ^{238}U $\geqslant$ ^{230}Th $\geqslant$ ^{226}Ra indicates a young accumulation of epigenetic U;
(2) ^{234}U $\geqslant$ ^{230}Th $\simeq$ ^{226}Ra $\geqslant$ ^{238}U indicates an older accumulation of epigenetic U than in (1) above;
(3) ^{230}Th $\geqslant$ ^{226}Ra $\geqslant$ ^{238}U $\geqslant$ ^{234}U indicates a recent partial leaching of an old U deposit; and
(4) ^{230}Th $>$ ^{226}Ra $\gg$ ^{234}U $>$ ^{238}U indicates recent leaching but more severe than in (3) above.

With this brief historical summary now completed it remains to discuss the modern use and interpretation of U-series disequilibrium data in exploration for radioactive ores, specifically U. Numerous nuclides have been used for such purposes. These are usually the long-lived nuclides such as ^{234}U, ^{230}Th, and ^{226}Ra, or those which have some special property of value, such as ^{222}Rn

Table 14.1. Types of radioactive disequilibrium found in present day ore deposits

Type	Relative activities	Comments
1	$^{238}U > ^{231}Pa > ^{230}Th > ^{226}Ra$	Indicates that leaching of daughter products has taken place or U may have migrated to its present location in a time less than that required by its daughter products to reach approximate equilibrium, that is less than 300 ky. The former explanation is more probable in carnotite and other types of deposits where U fixative agents such as V or phosphate are present.
2	$^{238}U \gg ^{231}Pa > ^{230}Th > ^{226}Ra$	Shows considerable daughter product deficiency. The most probable explanation is a recent accumulation of U.
3	$^{238}U < ^{231}Pa > ^{230}Th < ^{226}Ra$	Disequilibrium is believed to be the result of rather recent deposition of U contaminated with significant amounts of daughter products other than ^{230}Th.
4	$^{238}U \ll ^{231}Pa > ^{230}Th > ^{226}Ra$	Disequilibrium found in samples taken from an oxidized environment, and is the result of U leaching. The high ^{231}Pa content and near equilibrium abundances of U, ^{230}Th, and ^{226}Ra is very significant for many U anomalies.
5	$^{238}U < ^{231}Pa > ^{230}Th \ll ^{226}Ra$	Disequilibrium occurs in pyritic ores or ore dumps, and is the result of differential leaching of all components.
6	$^{238}U = 0$; $^{231}Pa = 0$; $^{228}Ra < ^{226}Ra$	Disequilibrium is commonly associated with oil and gas field brines. Radioactive hot spring deposits also show this type of disequilibrium.

(Based on the classification of Rosholt 1959 and summarized by Ostrihansky 1976)

and ^{214}Bi. Other nuclides, such as ^{210}Po, ^{210}Pb, and ^4He will also be considered. It will be assumed that the reader is familiar with the physics, chemistry, and geochemistry of U-series nuclides as discussed in chapters 1 to 5. Further, only the ^{238}U series disequilibrium will be considered as all exploration methods are based on the detection of this nuclide or its daughter products owing to its much greater abundance than ^{235}U, and the assumption that the $^{238}U/^{235}U$ ratio is essentially constant in all ore deposits (that at Oklo, Gabon, being the obvious exception; see Lancelot *et al.* 1975). The two themes which will be stressed in this chapter are: (i) how disequilibrium affects prospecting methods (for example, borehole logging, gamma ray spectrometry); and (ii) how disequilibrium can assist in the location of U deposits.

14.2 Classification of U deposits

Hambleton-Jones (1978) pointed out that U responds readily to the environment within which it occurs, and therefore, the classification of U deposits is a useful precursor to U exploration programmes. The establishment of a classification system for uraniferous ore deposits should attempt to emphasize the common and distinguishing features which are instructive in understanding ore genesis, as such information is fundamental in determining which features are important in exploration for other deposits of similar kind.

Numerous types of classifications of U deposits have been proposed, most of which appear to have several of four essential themes (Hambleton-Jones 1978): (i) morphological relations to host rocks; (ii) genetic associations; (iii) metallotectonic controls; and (iv) element associations. However, the classification of mineral deposits of any kind is always a difficult and contentious exercise because most natural phenomena are not readily separable into unequivocally distinct subdivisions. For U deposits this problem is compounded by the fact that the deposits are commonly polygenetic, and they may have undergone a complex history of deposition, concentration, and remobilization prior to preservation as an ore deposit.

Of the numerous classifications which have been proposed, as well as discussions of classifications, that of Ruzicka (1970) is significant because he summarized the classifications of Russian, East European, and Canadian deposits. General classifications embracing world deposits have been proposed by Barnes and Ruzicka (1972) and by Ziegler (1974). The former developed a genetic classification in terms of the geochemical cycle, and modes of transportation and deposition. The latter compiled a metallotectonic classification based on his conclusion that U is associated preferentially with certain structural features in the earth's crust.

One of the most modern, practical, and at the same time, relatively simple classifications is that of McMillan (1978) presented in Table 14.2. Although it is based mainly on Canadian examples, most important deposits in the world are accommodated. Two main differences between the classification of Table 14.2 and earlier ones are the use of the term hydrogenic, and the distinction between classical and unconformity veins. Both of these points are important because: (i) they are based upon the recognition that hydrogenic processes of one type or another have played a major role in the formation of much of the known U ore reserves in the world; and (ii) many large recently discovered deposits in Saskatchewan, and elsewhere, are classified as unconformity veins (i.e. closely associated with stable peneplaned surfaces) and are also hydrogenic in origin.

The term hydrogenic, as used by McMillan (1978) as well as in this chapter, refers to deposits formed as precipitates from aqueous solutions; hydrothermal and supergene are two of several subdivisions recognized within the

Table 14.2. Classification of important uranium deposits (from McMillan 1978)

Genetic type of deposit	Structural or petrographic association	Characteristic elements	Characteristic minerals	Examples
Igneous	Carbonatite, Alkaline syenite	Nb, U, Th, Cu, P, Ti, Zr, REE	uranothorianite, pyrochlore, betafite, perovskite, niocalite, ilmenite, apatite, zircon	Prairie Lake, Ont; Nova Beaucage, Ont; Pocos de Caldas, Brazil
Metamorphic	Pegmatite (alaskite), Skarn	U, Th, Mo, REE, Nb, Ti	uraninite, uranothorite, molybdenite, betafite, fluorite, zircon	Bancroft, Ont; Rossing, S. W. Africa
Detrital	Pyritic quartz pebble conglomerate	U, Th, Ti, REE, Au, Zr, C	uraninite, brannerite, pyrite, monazite, native Au	Elliot Lake, Ont; Witwatersrand, S. Africa
	Volcanogenic	U, Th, REE, Mo, Cu, F, Sr,	uraninite, uranothorite, fluorite, celestite, pyrite	Rexspar, B. C.
Hydrothermal	Hydrothermal veins	U, Th, REE, Be, Nb, Zr	uraninite, brannerite, thorite, allanite, quartz, fluorite, carbonates, sulphides	Bokan Mt, Alaska

Genetic class	Deposit type	Elements	Minerals	Localities
Syngenetic	Shale, Phosphorite, Evaporitic Limestone, Duricrusts (Calcrete)	U, P, V, Cu, Co, Ni, As, Ag, C	pitchblende, carnotite, apatite, gypsum, carbon	Ranstad, Sweden; Kitts, Labrador; Todilto Limestone, New Mexico; Yeelirrie, W Australia
Hydrogenic	Sandstone (Tabular, Roll, etc.)	U, C, Cu, V, Se, Mo	pitchblende, pyrite, coffinite, carnotite	Colorado Plateau; Wyoming, Texas Basins
Hydrogenic	Channel conglomerate	U, C	pitchblende, marcasite, coffinite, autunite	Kelowna-Beaverdell District B. C.; Ninge Toge, Japan
Epigenetic	Lignite	U, C, Mo, V		Cypress Hills, Sask; Dakotas
	Classical veins	U, Cu, Ag, Co, V, Ni, As, Au, Mo, Bi, Se	pitchblende, pyrite, chalcopyrite, Ni-Co arsenides, native Ag, Au	Beaverlodge, Sask; Port Radium, N. W. T.; Schwartzwalder, Colo.
Syngenetic, Supergene, and Epigenetic	Unconformity veins	U, C, Cu, Ag, Co, Ni, As, V, Se, Au, Mo	pitchblende, pyrite, Ni-Co arsenides, chalcopyrite, native Au, Ag	Wollaston, Key and Cluff Districts, Sask; Midnite, Washington
Supergene	Cappings, Enrichments	U, Si, Ca, Cu, Ag, Ni, As	'gummite', uranophane, carnotite, coffinite	Bolger, Eldorado, Sask; Pocos de Caldas, Brazil; Rossing, S. W. Africa.

hydrogenic type. The use of the term hydrogenic is very appropriate in this chapter because it is now recognized that many important U deposits were (and are being) formed by such processes, and also because these same deposits may be simultaneously remobilized, modified, or even destroyed by essentially the same solutions which were responsible for their initial accumulation. During the course of such penecontemporaneous or post-depositional modifications, differences in the mobilities of U and its daughter nuclides will result in disequilibrium in the waters and in those radioactive components of the U deposit which remain. Such disequilibrium should be looked upon in a positive light from an exploration point of view because it may generate anomalous concentrations of U and its daughter nuclides at varying distances from U deposits in a variety of media (for example, water, stream sediments, soil gas) which permit the original source (deposit) to be located by means of suitable geochemical and/or geophysical procedures.

The classification of Table 14.2 shows that many geological environments are favourable for the formation of economic U deposits each with its own geochemical characteristics in the form of distinctive trace element associations and mineralogy. The elements characteristic of each type of deposit may be particularly useful when pathfinders are employed in the exploration and interpretive stages of geochemical exploration programmes (see Hambleton-Jones 1978 and Levinson 1980, for discussions on geochemical prospecting methods for U). Further details on element and mineral associations of twelve types of U deposits and occurrences, most of which are listed in Table 14.2, may be found in Boyle (1974).

14.3 Determination of U source rocks by disequilibrium methods

One of the major problems facing a prospector is the recognition of potential source areas for U of hydrogenic origin, such as those classified as epigenetic or syngenetic (see Table 14.2). If such areas can be delineated, he will first attempt to determine how far U removed from this source area may have travelled, and second speculate on possible environments for its precipitation.

Certain granites characterized by high silica (greater than 70 per cent), Th, and alkali contents, and by moderate contents of biotite (Stuckless and Nkomo 1978), have experienced large losses of U by leaching and, in some cases, the mobilized U has precipitated at some distant site in ore-bearing quantities. Uranium (averaging perhaps 20 ppm or more in fresh rock) leached from granites has been proposed as the primary source of U in the sandstone deposits in Wyoming, the vein type deposits in France, the calcrete deposits at Yeelirrie, Australia, and some newly discovered deposits in calcretes and other hosts in the very arid regions of Namibia and other parts of southern Africa. However, only in the case of the Wyoming deposits, as a result of work by members of the U.S. Geological Survey over the past decade based on disequilibrium studies, has it been possible to establish unquestionably that the source area for several major U districts, like Gas Hills and Shirley Basin, for example, was the Granite Mountains.

Early disequilibrium studies by Rosholt and Bartel (1969) showed that low $^{238}U/^{206}Pb$ values, coupled with high $^{206}Pb/^{204}Pb$ and $^{207}Pb/^{204}Pb$ ratios, indicated that the contents of U in near surface rocks of the Granite Mountains would have had to have been considerably greater than those presently observed to have generated the high radiogenic ^{206}Pb and ^{207}Pb values. They concluded that the amount of U leached from the granites and potentially available to the surrounding sedimentary basin was 1000 times the estimated reserves. Further studies (Rosholt *et al.* 1973), based on Pb isotope systematics, showed that about 75 per cent of the U required to produce the radiogenic Pb had been removed by leaching, and that the process was operative to at least 50 m below the present surface; Nkomo *et al.* (1978) found evidence that U had been mobilized (leached) to a depth of 122 m within the past 150 ky. However, there is no apparent loss of associated Pb or Th. Most of this U loss occurred during the Cenozoic and probably represents the major source of U that was subsequently accumulated in ore deposits in the adjacent sedimentary basins.

More recent studies by Stuckless and Nkomo (1978) pointed out that the timing of U loss from a source rock, as well as the amount of such loss, is an important factor in interpreting the relationship between potential source and host rocks. For example, they found that the timing for at least part of the U loss from the Granite Mountains agrees with the radiometric dates for the ores, whereas these ores are older than certain volcanic rocks which had been proposed erroneously as source rocks for the ore deposits. Based on similar disequilibrium studies in the Owl Creek Mountains and Laramie Range, Stuckless (1979) believes that much of central Wyoming has been a U province since Precambrian time. He also suggested the possible use of Th contents and Th/U ratios in exposed basement rocks as an indicator of U provinces, the U content itself being unsatisfactory owing to the often observed drastic depletion of U in surface rocks by leaching. Specifically, if high Th (greater than 40 ppm) values are accompanied by Th/U ratios greater than 5, as is the case in the Granite Mountains, significant U leaching may be postulated, because Th/U ratios in igneous rocks typically average about 3.5 (see chapter 2). This simplified approach is certainly worthwhile as a first step in U exploration especially since it is not economically feasible to determine accurately the state of disequilibrium in granites on a routine basis owing to the high analytical costs involved.

Uranium which is easily leached by water from rocks is called *labile* U, and it is the source for many ore deposits of hydrogenic origin, such as the sandstone deposits in central Wyoming. Labile U is assumed to be weakly bound either within the crystal structure of soluble minerals, or along grain boundaries, or it may become available during the alteration of rock-forming minerals, for example, biotite and epidote (Stuckless and Nkomo 1980). Labile U is particularly mobile if channels (for example, microfractures) are available for the penetration of water. The entire subject of labile U was reviewed by Stuckless and Ferreira (1976), who first proposed the term, although the fact that U can be easily leached from rocks by various dilute natural solutions has been known for a long time.

Experiments concerned with the leachability of U from potential source rocks have employed ammonium carbonate, sodium carbonate with hydrogen peroxide (Beus and Grigorian 1977, pp. 25 and 78), weak solutions similar in composition to local ground water, and others, as the leachate. Szalay and Samsoni (1969) conducted experiments using a sodium bicarbonate leach solution on potential U source rocks and found (as summarized by Hambleton-Jones 1978): (i) the solubility of U in granites is much greater than in other types of igneous rocks; (ii) the concentration of U in the solution is a function of the amount of leachable U present in the source rocks; and (iii) water circulating through granite detritus attains an equilibrium concentration of U after a period of 3 to 5 hours. If the water is removed and replaced by a fresh solution, a new but lower equilibrium concentration is attained and indicates that the amount of leachable U decreases. Further discussion on the leaching of U will be found in other sections of this book (see chapters 2 and 7).

14.4 Uranium ore exploration

14.4.1 Uranium (^{238}U and ^{235}U) in U exploration

In the previous sections it was shown that during chemical weathering U may be leached from rocks and minerals. The nuclides ^{238}U and ^{235}U in fluids, which may be either surface or ground waters, will be in the ratio of about 137 to 1 by weight, as it is in other natural materials, suggesting that they do not undergo fractionation during the weathering process. In this discussion of U both nuclides in the above proportion are implied even though the ^{238}U series is mentioned.

Generally in oxidizing aqueous environments U can be considered to be more mobile than any of its daughters (exceptions will be discussed below). Because of this inherent mobility, disequilibrium will exist between parent ^{238}U and its daughters. The mobility of U in the secondary environment is governed primarily by three factors (see chapter 2): (i) its variable oxidation state; (ii) its tendency to form complexes with a large number of naturally occurring anions; and (iii) its tendency to be adsorbed on organic matter and mineral surfaces. All of these points, which are essential to understanding and interpreting data from hydrogeochemical U surveys, involve a fluid medium.

In the secondary environment the fluid is either surface water or ground water. (Mechanical transport and concentration of U minerals, such as uraninite, into placers is not considered in this discussion.)

As already discussed in chapter 2, U can exist in a hexavalent (uranic, U^{6+}) state which, when it forms complexes, is much more soluble than tetravalent U (uranous, U^{4+}); for all practical purposes, U^{4+} can be considered to be insoluble. In fact, the high solubility of U in the surficial environment is due to the oxidation of U to U^{6+} as UO_2^{2+} (the uranyl ion) in aerated or near-surface oxidizing environments.

Uranium forms complexes with many components in natural aqueous systems. Langmuir (1978) has reported a total of 43: fifteen with water; ten with fluoride; nine with phosphate; four with sulphate; three with carbonate; and two with chloride. The importance of each will depend on the pH and on the activities of the various anions. The most important complexes are those of the carbonates (uranyl dicarbonate and uranyl tricarbonate) and several phosphates, particularly the uranyl acid biphosphate, $UO_2(HPO_4)_2^{2-}$. The importance of phosphate complexes has been recognized only recently and, in fact, they are probably more stable than the carbonate complexes under certain conditions (Langmuir 1978).

With respect to uranyl complexes, the following generalizations can be made:

(i) Carbonate complexes are very important in slightly acid to alkaline pH in the absence of phosphate, but their stability gradually decreases with increasing temperature because the solubility of CO_2 (and other volatiles) decreases with increasing temperature resulting in less CO_2 to complex with U. Hence, boiling of thermal waters is a good mechanism for the deposition of U, as the volatiles are partitioned into the vapor phase.

(ii) In the presence of phosphate, even as little as 0.1 ppm PO_4^{3-} at low temperatures, phosphate complexes dominate the slightly acid to slightly alkaline pH range. Therefore, U is generally transported as some uranyl complex, and deposition takes place on reduction to the U^{4+} state, on adsorption, or by the formation of certain minerals with low solubility (for example, carnotite).

Uranium is adsorbed by certain types of organic matter, such as found in bogs and organic-rich lake and ocean sediments. Following adsorption, the uranyl ion may be reduced to U^{4+} (uraninite, or some other reduced phase). Various other materials, such as zeolites, clays, and iron oxides and hydroxides, may also be adsorbers of U to varying degrees and under certain environmental conditions. Occasionally, adsorption may result in economic deposits being formed. Hambleton-Jones (1978), for example, reported that adsorption of U by clay minerals, particularly montmorillonite, is the mechanism by which U deposits in calcrete and gypcrete formed in South West Africa (Namibia). At this time, however, insufficient data are available to allow accurate prediction of the role of adsorption in low-temperature sedimentary environments (Langmuir 1978).

To this point the discussion has been concerned with the mobility of U. Now, it is appropriate to consider the mechanisms by which U is precipitated. In all cases, the U will be out of equilibrium because it is young.

The reduction of U^{6+} to U^{4+} will result in the formation of uraninite or other uranous phases. This mechanism, which usually involves the removal of O_2 from waters during the oxidation of organic matter, explains the occurrence of U in roll front deposits, bogs, uraniferous logs, and coal beds. Methane has been postulated as the reductant in the formation of the Rabbit Lake and other large deposits in Saskatchewan. Other mobile reductants of importance include hydrogen sulphide, other hydrocarbons, and hydrogen. The oxidation of Fe from the ferrous to ferric state may, in some instances, result in the reduction of U in some deposits.

Adsorption of uranyl complexes by organic matter and other materials, for example, clays and iron oxides, has been discussed above. These materials certainly reduce the mobility of U. Evidence appears to be emerging that following the initial adsorption of the uranyl complexes, there may be subsequent reduction and precipitation by mobile reductants, such as methane and hydrogen sulphide.

Uranyl ion (in an oxidizing environment) can be precipitated from solution when it combines with vanadate, phosphate, arsenate, silicate, carbonate or sulphate anions in association with Ca, Mg, K, Ba, Pb, Cu, and several other elements. Minerals such as carnotite, $K_2(UO_2)_2(VO_4)_2 \cdot 3H_2O$ and soddyite, $(UO_2)_5 Si_2O_9 \cdot 6H_2O$, are the result. Uranyl ions may also be precipitated from solution after the dissociation of uranyl carbonate complexes at elevated temperatures.

From the above discussion it is clear that in the surficial environment disequilibrium between U and its daughter nuclides can take place only in an aqueous environment (the gaseous migration of ^{222}Rn being the obvious exception). The extent of disequilibrium is a function of the relative mobilities of the various nuclides in the ^{238}U series in each specific aqueous environment (for example, oxidizing, reducing). In the case of U, it is the mobility of the uranic (U^{6+}) ion which is the factor controlling mobility because the uranous (U^{4+}) ion is immobile except in extremely acidic (pH ~ 1) solutions. Finally, it must be stressed that recently precipitated hydrogenic U, such as is found in organic lake or stream sediments, will require 10 half-lives of ^{230}Th (i.e. $t_{\frac{1}{2}} = 7.52 \times 10^4$ y) in a closed system for complete equilibrium to be established and for gamma ray methods of analysis based on ^{214}Bi to be reliable (an exception is considered in connection with Fig. 14.8). This point is discussed further in connection with ^{214}Bi disequilibrium in section 14.4.6.

14.4.2 ^{234}U in U exploration

Cherdyntsev (1971) reviewed the early work in the Soviet Union in which the $^{234}U/^{238}U$ activity ratio had been used for exploration purposes. With respect to this ratio he stated (pp. 110–11):

> Isotope techniques are increasingly often used in prospecting and their reliability keeps becoming more evident. In fact, a change in the isotopic composition yields more information than does a mere determination of the element concentration . . . Even a high uranium content in water cannot serve as a criterion in uranium prospecting, since waters of igneous rocks have a high ($n \times 10^{-5}$ to 10^{-4} grams/liter) uranium content, while uranium deposits, on the contrary, do not easily give up uranium to the percolating waters.

Very early studies by Cherdyntsev had shown that waters from rocks enriched in U (but not ore bearing) often contain excess ^{234}U, whereas extracts (i.e. various leachates) from U minerals typically contain ^{234}U and ^{238}U in equilibrium amounts. The application of this concept is illustrated in Fig. 14.1 in which the striking difference between the $^{234}U/^{238}U$ activity ratios of waters from U ore deposits are compared with those of waters from non-economic U occurrences. The former yield ratios of approximately unity whereas the latter generally have a distribution of between 2.5 and 6.5, and may have large isotope shifts as high as 12. This Cherdyntsev (1971, p. 82) explained on the basis of the location of the U atoms in crystal lattices. Specifically, in U

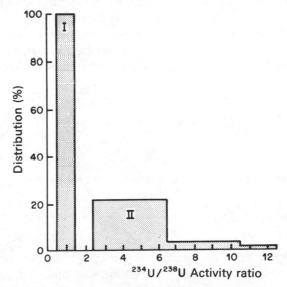

Fig. 14.1. Distribution of $^{234}U/^{238}U$ values in waters from: (I) economic uranium deposits; and (II) non-economic (disseminated) U occurrences. (After Cherdyntsev 1971, Fig. 24 based on work by Syromyatnikov).

minerals of economic deposits the ^{234}U recoil atoms are able to occupy the
positions of ^{238}U atoms and when the element is leached, both isotopes pass
into solution to about the same extent (activity ratio $\simeq$ 1). When the U atom is
an accessory or minor constituent in a mineral (for example, apatite, monazite)
or rock, and is uneconomic as an ore deposit, the ^{234}U recoil atom leaves its
original position, and is not able to find a suitable crystallographic site. It then
migrates to microfractures where it is more susceptible to leaching and thus
produces a high activity ratio in solution.

Syromyatnikov (1965) and Syromyatnikov *et al.* (1967) used the ^{234}U/^{238}U
ratio to distinguish between surface U anomalies in soils of ore and non-ore
origin in friable (unconsolidated, weathered) sandstones. The method employed involved the extraction with a 1 per cent sodium carbonate solution of
the mobile U in the soils and unconsolidated sandstones. The results, as
illustrated in Figs 14.2 and 14.3, show that for U anomalies of non-ore origin,
samples from both the background and anomalous (with U) concentrations
have the same ^{234}U/^{238}U isotopic ratios (Fig. 14.2). Those samples taken
from U anomalies associated with ore deposits have isotopic ratios deviating,
either higher or lower, from the surrounding background areas (see Fig. 14.3).
Such studies are likely to be particularly useful during the detailed or follow-
up stages of exploration as they should eliminate anomalies of low economic
potential from further consideration. Syromyatnikov, *et al.* (1967)
pointed out limitations with the procedure which include: (i) the
anomalous U values must be in contact with the original rock or ore; (ii) it is
difficult to determine beforehand what constitutes anomalous activity ratios;
and (iii) the method is designed primarily for arid regions. At this time there do
not appear to be any recent published studies confirming the use of this
concept.

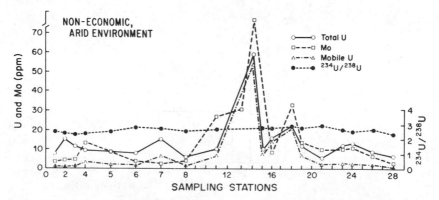

Fig. 14.2. Non-economic accumulation of anomalous U accompanied by Mo in an
argillaceous weathering zone at the contact of an amphibolite and granite. Note the
uniform ^{234}U/^{238}U activity ratio (approximately 2.5) of the mobile U in both the background and anomalous U areas. (After Syromyatnikov *et al.* 1967, Fig. 1.)

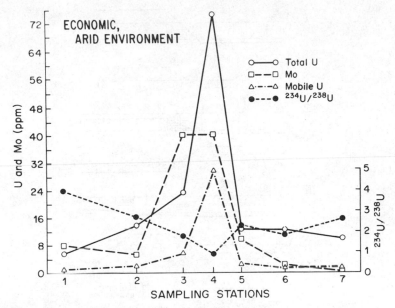

Fig. 14.3. Dispersion halos from economic mineralization from outcropping quartz porphyry rocks. Note the $^{234}U/^{238}U$ activity ratio of less than unity at station 4. (After Syromyatnikov *et al.* 1967, Fig. 3.)

Cowart and Osmond (1977) conducted one of the few detailed studies designed to investigate the use of the $^{234}U/^{238}U$ activity ratio in ground water for exploration purposes. These authors measured changes in the relative abundances of the U isotopes in ground water as it flows through various known sandstone-type U deposits in Wyoming and Texas. For a U accumulation still being formed, the down-flow total U content was found to be low, but it was relatively enriched in ^{234}U (see Fig. 14.4, case II). Sakanoue, Yoneda, Onishi, Koyama, Komura, and Nakanishi (1968, p. 79) observed similar results in their analysis of ground waters associated with a Japanese U ore body. Cowart and Osmond (1977) also considered two other conditions (see Fig. 14.4, cases I and III) likely to be found in sandstone-type U deposits. Thus, the measurement of the $^{234}U/^{238}U$ activity ratio, as well as total U concentration, in ground water can greatly enhance the effectiveness of hydrogeochemical exploration. In particular, it is an abrupt increase in the U activity ratio coupled with a decrease in concentration which may be the significant phenomenon for U ore exploration.

It must be noted that the results of Cowart and Osmond (1977) appear, superficially at least, to be possibly in disagreement with those described above from the Soviet literature, such as in Fig. 14.1. With regard to the Soviet data, it is difficult to determine the type of U mineralization, the hydrology,

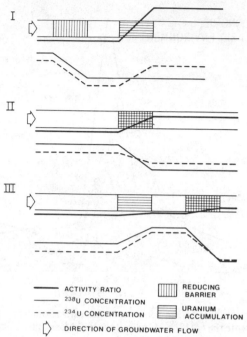

Fig. 14.4. Three configurations of a U precipitation–alpha recoil system in a confined aquifer. The relative positions of ^{234}U and ^{238}U concentrations and the concentration changes shown are schematic. Actual concentration, at equilibrium, of ^{234}U is about 5 orders of magnitude less than ^{238}U. I. Reducing barrier up-dip from main U accumulation. It is assumed that the up-dip migration was recent and, as yet, only a negligibly small quantity of U has precipitated at that location. II. Reducing barrier coincident with U accumulation. Precipitation of U isotopes is not preferential; input by alpha recoil from previously precipitated U is the cause of increase of ^{234}U concentration relative to ^{238}U concentration. III. Reducing barrier down-dip from previous U accumulation which is now in an oxidizing environment. It is, therefore, being dissolved, transported to the reducing barrier, and reprecipitated. Because the U concentrations in water are greater than in I and II above, the change in relative concentration between ^{234}U and ^{238}U is significantly less. It is assumed that the quantity of U in both accumulation zones of III is similar. (After Cowart and Osmond 1977.)

and the lateral scale (such information is not available for Figs. 14.2 and 14.3) upon which their conclusions are based. The conclusions with respect to Fig. 14.1, for example, may be absolutely true for waters in the case of a well crystallized detrital uraninite or brannerite ore (for example, Elliot Lake). However, based on the work of Cowart and Osmond (1977), they do not appear to be applicable for waters percolating through hydrogenic sandstone-type U deposits in Texas and Wyoming in which the uraninite (or other mineralogical phases) is relatively poorly crystallized, and from which ^{234}U is likely to be more easily moved into the aqueous phase. The low activity ratio illustrated in Fig. 14.3, which is related to economic mineralization, could well be explained by oxidation and dispersal of the ore body at the present time. This situation would be analogous to that illustrated in Fig. 14.4, case III. Cowart and Osmond (1980) discuss some recent experiences with respect to the $^{234}U/^{238}U$ ratio, and its use in exploration in sandstone-type deposits.

The use of ^{234}U in mineral exploration involves the determination of the $^{234}U/^{238}U$ activity ratio in waters, soils, and/or rocks. As the determination of this ratio by alpha spectrometry generally requires extensive chemical pretreatment and long counting times, the production rate is slow and, as a result, the data are expensive. Other techniques using gamma ray spectrometry (Ostrihansky 1976) or low energy photon spectrometry (Szoghy and Kish 1978) have been applied to whole rock samples but, again, the analyses are slow, especially for samples with low radionuclide content. It is not surprising, therefore, that the full potential of $^{234}U/^{238}U$ activity ratios has yet to be realized.

14.4.3 ^{230}Th in U exploration

The nuclide ^{230}Th is only occasionally determined for exploration purposes, and the few examples in which it has been reported for such purposes are mostly found in the Soviet literature. For example, Cherdyntsev (1971, pp. 111–13) showed how the activity ratio $^{230}Th/^{232}Th$ can be used to assist in the interpretation or corroboration of other isotopic data (particularly $^{234}U/^{238}U$ and $^{228}Ra/^{226}Ra$) collected in connection with U exploration. Specifically, the $^{230}Th/^{232}Th$ ratio in rocks of the earth's crust is approximately 0.8, and it is somewhat higher in natural waters. In the vicinity of economic U deposits, however, this ratio may become from ten to hundreds of times greater, as ^{232}Th is not normally associated with U deposits (except those in alkalic rocks).

The ^{230}Th is more often determined during the detailed stages of exploration of a deposit or occurrence when information on the age of mineralization, the state of equilibrium, and the extent and timing of U migration in near surface environments is needed. In such cases, the activity ratios $^{230}Th/^{234}U$, $^{230}Th/^{231}Pa$, and $^{226}Ra/^{230}Th$ are likely to be obtained. By determining first two of these ratios on samples from roll fronts, Rosholt, Tatsumoto, and Dooley (1965) were able to show the geometric relationship of U migration in certain roll front deposits in Colorado and Wyoming, as well as the fact that leaching of a deposit in the Powder River Basin of Wyoming took place over a period of at least 10^5y. Lively, Harmon, Levinson, and Bland (1979) used $^{230}Th/^{234}U$ and $^{226}Ra/^{230}Th$ activity ratios to study the extent of leaching and redeposition at the Yeelirrie calcrete-type deposit in Western Australia.

In general, fractionation of ^{230}Th from its parents (^{238}U and ^{234}U) and daughter (^{226}Ra) nuclides takes place in most environments (except deserts) because Th is extremely insoluble in aqueous media in comparison with U (averages for river water: [Th] = 0.1 ppb and [U] = 0.4 ppb). Therefore, U and Ra move away resulting in disequilibrium between ^{230}Th, ^{226}Ra, and ^{234}U so that in an active hydrological system, unsupported ^{230}Th can result and, indeed, samples in this state have occasionally been mentioned in the Soviet literature. It is worthwhile noting that Titayeva, Filonov, Ovchenkov, Veksler, Orlova, and Tyrina (1973) have reported on the mobility of U and Th isotopes during the interaction of surface waters with crystalline rocks in a cold, wet environment in the Urals. They observe that although the waters interacting with the rocks in this region have low salt contents and are at only 2 to 5° C, they extract appreciable amounts of Th as well as U. In fact, in their study they found that initially the entry of U into the liquid is only about twice as extensive as that of Th. However, based on study of spring waters with high (1 to 5) Th/U ratios, they observed that after some tens of metres the ratio falls substantially owing to the general insolubility of Th in most natural waters (except those with a low pH; pH of hydrolysis of Th is 3.5). In other words, Th nuclides yield very short hydrogeochemical dispersion patterns in waters and are, effectively, of value in hydrogeochemical surveys only in a limited number of special situations.

14.4.4 ^{226}Ra in U exploration

Except for U itself, ^{226}Ra is the most important nuclide in the ^{238}U decay chain as far as exploration for U ore deposits is concerned. This generalization is valid because:

(i) The ^{226}Ra is the parent of all the short-lived nuclides further down the decay chain upon which most instrumental exploration techniques are based, for example, the techniques to detect ^{222}Rn, ^{214}Bi, and ^{210}Po. Therefore, the instrumental methods which detect these nuclides are, in effect, detecting ^{226}Ra.

(ii) The ^{226}Ra may be mobile in some surficial environments, specifically in the reducing environment, in which U itself is immobile. Hence, this relatively long-lived nuclide ($t_{\frac{1}{2}} = 1602$ y) is a potential pathfinder for U ore deposits even though it is usually out of equilibrium with its parents in aqueous surficial environments (except in the case of detrital minerals). In oxidizing environments, Fe and Mn oxides tend to adsorb Ra thus reducing its mobility; this is discussed in greater detail below.

(iii) The ^{226}Ra forms many false anomalies (for example, in spring waters and their precipitates) originating from sources, such as the weathering of felsic igneous rocks, which are in themselves non-economic.

Although this has been discussed in chapter 2, it is now appropriate to consider the geochemical characteristics of Ra in the surficial environment. In this regard, it is important to recognize that Ra is an alkaline earth element and chemically it is similar to Ba. Some compounds of Ra, such as sulphate, have very low solubilities (i.e. the solubility of $RaSO_4$ is about 20 ppb at 25°C). Nevertheless, the extremely low abundance of Ra (each ton of U in equilibrium with its daughters will contain only 0.334 g of ^{226}Ra) permits all that is produced by decay of ^{230}Th to be mobile (i.e. the mobility of Ra is not limited by its solubility product). Radium also forms soluble complexes with the chloride ion as evidenced by the high Ra content of deep, chloride type brines. However, even though Ra can be leached from its parent ^{230}Th, there are several natural mechanisms by which its mobility can be reduced or stopped, at least temporarily, in the waters of the surficial environment: co-crystallization, co-precipitation, inorganic adsorption, and biological absorption.

Co-crystallization occurs when Ra is able to substitute for Ba in barite, resulting in the so-called radiobarite. As neither U nor Th can occupy such structural positions in natural situations, the Ra is unsupported, and all evidence of ^{226}Ra will disappear within 10 half-lives, or about 1.6×10^4 y. The presence of radiobarite must indicate recent geological processes, such as active leaching of pyrite-containing U ores. In such cases, sulphuric acid is formed, U is leached away as a soluble uranyl sulphate complex, and Ra will co-crystallize with any available Ba as radiobarite (as in the uraninite deposit at Blind River, Ontario).

Co-precipitation scavenging of Ra may occur where Fe-Mn oxides and hydroxides are being precipitated, such as at the orifices of springs. Radium, and U if present in spring water, may also be co-precipitated with Fe-As precipitates (Cadigan and Felmlee 1977). In practice, it is difficult to determine whether Ra is co-precipitated or adsorbed by Fe-Mn oxides.

Adsorption of Ra may be effected by Fe-Mn oxides, clay minerals or organic matter. Clay minerals preferentially adsorb divalent, alkaline earth cations, including Ra. Differential adsorption of Ra and U is illustrated in Fig. 14.5 based on an example from the Soviet literature (Perel'man 1977). This diagram shows the paths taken by U and Ra as they are leached from three U-bearing veins that have been weathered to depths of several tens of metres below the surface in a humid environment; presumably ^{230}Th would remain immobile in the weathered zone of the ore veins. Radium is adsorbed by clays in the glacial moraine, whereas U is mobile until it is adsorbed and/or reduced by the organic matter in the peat bog.

Ground water in many springs and wells has high contents of Ra. Manganese oxides, locally called reissacherite in the Badgastein area of Austria (sometimes called radioactive wad), are examples of spring pre-cipitates which may contain over 400,000 pCi/g of ^{226}Ra. The Ra has been adsorbed from the spring waters by the Mn oxides over an extended period of time, and is the main source of the Rn and other daughter nuclides in the area. Other less spectacular concentrations of Ra-containing Mn and other types

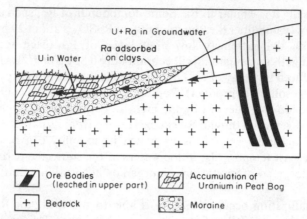

Fig. 14.5. Differential dispersion of U and Ra from a chemically weathered U deposit in a taiga landscape. (After Levinson 1980, but based on an example from Perel'man 1977.)

of precipitates from springs are known from many localities, such as in the western United States and more recently Saskatchewan. On occasion, the adsorption process may be so effective that it may result in Ra being essentially immobile. Rose and Korner (1979), for example, observed such a phenomenon in Pennsylvania in which Fe oxide is the adsorber. The environment in such situations is always oxidizing (the Fe oxides are insoluble; in reducing environments the Fe will be in the soluble ferrous state). Thus, in some, but not all oxidizing environments, ^{226}Ra released by the decay of U will be adsorbed or co-precipitated and another nuclide (for example, ^{222}Rn or U itself), may have to be used in hydrogeochemical surveys.

Biological absorption of Ra by vegetation exists because plants have no physiological barriers to Ra. This is because plants do not differentiate between the alkali earth elements or, in other words, physiologically Ra is similar to Ca or Ba as far as plants are concerned. As almost all plants die within the 1600 year half-life of ^{226}Ra, and ^{226}Ra is thus released during subsequent biological decay, biological absorption can be considered as only a temporary hindrance to the mobility of ^{226}Ra. Autoradiographs showing that plants absorb Ra have been presented by Levinson (1980).

In connection with radioactive spring and well waters (usually containing ^{226}Ra and/or ^{222}Rn) it is important to note that they frequently contain little or no U and, therefore, the Ra and Rn are unsupported. In most cases, U deposits have not so far been found in association with these nuclides, hence the springs yield false anomalies. Apparently, warmed meteoric waters have leached large volumes of granitic rock, which may or may not contain high contents of U, and these waters are then channelled to specific orifices. It has been suggested that at the elevated temperatures characteristic of thermal springs, Ra carbonate is soluble whereas the uranyl carbonate complex is unstable and dissociates, following which the U is removed from solution (Hambleton-Jones 1978).

According to Eisenbud (1977), the most common of the radioactive anomalies are the thousands of mineral springs that are known to have extraordinarily high levels of radioactivity. In a typical aquifer, ground water movement is slow, the amount of ^{222}Rn removed by springs and wells is small (compared with the amount left in the aquifer), and the distance that ^{222}Rn migrates is short. Under such conditions, it is possible that (approximate) radioactive equilibrium exists among ^{226}Ra, ^{222}Rn, and other decay products, so that the source of the total radioactivity may be several isotopes lower in the decay chain (but chiefly Ra and Rn). Water drawn from wells and springs in the St Peter Sandstone in various parts of Illinois and Iowa is elevated in Ra (and Rn) which may possibly be attributed to the fact that the aquifer is in contact with non-economic black shales (hence the anomalies are false or non-significant). One of the more remarkable discoveries is that some ground water in the Helsinki area of Finland contains substantial quantities of ^{222}Rn, ^{226}Ra, and U (Asikainen and Kahlos 1979). The Rn content has been found to be as high as 880,000 pCi/l. Other values include ^{226}Ra as high as 256 pCi/l, and U as high as 15 ppb in some of the wells.

Cadigan and Felmlee (1979) determined the concentration of Ra, Rn, and U in springs from a hydrothermal area northeast of Great Salt Lake, Utah and they estimated the U potential of the area. By determining Ra and Rn values in excess of equilibrium amounts with the U, they calculated that the hydrothermal conduit system needed approximately 170,000 metric tons of U to support such radioactivity. Of course, such calculations do not define an exploration target, rather they identify the magnitude of the U sources in the area. Whether U can be extracted economically depends on the degree of dispersal (for example, whether the U occurs within the structure of rock forming minerals or as a concentration of U minerals), and the depth to the U mineralization (if it does occur). Both of these factors are unknown in the area studied. Further, it must be recognized that thermal springs are frequently located along fault zones where extensive fracturing, brecciation, mylonitization, etc., have occurred. Under these conditions, U may be readily available for leaching and a high radioactivity in a thermal spring need not constitute a potential deposit.

If the dissolved Ra content of large quantities of spring water can be concentrated, the activity of the ^{223}Ra ($t_{\frac{1}{2}} = 11.68$ d) isotope may reveal the average distance of the source of dissolved Ra. As this isotope constitutes only 4.6 per cent of the total Ra activity close to the source, such measurements require high resolution alpha spectrometry and careful analysis to avoid interference from ^{224}Ra ($t_{\frac{1}{2}} = 3.64$ d) activity. This procedure is currently under development by the authors.

14.4.5 ^{222}Rn in U exploration

This nuclide is useful in exploration for U because it is an inert gas and, therefore, it has the ability to migrate through porous media without forming compounds. However, its high atomic weight and short half-life ($t_{\frac{1}{2}} = 3.8$ d) are limiting factors as to the distance it can migrate through rock materials.

It is essential to emphasize that as the immediate parent of ^{222}Rn is ^{226}Ra, the detection of ^{222}Rn only guarantees the presence of ^{226}Ra and not U unless the system is in equilibrium. This point is illustrated in Fig. 14.6 which is based on an example from an arid region in southern Africa, where the position of the water table has fluctuated in recent geological times. This has resulted in disequilibrium, as ^{226}Ra has been moved to point A (alternatively, movement of ^{234}U may be involved, as discussed in connection with Fig. 14.8). A ^{222}Rn anomaly detected at the surface at the present time represents only ^{226}Ra, and not U. As this ^{226}Ra is unsupported, it will have completely disappeared in about 1.6×10^4 y from the time of its original precipitation at A. The ^{222}Rn generated at B is too deeply buried to be detected at the surface.

Another point which merits comment here is the *emanation coefficient* defined as the ratio of liberated Rn to the Rn formed in a sample. Clearly, ^{222}Rn cannot migrate toward the surface if it cannot escape (emanate) from a sample. The emanation coefficient for most U ore samples only averages 21 per cent, but for rocks it is usually only a few per cent. The factors controlling the emanation of ^{222}Rn are complex with the most important variables being the mineralogy of the ore, the particle size of the U (Ra-containing) minerals, and the porosity and permeability of the ore (see Austin (1975) and Barretto

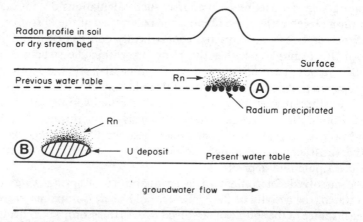

Fig. 14.6. Idealized diagram showing a ^{222}Rn anomaly over ^{226}Ra at A, which was precipitated when the water table was higher. The U deposit is located at B where the Rn anomaly presently being generated does not reach the surface. Such situations, or variations thereof, are believed to be common in certain desert environments, such as in southern Africa. (After Levinson and Coetzee 1978.)

(1975) for further details on Rn emanation). Poorly crystallized, very fine-grained samples are likely to have a higher emanation coefficient compared to samples with well-crystallized, coarse grained U minerals. The variability of the emanation coefficient in uraniferous materials has been discussed by Dyck (1978), and by Levinson and Bland (1978). From their reports it is obvious that because of the variable and unpredictable amounts of ^{222}Rn emanation, Rn surveys can only be considered qualitative at best. Further, the amount of ^{222}Rn that does emanate, and which may be detected at the surface is a reflection of ^{226}Ra and not necessarily U (see Fig. 14.6).

The matter of Rn migration from the point of view of exploration has been reviewed by Smith, Barretto, and Pouris (1976). They point out that Rn is considered to migrate by two methods: diffusion (such as through rocks), and transport (such as movement in ground water). With respect to diffusion, assuming reasonable geological conditions of rock permeability and porosity, they calculated that the mean migration distance of ^{222}Rn before decay occurs is about 1.6 m. In areas with highly fractured rocks, the migration distance will certainly be greater. Migration by transport in ground water is the mechanism by which ^{222}Rn can migrate over larger distances (turbulence in surface waters results in a rapid loss of dissolved gases). Assuming a flow rate of 1.5 metres per day, ^{222}Rn could migrate 60 to 70 m before its concentration would be reduced by decay to below the detection limit (Smith *et al.* 1976). Considering all the variables and limitations of Rn decay (for example, limited emanation and diffusion factors, short half-life), rather than postulate great upward movement of Rn from a U ore body (up to several hundred metres have been claimed) many investigators recognize that it is much more reasonable to assume that fluctuations in the water table over periods of thousands of years result in the differential migration (disequilibrium) between ^{238}U, ^{234}U, ^{230}Th, and ^{226}Ra, and they interpret ^{222}Rn anomalies as representing ^{226}Ra anomalies. There are, however, some who insist that ground water can move much faster than commonly believed, especially in good aquifers, and they reject the above argument. Radioactive springs high in ^{222}Rn, such as those mentioned in section 14.4.4, are examples of ground waters in which ^{222}Rn apparently has moved large distances; alternatively, high ^{222}Rn activities may represent local concentrations of ^{226}Ra. Clearly, the problem of how far Rn can migrate is very pertinent to U exploration (see Fleischer and Mogro-Campero 1978 for a detailed review). Those inclined toward the theoretical generally favour a limited migration of ^{222}Rn, whereas others are likely to be considerably less conservative on this point.

For the purposes of prospecting, Rn can either be determined in the field in soil gas or in water (for example, ground water or lake water), using any one of several techniques currently available commercially. Hambleton-Jones (1979) summarized these techniques as falling into five categories: (i) the emanometer group; (ii) the alphameter; (iii) the track-etch method; (iv) the thermoluminescence method; and (v) the RAOC (Rn adsorption on charcoal) method. They have also listed the advantages and disadvantages of each type.

Radon surveys must be conducted with caution and, in fact, it is often very difficult, if not impossible, to obtain identical results when a Rn survey is repeated (this is characteristic of most vapour surveys used for exploration). One of the problems involves changes in atmospheric pressure which affect the ability of ^{222}Rn to migrate in soil gas (see Levinson 1980, p. 735). The ^{220}Rn (thoron, $t_{\frac{1}{2}} = 56s$), a product of the decay of ^{232}Th, also must be considered (see Fleischer and Mogro-Campero 1979) during field determinations as it can add to the total Rn count (most field instruments are capable of discriminating between the Rn nuclides). The greater problem, by far, with Rn surveys is the large number of anomalies which are mostly false as they may frequently not reflect mineralization. These occur with great frequency in all types of waters (spring, ground, lake) and in soils. Although the reasons for these false anomalies are usually not determined, they do represent disequilibrium in some form.

In the last few years, efforts have been made to understand the accumulation of Rn in buildings built in the vicinity of U mines (for example, at Port Hope, Ontario). Despite enormous expenditures of time and money, no simple method has been devised to exclude Rn from buildings and, although this work is not of direct interest in exploration, it does illustrate disequilibrium and the complexities of Rn migration. Cliff (1978a,b) has provided excellent material on this subject; his papers are highly recommended to those interested in a thorough understanding of Rn measurements.

14.4.6 ^{214}Bi in U exploration

The nuclide ^{214}Bi decays by beta emission into ^{214}Po which is left in one of several excited states thus giving rise to characteristic gamma radiation at 1.76 MeV. It is this radiation which is detected by scintillation counters, and by gamma ray spectrometers with a U-channel designed for U exploration. Therefore, it must be stressed that such instruments detect ^{214}Bi and not U. It should be recognized that the presence of ^{214}Bi indicates proximity to ^{222}Rn (and hence ^{226}Ra) because, on a practical basis, ^{222}Rn is the parent of ^{214}Bi owing to the short half-lives of the intermediate nuclides ^{218}Po and ^{214}Pb (exceptions might be found where strong atmospheric circulation might occur as in a well-ventilated mine, where waters containing Rn are turbulent and free-flowing, or where rocks have a very high permeability). Thus, with reference to Fig. 14.6 a gamma ray spectrometer survey should record anomalous values in those same areas where the high Rn anomalies were detected. However, there is no U occurrence corresponding to anomaly A as there is at B (see Fig. 14.6) as a result of disequilibrium between Ra and U at location A. The apparent amount of U calculated from the gamma ray emission of ^{214}Bi is called equivalent U (often abbreviated eU, or in the oxide

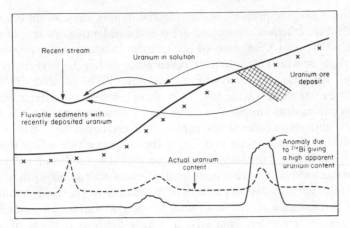

Fig. 14.7. Diagrammatic representation of the relationship between actual U and apparent (equivalent) U (eU based on gamma radiation from [214]Bi) in an environment with chemical weathering. (After Hambleton-Jones 1978.)

form as eU_3O_8), and is based on the assumption that equilibrium conditions prevail, which is not always the case.

In many instances the relation between actual U and equivalent (apparent) U shows marked deviations (i.e. disequilibrium), similar to that in Fig. 14.6. Figure 14.7 illustrates the point further, with plots of the apparent (equivalent) U from [214]Bi, and actual U (which can be determined by fluorometry, delayed neutron counting, or X-ray fluorescence). In Fig. 14.7 soluble U is leached from a deposit, transported as a complex such as uranyl dicarbonate, and deposited in stream sediments. Residual [226]Ra, in part generated from insoluble [230]Th in the ore deposit, will give an apparent U anomaly determined by gamma ray methods significantly above the actual U content. The actual U content of the stream sediments in Fig. 14.7 may be high, but since there has been insufficient time for [214]Bi to have been produced, the gamma ray activity is negligible. In an analogous situation, Levinson, Bland, and Parslow (1978) showed that U transported in solution into organic-rich lake sediments in the Seahorse Lake area, near the large Key Lake U deposit, Saskatchewan, yielded apparent U values of only 3 to 8 per cent of the actual U present in most samples owing to the young age of the U (the area was glaciated during the Pleistocene, and the lakes are believed to be no older than 10^4 y). Any clastic U minerals in the stream sediments (see Fig. 14.7) or lake sediments will be in equilibrium and equivalent U will correspond to the measured U content.

Numerous other examples (such as the sandstone deposits of the United States) could be cited in which disequilibrium exists between [214]Bi (apparent U) and the true U content, at least on a local basis. The problem can take on significant economic proportions during the mining stage of an ore deposit especially if the extent of disequilibrium is of the order of a few tens of metres. Thus, in all types of U deposits or occurrences in which water movement is taking place, or where it could have taken place within the past few hundred thousand years (governed by the half-life of ^{230}Th, or that of ^{234}U as discussed in connection with Fig. 14.8 below), radiometric determinations based on the gamma emission of ^{214}Bi may be misleading.

A very important concept for exploration geologists to consider in the context of the disequilibrium problem is the volume within which this effect occurs. If smaller than the sampling volume, then disequilibrium will be of no practical significance. A case in point would be an airborne survey in which the movement of the nuclides may be tens of metres but where the disequilibrium problem is minimized by the large sample volume examined by an airborne gamma ray spectrometer. Similarly, if a bulk analysis is made of a hand specimen weighing several kilograms, and the isotopic movement were a few centimetres, the radioactive decay series will appear in equilibrium. On the other hand, if movement were a few centimetres but only one gram of a sample was analysed, then extensive disequilibrium may be observed.

A voluminous literature exists on the application and interpretation of gamma ray spectrometric methods used in U exploration. The review of this subject by Killeen (1979), in which the ramifications of radioactive disequilibrium are considered, may be recommended highly.

Earlier in this chapter, it was shown that the ^{234}U/^{238}U activity ratio in waters is likely to deviate considerably from unity (see Figs. 14.1 to 14.4). Hambleton-Jones (1978) is the first to point out the importance of this from the point of view of gamma ray spectrometry and its use in exploration. The following four paragraphs are a direct quotation from Hambleton-Jones (1978, p. 189) with only changes in the figure numbers and references for consistency herein.

If the uranium dissolved in the subsurface water has a ^{234}U/^{238}U activity ratio of 1.0 and is precipitated from solution in the same ratio, then the growth of ^{226}Ra would follow the line R = 1.0 in Fig. 14.8 where the activity ratio ^{226}Ra/^{238}U, would attain equilibrium at 10^6 years. However, the ^{234}U/^{238}U activity ratio for subsurface waters is usually greater than unity. Curves for R = 1.2, 1.5 and 2.0 are also plotted. It is noted that the curves go above the equilibrium line, ^{226}Ra/^{238}U = 1, and return back to that line at approximately 2×10^6 years. The reason for this peak is that there is ^{226}Ra accumulating from the unsupported ^{234}U. Therefore, this implies that there is a 'mixture' of ^{226}Ra's, some from the unsupported ^{234}U and the remainder from the supported ^{234}U. The unsupported ^{226}Ra differentially grows and finally decays with the unsupported ^{234}U. The time of equilibrium to be reached for the R = 2.0 case is approximately double that of the R = 1 case.

The implication for this situation is very important and in particular for those uranium accumulations whose ages are less than 2×10^6 years. Consider the situation where the uranium in the groundwater at R = 1.5 is being deposited. There are two ages where the activity ratio of ^{226}Ra/^{238}U is at unity: 1.4×10^5 years and

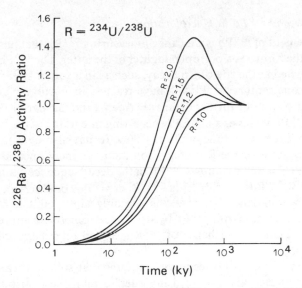

Fig. 14.8. The growth of ^{226}Ra from its parent ^{234}U is shown. When U is precipitated from solution with an activity ratio R = ^{234}U/^{238}U = 1, then secular equilibrium will be attained in 10^6 y. If R > 1, then, there is an amount of unsupported ^{234}U, and secular equilibrium is attained at a greater age. The relative gamma ray activities of the cases where R > 1 will be higher than where R = 1, which may then lead to errors in ore reserve estimation. (After Hambleton-Jones 1978.)

1.6×10^6 years. This does not mean that there are two equilibrium ages, which is not possible, but that the first age is an apparent equilibrium age, brought about by there being 50 per cent unsupported ^{234}U. Furthermore, if a uranium deposit is being explored using a calibrated γ-spectrometer (either the portable or borehole types) and has an actual age of, say, 2×10^5 years, there will be an overestimation of the equivalent uranium by an amount of 30 per cent i.e., difference between the value at R = 1.5 and R = 1.0. Conversely, assume that ^{226}Ra/^{238}U = 0.8, then the apparent discrepancy between the ages for the case R = 1 and 1.5 will be 8.75×10^4 years.

Therefore, if the uranium flux has a value R > 1 and part of that uranium is being precipitated then the grade of uranium will be overestimated if the age of the deposited uranium is less than a few million years. If the exact age of the deposit is known and is less than a few million years, a good estimate of the amount of disequilibrium may be calculated. However, age determinations within this time scale have been attempted but in general have not been found to be entirely satisfactory.

The radiometric technique whereby the degree of disequilibrium can be determined using the gamma energies of the daughter products of ^{226}Ra in the (4n + 2) decay chain may lead to severe errors. If R $\geqslant$ 2 then the error will be so large (greater than 100 per cent) that any results will become virtually meaningless for those deposits younger than 2×10^6 years.

14.4.7 ^{210}Pb and ^{210}Po in U exploration

The measurement of ^{210}Pb is a means of measuring ^{222}Rn that has decayed in a sample as the former is a prompt daughter of the latter. The ^{210}Pb ($t_{\frac{1}{2}} = 22$ y) will be deposited at the site of Rn decay, such as in a soil or stream sediment. After about one century, ^{210}Pb will have reached its maximum activity with unsupported ^{226}Ra (for further details see Adams and Gasparini 1970, p. 124). Therefore, ^{210}Pb serves as a naturally occurring integrator of the Rn flux in an area with an integration time of about 100 y. It may also be considered as a substitute for some of the Rn integration methods mentioned above.

The ^{210}Po ($t_{\frac{1}{2}} = 138$ d) follows ^{210}Pb in the decay sequence being separated from it only by ^{210}Bi($t_{\frac{1}{2}} = 5$ d). Either ^{210}Bi or ^{210}Po could serve as a measure of the ^{210}Pb abundance but ^{210}Po is the only alpha emitter of the three nuclides. Thus, by measuring ^{210}Po by alpha spectrometry an indirect measure of ^{210}Pb is obtained, and hence of ^{222}Rn, in soil or stream sediment (see section 5.5.5).

Application of the ^{210}Po technique requires that soil or stream sediment samples be collected and analysed in order to map the distribution of Rn emanation. Rainfall will deposit ^{210}Pb and ^{210}Po attached to aerosols. However, ^{210}Pb and ^{210}Po deposited from the atmosphere are widely and uniformly distributed and present no unusual problems in exploration. Further, ^{210}Pb is very effectively sequestered by organic material in the top soil. Sampling from a few centimetres beneath the surface should remove contributions from the surface which may fluctuate with the season and weather. Although the analysis of ^{210}Po (or ^{210}Pb) alleviates some of the major problems associated with Rn surveys (for example, variations due to atmospheric changes), it does not eliminate all limitations of the Rn method (for example, the emanation factor must still be considered). It also is not applicable to Rn surveys in waters.

Studies by ScienTerra Inc. Spokane, Washington (courtesy of Dr. D. A. Hansen) have established that the ^{210}Po and the ^{210}Po/U ratio are both useful exploration parameters, the latter being a measure of the excess Rn integration with respect to the Rn emanating power at the sample site. Other studies by Scien Terra Inc. have shown that, assuming a uniform Rn flux from any source, the average migration time through soils will be greater through clayey soils than through sandy soils owing to the lower permeability and interstitial porosity of the former. Therefore, ^{210}Pb and ^{210}Po concentrations in clayey soils will be greater than in sandy soils, other factors such as Rn flux being equal. By normallizing the ^{210}Po (or ^{210}Pb) values to the surface area of the sample, it is possible to identify those anomalous features that may be due to the difference in the migration rate of the Rn through soils of varying type. To date, it has been possible to identify anomalous responses to ^{210}Po surveys as originating from three different sources: (i) U mineralization at depth; (ii) anomalies due to decreased rate of Rn migration in clayey soils; and (iii) minor concentrations of U in the sample itself (i.e. false anomalies due to different U contents in different soils). As in the case of ^{222}Rn, true anomalies of ^{210}Pb or ^{210}Po effectively represent Ra dispersion halos.

Certain plant species such as mosses and lichens derive nutrients from atmospheric moisture (see Shacklette and Connor (1973) for a review of this subject). Such species are long-lived (some lichens live several centuries!) and thus may provide an excellent monitor of atmospheric ^{210}Pb and while not identifying the Rn halo in the soil beneath, analysis of these materials may be used to determine areas of enhanced activity. Levels of ^{210}Po in lichens growing in uraniferous regions may exceed 100 pCi/g on a dry weight basis (it would be significantly higher on a dry ash basis).

14.4.8 Stable nuclides (^{4}He and ^{206}Pb) in U exploration

Although ^{4}He and ^{206}Pb are stable nuclides and, therefore, do not fit exactly within our framework of disequilibrium, they are generated during radioactive decay in fixed and predictable amounts, and used in exploration often in conjunction with other nuclides, particularly ^{222}Rn. Accordingly, they will be considered here briefly.

As already discussed in chapter 2 He is a by-product of radioactive decay. It is this fact that makes it a potential pathfinder for U and Th deposits. Other uses include exploration for geothermal resources, locating faults which may be the locus of many types of mineralization, and in petroleum exploration. Dyck (1976, 1979) has reviewed the uses of He in locating mineralization, with many references to Soviet experiences in this field, and the reader is referred to his works for more detail.

Alpha particles consist of two protons and two neutrons, i.e. a He atom minus two electrons (see chapter 1). Being positively charged, these particles readily pick up available free electrons in the subsurface environment and become gaseous, inert He atoms. Eight ^{4}He atoms are produced in the decay of ^{238}U to ^{206}Pb, seven in the decay of ^{235}U to ^{207}Pb, and six in the decay of ^{232}Th to ^{208}Pb. The constant generation of non-radioactive He (^{4}He) gas from U and its daughter products provides a potentially useful means of remote detection of buried radioactive deposits. One major advantage of the use of He in exploration is that it is a stable nuclide and, therefore, there is no time restriction on its migration as, for example, there is with ^{222}Rn ($t_{\frac{1}{2}} = 3.8$ d). Also, because it is an inert, very light and small atom, it has a much greater ability to diffuse greater distances through rocks than does Rn and other gases.

Numerous studies (see Dyck 1976, 1979; De Voto et al. 1980) have shown that: (i) He anomalies in the atmosphere are essentially undetectable even above known mineralization; (ii) He anomalies in soil gas can be extremely subtle and difficult to detect (De Voto et al. (1980) found that instantaneous gas samples collected from soils over known U deposits in the Grants District, New Mexico, buried 15 to 250 m below the surface contained a maximum of 0.26 ppm He above the background of 5.24 ppm); and (iii) the He contents of ground waters show distinctly anomalous relationships to buried mineral deposits. Therefore, based on the present state of the art, the use of He surveys in ground water should be emphasized.

Although the amount of He emanating from a U deposit may be substantial (for example, 1 g of ^{238}U in equilibrium with its decay products will produce 11.0×10^{-8} cm^3 of He per year), the great mobility of this element can frequently be detrimental. Dyck and Jonasson (1977) have discussed some aspects of the nature and behaviour of He and other gases in natural waters with particular emphasis on exploration. They concluded that Rn and He are clearly related to the presence of U-enriched rocks. They further concluded that Rn can be used as a tracer to such mineralization but the extreme mobility manifest in He, limits its usefulness in this regard. Dyck (1979) noted that most He observed in ground waters and springs has escaped into the water systems from rocks and minerals, particularly the U and Th enriched basement rocks. Thus, most regional He anomalies will reflect fault and fracture zones rather than U ore deposits.

Another problem apparently not previously mentioned in the literature, which may result in a false anomaly, may be illustrated with reference to Fig. 14.6. At position A characterized by unsupported ^{226}Ra, high He values may be expected (along with high ^{222}Rn and ^{214}Bi) because five out of the total of eight alpha particles are released between the nuclides ^{226}Ra and ^{206}Pb (all of which have relatively short half-lives).

Notwithstanding possible pitfalls, He has great potential in U exploration in view of its stability, inertness, and mobility. Dyck (1979) concluded that tests of the usefulness of ^{3}He/^{4}He ratios as a means of differentiating between He from U deposits and basement rocks, and ^{3}He measurements as a means of determining residence times of water reservoirs, are urgently needed to evaluate the He method as a U prospecting tool for buried deposits.

The use of the stable isotope ^{206}Pb (also ^{207}Pb), particularly involving ratios such as ^{238}U/^{206}Pb, have been discussed above in connection with the determination of U source rocks by disequilibrium methods. However, Pb isotopes also have been used to evaluate other types of mineral deposits. For example, Doe (1979) has shown that the Pb isotopic composition in all large mining districts in the western United States (magmatothermal types) is characterized by ^{206}Pb/^{204}Pb ratios of less than 18; this ratio exceeds 20 where the ore potential is very minor. One advantage of these types of studies is that only one or a very few samples need be analysed because there is very little variation in the Pb isotope ratios of specimens from the same district.

14.5 Environmental health aspects of U-series disequilibrium

Exposure to ^{222}Rn and ^{220}Rn and their daughters constitutes the most significant radiological hazard, primarily in U mining, but also in some non-U mines and caves in which Rn can be found (the latter include Fe, Sn, and fluorite mines in Great Britain, Fe mines in Sweden, fluorite mines in Newfoundland, and the large, commercial caves of the United States). Busigin, Van der Vooren, Lin Pai, and Philips (1979) summarized the early history of such problems, and note that the high incidence of lung cancer and respiratory ailments among U miners has been studied for about 100 years. It was not until 1921, however, that it was suggested that the high incidence might be attributable to ionizing radiation. In 1924 Rn was implicated as the responsible nuclide. In 1954 it was established that the major dose came from inhaled and subsequently deposited alpha-emitting Rn daughters. Since that time, many detailed epidemiological studies in the health physics and mine safety literature have established a definite relationship between cumulative exposure to Rn daughters in air and excess lung cancer deaths. High levels of Rn have also been encountered in some caves where it is unrelated to mineralization; its occurrence in buildings was mentioned in section 14.4.5.

Potentially dangerous amounts of radioactive solid and liquid wastes are produced at U (yellow cake) extraction plants. The ^{226}Ra is the isotope which is usually given the most attention and generally is removed by the addition of barium chloride and the retention of the resultant precipitate ($BaSO_4$) in large settling ponds. Other nuclides, such as ^{230}Th, ^{210}Pb, and ^{210}Po, are also monitored in the vicinity of U processing plants.

Interesting examples of disequilibrium conditions are often observed in those abandoned tailings subjected to acid leaching (for example, by the oxidation of pyrite). When the pH drops to values of about 2 to 3, Th and Ra are removed by leaching over a number of decades. Protactinium, however, is highly insoluble, and thus, remains in the tailings. The ^{231}Pa (in the ^{235}U chain) generates the short-lived ^{227}Th and ^{223}Ra. Whereas in most situations the activities of these nuclides are considered negligible (compared to the ^{238}U series) in conditions of strong acid leaching the activity ratio of say, ^{223}Ra to ^{226}Ra, may rise to unity or more.

Cliff (1978a, b) and O'Riordan (1978) have written papers concerned with the environmental aspects of U-series disequilibrium which will be of interest to those concerned with exploration and mining.

14.6 Concluding remarks

In exploration for U ores disequilibrium can often result in many difficult and frustrating situations as some of the examples presented in the above discussions have shown. However, it is well worth emphasizing that dis-equilibrium must not be viewed only in a negative light. The following points, if thoroughly understood, can result in disequilibrium being turned into an advantage during exploration.

In the search for U ore deposits, it must be recognized that a multi-element (for example, U, Ra, Rn) system is involved. Each element in the U decay series has its own chemical and physical properties which may result in different mobilities, different sampling media being used, and different detection methods (for example, gamma or alpha detectors) being required. Certain nuclides (for example, ^{234}U, ^{226}Ra, and ^{222}Rn), have significant chemical mobility under specific environmental conditions and these nuclides may be considered as pathfinders in such environments. Several of these nuclides (for example, ^{234}U, ^{230}Th, and ^{226}Ra) have half-lives of 1.6×10^3 to 2.5×10^5 y, long enough for a great deal of dispersion, with resultant disequilibrium, to take place. Young U ores have few or no radiogenic pathfinders because daughter nuclides will not have grown in. In the chemical weathering environment, where water transport is required, disequilibrium may occur; one practical aspect of this phenomenon is that U and certain pathfinder nuclides (for example, ^{226}Ra) may be leached from older ores.

Each instrument used for exploration purposes (for example, alpha and gamma detectors) generally records a different pathfinder nuclide (for example, gamma detectors detect ^{214}Bi) and not U itself. In effect, however, most of these instruments (and techniques in the case of some of the Rn detection schemes) determine ^{226}Ra because they detect the presence of one of the essentially prompt daughters (^{222}Rn, ^{214}Bi, and ^{210}Po) of this nuclide.

Two basic types of disequilibrium, both involving transport in solution, can be recognized and both are of importance in exploration: residual, that is to say, where U is leached and certain daughter nuclides (for example, ^{230}Th) remain, and, new, that is to say, where hydrogenically-formed U occurrences are young, and equilibrium between U and its daughters has not been established. In view of the possibility of disequilibrium, and its effect on the interpretation of gamma ray spectrometry, and on Rn and Po data, ideally it is advisable to analyse important samples (sediments, rocks, waters) for the following: ^{238}U, ^{234}U, ^{230}Th, ^{226}Ra, and ^{214}Bi. Such data can be used to determine the state of equilibrium, the nature and age of leaching, and whether an ore is accumulating or leaching.

Uranium is a ubiquitous element occurring in many places and forms, and to date, there is no single technique, whether geophysical or geochemical, that can solve all the exploration problems associated with it. More often than not combinations of techniques are applied, and in some cases it is just pure intuition and luck that makes the ultimate strike (Hambleton-Jones 1978, p. 296).

Acknowledgements

Dr B. B. Hambleton-Jones (Atomic Energy Board, Pretoria, South Africa), Dr D. A. Hansen (ScienTerra Inc., Spokane, Washington) and Dr J. S. Stuckless (U. S. Geological Survey, Denver, Colorado) kindly reviewed this chapter. We sincerely thank them for their suggestions and comments which have added considerably to the clarity of presentation and to the technical content.

15 THE OCEANIC CHEMISTRY OF THE U- AND Th-SERIES NUCLIDES
J. K. Cochran

15.1 Introduction

In the marine environment, the various nuclides of the U- and Th-decay series can be classified into two groups: (i) those which remain stably dissolved in sea water; and (ii) those which are particle reactive and removed from sea water by adsorption, coprecipitation, or biological processes. In the first group are the isotopes of U (^{238}U, ^{235}U, ^{234}U), Ra(^{226}Ra, ^{228}Ra), and ^{222}Rn, although the Ra isotopes are to some extent involved in biological cycling. Particle-reactive nuclides include the Th isotopes (^{232}Th, ^{234}Th, ^{230}Th, ^{228}Th), ^{231}Pa, ^{210}Pb, and ^{210}Po. This disparity in chemical behaviour allows radioactive disequilibria to be established in the various phases of the marine system (for example, sea water, suspended particles, and bottom sediments). As the radionuclides contained in any given phase approach a state of radioactive equilibrium through ingrowth or decay, they provide useful chronometers for determining the rates of a variety of marine processes. In this context, the present chapter describes the distributions of the U- and Th-series nuclides in the oceanic water column, discusses their rates of removal, and evaluates nuclide mass balances for the marine system as a whole.

15.1.1 Input and removal of U- and Th-series nuclides in the oceans

The possible ways in which the U- and Th-series nuclides are cycled in the marine environment are illustrated schematically in Fig. 15.1. Four main pathways provide for addition of these nuclides to the oceans.

(1) Rivers supply dissolved U and Ra. In addition, these nuclides and their daughters, as well as ^{232}Th and its daughters, are carried in the particulate load of rivers. Reactions occurring in river or estuarine environments may modify the fluxes of nuclides to the oceans either through uptake or release.

(2) Nuclides such as ^{226}Ra, ^{228}Ra, and ^{222}Rn which are produced in marine sediments by radioactive decay of longer-lived parents can be mobilized to sediment pore water by recoil associated with their production or by subsequent leaching by the fluid phase. Depending on the extent of their

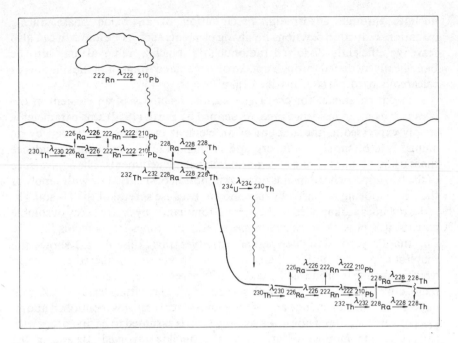

Fig. 15.1. Schematic diagram showing oceanic cycles of selected members of the U and Th decay series. Solid horizontal arrows correspond to radioactive decay characterized by a rate constant. Solid vertical arrows denote fluxes across the sediment–water interface and wavy vertical arrows represent removal from the atmosphere (for ^{210}Pb) or oceans through chemical scavenging and particle settling. (After Cochran 1980a.)

interaction with the sediment particles, appreciable quantities of these nuclides diffuse out of bottom sediments into the overlying water.

(3) Decay of the relatively soluble isotopes of U, Ra, and Rn in coastal and open ocean waters provides a mechanism for *in situ* production of daughter nuclides in the oceans. Introduction of ^{234}Th, ^{230}Th, ^{228}Th, ^{231}Pa, ^{210}Pb, and ^{210}Po occurs in this way.

(4) A fourth pathway by which nuclides enter the ocean is through the atmosphere. Wind-blown particles and their associated radionuclides can enter the ocean in this way. Of perhaps greatest importance is the atmospheric supply of ^{210}Pb to the oceans. This occurs because the noble gas ^{222}Rn emanates from continental rocks and soils to the atmosphere where it decays to ^{210}Pb. The ^{210}Pb so produced is scavenged from the atmosphere by precipitation and deposited onto the surface ocean.

Removal of U- and Th-series nuclides from the oceans is accomplished by several mechanisms. Radioactive decay is perhaps the simplest of these. As noted above, this process results in a production or input of the daughter nuclide. Particles settling through the water column can remove dissolved nuclides through adsorption. Precipitation of an oxide phase during oxidation–reduction reactions involving elements such as Fe and Mn can also scavenge efficiently dissolved radionuclides. Nuclide removal may also be biologically mediated through uptake on organic coatings or in the siliceous or calcareous hard parts of marine organisms.

A useful parameter for characterizing the reactivity of an element in the oceans is its mean residence time. As defined by Barth (1952), this parameter is simply expressed as the amount of an element in the ocean divided by its annual rate of supply. Goldberg and Arrhenius (1958) calculated residence times using the annual rate of removal into pelagic sediments. The equivalence of the two approaches depends on an assumption of steady state with supply of the element being balanced by its removal. Because several of the U- and Th-series nuclides are supplied to the oceans dominantly by *in situ* decay of soluble parents, it is possible to calculate mean residence times for elements like Th, Pb, and Po from daughter/parent activity ratios. Figure 15.2 shows the simplest approach. In this scheme a parcel of water is considered as a box in which a nuclide is produced by decay of its parent. The nuclide is removed by radioactive decay and 'chemical scavenging.' The latter term is not well understood but can include any of the removal mechanisms mentioned above. For convenience the chemical scavenging term is assumed to be first order with respect to the number of atoms of the nuclide involved. Balancing the production and removal terms gives:

$$\lambda_P N_P = \lambda_D N_D + k_D N_D, \tag{15.1}$$

where the subscripts P and D denote the parent and daughter, respectively, and λ the decay constant, N the number of atoms, and k the first order rate constant for removal processes other than decay. Multiplying by λ_D, rewriting eqn (15.1) in terms of activities, and solving for k_D yields:

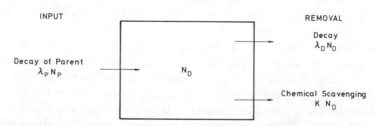

Fig. 15.2. Simple box model for describing U- and Th-series disequilibrium. The subscripts P and D denote the parent and daughter for a radioactive parent decaying to a reactive radioactive daughter; N refers to the number of atoms; λ is the radioactive decay constant; and k is the first order rate constant for processes other than radioactive decay (for example, chemical scavenging). The reciprocal of k is τ, the mean residence time of the daughter in the box relative to non-decay removal.

$$k_D = \lambda_D \left(\frac{A_p}{A_D} - 1 \right). \tag{15.2}$$

Thus, a measurement of the parent/daughter activity ratio in the water sample suffices to determine k, the inverse of which is the mean residence time of the nuclide (often designated as τ). The mean residence time is essentially a measure of the time an atom of the given nuclide spends in the water parcel and is distinct from the rates of reactions which effect its removal.

This simple model is an appropriate first attempt at determining the mean residence time when sample coverage is sparse or the system is sufficiently homogeneous. As will be seen in subsequent sections, the latter is frequently not the case. Significant horizontal and vertical variations in daughter/parent activity ratios can occur in the ocean. As sampling becomes more widespread, it is more appropriate to apply one-dimensional advection diffusion models expressed by the relationship:

$$\frac{\partial N}{\partial t} = K \frac{\partial^2 N}{\partial z^2} + w \frac{\partial N}{\partial z} - \lambda N + J + P, \tag{15.3}$$

where K is the eddy diffusion coefficient, w the advection velocity, J the input or removal (for example kN) term, P the radioactive production from decay of parent, and z the linear distance, frequently depth in the water column.

Models such as described by eqn (15.3) assume that transport of the tracer in the water column occurs by eddy diffusion and advection, and have the advantage that the removal (J) and radioactive production (P) terms can be made functions of location. In solving eqn (15.3) for a particular tracer, steady state is usually assumed ($\partial N/\partial t = 0$). The ratio K/w is obtained from non-radioactive conserve tracers like salinity or temperature and, using this result, J is calculated for the nuclide of interest.

A useful test of the steady-state assumption invoked in calculating removal rates of a given nuclide is whether or not a satisfactory mass balance can be constructed. Such a balance simply matches the total inputs against the outputs. Success depends critically on knowing and evaluating the sources and sinks. An apparent imbalance between inputs and outputs can be useful in focussing attention on the ways in which removal or supply takes place.

15.2 Uranium

15.2.1 Uranium input to the oceans

The naturally occurring U isotopes ^{238}U, ^{235}U, and ^{234}U are dissolved during chemical weathering and form the stable uranyl carbonate species $UO_2(CO_3)_3^{4-}$ in oxidizing aqueous environments (pH approximately 6 or greater). The flux of U to the oceans can be determined from estimates of the concentration of ^{238}U in river water, assuming conservative behaviour in the estuarine regime (see section 15.2.2). Although the U concentration in rivers varies considerably, Bertine *et al.* (1970) determined an average value of approximately 0.3 µg U/l.

Turekian and Chan (1971) and Spalding and Sackett (1972) suggested that the higher values might be due to contamination from phosphate fertilizers and U mining activities. Arguing against this as a general phenomenon are the values of about 0.6 µg U/l for the MacKenzie River (Turekian and Chan 1971) and about 7 µg U/l for the upper Ganges (Bhat and Krishnaswami 1969). Both rivers should have little fertilizer U contribution, yet their ^{238}U concentrations are higher than the average.

Broecker (1974) suggested that because U forms a stable carbonate complex, the U variation in rivers may be due to a variation in carbonate concentration. Indeed, Mangini *et al.* (1979), in a study of ground water and river samples from heavily fertilized areas of Germany, noticed a positive correlation between U and HCO_3^- concentrations. Other rivers such as the Amazon (low U, about 0.04 µg/l, Moore 1967) and the Texas rivers studied by Spalding and Sackett (1972) also exhibit a correlation of U with HCO_3^-. Turekian and Cochran (1978) pointed out that dissolved U also correlates with total dissolved solids, and that the control on U content may be simply the extent of chemical weathering. Borole, Krishnaswami, and Somayajulu (1982) used the relationship between U and total dissolved solids in world rivers coupled with the total dissolved solids flux to the oceans to deduce a U input to the oceans of about 1×10^{10} g/y. This value is in substantial agreement with an average U content of rivers of 0.3 µg/l and a rate of water discharge to the oceans of about 3.5×10^{16} l/y (Livingstone 1963; Alekin and Brazhnikova 1961). In a compilation of U concentrations from world rivers, Sackett *et al.* (1973) obtained an average of 0.6 µg U/l, a value which has been used in a recent paper on the oceanic U balance (Bloch 1980b). Thus, for an average river U concentration of 0.3 to 0.6 µg/l, the U input to the oceans is 1 to 2×10^{10} g/y.

During weathering ^{238}U and ^{235}U are taken in aqueous solution in the constant ratio in which they occur in nature (137.88:1). Because ^{234}U is produced from ^{238}U, mobilization processes such as recoil which are associated with its production may lead to preferential dissolution of ^{234}U (see for example, Kigoshi 1971; Fleischer and Raabe 1978; Osmond and Cowart 1976a). The activity ratio of ^{234}U to ^{238}U in natural waters is commonly greater than 1.0 (river waters generally fall in the range of 1.20 to 1.30) and gives important information about the hydrological regime through which the water passes (see chapters 8 and 9 for a fuller discussion).

15.2.2 Uranium in the coastal ocean

The behaviour of U in the coastal zones and in particular, during estuarine mixing, is not well understood. The concentration of U in nearshore and coastal waters is variable and must in part be related to the inherent variability of U concentration in rivers entering the coastal zone. Values greater than the open ocean value of 3.3 µg/l have been noted by Miyake *et al.* (1964) for the north-west Pacific and Sackett and Cook (1969) for the Gulf of Mexico. Elsewhere U concentrations lower than the open-ocean value have been reported (Blanchard and Oakes 1965; Bhat *et al.* 1969; Noakes, Kim, and Supernaw 1967; Aller and Cochran 1976; Borole *et al.* 1977; Martin, Nijampurkar, and Salvadori 1978).

Figure 15.3 shows plots of U against chlorinity for several estuaries. Most of the patterns are consistent with a conservative mixing trend of river water with sea water. In the case of the Charente estuary, Martin, Nijampurkar, and Salvadori (1978) document an anomalously high U input due to the effluent from phosphate processing plants. Using the highest observed U concen-

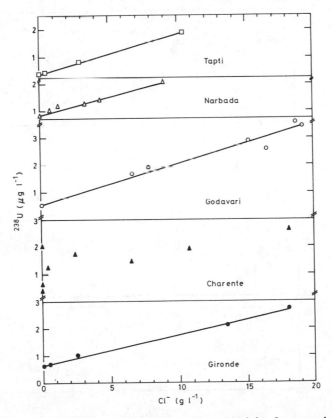

Fig. 15.3. The plot of ^{238}U concentrations against chlorinity for several estruaries. Data for the Tapti, Narbada, and Godavari are from Borole *et al.* (1977), and for the Charente and Gironde from Martin, Nijampurkar, and Salvadori (1978).

tration as the river water end-member, the plot of U against chlorinity appears to show removal of U during estuarine mixing. Martin, Nijampurkar, and Salvadori (1978) also observed high variability in the low-chlorinity water of the Charente, and an alternate explanation is that U is essentially conservative with a temporally variable river end-member.

An alternative approach to the behaviour of U in estuaries is to plot the $^{234}U/^{238}U$ activity ratio against the reciprocal ^{238}U concentration (see Fig. 15.4). Conservative mixing between two water types produces a linear

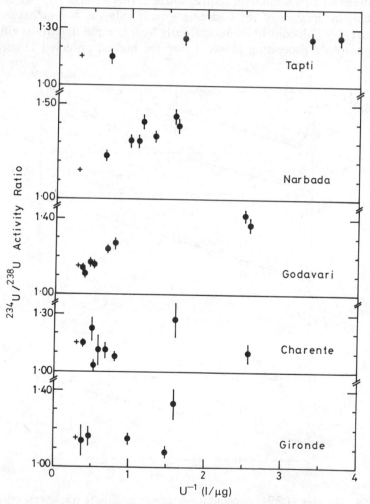

Fig. 15.4. The plot of $^{234}U/^{238}U$ activity ratio against reciprocal U concentration for several estuaries. (See caption to Fig. 15.3 for sources of data.)

relationship between these variables (see Osmond and Cowart 1976a). Figure 15.4 shows that for the estuaries considered above, the uncertainty of the data is, in general, too great to determine whether U is being released or removed during mixing.

15.2.3 Uranium in the open ocean

The distribution of ^{238}U, ^{235}U, and ^{234}U in the oceans is of considerable importance because these nuclides produce the more particle-reactive nuclides ^{234}Th, ^{231}Pa, and ^{230}Th.

The U concentration of sea water has been studied by a number of workers, and the literature recently has been reviewed by Burton (1975) and Ku *et al.* (1977). Values for open-ocean waters range from 1 to 5 μg/l. For oxidizing sea water of about 35 per mil salinity, it is doubtful if this variation is real. Indeed, several detailed studies of depth profiles of U in the open ocean reveal that U is essentially conservative, varying only with salinity.

Turekian and Chan (1971) found an average U concentration of 3.16 $\pm$ 0.13 μg/l and average U/salinity ratio of $(9.21 \pm 0.04) \times 10^{-8}$ g/g for a depth profile (10 to 4000 m) in the North Pacific. Ku *et al.* (1977) analysed samples from nine depth profiles in the Atlantic, Pacific, Arctic, and Antarctic Oceans observing that the U content of 35 per mil salinity water is 3.3 ± 0.2 μg/l with a U/salinity ratio of $(9.34 \pm 0.56) \times 10^{-8}$ g/g (see Fig. 15.5). These results are substantially in agreement with the early work of Rona *et al.* (1956) and Wilson, Webster, Milner, Barnett, and Smales (1960) who found sea water U concentrations of about 3.3 μg/l.

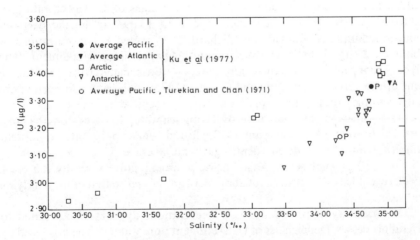

Fig. 15.5. The plot of U concentration against salinity for open sea waters. (Data from Turekian and Chan 1971; Ku *et al.* 1977).

The preferential mobilization of ^{234}U during chemical weathering which gives rise to a ^{234}U/^{238}U activity ratio greater than unity in rivers implies that the ^{234}U/^{238}U activity ratio in the oceans will be greater than unity as well. This fractionation was first documented by Cherdynstev et al. (1955) and by Thurber (1962) who found that ocean water has a ^{234}U/^{238}U activity ratio of 1.15. Subsequent work has revealed little variation in this value (Koide and Goldberg 1964; Veeh 1968; Sackett and Cook 1969). Although Miyake, Sugimura, and Uchida (1966) report values as low as 1.09, the work of Ku et al. (1977) gives an average value of 63 samples of 1.14 ± 0.03.

The flux of ^{238}U to the oceans from rivers calculated in section 15.2.1, coupled with the U concentration of sea water, can be used to calculate the mean residence time of U in the oceans (see section 15.1). For a mean U concentration of 3.3 µg/l and an oceanic volume of 1.37×10^{21} l, the total U content of the oceans is 4.5×10^{15} g, and hence, the residence time is approximately $2 \pm 0.4 \times 10^5$ y. In constructing a mass balance for U, any expectation of having input balance output rests on the assumption of a steady state, at least for a period of several mean residence times (about 10^6 y). Indirect evidence that this is the case comes from the U/Ca ratio of corals through geological time. Broecker (1974) finds essentially no change in the U/Ca ratio of corals from the present through at least the Pliocene (2 to 5 my). Assuming the Ca content of the oceans has not changed markedly in that time, the coral data imply that neither has the U content.

15.2.4 Uranium in sediment pore water

There have been relatively few studies of U content in deep-sea sediment pore waters. Baturin and Kochenov (1973) found ranges of U concentration of 1.3 to 65 µg/l in Atlantic sediments and 2 to 13 µg/l in Pacific deep-sea sediments. Similarly, Boulad and Michard (1976) noted U concentrations of about 3 to 27 µg/l in the pore waters of South Atlantic deep-sea sediments. The higher U concentrations in these pore waters were associated with the reducing zone of the cores. Somayajulu and Church (1973) analysed a composite pore water sample from a 2 m North Equatorial Pacific core and found about 2 µg U/l, somewhat less than in the overlying sea water. In cores from the same area, Cochran and Krishnaswami (1980) found similar pore water U content samples from 10 to 40 cm depth gave an average U concentration of 2.5 ± 0.2 µg/kg. Analyses of East Pacific sediment pore water by Ku et al. (1977) reveal U concentrations of about 3.3 µg/l, similar to that of the overlying sea water.

The variability of pore water U concentrations appears to be related to several processes. Deficiencies of U in sediment pore water can be produced by uptake of U by authigenic phases (for example, Mn nodules) forming in the sediment. Evidence for this process is presented by the data of Immel and

Osmond (1976) who found ^{238}U concentrations of 0.5 to 6.5 ppm in micronodules from Antarctic cores.

Buildup of recoil ^{234}U in pore water can lead to activities of this nuclide which are higher than sea water. This process was documented by Cochran and Krishnaswami (1980) who found activity ratios as high as 1.23 ± 0.03 in Pacific sediment pore water. High U concentrations in reducing sediment pore water are difficult to explain based on the chemical behaviour of U. These observations will have to be confirmed in situations in which there is greater control over the separation of sediment and pore water.

15.2.5 Removal of U from the oceans

Table 15.1 summarizes the balance for U in the oceans. The sinks described below account for a large fraction if not all of the input of U by rivers. Within the uncertainties of input and output, it is possible to say that the oceans are in balance with respect to U.

Table 15.1. The oceanic U balance

Supply (10^{10}g U/y)	1–2
Removal (10^{10}g U/y):	
Carbonate deposits	
Calcareous oozes	<0.04
Corals and molluscs	≈0.08
Siliceous oozes (deep sea)	0.03
Pelagic clay	0.01
Organic-rich anoxic sediments	0.20
Metalliferous sediments and	
ferromanganese nodules	0.14
Uptake during rise crest	
hydrothermal circulation	0.9 ± 0.2
Low temperature weathering of	
oceanic basalts (500 m thick)	0.8
Total removal	2.2

(i) *Marine sediments.* Uranium is removed from the ocean by organisms which secrete a calcareous or siliceous test. These are calcareous plankton (for example, foraminifera, coccoliths, pteropods), corals, and molluscs in the first instance, and diatoms and radiolaria in the latter. Direct evidence for this comes from the good correlation of U with biogenic material observed by Brewer, Nozaki, Spencer, and Fleer (1980) in sediment trap experiments. The U content of calcareous plankton is low. Ku (1965) determined that coccoliths have less than 0.1 ppm U and foraminifera have 0.025 ppm. Higher values for foraminifera (0.3 to 1.2 ppm) and pteropods (less or equal to 2.7 ppm) from the Gulf of Mexico were reported by Mo, Sattle, and Sackett (1973). Using carefully cleaned and dissolved samples, Delaney (1980) determined a U content for foraminifera of 0.027 ± 0.008 ppm U, in close agreement with the results of Ku (1965). The higher values observed by Mo *et al.* (1973) probably reflect U associated with Mn or Fe oxyhydroxides or possibly residual organic matter in the carbonate tests. Formation of an Fe-Mn oxide coating may happen during settling through the water column or during diagenesis. Assuming that the high U contents of deep-sea calcareous sediments represents U removal from sea water, and using a concentration of 0.3 ppm U for these deposits, a U removal flux of 0.04×10^{10} g/y is obtained using the $CaCO_3$ accumulation rates of Turekian (1976). Sackett *et al.* (1973) estimate approximately 0.08×10^{10} U g/y is removed by corals and molluscs, bringing the removal in calcareous deposits to about 10 percent of the total input.

The U content of deep-sea siliceous oozes is about 1 ppm based on the data of Scott, Osmond, and Cochran (1972) for Indian–Antarctic deposits and of DeMaster (1979) for South Atlantic–Antarctic siliceous oozes. When multiplied by the total removal of silica in such deposits (about 2.9×10^{14} g SiO_2/y) the deposition of U in siliceous sediments is about 0.03×10^{10} g U/y. This is probably an upper limit because no attempt is made to correct for the fraction of the total sediment which is biogenic silica (about 50 per cent) or possible diagenetic effects.

Removal of U in deep-sea clays is difficult to estimate because of the large detrital U component. Krishnaswami (1976) has estimated the authigenic U concentration of Pacific pelagic clays by assuming that the accumulation rate of authigenic U is the same in Fe-Mn nodules as in deep-sea clays. Using this approach the accumulation rate of authigenic U in Pacific clay is calculated to be 0.04 μg/cm^2 10^3 y. Extrapolating this to the world ocean gives a removal rate of 0.01×10^{10} g/y, only about 1 per cent of the river U input.

Removal of U from sea water also occurs in association with reducing sediments. In such sediment U^{6+} can be reduced to the relatively insoluble U^{4+}. In organic-rich sediments accumulating below an anoxic water column, at least part of the uptake of U may be related to the adsorption or complexing of U with organic matter. Accumulation of U in organic-rich anoxic sediments has been documented for fjords (Veeh 1967; Kolodny and Kaplan 1973) and other anoxic basins (Mo *et al.* 1973) as well as for the phosphorite-rich sediments associated with upwelling areas off Peru (Veeh, Burnett, and Sontar 1973) and Africa (Veeh *et al.* 1974). A detailed accounting of the U removal in these sediments is given in Table 15.2 and the total is about 16 per cent of the U input to the oceans.

A category of sediments not documented in Table 15.2 is the sediments accumulating in estuaries and on the continental shelf where the water column is well oxygenated. Such sediments frequently become anoxic below the top few centimetres due to the effects of diagenesis. Removal of U in these sediments was postulated by Koczy (1963). More recently Thomson *et al.*

Table 15.2. Uranium accumulation rates in organic-rich anoxic sediments

Location	Area (cm^2)	Sediment accumulation rate (g/cm^2 y)	Authigenic U (µg/g)	U removal rate (µg/cm^2 ky)	(10^{8}g/y)
Peru–Chile	1.9×10^{15}(1)	0.024 (1)	11 (2)	260	5.0
Gulf of California	7.6×10^{14}(1)	0.063 (1)	4 (2)	250	1.9
N. American west coast basins:					
S. Barbara		0.075 (1)	2 (2)	150	
Saanich Inlet		0.13 (3)	5 (4)	650	
Total area	$\leq 8 \times 10^{14}$(1)			~ 200	1.6
Black Sea	2.7×10^{15}(5)	—	—	145 (5)	3.9
Sea of Azov	0.4×10^{15}(5)	—	—	129 (5)	0.5
Baltic Sea	3.8×10^{15}(7)	—	—	40–200 (6)	~ 3.8
SW African Shelf (Walvis Bay)	2.5×10^{14}(1)	—	10–55 (8)	~ 500 (8)	~ 1.3
TOTAL	10.6×10^{15}				18×10^8 gU/y

References
(1) DeMaster (1979).
(2) Veeh (1967)
(3) Bruland (1974).
(4) Kolodny and Kaplan (1973)
(5) Nikolaev, Lazarer, Korn, and Drozhzhin (1966)
(6) Koczy et al. (1957)
(7) Menard and Smith (1966)
(8) Veeh et al. (1974)

(1975) and Aller and Cochran (1976) have found evidence for U uptake in the reducing sediments of Long Island Sound. Because the areal extent of the continental shelves is significant, this could be an important sink for U. However, the rate of renewal of U in these areas is not likely to be as high as in organic-rich anoxic sediments accumulating under an anoxic water column.

(ii) *Authigenic deposits.* Non-biogenic marine deposits which form from sea water also are sinks for U. Two important types of such deposits are the metalliferous sediments associated with spreading centres and Fe-Mn nodules.

Metalliferous deposits associated with ridge crest hydrothermal activity include both Fe-Mn rich sediments (Boström and Peterson 1969) and Mn-rich crusts (Scott, Scott, Rona, Butler, and Nalwalk 1974). Removal of U has been documented in both types (Fisher and Boström 1969; Scott et al. 1974), and in general the U appears to be of sea water origin (Veeh and Boström 1971; Bender, Broecker, Gornitz, Middler, Kay, Sun, and Biscaye 1971, Scott et al. 1974). The incorporation of U in such deposits probably occurs through adsorption or coprecipitation with Fe-Mn oxides. There may also be a contribution from carbonate-rich hydrothermal solutions, although sulphide-rich hydrothermal solutions from the Galapagos rise are depleted in U as they emerge at hot springs (Edmond, Measures, Mangum, Grant, Selater, Collier, Hudson, Gordon, and Corliss 1979, see paragraph (iii) below).

Bloch (1980b) has attempted to calculate the U removal in ridge crest metalliferous sediments by assuming all the U is scavenged by hydrothermally supplied Fe. Using estimates for Fe input from hydrothermal activity and from cooling igneous bodies (Wolery and Sleep 1976) together with the U/Fe ratios of some East Pacific Rise metalliferous sediment, he estimates the removal of U by this mechanism to be about 1.4×10^9 g/y.

Slowly accumulating Fe-Mn nodules and concretions also concentrate U from sea water. Ku and Broecker (1969) report that U in such nodules varies from 4 to 13 ppm with an average of 9 ppm. For a typical nodule accumulation rate of $3 \, \text{mm}/10^6$ y and an *in situ* dry bulk density of $2 \, \text{g/cm}^3$, the accumulation rate of U in nodules is $5.4 \times 10^{-12} \, \text{g/cm}^2$ y. If 10 per cent of the deep ocean floor is covered with nodules, then they remove $1.9 \times 10^6 \, \text{g U/y}$, a relatively small fraction of the total U input.

(iii) *Hydrothermal circulation.* Edmond, Measures, Mangum, Grant, Selater, Collier, Hudson, Gordon, and Corliss (1979) have shown that the water debouching from hot springs at the Galapagos spreading centre is depleted in U. Such a depletion is probably caused by reduction of U to insoluble UO_2 during sea water circulation through the vent system. The approach used by Edmond, Measures, McDuff, Chan, Collier, Grant, Gordon, and Corliss (1979) in evaluating the importance of hydrothermal circulation as a sink for U is to relate it to the flux of heat through the ridges. Silica shows a good positive correlation with temperature and is thus used as a normalizing variable. Determination of the heat flux can be approached in several ways. Edmond, Measures McDuff, Chan, Collier, Grant, Gordon, and Corliss (1979) use the approach of Jenkins, Edmond, and Corliss (1978) relating the heat flux to that of dissolved ^{3}He. This estimate gives a hydrothermal heat flux from ridge axes of $4.9 \pm 1.2 \times 10^{19}$ cal/y. The calculation assumes a ^{3}He/temperature correlation similar to that of the Galapagos for the entire ridge system, but on the whole, there is general agreement with geophysical estimates (Wolery and Sleep 1976).

The plot of U against silica of Edmond, Measures, Mangum, Grant, Selater, Collier, Hudson, Gordon, and Corliss (1979) is shown in Fig. 15.6. The removal of U is indicated from the negative slope of the correlation and multiplication by the silica/temperature relationship, and the heat flux gives a U removal of $0.9 \pm 0.2 \times 10^{10} \, \text{g U/y}$. This is a substantial fraction of the input by streams, and represents a major U sink. It is important to note that not all of the vents studied by Edmond, Measures, Mangum, Grant, Selater, Collier, Hudson, Gordon, and Corliss (1979) showed a marked U removal. Moreover, carbonate-rich hydrothermal solutions containing U have been postulated to explain the anomalously high U contents of East Pacific Rise crest sediments

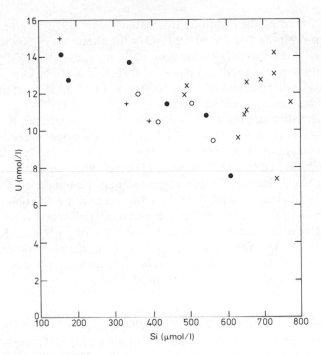

Fig. 15.6. Plot of U concentration against silica concentration in four hydrothermal vent fields at the Galapagos spreading centre (from Edmond, Measures, Mangum, Grant, Selater, Collier, Hudson, Gordon, and Corliss 1979). The fieds are: solid circles = Clambake; crosses = Garden of Eden; pluses = Dandelions; open circles = Oyster Beds. Excluding the Garden of Eden samples, the slope of the U plot against Si trend is approximately -0.01 nmol U/μmol Si or, using the silica–temperature correlation given in Edmond, Measures, McDuff, Chan, Collier, Grant, Gordon, and Corliss (1979), -0.8 n mol U/k cal.

(Rydell, Kreamer, Boström, and Joeusuu 1974). Thus the above calculation points out the possible importance of hydrothermal circulation as a removal mechanism for U, but further work is needed to confirm the magnitude of this effect.

(iv) *Weathering of marine basalts.* Aumento (1971) and MacDougall (1977) have identified the uptake of U during the weathering of marine basalts as a sink for U. MacDougall (1977) noted that the concentration of U in altered basalt was about an order of magnitude greater than initial content as represented by the glassy margins of the same basalts. He estimated that each gram of altered basalt gained about 0.4 μg U during weathering. Aumento, Mitchell, and Fratta (1976) found about 0.2 μg U increase per gram of basalt. The total U uptake by such a process can be determined if the time scale and depth to which oceanic basalt is altered is known. Bloch (1980b) has made such a calculation using the hydration rate data of Hart (1973) as given in Wolery and Sleep (1976) (1.5 to 6×10^{14} g H_2O/y for alteration of 500 to 2000 m of basalt) and Aumento's (1971) estimate of U uptake of 5×10^{-5} g U/g H_2O. These values give a total U uptake of 0.75 to 3×10^{10} g U/y. Thus, for a weathered basalt thickness of about 1000 m, virtually all U input to the oceans could be balanced by this sink alone.

An alternative approach to the problem is through the K/U ratio in weathered basalts. Aumento's (1971) data, based on slopes of correlation lines of U plotted against K_2O, generally range from 1.5×10^3 to 26×10^3 g K/g U. The data presented by MacDougall (1977) average to the latter value, about 28×10^3 g K/gU. Hart (1973) has suggested that the uptake of K during the weathering of oceanic basalts accounts for a significant fraction of the river supply of K to the oceans. Bloch and Bischoff (1979) have estimated the K uptake of weathering 600 m basalt to be about 2×10^{13} g/y (for a K content of 2 per cent). The increase in U corresponding to this K uptake ranges from 0.07×10^{10} to 1.3×10^{10} g/y. The latter value is in general agreement with that obtained from the hydration rate method.

15.2.6 Sources and sinks of ^{234}U in the oceans

The U sinks discussed above, although formulated for ^{238}U, apply to ^{234}U as well. A number of workers have evaluated the ^{234}U balance in the oceans (Ku 1966; Veeh 1967; Bhat and Krishnaswami 1969; Ku et al. 1977; Turekian and Cochran 1978; Borole et al. 1982). The oceanic parameters entering into this balance are the ^{238}U concentration (3.3 µg/l) and the $^{234}U/^{238}U$ activity ratio (1.14). The ^{234}U entering the oceans from rivers can be obtained from the ^{238}U concentration of river water and the $^{234}U/^{238}U$ activity ratio of streams. There is considerable uncertainty in both these parameters, but estimates generally range from 0.3 to 0.6 µg/l as the average ^{238}U content of streams (see section 15.2.1) and 1.20 to 1.30 for the average $^{234}U/^{238}U$ activity ratio. Higher concentration of ^{238}U in rivers (about 0.6 µg/l) produces a ^{234}U balance between removal and supply if stream $^{234}U/^{238}U$ activity ratios are in the range 1.20 to 1.30. If, however, the average ^{238}U content of streams is on the low side (0.3 µg/l or less), then an additional source of ^{234}U to the oceans is required if river water $^{234}U/^{238}U$ activity ratios are 1.2 to 1.3. Ku (1965, 1966) has suggested that recoil and mobilization of ^{234}U in deep-sea sediments can supply ^{234}U to the oceans and provide a ^{234}U balance. Ku (1965) observed a flux of ^{234}U out of deep-sea pelagic clays, while Cochran and Krishnaswami (1980) found that sediments of the North Equatorial Pacific constituted a sink for U. Thus, as in the case of the ^{238}U balance, better knowledge of the input of ^{234}U to the oceans from rivers is needed to assess accurately the ^{234}U budget.

15.3 Thorium

15.3.1 Scavenging of Th in the deep sea

The primary characteristic of Th in the marine environment is its high particle reactivity. Indeed, ^{232}Th enters the oceans primarily in detrital form, although Scott (1968) has shown that there is some mobilization of Th during weathering and soil formation. Unlike U which can be dissolved during this process, Th is largely retained on particle surfaces. The naturally occurring Th isotopes ^{234}Th, ^{230}Th, and ^{228}Th have *in situ* sources in sea water resulting from the decay of U and Ra parents, as does ^{231}Pa produced from ^{235}U. This latter nuclide will be considered in this section because of the general similarity of Pa to Th in geochemical behaviour, although there may be differences in the rate of removal of Th and Pa from the oceans. Table 15.3 indicates that the Th (^{232}Th) concentration of sea water is quite low. Although estimates vary by

Table 15.3. The ^{232}Th, ^{230}Th, and ^{231}Pa concentrations in open ocean waters

Sample location	Water depth (m)	^{232}Th (μg/1000 *l*)	^{230}Th (dpm/1000 *l*)	^{231}Pa (dpm/1000 *l*)	Reference
N. Atlantic*	Surface	0.64 ± 0.20	0.44 ± 0.11	0.13 ± 0.13	Moore and Sackett (1964)
	4500	4.5 ± 0.8	1.6 ± 0.2	—	
Caribbean*	Surface	0.64 ± 0.20	0.43 ± 0.11		
	800	0.36 ± 0.17	0.31 ± 0.12	0.22 ± 0.24	
N. Pacific	Surface	0.33 ± 0.04	0.55 ± 0.10	—	Somayajulu and Goldberg (1966)
	2500	0.20 ± 0.04	1.6 ± 0.3	—	
	2500	0.65 ± 0.09	0.52 ± 0.16	—	
Pacific	Surface	1.6	29	5.4	Kuznetsov, Simonyak, Elizarova, and Lisitsyan (1966)
	Surface	7.0	62		
	1000–5000 average	4.4 (1.0–7.9)	4.3 (12–94)	< 3.22 (4000m)	
Composite	Surface	< 0.07	—	—	Kaufman (1969)
N. Pacific	3500	< 0.08	0.5	—	Krishnaswami *et al.* (1972)
NW. Pacific	Surface	2.4 (0.1–7.8)	1.3 (0.09–3.2)	—	Miyake *et al.* (1970)
	> 1000	6.8 (0.7–28)	1.4 (0.6–3.5)		

* Samples were centrifuged to eliminate particles. All other analyses represent total water sample.

nearly two orders of magnitude, several workers report upper limits of about 0.08 µg Th/1000 l. This value agrees well with the 0.07 µg Th/1000 l calculated by Turekian et al. (1973) from the Th/La ratio in pteropods. Higher values probably represent the contribution of particulate Th to the samples.

Table 15.3 also gives ^{230}Th and ^{231}Pa activities of sea water. These values may be compared directly to the respective parent activities of ^{234}U (2.8 dpm/l) and ^{235}U (0.11 dpm/l), and yield a sea water ^{230}Th/^{234}U activity ratio of about 6×10^{-4} or less and a ^{231}Pa/^{235}U activity ratio of about 2×10^{-3} or less. These ratios, when compared with the equilibrium value of 1.0, indicate that removal of ^{230}Th and ^{231}Pa is rapid relative to their half-lives.

Because the activities of ^{230}Th and ^{231}Pa in sea water are low and their measurement is subject to some uncertainty, scavenging rates of Th and Pa are perhaps best determined by analysing the particles onto which the nuclides are scavenged. Settling particles have been sampled by sediment traps (Spencer, Brewer, Fleer, Honjo, Krishnaswami, and Nozaki 1978; Brewer et al. 1980) and by in situ pumps which filter particles from large volumes of sea water (Krishnaswami, Lal, and Somayajulu, 1976; Krishnaswami, Lal, Somayajulu, Weiss, and Craig 1976).

Sediment traps placed at several depths at stations in the Atlantic were analysed for Th isotopes by Spencer et al. (1978) and Brewer et al. (1980). They observed that the ^{234}Th activity of the particles was essentially constant with depth, indicating that the scavenging of ^{234}Th onto the particles was balanced by its decay. The activity of ^{230}Th on the particles increased with depth (see Fig. 15.7). This is a direct consequence of the constant U concentration of sea water and the fact that as the particles sink they scavenge progressively more ^{230}Th, there being no time for ^{230}Th decay during particle settling. The relationship between ^{230}Th flux and water depth is, in simplest terms, dependent solely on the ^{234}U activity of the overlying water column and is equivalent to $2.6 D$ dpm/cm^2 y where D is the water depth in kilometres (Cochran and Osmond 1976). The increase in ^{230}Th flux with depth noticed by Brewer et al. (1980) was actually about 74 per cent of the predicted value, which they attributed to inefficient trapping of all the settling particles. Brewer et al. (1980) modelled the variation in particulate ^{230}Th and ^{228}Th with depth by a vertical settling and scavenging model:

$$S \frac{\partial \chi}{\partial z} + \psi C + \lambda_p \chi_p - \lambda \chi = 0 \qquad (15.4)$$

where S is the mean settling velocity of the particle, ψ the first order scavenging rate constant, χ the concentration of radionuclide in particulate phase, and C the concentration of radionuclide in solution.

The $\lambda_p \chi_p$ term denotes production of the nuclide in the particulate phase from decay of its parent. Although the formulation of eqn (15.4) in terms of vertical scavenging is required by the nature of the data, Brewer et al. (1980)

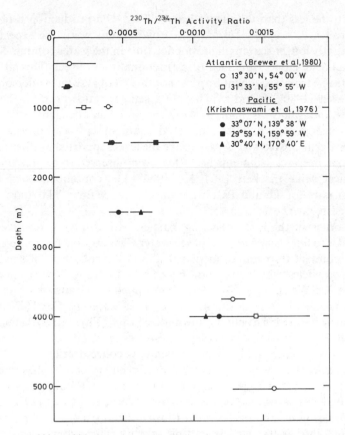

Fig. 15.7. The ^{230}Th/^{234}Th activity ratio in particles as a function of depth in the water column. The particles were collected in traps (Brewer *et al.* 1980) and on filters (Krishnaswami, Lal, Somayajulu, Weiss, and Craig 1976).

point out that there may be horizontal scavenging and transport processes which contribute to removal of nuclides from the water column.

In applying eqn (15.4) to the Th isotope data, Brewer *et al.* (1980) used the ^{228}Th/^{234}Th activity ratio on the particles to calculate best-fit values of 21 m/d for S and 27 y for the scavenging residence time $(1/\psi)$ of Th. These values were consistent with the ^{230}Th/^{234}Th data as well. The value of S determined from the Th data and applied to the ^{231}Pa data yielded a scavenging residence time for Pa of 31 y.

The difference between Th and Pa residence times in the oceans, if real, implies that there can be chemical fractionation between these two elements during their removal from sea water. The water ^{230}Th/^{231}Pa activity ratios obtained from Table 15.3 suggest that this is occurring, in that the values appear to be less than the theoretical ^{230}Th/^{231}Pa production ratio of 10.6 (equal to $[\lambda_{230}/\lambda_{231}]\cdot[^{234}U/^{235}U]$). Lower ratios would be expected if Pa has a slightly longer scavenging residence time in the water column. Sediment trap data also support the idea of a fractionation between Th and Pa. In a depth profile from a sediment trap site south-east of Hawaii, Anderson, Bacon, and Brewer (1980a) found ^{230}Th/^{231}Pa activity ratios of 35 in the settling particles and values as high as 47 were noted in the equatorial Atlantic (Anderson et al. 1980b). Indeed the activity ratio of sea-water-produced ^{230}Th to ^{231}Pa (termed 'excess' activities above the level supported by their respective parents) in deep-sea sediments has often been observed to be greater than the theoretical value (Sackett 1964; Ku 1966). The consequences of different removal rates for Th and Pa on the mass balance for ^{230}Th and ^{231}Pa are discussed further in section 15.3.2.

The other method of collecting particles for studying reactive nuclide removal is to filter large volumes of sea water. This approach is complementary to the sediment trap studies in that it allows for collection of small, slowly sinking particles which may not be caught in traps. Krishnaswami, Lal, Somayajulu, Weiss, and Craig (1976) determined ^{230}Th and ^{234}Th activities in particles collected from *in situ* filtration of sea water. As Fig. 15.7 shows, the depth increase in ^{230}Th activity (normalized to ^{234}Th) at their Pacific stations is nearly identical to that observed by Brewer et al. (1980) for their Atlantic sediment trap stations. Given the constant U concentration and ^{234}U/^{238}U activity ratio of sea water this is sensible agreement and implies that, even if different types of particles are sampled by the two methods, they scavenge Th similarly. Krishnaswami, Lal, Somayajulu, Weiss, and Craig (1976) modelled their data using an approach similar to that of eqn (15.4) and obtained a value of about 2 m/d as the particle settling velocity. This value is consistent with sediment accumulation rates in this area, but is about a factor of 10 lower than that calculated by Brewer et al. (1980). The difference may reflect the possibility that spatial differences exist in particle fluxes in the oceans or that sediment traps collect more rapidly sinking (hence larger) particles than *in situ* filtration.

15.3.2 The ^{230}Th and ^{231}Pa balance

^{230}Th is removed from sea water rapidly relative to its half-life, and so a first attempt at constructing its balance may be made by comparing the total activity for excess ^{230}Th in a sediment column (dpm/cm^2) with that produced from ^{234}U decay in the immediately overlying water column. If transport of Th on sinking particles is dominantly vertical, then these quantities should balance. In the deep sea such a balance is often achieved so that deviations from it can be attributed to post-scavenging processes like sediment focussing, winnowing, or bottom transport of sediment. Downslope transport of sediment from topographic highs has been inferred from ^{230}Th-excess accumulation patterns in the Tasman Basin (Cochran and Osmond 1976) and the Bauer Deep (Dymond and Veeh 1975). Despite local variability, ^{230}Th in the sediments of the Tasman Basin is in balance with its production in the overlying water (Cochran and Osmond 1976). On a larger scale, a similar conclusion has been reached by Krishnaswami (1976) and Kadko (1980a). Kadko (1980a) considered the effect of mixing of the upper few centimetres of sediment on the ^{230}Th-excess activity at the top of a core, and showed that reasonable agreement between calculated and observed activities resulted for a mixed depth of 10 cm.

Although early studies on the use of ^{230}Th and ^{231}Pa as chronometer in deep-sea sediments (Sackett 1960; Rosholt et al. 1961) argued that these nuclides behave identically with respect to removal from sea water, the sediment trap studies of Anderson et al. (1980a, b) show that fractionation of ^{230}Th and ^{231}Pa takes place. This shows up in departure of the ^{230}Th/^{231}Pa activity ratio from the value of 10.6 set by instantaneous production from U in the water column. The ^{230}Th/^{231}Pa activity ratios of sediment core tops is often greater than 10.6 (Sackett 1964; Ku 1966) and less than 10.6 in Fe-Mn nodules (Sackett 1966).

Indeed, Turekian and Chan (1971) tried to take this fact into account in calculating oceanic mass balances for ^{230}Th and ^{231}Pa. They concluded that coverage of 25 per cent of the sea floor was required to provide a significant sink for Pa. In fact greater than 25 per cent is required if allowance is made for the fact that nodules contain only 5 to 10 per cent of the ^{230}Th and ^{231}Pa produced in the overlying water column.

Although at present, there are insufficient data on deep-sea cores to construct a quantitative balance it is possible to delineate areas which are sinks for Pa by comparing the ^{230}Th/^{231}Pa activity ratio at the core top with the production ratio of 10.6. In doing this it is necessary to take into account the fact that ^{230}Th/^{231}Pa ratios increase down a core as a function of age (due to the different half-lives) and that continuous mixing of the upper few centimetres by currents or organisms will raise the ^{230}Th/^{231}Pa activity ratio at the core top above this value for freshly deposited sediment. If the upper mixed zone of deep-sea sediments is considered to be rapidly mixed to a depth L such that constant ^{230}Th and ^{231}Pa activities result, then the ^{230}Th/^{231}Pa activity ratio in the mixed zone is given by the expression:

$$\left(\frac{^{230}\text{Th}}{^{231}\text{Pa}}\right)_L = \left(\frac{S + \lambda_{231} L}{S + \lambda_{230} L}\right)\left(\frac{^{230}\text{Th}}{^{231}\text{Pa}}\right)_0 \qquad (15.5)$$

where $\left(\dfrac{^{230}\text{Th}}{^{231}\text{Pa}}\right)_L$ is activity ratio in mixed zone, $\left(\dfrac{^{230}\text{Th}}{^{231}\text{Pa}}\right)_0$ activity ratio in

sediment being deposited on the sea floor, and S is sediment accumulation rate.

Equation (15.5) results from balancing the flux of ^{230}Th or ^{231}Pa into the mixed zone by deposition of fresh sediment with the removal by radioactive decay and sediment accumulation. From the observed variation of ^{230}Th/^{231}Pa ratio with sediment accumulation rate, the ratio in freshly deposited sediment can be obtained.

An attempt to do this with existing data is shown in Fig. 15.8. The cores plotted were carefully collected and should represent *in situ* conditions. However, for a mixed zone of about 10 cm the pattern will not change seriously even if a few centimetres of sediment were lost during the coring operation. The ^{230}Th/^{231}Pa activity ratios in the mixed zone range from 35 to 4 and are higher in more slowly accumulating red clay sediments from the North Pacific Ocean. Allowing for mixing to 10 cm, these high values are consistent with an input ^{230}Th/^{231}Pa ratio of approximately 22, still significantly larger than 10.6 (see upper solid curve in Fig. 15.8). Of particular interest in Fig. 15.8 are the cores in which the ^{230}Th/^{231}Pa ratio is less than the theoretical value. Two such environments are seen from the figure: the ratio is low on the crest of the mid-ocean ridges based on one value for the East Pacific Rise (Kadko 1980b) and one value for the Mid-Atlantic Ridge (Cochran 1979). The latter core is complemented by a suite of cores from the Mid-Atlantic Ridge flank in the same area (Ku *et al.* 1972). Away from the rise crest, the ratio is greater than the theoretical value. The other environment with the lowest ^{230}Th/^{231}Pa ratio is South Atlantic–Antarctic where siliceous sediments are rapidly accumulating (De Master 1979). Not only is this area characterized by anomalously low ^{230}Th/^{231}Pa ratios, but the inventories of both Th and Pa are substantially greater than production in the overlying water column.

The reasons for the pattern observed in Fig. 15.8 can only be speculated

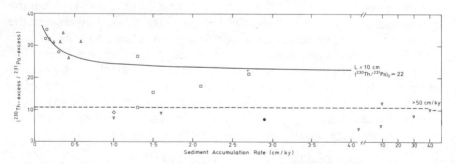

Fig. 15.8. The plot of ^{230}Th-excess/^{231}Pa-excess activity ratio in the mixed zone of the deep-sea sediment column against sediment accumulation rate. The dashed line is the theoretical instantaneous production rate value based on dissolved U in the water column. The solid curved line is calculated based on a mixing depth of 10 cm and a ^{230}Th/^{231}Pa activity ratio of 22 for freshly deposited sediment. (See text for discussion). Sources of data: triangles—Mangini and Sonntag (1977), Mangini *et al.* (1980), unpub. MS., and Müller and Mangini (1980); circles—Cochran (1979); diamond—Kadko (1980b); inverted triangles—DeMaster (1979); squares—Ku *et al.* (1972).

upon at present. The cores from the Pacific suggest a ^{230}Th/^{231}Pa activity
ratio of about 22 in freshly deposited sediment, a value consistent with high
values found in open ocean sediment trap samples (Anderson *et al.* 1980*a*, *b*).
This effective hold-up of Pa in the water column implies that proportionately
more Pa than Th can be transported through oceanic mixing processes to areas
of more rapid Pa (and probably Th) removal (DeMaster 1979; Anderson *et al.*
1980*b*). Enhanced removal in the vicinity of ridge crests may be related to
effective scavenging by precipitation of hydrothermally supplied Mn
(Klinkhammer, Bender, and Weiss 1977; Weiss 1977). The effect in the rapidly-
accumulating sediments of the South Atlantic–Antarctic may be one of high
particle flux to the bottom.

15.3.3 Thorium removal from the coastal and surface ocean

The shorter lived Th isotopes ^{234}Th ($t_{\frac{1}{2}} = 24$ d) and ^{228}Th ($t_{\frac{1}{2}} = 1.9$ y) show
marked variation in their sea water concentrations which are related to varying
rates at which Th is removed from sea water. Bhat *et al.* (1969) showed that the
^{234}Th/^{238}U activity ratio increases with distance from a coast. This suggests
that Th is removed more rapidly in nearshore waters with higher suspended
particle concentrations than the open ocean. Bhat *et al.* (1969) also studied the
depth distribution of ^{234}Th in the top 200 m of the oceanic water column. They
observed ^{234}Th/^{238}U activity ratios less than 1.0 in the top 100 m with an
approach toward equilibrium with depth.

Matsumoto (1975) analysed samples from the upper 200 m of profiles taken
in the North Pacific. He found ^{234}Th/^{238}U ranging from 0.4 to 1.2 with a
median value of 0.80 corresponding to a scavenging mean residence time for
Th of 0.4 y. In near-bottom waters of the South Pacific, Amin, Krishnaswami,
and Somayajulu (1974) observed no disequilibrium between ^{234}Th and ^{238}U.
This result is consistent with a residence time of ^{234}Th greater than several
months, but because of the absence of a nepheloid layer, may not be typical of
all near-bottom areas.

Like ^{234}Th, ^{228}Th shows deficiencies in the surface ocean relative to its
parent ^{228}Ra. Broecker *et al.* (1973) found a median ^{228}Th/^{228}Ra activity ratio
of 0.21 for open-ocean sea water. They also noticed that the ^{228}Th/^{228}Ra ratio
decreased as a coast was approached. Using the open-ocean data, Broecker
et al. (1973) calculated a mean residence time of approximately 0.7 y for Th in
open-ocean surface water. Li *et al.* (1979) examined the pattern of
^{228}Th/^{228}Ra disequilibrium over the continental shelf in the New York Bight
in detail and found a progressive decrease over a salinity range of 35 to 31 per
mil. Mean residence times for Th in this region varied from about 10 to 70 d,
with a suggestion of a slightly longer residence time for Th during the winter in
the inner shelf region.

Kaufman, Li, and Turekian (1981) used both ^{228}Th/^{228}Ra and ^{234}Th/^{238}U
activity ratios to calculate removal rates for Th in the New York Bight.

Residence times varied from about 10 to 90 d, in agreement with the results of Li *et al.* (1979), and values calculated with the two isotopes were generally concordant. Samples in which the residence time based on ^{228}Th/^{228}Ra differed from that based on ^{234}Th/^{238}U were shown to have anomalously high ^{228}Th activities. A possible explanation for the discrepancy is that the water has circulated from a region of long Th residence time into a region of more rapid Th removal because ^{234}Th, with its shorter half-life will reach a new steady state more rapidly than will ^{228}Th. The general trend of decreasing ^{234}Th/^{238}U and ^{228}Th/^{228}Ra activity ratios toward shore also has been observed in California coastal waters (Knauss *et al.* 1978) and similar residence times (approximately 0.3 y) have been calculated.

In the estuarine environment, removal rates for Th are quite rapid. In Narragansett Bay, Santschi *et al.* (1979) have shown concordancy for Th removal rates calculated using ^{234}Th/^{238}U activity ratios (values less than 0.30) and ^{228}Th/^{228}Ra activity ratios (values less than 0.02). The mean residence time for Th obtained from these data is less than 10 d. This agrees with the result of Aller and Cochran (1976) of less than 2 d based on ^{234}Th/^{238}U disequilibrium in Long Island Sound. Santschi *et al.* (1979) also demonstrated an inverse correlation between the Th removal rate (from ^{234}Th measurements) and the rate of sediment resuspension. When sediment resuspension is high, as in the summer, Th is rapidly scavenged from the water column.

In an effort to better understand the controls on Th removal from coastal waters, Santschi, Li, and Carson (1980) and Santschi, Adler, Admurer, Li, and Bell (1980) have performed tracer experiments in microcosms designed to simulate Narragansett Bay. Such studies have shown that Th is removed from the tanks with a half-removal time of days, similar to Th in Narragansett Bay. Other elements, notably Fe, $Cr^{(3+)}$, Pb, Po, Am, and Hg behave similarly leading these authors to conclude that adsorption of Th onto iron hydroxide coatings on particles can account for the similar removal behaviours of Fe and Th, as well as the other easily hydrolysed reactive metal ions. The seasonal cycle of Th removal rates observed in Narragansett Bay is duplicated in the tank experiments, with the removal rate for Th being shortest during high temperature periods with high sediment resuspension rates. In Long Island Sound, Aller (1977) has shown that the highest flux of Mn from the bottom sediments occurs during the summer, and this may enhance the scavenging of particle reactive nuclides like Th.

Thus, the general increase in Th removal rates in the nearshore environment may be attributed to increased suspended particle concentrations as well as to more efficient oxidation–reduction cycling of Fe and Mn. To a certain extent this concept of 'boundary scavenging' (see for example Bacon *et al.* 1976) can be generalized to the deep sea as well. Sarmiento (1978) has presented data which show that the ^{228}Th/^{228}Ra ratio is as low as 0.30 near the sediment–

water interface and increases toward equilibrium with increasing distance above the bottom. Such effects are observed in both Atlantic and Pacific Oceans, but as with the ^{234}Th/^{238}U data of Amin *et al.* (1974), are not ubiquitous. The absence of enhanced scavenging of ^{228}Th and ^{234}Th near the sea floor in certain areas may be associated with the local absence of a nepheloid layer or appreciably higher suspended particle concentrations near the bottom.

The rapid removal of ^{234}Th from the water column in near shore areas implies that the sediments should contain an inventory equal to the rate of production from dissolved U. Aller, Benninger, and Cochran (1980) demonstrate this to be the case in Long Island Sound. While a balance exists on a basin-wide scale, there are smaller scale perturbations caused by horizontal transport, variations in resuspension and types of particles over the Sound, and variations in the rate and nature of bioturbation in the bottom sediments. Inventories of ^{234}Th-excess in New York Bight sediments are also comparable to production in the overlying water column (Cochran and Aller 1979).

15.4 Radium

15.4.1 Radium-226

Within a decade of the discovery of ^{226}Ra (Curie, Curie, and Bémont 1898) Joly (1908b) had observed that concentrations of this nuclide in surface deep-sea sediments and Fe-Mn nodules were considerably higher than in nearshore sediments and sedimentary rocks. To explain these results, he argued that Ra was being removed from the water column to the underlying sediments. Early analyses of ^{226}Ra in sea water showed that it had low but measurable concentrations (Joly 1908a, 1909; Eve 1907, 1909; Satterly 1911). The high background of the relatively simple detection techniques used yielded sea water ^{226}Ra concentrations that have since been shown to be too high. It was not until some 25 years later that more accurate values were obtained (Evans, Kip, and Moberg 1938) who measured ^{226}Ra concentrations from several depths in the Pacific documenting an increase of ^{226}Ra concentration with depth. At the same time, the first U analyses of sea water (Hernegger and Karlik 1935) suggested that the ocean was depleted in Ra relative to that expected if secular equilibrium with U had been established. Since no measurements of the ^{230}Th concentration of sea water had yet been made, these observations led Evans and Kip (1938) to reaffirm Joly's (1908b) contention that Ra was being removed from the oceanic water column to the sea floor.

In a departure from this line of thinking, Petterson (1937) hypothesized that the high concentrations of ^{226}Ra in pelagic sediments resulted from the removal of its parent, ^{230}Th, from sea water. The predicted pattern of ^{226}Ra with depth in a deep-sea core corresponded to that of a daughter growing into equilibrium with its longer lived parent (i.e. low values near the core top, increasing to a maximum corresponding to equilibrium with ^{230}Th, followed by a decrease with depth to U-supported activities of ^{230}Th and ^{226}Ra.) Such a pattern was determined by Piggot and Urry (1941, 1942) in several deep-sea cores, and this led to widespread attempts to use ^{226}Ra as an index of ^{230}Th to determine sediment accumulation rates over time spans of about 4×10^5y or about five half-lives of ^{230}Th (Pettersson 1951, 1953; Kroll 1954, 1955; Volchok and Kulp 1955).

However, doubt was subsequently cast on the validity of this approach by the detailed studies made on cores taken on the Swedish Deep Sea Expedition (Pettersson 1951; Kroll 1954, 1955). Irregularities including multiple maxima of ^{226}Ra were observed in some cores. Pettersson (1951) and Kroll (1953) suggested that one explanation was diffusion of ^{226}Ra after production from ^{230}Th in the sediment. Further evidence of mobilization of Ra was Arrhenius and Goldberg's (1955) observation of high radioactivity in authigenic phillipsite separated from Pacific sediment as well as Pettersson's (1955) finding of higher Ra concentrations on the bottom sides of Fe-Mn nodules which were in contact with the sediment than on the top sides which were in contact with sea water. The development of analytical methods to measure ^{230}Th directly (Issac and Picciotto 1953; Goldberg and Koide 1958) not only confirmed disequilibrium between ^{226}Ra and ^{230}Th in surface sediments, but made ^{226}Ra determinations no longer necessary when used for chronometric purposes alone.

Another possibly important source of ^{226}Ra to the oceans is rivers. Table 15.4 gives the Ra content in a number of world rivers. Taking the flux of river water to the oceans to be 10 l/cm^2 1000 y (as summarized by Turekian 1971) and an average ^{226}Ra concentration of 0.10 dpm/l (see Table 15.4) yields a flux of 10^{-3} dpm ^{226}Ra/cm^2 y. If ^{226}Ra is removed from the ocean only by radioactive decay, then the average ^{226}Ra concentration (for a well-mixed ocean of 4 km depth) should be about 0.6 dpm/100 l. This is less than 10 per cent of the ^{226}Ra content in sea water even as measured by Evans et al. (1938). Based on more recent GEOSECS data (see below), rivers can account for only about 3 per cent of the oceanic ^{226}Ra. Moore (1967) reached a similar conclusion based on his study of Ra in water and suspended sediment of the Amazon and Mississippi Rivers.

For ^{228}Ra the situation is much the same. The river input of ^{228}Ra to the surface ocean is 10^{-3} dpm/cm^2 y (for an average ^{228}Ra river concentration of 0.1 dpm/l). If this ^{228}Ra is supplied to the upper 100 m of the surface ocean and removed only by decay, then a steady state surface concentration of 0.08 dpm/100 l is expected. The river input of ^{228}Ra should be greatest in the Atlantic, but surface waters there have at least 10 times the calculated

Table 15.4. Radium concentrations in rivers

River	^{226}Ra	^{228}Ra	^{238}U	References
	(dpm/l)			
Amazon	0.02	0.012	0.031	Moore (1967, 1969a)
Mississippi	0.07	—	0.70	Moore (1969a)
Hudson	0.07	—	—	Rona and Urry (1952)
	0.01–0.10	—	—	Li et al. (1977)
St Lawrence	0.05	—	—	Rona and Urry (1952)
Ganges	0.20	0.16	1.3	Bhat and Krishnaswami (1969)
Other Indian rivers				
Godavari	0.05	—	0.51	Bhat and Krishnaswami (1969)
Krishna	0.05	—	0.80	Bhat and Krishnaswami (1969)
Sabarmati	0.09	0.31	2.6	Bhat and Krishnaswami (1969)

concentration (Moore 1969b; Kaufman et al. 1973). This supports the idea that the primary flux of Ra to the oceans is from bottom sediments.

Further support for this hypothesis appeared to come from the work of Koczy (1958) on ^{226}Ra profiles in the ocean. In all cases an increase with depth was noted, as would be expected from a bottom source. However, biological recycling also causes concentrations of elements like Si, P, and N to increase with depth. This effect must be taken into account if, as suggested by Koczy, ^{226}Ra is to prove a useful tracer in studying oceanic mixing processes.

Following Koczy's work, most of the effort toward understanding the marine chemistry of Ra centred around its distribution in the water column. This involved a detailed determination of the distribution of ^{226}Ra in the world oceans. Such studies (Koczy and Szabo 1962; Broecker 1965; Broecker, Li, and Cromwell 1967; Szabo, Koczy, and Östlund 1967; Szabo 1967) showed that the ^{226}Ra concentrations in the surface waters of the Atlantic and Pacific were nearly the same, but that the Ra increase with depth was considerably greater in the Pacific. Broecker et al. (1967) showed, on the basis of mixing rates derived from ^{14}C, that such changes were not due to mixing alone, and Chow and Goldberg (1960) suggested, by analogy with the distribution of Ba in the oceans, that the pattern was generated by biological effects. They argued that Ba could be considered a stable analogue of Ra and thus could be used to account for non-mixing effects. The first such attempt was made by Turekian and Johnson (1966), who noted a correlation between ^{226}Ra and Ba for the Caribbean but not for the East Pacific.

The GEOSECS programme made possible accurate measurement of ^{226}Ra in a large number of depth profiles throughout the oceans (Broecker, Kaufman, Ku, Chung, and Craig 1970, Chung, Craig, Ku, Golder, and

Broecker 1974; Broecker *et al.* 1976; Ku and Lin 1976; Chung 1976). Simultaneous measurements of Ba would, in principle, allow correction for biological effects (Chan, Edmond, Stallard, Broecker, Chung, Weiss, and Ku 1976). The salient features of the ^{226}Ra distribution in the oceans are shown in Fig. 15.9 based on the GEOSECS data. Surface water has a nearly constant ^{226}Ra concentration of about 7.4 dpm/100 kg in the Atlantic (Broecker *et al.* 1976) and 6.4 dpm/100 kg in the Pacific (Chung 1976). Turekian and Cochran (1978) have discussed the possible explanations for this similarity despite the large difference in bottom water ^{226}Ra between the two oceans. These include: (i) rivers supply more ^{226}Ra to the surface Atlantic than to the surface Pacific; (ii) the biological removal of ^{226}Ra from surface water is more efficient in the Pacific than in the Atlantic; or (iii) the surface waters of the Atlantic circulate through the high ^{226}Ra water around Antarctica faster than do those of the

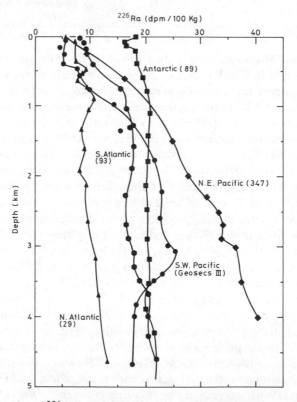

Fig. 15.9. The plot of ^{226}Ra activity against depth in the water column. The data were collected during the GEOSECS expeditions, and the number in parenthesis is the GEOSECS station number. Sources of data: Stn 29—Broecker *et al.* (1976); Stns 83 and 93—Ku and Lin (1976); GEOSECS III—Chung *et al.* (1974); Stn 347—Chung (1976).

Pacific. In the deep water, there is a progressive increase in ^{226}Ra content from the North Atlantic to the North-east Pacific (Broecker et al. 1967; Chung and Craig 1973). Preliminary results from the 1978 GEOSECS Indian Ocean expedition, as well as the previous work of Koczy and Szabo (1962), indicate that deep water of the Indian Ocean is intermediate between the Atlantic and Pacific. Bottom water formation in the oceans takes place in the North Atlantic and around the Antarctic continent, and in the Southern Ocean ^{226}Ra profiles show nearly constant values of about 22 dpm/100 kg below 2 km depth (Ku, Li, Mathieu, and Wong 1970; Chung 1974a; Ku and Lin 1976, see Fig. 15.9). Broecker et al. (1976) have shown that the ^{226}Ra increase from north to south in the Atlantic is not the result of conservative mixing of two water types with origins in the North Atlantic and Antarctic. The progressive increase in ^{226}Ra in the deep water from Atlantic to Pacific via Antarctic waters can be shown to be caused primarily by biological cycling rather than input from the bottom sediments. This follows from the good correlations of ^{226}Ra with Ba in the Atlantic and much of the Pacific (Li et al. 1973; Chung 1974a; Chan et al. 1976). In the East Equatorial and North-east Pacific, however, the Ra increase in the near-bottom waters is not matched by a corresponding increase in Ba (Chung 1974a; Chan et al. 1976), and the bottom water of this area appears to reflect a strong primary input of ^{226}Ra from the sediments. Indeed, this input is traceable as an advective core of high ^{226}Ra which extends westward from about 130° W in the North-east Pacific (Chung 1976). Chung (1976) concluded from these data that the flux of ^{226}Ra from the bottom varies locally as a function of sediment type. Cochran (1980b) used the patterns of ^{226}Ra/^{230}Th disequilibrium in ten deep-sea cores to determine the flux of Ra from the bottom. Variation over two orders of magnitude (0.002 to 0.2 dpm/cm^2 y) was observed with the highest fluxes occurring in the North-east Pacific.

Attempts to use the Pacific ^{226}Ra pattern as an oceanic tracer have been made by Chung and Craig (1980) and Ku et al. (1980). Chung and Craig (1980) considered mixing along an east–west transect in the North Pacific away from the zone of high bottom flux of ^{226}Ra in the NE Pacific. They argued that the relatively low value of the horizontal eddy diffusion coefficient determined from a steady-state one-dimensional diffusion-decay model implied eastward advection along the trajectory. Chung and Craig (1980) also considered mixing along the benthic front in the Western boundary of the Pacific. The pattern of ^{226}Ra along the trajectory is consistent with the front as a surface of no motion separated by northward-flowing Antarctic Bottom Water and southward-flowing Pacific Deep Water. The advective flow velocities deduced from the ^{226}Ra gradients are consistent with independent estimates of this parameter.

Ku, Huh, and Chan (1980) applied a two dimensional vertical diffusion advection model to a north–south ^{226}Ra in the eastern Pacific along 125° W.

While they derived reasonable estimates for horizontal and vertical transport ratio, they noted that the ^{226}Ra balance for the area is only satisfied with a relatively small flux of ^{226}Ra from the bottom. As this is contradicted by bottom water ^{226}Ra concentrations (Chung 1976; Chung and Craig 1980; Ku *et al.* 1980) as well as fluxes inferred from sediment data (Cochran 1980*b*), it was concluded that vertical transport through the thermocline was low in the Eastern Pacific.

The relatively large flux of ^{226}Ra from bottom sediments implies that ^{226}Ra is present in sediment pore water. A direct test of this was made by Somayajulu and Church (1973) who analysed a composite pore water sample from a North Equatorial Pacific core and found a ^{226}Ra activity of about 2.5 dpm/kg or about five times the bottom water ^{226}Ra content. This means that ^{226}Ra can diffuse into the overlying water, but Somayajulu and Church

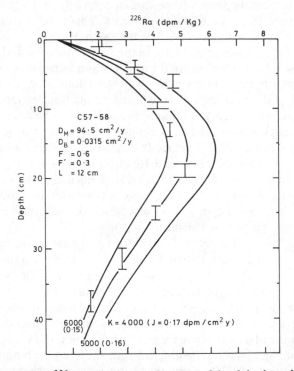

Fig. 15.10. Pore water ^{226}Ra activities as a function of depth in the sediment column for a N. Pacific deep-sea clay–siliceous ooze. The solid curves are model fits to the data and result from varying the adsorption coefficient, K; D_m and D_B are the molecular diffusion and biological mixing coefficients, respectively. Bioturbation is assumed to operate to depth of L cm, and F and F' refer to the fraction of Ra nuclei which have recoiled from the solid phase, where they are produced, into the pore water. The values in parenthesis are the fluxes of ^{226}Ra out of the sediment. (See Cochran and Krishnaswami 1980 for further discussion.)

(1973) could not adequately characterize what controls this diffusive flux. More recently, Cochran and Krishnaswami (1980) measured pore water ^{226}Ra activities as a function of depth in several deep-sea cores from the North Pacific (see Fig. 15.10). Their data suggest that Ra is adsorbed onto sediment particles from the sediment pore water. The mechanism by which ^{226}Ra is introduced to the pore water is thought to be its recoil after production from ^{230}Th decay (see, for example, Kigoshi 1971). Leaching by the pore fluid may enhance the transfer of Ra from the solid phase (Fleischer and Raabe 1978).

The possible importance of the nearshore regime in supplying ^{226}Ra to the oceans was recognised by Li *et al.* (1973). On the basis of a calculated mass balance for ^{226}Ra in the oceans, they suggested that the flux of ^{226}Ra from continental-shelf sediments is of comparable magnitude to that from deep-sea sediments. In a later study of the Hudson River Estuary, Li *et al.* (1977) showed that ^{226}Ra behaved in a non-conservative manner over the zone of fresh water/sea water mixing. They explain this as a desorption of Ra from suspended sediment. Contributing effects may also include the dissolution of a Ra-bearing phase such as diatoms (Dion 1979) or a flux of Ra from bottom sediments (Cochran 1979).

Other recent work on the distribution of ^{226}Ra in marine sediments has shown that in some parts of the ocean, particularly in areas of coastal upwelling, Ra removal from the surface water is intensive enough to be seen as excess ^{226}Ra in the underlying sediments (Koide, Bruland, and Goldberg 1976; DeMaster 1979). If the removal rate has been constant with time, excess ^{226}Ra may then be used as a chronological indicator in its own right.

15.4.2. Radium-228

The detection of ^{228}Ra is usually accomplished through beta analysis of its daughter ^{228}Ac$(t_{\frac{1}{2}} = 6.1\ \text{h})$ or alpha spectrometry of its granddaughter ^{228}Th$(t_{\frac{1}{2}} = 1.9\ \text{y})$ (see section 5.5.3) since ^{228}Ra itself is only a weak beta emitter $(t_{\frac{1}{2}} = 5.75\ \text{y})$. These analytical techniques were not available as early as those for ^{226}Ra. The first evidence of disequilibrium between ^{228}Ra and ^{232}Th in the oceans came from the work of Koczy *et al.* (1957), who noticed an excess of ^{228}Th relative to ^{232}Th in coastal waters off Sweden. They interpreted this as due to an excess of ^{228}Ra in the water and argued, by analogy with ^{226}Ra, that excess ^{228}Ra was supplied to the oceans from rivers or from bottom sediments. Further, they suggested that because of its short half-life, ^{228}Ra would be useful in studying mixing away from the shore or near the bottom. Subsequently, this was confirmed by Moore and Sackett (1964) and Somayajulu and Goldberg (1966). They analysed several open-ocean samples by more precise alpha spectrometeric techniques and also found high ^{228}Th/^{232}Th ratios. Later, Moore (1969, 1969a) made the first direct determinations of ^{228}Ra in the oceans, confirming its excess relative to ^{232}Th and calculated the amount of ^{228}Ra which could be supplied to the surface oceans by rivers (see section 15.4.1) or by ^{232}Th contained in suspended particles. He found these sources to be less than 2 per cent of measured ^{228}Ra content and thus concluded that ^{228}Ra was supplied from bottom sediment to sea water in contact with it (i.e. in the shallow coastal region and near the bottom of the deep sea). Moore (1969b) also demonstrated how oceanic mixing affects the ^{228}Ra distribution, documenting that ^{228}Ra decreased markedly away from the coasts and away from the bottom in the deep sea. Intermediate waters have low ^{228}Ra content because the oceanic circulation (i.e. eddy diffusion away from the bottom) takes places on a time-scale greater than the mean life of ^{228}Ra. This general pattern of ^{228}Ra distribution was verified in greater detail by Kaufman et al. (1973). They found ^{228}Ra concentrations of 10 to 20 dpm/100 kg in shallow coastal waters.

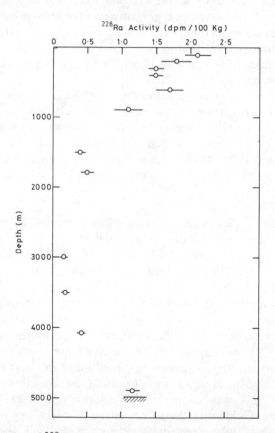

Fig. 15.11. The plot of ^{228}Ra activity against depth in the water column. The profile was taken at the second GEOSECS intercalibration station in the North Atlantic. (After Trier, Broecker, and Feely 1972.)

Concentrations in the centre of oceanic circulation gyres were about 1 dpm/100 kg in the Atlantic and as low as 0.2 dpm/100 kg in the Pacific. Values of ^{228}Ra near the sediment–water interface ranged from 0.5 to 1 dpm/100 kg; these features are illustrated in Fig. 15.11 which shows a depth profile of ^{228}Ra from the North Atlantic Ocean (Trier, Broecker, and Feely 1972).

The relatively short half-life of ^{228}Ra makes it a useful tracer for ocean mixing in surface and near-bottom waters. Because ^{228}Ra is added to the oceans from bottom sediments, the effect of this source will decrease with increasing distance from the sediment water interface. This is seen in the decrease in ^{228}Ra with distance away from shore (Kaufman *et al.* 1973; Knauss *et al.* 1978), with depth in the thermocline (Moore 1972; Trier *et al.* 1972) or with height above the bottom in the deep ocean (Moore 1976; Sarmiento, Feely, Moore, Bainbridge, and Broecker 1976; Sarmiento 1978). In the simplest case the ^{228}Ra distribution can be treated as a balance between eddy diffusion and radioactive decay of ^{228}Ra.

Thus, for steady state:

$$K \frac{\partial^2 C}{\partial z^2} - \lambda C = 0, \qquad (15.6)$$

where K is the eddy diffusion coefficient (vertical or horizontal), z is linear distance away from interface, C is concentration of dissolved ^{228}Ra, and λ is radioactive decay constant.

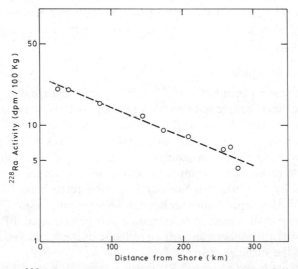

Fig. 15.12. The ^{228}Ra activity in surface waters as a function of distance from the shore of Long Island, USA. The dashed line corresponds to a horizontal eddy diffusion coefficient of 10^6 cm^2/s. See text for further discussion. (After Kaufman *et al.* 1973.)

The solution to eqn (15.6) for the boundary conditions: $C = C_0$ at $z = 0$, $C \rightarrow 0$ as $z \rightarrow \infty$ is:

$$C = C_0 \exp\left(-Z(\lambda/k_z)^{\frac{1}{2}}\right) \tag{15.7}$$

Figure 15.12 shows an application of eqn (15.7) to ^{228}Ra content in a transect off the coast of the north-eastern U.S.A. (Kaufman *et al.* 1973). In this case K corresponds to a horizontal eddy diffusion coefficient. In applying models such as eqn (15.6) to tracer distributions in the oceans, the validity of the assumptions made must be checked. For ^{228}Ra, the validity of steady state in surface waters has been questioned by Reid *et al.* (1979) who found temporal variability in the ^{228}Ra activities of the Gulf of Mexico.

15.5 Radon

The decay of ^{226}Ra produces ^{222}Rn which, because of its short half-life (3.82 d), is generally in equilibrium with Ra in the oceans. Broecker (1965) noted that significant exceptions to this pattern occur near the air–sea and sediment–water interfaces. In surface waters, ^{222}Rn is deficient with respect to ^{226}Ra (the activity ratio ^{222}Rn/^{226}Ra is less than unity) because it is lost to the atmosphere. The pattern near the sediment–water interface is one of excess Rn which is created by ^{222}Rn diffusing out of bottom sediments. The mobilization of ^{222}Rn to sediment pore water occurs in a manner analogous to that in the Ra isotopes (see section 15.4). However, because it is chemically less reactive than ^{226}Ra, the activity of ^{222}Rn in near-bottom waters is generally greater. The patterns of ^{222}Rn/^{226}Ra disequilibrium in these two regions give tracers for studying gas exchange of the surface ocean with the atmosphere and near-bottom mixing.

15.5.1 Radon in surface waters

Broecker's (1965) suggestion that ^{222}Rn deficiencies relative to ^{226}Ra should be observed in near-surface waters was confirmed by Broecker *et al.* (1967) for samples taken in the North-west Pacific. Since then, a number of surface Rn profiles have been obtained, many as a result of the GEOSECS programme (Broecker and Peng 1974). An example of such a profile is seen in Fig. 15.13. The upper 40 m of this station correspond to the mixed layer and are characterized by a constant deficiency of ^{222}Rn relative to ^{226}Ra. In the thermocline, ^{222}Rn rapidly approaches equilibrium with ^{226}Ra. From such a profile two important pieces of information can be extracted: (i) the evasion rate of Rn to the atmosphere; and (ii) an estimate of the vertical eddy diffusion coefficient in the surface ocean.

These workers have calculated Rn evasion rates by a model in which Rn diffuses through a surface boundary layer or 'stagnant film.' The diffusive flux across this boundary layer may then be equated with that required to support

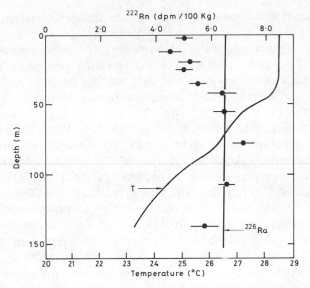

Fig. 15.13. Dissolved ^{222}Rn activity as a function of depth in the surface ocean. The profile was taken at GEOSECS station 263 in the tropical Pacific Ocean. The temperature and ^{226}Ra–depth profiles are indicated by the solid line. The ^{222}Rn deficiency relative to ^{226}Ra in the upper 50 m is caused by loss to the atmosphere.

the integrated ^{222}Rn deficiency in the mixed layer. Thus,

$$D\left(\frac{C_s - \alpha P^{atm}}{z}\right) = \lambda_{228} \int_0^\infty (C_E - C_x)\,dx, \qquad (15.8)$$

where D is molecular diffusion coefficient of Rn in sea water, α is solubility of Rn in surface sea water, P^{atm} is partial pressure of Rn in air, z is thickness of the stagnant film, C_s is Rn activity at the base of the boundary layer, C_E is Rn activity in equilibrium with dissolved ^{226}Ra, and C_x is Rn activity at depth x in the surface ocean.

The right hand side of eqn (15.8) may be written as

$$\lambda_{228} \int_0^\infty (C_E - C_x)\,dx = \lambda_{228}(C_E - C_s)h, \qquad (15.9)$$

where h is the characteristic depth, that is the depth to which the surface Rn deficiency $(C_E - C_s)$ extends.

Because C_s is much greater than αP^{atm}, eqn (15.8) coupled with eqn (15.9) yields an expression for the stagnant film thickness (z) of:

$$z = \frac{DC_s}{\lambda_{228}\,h(C_E - C_s)}. \qquad (15.10)$$

The parameter D/z is referred to as the gas transfer coefficient or piston velocity.

The model assumes that only vertical exchange processes are operating and that the ^{222}Rn profile is in a steady state. The latter assumption is probably poor inasmuch as the gas exchange rate depends on local meteorological conditions (for example, wind speed) which may change on a time-scale comparable to the half-life of ^{222}Rn.

Prior to the GEOSECS programme, Broecker and Peng (1971) determined a boundary film thickness of about 60 µm (piston velocity of approximately 2 m/d) in the North Atlantic, and Peng et al. (1974) found a film thickness of 20 µm (piston velocity of 3.6 m/d) in the North Pacific. The GEOSECS results are summarized in Peng, Takashi, and Broecker (1979). The average film thickness from these data is 36 µm corresponding to a piston velocity of about 2.8 m/d. Considerable zonal variation was noted, with the equatorial zone of both Atlantic and Pacific having a lower gas exchange rate than higher latitudes. Peng, Broecker, Mathieu, Li, and Bainbridge (1979) noted no clear relation between Rn exchange rate and wind speed, in contrast to laboratory experiments predicting that exchange rate varies as the square of wind velocity (Kanwisher 1963).

Using the Rn exchange rates and the good approximation that the rate is the same for CO_2, Peng et al. (1979) calculated the global average CO_2 exchange rate for the surface ocean. The determined value of 16 ± 4 moles/m^2 y is about 80 per cent of that required by the bomb ^{14}C distribution in the upper water column. The difference, if significant, may be explained by enhanced rate of conversion of CO_2 to HCO_3^- in the surface ocean or by the possibility that the Rn exchange rates are reflecting a mean wind speed greater than the global average.

In addition to providing a means of determining the exchange rate of Rn with the atmosphere, the shape of the surface Rn profile can be used to estimate vertical eddy diffusion coefficients in the surface ocean. Assuming no vertical advection and a steady state:

$$K_z \frac{\partial^2 C}{\partial z^2} - \lambda_{222} C + \lambda_{222} C_E = 0, \qquad (15.11)$$

where K_z is the vertical eddy diffusion coefficient, z is the depth in the water column, C is the ^{222}Rn activity, and C_E is the ^{226}Ra activity.

The example given in Fig. 15.13 shows essentially constant ^{222}Rn activity throughout the mixed layer. Such a profile allows only a minimum value of K_z (of approximately 200 cm^2/s) to be calculated for the mixed layer.

15.5.2 Near-bottom ^{222}Rn as a tracer for vertical mixing

The usefulness of the near-bottom excess ^{222}Rn to determine vertical mixing rates in the deep sea was tested by Broecker, Li, and Cromwell (1967) and

Broecker, Cromwell, and Li (1968). They showed that if vertical eddy diffusion is responsible for mixing near the deep sea sediment–water interface, then eqn (15.6) may be used to evaluate near-bottom ^{222}Rn profiles such as Fig. 15.14. Thus, under steady state conditions:

$$K_z \frac{\partial^2 C}{\partial z^2} - \lambda_{222} C = 0 \qquad (15.12)$$

where K_z is the vertical eddy diffusion coefficient, z is the depth above the bottom, and C is the excess ^{222}Rn concentration.

The solution of eqn (15.12) is eqn (15.7) (see section 15.4.2). It predicts that the excess ^{222}Rn activity should decrease exponentially with distance above the bottom. Figure 15.14 shows that, in some instances, the predicted quasi-exponential decrease is observed.

Broecker, Cromwell, and Li (1968) used such profiles to estimate values of K_z ranging from 2 to 50 cm^2/s. In addition to the seemingly exponential Rn

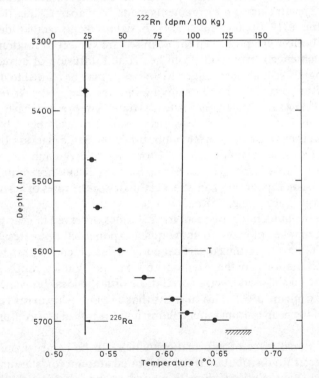

Fig. 15.14. Dissolved ^{222}Rn activity as a function of depth near the bottom. The profile is from GEOSECS station 263 (Pacific Ocean) and the ^{222}Rn excess relative to ^{226}Ra near the bottom is caused by Rn diffusing out of sediments. (After Sarmiento et al. 1976.)

profiles which can be fitted by eqn (15.7), Chung and Craig (1972) observed more complicated shapes, some with several maxima and minima. They attempted to relate Rn profiles to temperature profiles near the bottom, and observed the highest model eddy diffusivities (about 200 cm^2/s) in cases where the temperature gradient was adiabatic.

The validity of the steady-state assumption in treating near-bottom Rn profiles has been examined by Broecker and Kaufman (1970) and Chung (1973, 1974b). At a station in the North Pacific (GEOSECS-I), Broecker and Kaufman (1970) observed that two near-bottom Rn profiles taken 5 days apart were remarkably similar. Significantly, the Rn decrease above the bottom was not the simple expected exponential decrease, but instead exhibited a secondary maximum at about 50 m above the bottom. These authors then argued that lateral mixing from adjacent topographic highs could produce such an effect, and that the influx was approximately steady state. Chung (1974b) reported three additional profiles of near-bottom excess Rn at the same GEOSECS-I station. These profiles showed significant variation over a period of 2.5 y (including measurements made 1.5 d apart). Thus, the Broecker and Kaufman (1970) and Chung (1974b) data, taken together, demonstrate that near bottom Rn profiles in the deep sea are subject to random transient effects. Such effects may result from horizontal diffusion or advection from local topographic highs, as mentioned above, or may be related to detachment of the bottom mixed layer and its mixing into the deep water, as observed by Armi and D'Asaro (1980). Non-steady-state Rn profiles have also been observed on a smaller scale in offshore basins. For example, Chung (1973) found a transient near-bottom Rn minimum in the Santa Barbara Basin which he attributed to a turbidity current. Moreover, Lietzke and Lerman (1975) showed that in such cases as the Santa Barbara Basin, horizontal diffusion related to basin topography can affect estimates of the rates of vertical mixing derived from tracers like ^{222}Rn.

Despite the moderately frequent irregularities observed in near-bottom Rn profiles, a number of cases do show quasi-exponential decreases. Sarmiento et al. (1976) have determined vertical eddy diffusion coefficients for 14 such GEOSECS stations in the Atlantic and Pacific. Values range from 5 to 440 cm^2/s and show an inverse correlation with the local buoyancy gradient (see also Sarmiento 1978). This suggests that a relationship exists between the stability of the near-bottom water column and the rate of eddy diffusion in this region.

An additional piece of information which can be gained from the near-bottom excess Rn distribution is the integrated amount (or 'standing crop') of excess Rn. This quantity is directly related to the flux from the underlying sediments, as well as to any horizontal flux of Rn. Broecker, Cromwell, and Li (1968) found that the excess Rn inventory varied from less than 0.2 to about 20 dpm/cm^2. The highest values were found over slowly accumulating sediments in the Pacific. Kadko (1980a) further examined the realtionship between the Rn flux from the bottom, the sediment accumulation rate, and the ^{226}Ra activity of the underlying sediments. He noted that slowly accumulating sediments have, in general, more ^{230}Th and ^{226}Ra than do rapidly accumulating sediments, and hence a higher Rn flux from the bottom. Key et al. (1979) attempted to balance the measured excess Rn inventory in near-bottom water with a measurement of the Rn deficiency in the sediments from the same location. They found Rn deficiencies to depths of about 4 cm in the deep sea sediment column and, in general, reasonable agreement between Rn deficiency in the sediment and surplus in the bottom water.

Other marine-deposits also serve as sources of excess ^{222}Rn to bottom water. Krishnaswami and Cochran (1978) noted deficiencies of Rn (inferred from ^{210}Pb data) in Fe-Mn nodules from the North Equatorial Pacific. Their data show that in areas where nodules are abundant they can form an important source of Rn to the overlying water column. Sarmiento et al. (1978) also suggested Fe-Mn nodules and encrustations as the source of the high Rn inventories they observed in deep ocean passages.

In nearshore and estuarine waters, Rn has been used as a tracer to monitor the exchange between the pore water and overlying water. In the Hudson River estuary, Hammond, Simpson, and Mathieu (1977) found that molecular diffusion accounts for only about 40 per cent of the total Rn input from sediments. They argued that stirring of the surficial sediments, either by currents or organisms, could enhance the Rn flux to produce the observed Rn inventories in the overlying water column. At Cape Lookout Bight, North Carolina, an environment not affected by physical of biological mixing, Marteus, Kipphut, and Klump (1980) have found that enhanced fluxes of Rn out of the sediments can be explained by a biogenic formation of methane bubble tubes.

15.6 Lead-210

15.6.1 Supply of ^{210}Pb to the oceans

A significant source of ^{210}Pb to the surface oceans is the atmosphere, where it is produced from ^{222}Rn decay. The emanation of Rn from continental rocks and soils (the Rn flux from the surface ocean accounts for only about 2 per cent of the total atmospheric ^{222}Rn; Wilkening and Clements 1975) decays in the atmosphere to ^{210}Pb which falls onto continent and ocean in a component of precipitation. The supply of excess ^{210}Pb delivered from continents by rivers is largely in the particulate phase. Rama et al. (1961) and Lewis (1977) have shown that, except under conditions of low pH(less than 4), ^{210}Pb is rapidly removed from solution onto detrital particles. The ^{210}Pb which falls onto soils is trapped efficiently, and indeed, the mean residence time of ^{210}Pb in soil with respect to erosion is greater than 1×10^3 y (Benninger et al. 1975; Lewis 1977). The observations indicate that ^{210}Pb falling onto the continents is trapped in soils or scavenged from rivers onto particulate matter, and that the river-derived flux of ^{210}Pb to the surface oceans is considerably less than that derived from direct atmospheric input.

The supply of ^{210}Pb to the surface ocean varies with distance from the continental source, because the continents form the dominant source of Rn to the atmosphere. Turekian, Nozaki, and Benninger (1977) have recently reviewed the evidence for this variation and the factors controlling it, modelling the ^{210}Pb flux as a function of latitude and longitude. With a negligible Rn flux from the surface ocean, the prevailing west to east circulation of air implies that the primary control on latitudinal variation in the ^{210}Pb flux to the surface ocean is continental areas. Longitudinal variation in the flux is controlled by the air mass transit time (which affects the total production of ^{210}Pb as a function of distance away from the continent) and the aerosol mean residence time (which governs the rain-out rate of ^{210}Pb). The model developed by Turekian et al. (1977) indicates that, from 15° to 55° latitude in the Northern Hemisphere, the flux of ^{210}Pb to the surface ocean decreases by a factor of 3 from west to east in the Pacific and a factor of about 2 from west to east in the Atlantic.

An additional source of ^{210}Pb in the oceans is decay of dissolved ^{226}Ra (through ^{222}Rn). This source, if a function of the integrated ^{226}Ra activity in the water column, and in the surface and coastal ocean, at depths less than 100 m, is not as important as the atmospheric ^{210}Pb from flux. In the open ocean, the production of ^{210}Pb from ^{226}Ra in a 4 to 5 km deep water column can be comparable to the atmospheric flux.

15.6.2 The ^{210}Pb distributions in surface waters

The ^{210}Pb activity of surface waters varies considerably with geographic location. The observed concentration of ^{210}Pb is affected by inputs from ^{226}Ra decay, the atmosphere, and possibly upwelling as well as by removal onto particulates and radioactive decay. Nearshore waters have low ^{210}Pb concentrations and ^{210}Pb/^{226}Ra activity ratios. Bruland et al. (1974) observed ^{210}Pb/^{226}Ra activity ratios ranging from 0.10 to 0.39 in the mid-Gulf of California, and Krishnaswami, Somayajulu, and Chung (1975) reported a range of values from 0.2 to 0.6 for the Santa Barbara Basin. These low ^{210}Pb/^{226}Ra ratios suggest rapid removal of ^{210}Pb onto sinking particles. Indeed, Benninger (1978) found that all the ^{210}Pb in the waters of Long Island Sound was well correlated with the suspended particle concentration, and that there was effectively no dissolved ^{210}Pb. This observation agrees with the estimate of Krishnaswami et al. (1975) that particulate ^{210}Pb accounted for 50 to 70 per cent of the total ^{210}Pb in unfiltered water samples from the Santa Barbara Basin.

As the open ocean is approached, ^{210}Pb concentrations in surface waters increase and, owing to the atmospheric flux of ^{210}Pb to surface waters, the ^{210}Pb/^{226}Ra activity ratio can exceed unity (Rama et al. 1961). Nozaki and Tsunogai (1973b) demonstrated a latitudinal variation of ^{210}Pb in Pacific surface water with a maximum value of about 19 dpm/100 kg relative to ^{226}Ra, about 7 dpm/100 kg at around 30°N and 20°S (Tsunogai and Nozaki 1971). They also pointed out that this latitudinal ^{210}Pb pattern is well correlated with the geographical distribution of arid land area, which should have a high ^{222}Rn flux. The later ^{210}Pb data of Nozaki and Tsunogai (1973b) do not show any longitudinal variation as might be expected if the transit time for air to cross the Pacific were comparable to the half-life of ^{222}Rn ($t_{\frac{1}{2}} = 3.825$ d).

The pattern of high ^{210}Pb concentrations at the centre of the hemispheric gyres with lower values at the edges was mapped in more detail by Nozaki, Thomson, and Turekian (1976) and is shown in Fig. 15.15. They modelled the ^{210}Pb variation taking into account: (i) the variation in ^{210}Pb flux from the atmosphere; (ii) the variation in ^{210}Pb scavenging efficiency (greater at the continental margins and the productive equatorial regions); (iii) the production of ^{210}Pb from decay of ^{226}Ra (^{222}Rn); and (iv) horizontal diffusion and advection. Although Nozaki et al. (1976) did not find a significant difference for the model fit to the North Pacific data for cases of constant scavenging efficiency and for higher efficiency near the edges of the gyre, the latter seems a closer approximation to the actual behaviour of ^{210}Pb in nearshore waters. In addition to the greater scavenging efficiency of ^{210}Pb near the ocean boundaries, a contributing factor toward low ^{210}Pb activities relative to ^{226}Ra in these regions is upwelling. This was shown by Thomson and Turekian (1976) and Turekian and Nozaki (1980) for the upwelling area off Peru. Stations with the lowest surface ^{210}Pb concentrations (^{210}Pb/^{226}Ra activity ratios less than unity) were dominated by upwelling of ^{210}Pb-deficient water combined with more efficient scavenging of ^{210}Pb from the surface water due to the associated higher productivity.

Patterns of ^{210}Pb/^{226}Ra disequilibria in the surface equatorial North Atlantic are similar to those of the Pacific. Bacon et al. (1976) report average ^{210}Pb/^{226}Ra activity ratios of about 2 from eight Atlantic stations. Moreover, for the North Atlantic, dissolved ^{210}Pb concentrations are highest in the centre of the gyre and decrease toward the edges, similar to the pattern noted in the Pacific.

15.6.3. The behaviour of ^{210}Pb in the deep-sea

Early measurements of ^{210}Pb concentrations in deep water suggested secular equilibrium with its parent (Rama et al. 1961). Removal of ^{210}Pb from the deep sea was documented by Craig et al. (1973). They found ^{210}Pb/^{226}Ra activity ratios of 0.25 to 0.80 (average 0.50) in North Atlantic and North

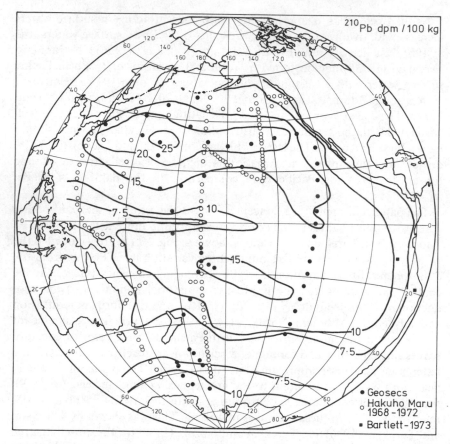

Fig. 15.15. The ^{210}Pb concentrations (in activity units) in surface waters of the Pacific Ocean. (From Nozaki *et al.* 1976.)

Pacific deep water and concluded that the mean residence time of ^{210}Pb in the deep ocean is about 50 y. More recent work, some of which is illustrated in Fig. 15.16, has demonstrated that ^{210}Pb/^{226}Ra disequilibrium is a ubiquitous occurrence in the deep sea. As shown in Fig. 15.16, the depth distribution of ^{210}Pb activities in the open ocean water column is characterized by three regions:

  (i) surface waters which frequently show excesses of ^{210}Pb over ^{226}Ra due to atmospheric input (see section 15.6.2);

 (ii) mid-depth waters which are characterized by a fairly constant ^{210}Pb/^{226}Ra activity ratio (of 0.6 to 0.8); and

(iii) bottom water in which the ^{210}Pb/^{226}Ra activity ratio continues to decrease to values as low as about 0.3.

Somayajulu and Craig (1976) examined the partitioning of ^{210}Pb between

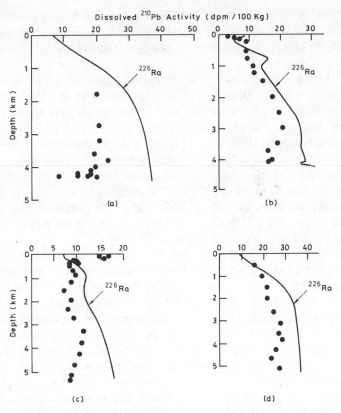

Fig. 15.16. The plot of dissolved ^{210}Pb against depth in the water column. Solid curves represent ^{226}Ra depth profiles: (a) NE. Pacific (Craig *et al.* 1973); (b) S. Pacific (Thomson and Turekian 1976); (c) N. Atlantic (Bacon 1976), (d) N. W. Pacific (Nozaki and Tsunogai 1976). (From Turekian and Cochran 1978.)

soluble and particulate (greater than 0.4μm) phases at two stations from the South Atlantic. In the equatorial South Atlantic, particulate ^{210}Pb is less than 1 per cent of the total ^{210}Pb in the surface water and this fraction increases to about 15 per cent in the bottom water. Using a box model, they estimated a mean sinking velocity of 440 m/y for the scavenging particles, corresponding to a Stokes diameter of 4 μm.

In contrast, Bacon *et al.* (1976) studied dissolved and particulate ^{210}Pb activities in the North Atlantic and found little gradient in particulate ^{210}Pb with depth. To fit their data, particulate sinking velocities as high as 1.1×10^3 m/y were required. Bacon *et al.* (1976) concluded that vertical scavenging by settling particles was responsible for only about 10 per cent of the ^{210}Pb removal from the deep water column. They then proposed that enhanced scavenging at topographic boundaries such as ocean margins, coupled with horizontal eddy diffusion, may be an important process in removing ^{210}Pb from the deep sea. This explanation is consistent with the horizontal distribution of deep-water ^{210}Pb/^{226}Ra activity ratios, which increase from values of about 0.3 near ocean margins to about 0.7 in the open oceans.

Much the same patterns of ^{210}Pb/^{226}Ra disequilibrium are observed in the Pacific Ocean as in the Atlantic. The atmospheric input of ^{210}Pb shows up in the central North Pacific in excesses of ^{210}Pb in the surface waters (Nozaki, Turekian, and von Damm 1980). As shown in Fig. 15.16, upwelling and enhanced scavenging associated with the high productivity in the eastern South Pacific results in surface water deficiencies of ^{210}Pb relative to ^{226}Ra (Thomson and Turekian 1976; Turekian and Nozaki 1980). In mid-depth waters of the North Pacific the ^{210}Pb/^{226}Ra activity ratio ranges from 0.77 to 0.94 (Nozaki and Tsunogai 1976; Nozaki et al. 1980). Nozaki et al. (1980) also document lower ^{210}Pb/^{226}Ra activity ratios in the bottom water, consistent with more rapid removal of ^{210}Pb in this region.

The patterns of ^{210}Pb/^{226}Ra disequilibrium in surface and bottom waters provide possible tracers for vertical mixing processes in these regimes. This is shown in Fig. 15.17 from Nozaki et al. (1980). In the top 2000 m of the water column, the ^{210}Pb excess (relative to ^{226}Ra) decreases from a nearly constant value in the approximately 100 m thick mixed zone to low values from 1500 to 2000 m. Using a vertical eddy diffusion–decay model, Nozaki et al. (1980) fit the profiles in Fig. 15.17 with a vertical eddy diffusion coefficient of 3 cm^2/s. However, Nozaki et al. (1980) note that it is difficult to compare this value with that obtained from other tracers because no account was taken of vertical or horizontal advection or of particulate transport and dissolution. In particular the latter must influence the near-surface ^{210}Pb profile because the residence time of ^{210}Pb in the mixed layer is sufficiently short (1 to 2 y) to remove all ^{210}Pb from the surface water.

In near-bottom waters of the mid-North Pacific, Nozaki et al. (1980) found a nearly constant deficiency of ^{210}Pb relative to ^{226}Ra over a distance of about 1000 m from the bottom. This constancy of ^{210}Pb/^{226}Ra ratio implies rapid vertical diffusive mixing in this region. Between 1000 and 2000 m above the bottom the ^{210}Pb deficiency decreases toward the mid-water value (see Fig. 15.17). Again, using a vertical diffusion–decay model (see eqn 15.6) Nozaki et al. (1980) calculated an apparent vertical eddy diffusion coefficient of about 3 cm^2/s. Thus, the near-bottom mixing pattern which emerges from these results is one of fairly rapid mixing in the bottom 1000 m with a change to a more slowly mixed region extending to at least 2000 m.

15.7 Polonium-210

Despite its relatively short half-life ($t_{1/2} = 138$ d) significant radioactive disequilibrium exists between ^{210}Po and ^{210}Pb in the Ocean. Shannon,

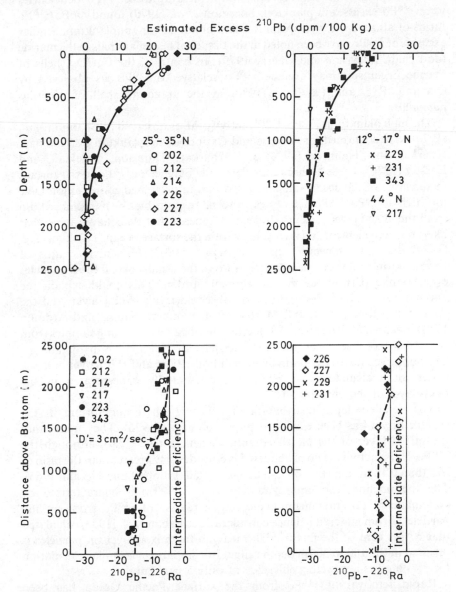

Fig. 15.17. (Top) Depth profiles of excess ^{210}Pb activity (per mass of sea water) from the N. Pacific. The value labelled 'O' is the mid-depth ^{210}Pb deficiency relative to ^{226}Ra. (Bottom) ^{210}Pb deficiency as a function of distance above the bottom in the N. Pacific. The deficiency is expressed as the difference between the ^{210}Pb activity and the ^{226}Ra activity. (From Nozaki *et al.* 1980.)

Cherry, and Orren (1970) found $^{210}Po/^{210}Pb$ activity ratios of about 0.5 in surface sea water around Africa. Complementing these ^{210}Po deficiencies were ^{210}Po excesses in plankton. Shannon et al. (1970) found $^{210}Po/^{210}Pb$ ratios of about 3 for phytoplankton and about 12 for zooplankton. Similar excesses of ^{210}Po have been noted in the tissues of higher animals in the marine food chain (Heyrand and Cherry 1979). In contrast, the $CaCO_3$ shells of marine organisms may exclude ^{210}Po relative to ^{210}Pb, as observed by Cochran, Rye, and Landman (1980) in the aragonite shell of Nautilus pompilius.

The high planktonic $^{210}Po/^{210}Pb$ activity ratios observed by Shannon et al. (1970) were confirmed for Atlantic and Caribbean zooplankton by Turekian, Kharkar and Thomson (1974) and Kharkar, Thomson, Turekian, and Forster (1976). These authors considered the balance of ^{210}Po in the surface mixed layer of the ocean and concluded that unprocessed zooplankton have too high a $^{210}Po/^{210}Pb$ activity ratio to be responsible for removing ^{210}Po from the mixed layer to depth. Indeed, it appears that Po behaves somewhat like a nutrient element and is recycled within the surface ocean. Turekian et al. (1974) did note, however, that the high $^{210}Po/^{210}Pb$ activity ratios of zooplankton could provide a carrier of Po to the atmosphere if these particles ever became part of the marine aerosol burden. This could explain the anomalously high $^{210}Po/^{210}Pb$ ratios observed in air over Hawaii and the Antarctic (Moore, Poet, and Martell 1974; Lambert, Sanak, and Ardouin 1974). Bacon and Elzerman (1980) tested this idea by examining samples from the sea-surface microlayer (the upper $300\,\mu m$) for $^{210}Po/^{210}Pb$ disequilibrium. They observed microlayer enrichments of both ^{210}Po and ^{210}Pb over bulk sea water and calculated that ^{210}Po is preferentially concentrated in the microlayer of the open ocean.

The detailed depth distribution of $^{210}Po/^{210}Pb$ disequilibrium in the surface ocean has been examined by Bacon et al. (1976). They found that the mixed layer of the North Atlantic is deficient in ^{210}Po, although the $^{210}Po/^{210}Pb$ activity ratio of about 0.5 is considerably greater than the ratio of less than 0.10 in rain (Burton and Stewart 1960; Lambert and Nezami 1965). The thermocline from approximately 100 to 300 m is characterized by systematic ^{210}Po enrichment, indicating a release of ^{210}Po from sinking particles. From material balance considerations Bacon et al. (1976) calculated that 67 per cent of the flux of ^{210}Po leaving the mixed layer on particles is released in the 100 to 300 m depth range. In comparison a similar calculation for ^{210}Pb gives a recycling efficiency of only 5 per cent.

Rapid removal of ^{210}Po from the surface Pacific Ocean has been demonstrated by Nozaki et al. (1976), and their data are shown in Fig. 15.18. The $^{210}Po/^{210}Pb$ activity ratio of the surface water is highest near the centre of the gyres due to a longer residence time of ^{210}Po in these areas. The situation is complicated by the fact that upwelling can cause near unity values of the

Fig. 15.18. Distribution of ^{210}Po/^{210}Pb activity ratios in the surface waters of the Pacific Ocean. (From Nozaki *et al.* 1976.)

^{210}Po/^{210}Pb ratio even at the gyre edges. From the data presented in Fig. 15.18 Nozaki *et al.* (1976) calculated a mean residence time of ^{210}Po relative to removal from the mixed layer on particles of about 0.6 y.

The effect of upwelling on ^{210}Po/^{210}Pb disequilibrium was further examined in the eastern South Pacific off Peru by Thomson and Turekian (1976). They found that no measurable ^{210}Po/^{210}Pb disequilibria could be detected from the surface to about 1500 m depth. Samples from 1500 to 400 m showed a pronounced ^{210}Po deficiency. A re-examination of new samples from the same area revealed ^{210}Po in equilibrium with ^{210}Pb at all depths (Turekian and Nozaki 1980), suggesting sampling difficulties as the probable cause for the initial results. The conclusion from this work is that in this area upwelling imprints even the surface water with a higher ^{210}Po/^{210}Pb activity ratio (and a lower $^{210}Pb/^{226}Ra$ activity ratio, see section 15.6.2), despite the high productivity which would enhance ^{210}Po removal.

In the deep ocean, disequilibrium between ^{210}Po and ^{210}Pb is much less striking. For example, Bacon et al. (1976) found a slight (about 12 per cent) deficiency of ^{210}Po to ^{210}Pb in the dissolved phase and an enrichment in the particulate phase. They attributed their observations to rapid in situ scavenging of ^{210}Po. For the water column as a whole, however, total ^{210}Po and ^{210}Pb, obtained by summing dissolved and particulate analyses, are in balance. This suggests that the mean sinking velocity of the particles responsible for scavenging ^{210}Po is not large enough to establish any total disequilibrium between ^{210}Po and ^{210}Pb.

Acknowledgments

I am indebted to K. Turekian, A. Mangini and S. Krishnaswami for discussions and helpful comments on portions of this manuscript and to Ann Phelps for typing the first draft. Support of the US Department of Energy (Grant DE-ACO2-76-EV13573 to K. K. Turekian, principal investigator) is gratefully acknowledged.

16 SEDIMENTS AND SEDIMENTATION PROCESSES
C. Lalou

16.1 Introduction and historical perspectives

The ^{238}U, ^{235}U, and ^{232}Th decay series provide a useful mechanism for studying sedimentological processes because of differences in the physico-chemical behaviour of the actinides in the marine environment. Thus, daughter products within each decay series can be separated from their parent nuclides causing a disequilibrium by which it is possible to study accumulation rates, solid–fluid interactions, and determine absolute ages for marine sediments.

The beginning of deep sea sedimentology dates from the Challenger expedition of 1872–76, and ideas to determine the absolute age of these deposits quickly followed. Joly (1908b) measured Ra contents in different types of marine sediments. He observed that deep sea sediments had higher Ra concentrations than sediments of the continental shelf and considered that this Ra was directly scavenged from sea water by descending detrital sediments. Although this hypothesis has proved to be incorrect, the work of Joly developed what is the classical method for determining sedimentation rates and formed the basis for all later work in the field. Pettersson (1937) attributed correctly the Ra excess in pelagic sediments to the scavenging of ^{230}Th and Piggot and Urry (1939, 1942) showed that variations in ^{226}Ra with depth in deep sea sediment cores effectively represented its parent ^{230}Th. With this work a method for directly measuring the absolute age of deep sea sediments was born. Subsequent advances over the past forty years have been made by the development of alpha spectrometry which provided a means of measuring directly the activity of individual nuclides and by evolution in our understanding of the general geochemistry of the actinides in the marine environment.

The initial reason for using U-series disequilibrium to investigate deep sea sediments was to establish a global budget between the processes of erosion and sedimentation and more recently to study the complex oceanic geochemical system. Because different time-scales are required to study the various oceanic sub-systems a variety of radiometric clocks are required. For example, a longer time-scale is needed to study climatological fluctuations during the Pleistocene ice ages than to trace the path of heavy metal pollutants in the ocean on a human time-scale.

While the ^{14}C dating method was developed by Libby in 1952, permitting absolute dating of marine sediments younger than about 30 ky, within a year Issac and Picciotto (1953) suggested the ^{230}Th method thus introducing the possibility of extending an absolute chronology back to about 300 ky. This method, utilizing the direct measurement of ^{230}Th by alpha spectrometry or more recently by high resolution gamma ray spectroscopy (Yokoyama and Nguyen 1980) or indirect measurement of ^{230}Th by ^{214}Bi by traditional gamma ray spectroscopy (Osmond and Pollard 1967; Yokoyama et al. 1968), found extensive application to deep sea sediment chronology and sedimentation budgets and established that the geochemistry of ^{230}Th and daughter nuclides is not as simple as had been generally presumed and, in so doing, laid the foundation for the application of new methods based upon ^{234}Th and ^{210}Pb to date rapid or more recent phenomena.

This general picture is typical of developments and evolution in the field of geochronology. An initial idea is proposed and successfully applied to a particular problem. Subsequently problems are encountered and a study of the geochemistry of an element undertaken and budgets investigated in order to understand the original problem. In this chapter the reverse path will be followed. First, our present knowledge about the abundance and distribution of the actinide elements in deep sea sediments will be presented. Based upon this, possible application of various U-series nuclides as chronometers will be discussed. Finally, this information will be synthesized and a review of the geochemistry and element budgets presented.

16.2 Concentration and distribution of actinides in sediments and pore waters

Before making an inventory of the actinides in sediments and pore waters, it is necessary to define first the main components of deep sea sediments. The actinide content of a sediment will depend primarily on the relative proportions of each constituent. Deep sea sediments comprise essentially four components:

(i) a detrital component, essentially made of clays or minerals derived from the weathering of continental rocks;

(ii) a mineral biogenic component which is calcareous in the middle and equatorial latitudes, and siliceous in high latitudes; the calcareous component being essentially foraminifera tests and coccoliths, the siliceous one being diatoms and radiolarian skeletons, although other mineral biogenic elements such as pteropods, coral debris, sponge spicules and fish teeth, are present in minor quantities;

(iii) an organic biogenic component which, in some instances, may play an important role in establishing the chemical characteristics of the sediment; (iv) an authigenic phase including such components as Fe and Mn oxides, (from diffuse oxihydroxide precipitates to Mn nodules) metalliferous sediments, and authigenic clays whose origin, chemical composition, and abundance may be greatly variable.

The general U and Th contents of the various components of marine sediments are summarized in Table 16.1 with details discussed below.

16.2.1 Uranium

Clay minerals are the principal detrital component of pelagic sediments on the abyssal plains. The U content of such clays is about 2 to 3 ppm, slightly lower than the mean concentration in continental rocks as a result of a small amount of U loss during erosion and transport (Ku 1966). The addition of U-poor calcareous detritus on the continental shelf causes these sediments to have a slightly lower U content than their deep sea counterparts. Near the continent resistate phases high in U, such as zircon, or low in U, such as quartz, may be present in minor amounts.

The biogenic mineral component comprises the majority of most deep sea sediments. Above the carbonate compensation depth it represents more than 80 per cent of the sediment occurring as foraminifera tests or coccolith which have U contents generally in the range of 0.11 ± 0.04 to 0.13 ± 0.08 ppm (Holmes et al. 1968). However, Mo et al. (1973) have noted some exceptions such as in the Gulf of Mexico where the greater than $88 \mu m$ carbonate fraction, essentially planktonic foraminifera, had U contents up to 1.2 ppm. Because $CaCO_3$ content may be variable within a core, either due to difference in biological productivity related to climatic variations or to variations in the depth of the lysocline, it is common practice to express the concentration of U (and other actinides) in a sediment on a carbonate free basis. This normalization procedure helps to identify anomolies in U content. Biogenic silica is another important constituent of most deep sea sediments. No direct measurements have been made on the U content of this component, but Ku (1966), using the mean value of 2.6 ppm for the clay fraction in a core containing 50 per cent diatoms and radiolarians and 1.57 ppm U, concluded that the U content of biogenic silica in the core was about 0.5 ppm, substantially lower than that of the clay fraction. Other biogenic materials such as fish teeth with 21 to 49 ppm U (Bernat, Bieri,Koide, Griffin, and Goldberg 1970), coral debris with 2 to 4 ppm U, or pteropod skeletal material with more than 1 ppm U (Sackett and Cook 1969), frequently occur in marine sediments, but usually only in small amounts so that their overall contribution to total U in marine sediments is insignificant.

Uranium is present in normal oxidized sea water as the soluble hexavalent state, where it forms a highly soluble and mobile uranyl carbonate complex

Table 16.1. Some U and Th concentrations in different components of sediments

Origin	Nature	U (ppm)	Th (ppm)	Reference
Detritic	Clay	2–3		Ku (1966)
			9.1–18.4	Ku et al. (1968)
			10 ± 3	Heye (1969)
Biogenic (mineral)	Coccoliths and Foraminifera tests	0.1–0.13	< 0.2–2	Holmes et al. (1968)
			0	Heye (1969)
	Foraminifera Gulf of Mexico	0.27–1.20		Mo et al. (1973)
	Pteropods	> 1.0		Sackett and Cook (1969)
	Diatoms, radiolarian	< 0.5		Ku (1966)
	Fish teeth	21–49	63–230	Bernat et al. (1970)
Biogenic (organic)	Anoxic sediments, Gulf of California, fjords	4.8–39		Veeh (1967)
	Mid-Atlantic Ridge	40		Turekian and Bertine (1971)
	Mediterranean sapropels	47		Mangini and Dominik (1979)
	Reducing sediments (mean)	130		Baturin (1973)
Authigenic	*Metalliferous sediments*			Fisher and Boström (1969)
	East Pacific Rise	53	2.4	Boström and Rydell (1979)
	Red Sea brines	30–60	0.25	Ku (1969)
	Mid-Atlantic Ridge	38		Bertine et al. (1970)
	Galapagos Spreading Centre	up to 500		Lalou and Brichet (1980a)
	Phillipsite	0.25–0.69	13	Bernat et al. (1970)
	Barite	0.2–5	30	Church and Bernat (1972)
	Deep sea Mn nodules	4–13	100	Ku and Broecker (1969)
	Shallow Mn nodules	10–20	2.31–5	Ku and Glasby (1972)
	Hydrothermal crusts			
	Mn rich: TAG	9–16	2.34–3.95	Scott et al. (1974)
	FAMOUS	3–6.7	0.7–1.5	Lalou et al. (1977)
	Fe rich	0.25–1.25	—	Piper et al. (1975)
	Sulphur deposits 21° N East Pacific Rise	0.6–2.8	—	Lalou and Brichet (1980b)
	Oxide deposits 21° N	8.9–14.4	—	Lalou and Brichet (1980b)

(see chapter 2). As a result, oxidized sediments without biogenic silica, have between 2 to 3 ppm U on a carbonate free basis. This U has a $^{234}U/^{238}U$ ratio close to unity similar to that of the continental rocks from which it derived. Surficial sediments typically have $^{234}U/^{238}U$ ratios of about 0.9 as a result of ^{234}U leaching and ^{234}Th migration during particle transport. Although this disequilibrium slowly fades with depth, the evolution is not regular leading Ku (1965) to propose a hypothesis for the migration of labile ^{234}U in the sediment column and to calculate a diffusion coefficient of 10^{-8} cm^2/s to 10^{-9} cm^2/s. Subsequent work by Dysard and Osmond (1975) documented $^{234}U/^{238}U$ ratios of 1.02 to 1.59 in sediment pore waters substantiating this hypothesis. Sediments with greater U contents can be found in environments where the redox potential is sufficiently low to cause the soluble U^{6+} to be reduced to the poorly soluble U^{4+}. Such environments are usually associated with organic biogenic sediments.

Uranium is not concentrated in living tissues, but rather in regions of high productivity where planktonic organic matter accumulates (for example at the sediment–water interface). Thus in special hydrographic and topographic conditions anoxic basins are formed, and due to the high content of organic matter, reducing conditions develop. In such reducing environments, U may be scavenged from sea water and could be considered as authigenic. When deposited in this way, the initial $^{234}U/^{238}U$ activity ratio is 1.14 ± 0.02, characteristic of sea water (see chapter 15). Examples of this phenomenon have been documented in the Gulf of California and in Norwegian fjords by Veeh (1967) who calculated an accumulation rate for U of $100\,\mu g/cm^2/10^3$ y. Similarly, Mo et al. (1973) observed a strong correlation between the percentage of organic C and U content in the deeper part of the Gulf of Mexico, and considered that this enrichment occurred in a shallow anoxic coastal environment followed by a deposition in the deepest part of the Gulf. Baturin (1973) studied the correlation between organic C and U in more than 700 samples of sediments from around the world and concluded that localization of high concentrations up to 130 ppm depends mainly on the content of organic matter in the sediments. Rona and Joensuu (1974) documented that U enrichment is presently occurring in the Black Sea. The link between the accumulation of organic matter and development of anoxic conditions, leading to the scavenging of U from sea water has been demonstrated recently by Mangini and Dominik (1979). In sapropelic layers of Eastern Mediterranean, concentrations as high as 47 ppm were measured permitting the calculation of a deposition rate of U from sea water of $11\,\mu g/cm^2/10^3$ y. But the $^{234}U/^{238}U$ ratio, instead of being that of sea water varied in the same sapropel layer from 0.98 to 1.19. This is explained by an open system where in situ produced ^{234}U is able to diffuse out of the sapropel with a diffusion constant of about 10^{-11} to 10^{-12} cm^2/s, slightly lower than the one calculated by Ku (1965).

At the boundary between U associated with the organic biogenic component and U truly linked to authigenic phases, are the U enrichments linked to reducing environments, where organic C is suspected but not measured and where the reducing conditions are deduced from the presence of another metal. For example, high U concentrations along the mid-oceanic ridge areas have been attributed to local reducing environments by Turekian and Bertine (1971). These authors studied a core from the mid-Atlantic Ridge and observed a correlation between high U contents (up to 40 ppm on a $CaCO_3$ free basis) and high Mo concentrations at the same levels in the core. This was attributed to the formation of transient ephemeral basins along the ridge with high supply of organic material in a relatively shallow environment. Similarly, Bonatti et al. (1971), observed a relationship between the reducing conditions of hemipelagic sediments in an East Pacific core, as indicated by low Eh values, and significant enrichments of U as well as Cr, V, and S. These enrichments are attributed to post-depositional migrations of the elements from oxidized zones towards reducing ones through pore water. But, in this latter case, owing to the very low U concentration from 0.7 ppm in the oxidized zone to 2.74 ppm, $CaCO_3$ free, in the reducing one, there is no addition of U but only a redistribution.

The authigenic components of deep sea sediments may be classified into two groups: (i) diffuse phases, where U is related to scavengers as Fe or Mn hydroxides (Mn nodules excepted) ranging from slight enrichment in metallic elements to metalliferous sediments; and (ii) discrete mineral phases as zeolites (essentially phillipsite), barites, polymetallic crusts or encrustations, Mn nodules, or hydrothermal deposits which may be either depleted or enriched in U. (Phosphorites, another authigenic component, will be discussed in chapter 17.) In this authigenic component, U may derive either from sea water or volcanism. Generally, a sea water source is attributed to enrichments having a $^{234}U/^{238}U$ ratio consistent with that of modern sea water (1.14 ± 0.02), and a U with a higher isotopic ratio is attributed to hydrothermal leaching of basalts. This latter classification as to U source is largely theoretical because mixing of the two U sources must be commonplace as the hydrothermal fluid in the deep sea environment is sea water with a relatively high U concentration, and when this hydrothermal fluid discharges onto the sea floor, fresh sea water U may be scavenged by any hydrothermal precipitate which is formed. Therefore, it does not seem that a typical 'volcanogenic value' may be defined. The only published data concern the water issued from active volcanoes in USSR and indicate a mean $^{234}U/^{238}U$ ratio of 1.34, but with a range of 1.04 to 2.25 (Kuptsov and Cherdyntsev 1969). One of the first studies to define clearly the high U content of metalliferous sediments associated with mid-ocean ridges was that of Fisher and Boström (1969). Their measurements for a traverse across the East Pacific Rise show a significant increase of U up to 53 ppm (on a $CaCO_3$ free basis) just on the ridge axis as compared with the mean value of 2 ppm for the barren sediment of the adjacent abyssal plain. These authors also noted that there was a close association of high heat flow and enrichment of the crest sediments with certain trace elements of probable volcanogenic origin such as Fe, Mn, V, P, Ba, Hg, As, and Cd, thus suggesting a similar volcanogenic origin for the U. Similarly, Bertine et al. (1970) found U values as high as 38 ppm within discrete levels of a sediment core from the mid-Atlantic Ridge and concluded that the U enrichment was due to either episodic hydrothermal events or the periodic formation of ephemeral reducing basins along the ridge.

Both of these studies utilized fission track techniques for U analysis and thus could not determine the extent of any U-isotope disequilibrium. However, several alpha spectrometry studies have investigated this problem. Ku (1969) observed that Red Sea brines had U values of 30 to 60 ppm and $^{234}U/^{238}U$ ratios close to the mean sea water value of 1.14 and concluded that U in the brine was coprecipitated from Red Sea water with Fe hydroxides. Calculating a deposition rate for U of $450 \mu g/cm^2/10^3 y$ and assuming a residence time of 200 y for the waters, he further demonstrated that the U in the brines represented less than one third that contained in the one km water column above the brines. Similarly, Bender et al. (1971) conclude that the high U content of a core from the East Pacific Rise with a high Mn accumulation rate is of sea water origin as it exhibits a $^{234}U/^{238}U$ ratio of 1.16 ± 0.01, but also argue that submarine volcanism was responsible for the associated metal enrichment. Subsequently, Rydell and Bonatti (1973) questioned the extent to which $^{234}U/^{238}U$ ratios were a guide to the source of the U in metalliferous sediments. In a study of deposits of unequivocal hydrothermal origin these authors observed that $^{234}U/^{238}U$ ratios were close to the sea water value. They also observed that samples exhibiting an anomalously high $^{234}U/^{238}U$ ratio appear to have a relatively low U content. Finally, Lalou and Brichet (1980a) have found very high U concentrations of up to 500 ppm (on a $CaCO_3$ free basis) and high $^{234}U/^{238}U$ ratios in sediments deposited between 300 to 600 ky under the hydrothermal mounds near the Galapagos spreading centre which suggested a hydrothermal origin for this U.

In the authigenic formations, U concentration is rather variable. for example, phillipsite contains only 0.25 to 0.69 ppm U (Bernat et al. 1970) in a sediment where U concentration varies from 2.6 to 3.1 ppm. Moreover, an increase in quantity of phillipsite with depth is accompanied by a decrease in U concentration in the phillipsite, leading Bernat et al. (1970) to postulate a continuous growth of phillipsite in the sediment column. The precipitation of marine barite is another example of an authigenic component, U content ranging from 0.2 to 5 ppm (Church and Bernat 1972).

The other major authigenic component of marine sediments is represented by polymetallic nodules. Ku and Broecker (1969), observed that deep sea Mn nodules from different parts of the ocean floor had U concentrations ranging from 4 to 13 ppm, with an average value of 9 ppm. By comparison shallow water continental margin Mn nodules have a distinctly higher U content (10 to 20 ppm) which is attributed to their formation under more reducing conditions (Ku and Glasby 1972). Other Mn-rich marine encrustations usually attributed to hydrothermal activity, have U concentrations which are more variable. For example, on the mid-Atlantic Ridge, two areas of Mn deposits have been studied, the TAG hydrothermal field in which U content is comparable to deep sea Mn nodules, from 9 to 16 ppm with a $^{234}U/^{238}U$ ratio near the one of sea water (Scott *et al.* 1974), and the FAMOUS deposit where U is lower (3 to 6.7 ppm) with a $^{234}U/^{238}U$ ratio significantly higher than the sea water ratio (Lalou, Brichet, Ku, and Jehanno 1977). In Fe rich deposits from the Pacific, Piper, Veeh, Bertrand, and Chase (1975) found U values as low as 0.25 to 1.25 ppm with an activity ratio varying from 1.22 ± 0.01 to 1.29 ± 0.02, clearly higher than that of sea water. The recently discovered hydrothermal sulphide deposits at 21° N on the East Pacific Rise have very low U concentrations (0.6 to 2.8 ppm), whereas oxides of the same deposits are enriched in U (8.9 to 14.4 ppm). For both types U isotopic ratio is not distinguishable from the sea water value (Lalou and Brichet 1980*b*).

There is also much interest in the U content of interstitial waters due to the fact that they are the vehicle for migration of the more mobile actinides. Baturin and Kochenov (1973) found variable U contents, (1.3 to 650 ppb) in the interstitial waters from about 30 cores from Atlantic and Pacific Oceans and from the Black Sea documenting that diagenetic mobility of U is controlled by pH and *E*h of the water and by the U and organic carbon content of the sediment. The U content of interstitial water in a deep sea core from the Pacific Ocean, studied by Somayajulu and Church (1973), exhibits a lower value (1.9 ppb) than sea water (3 ppb). Similarly, Dysard and Osmond (1975) analysed a core from the Southern Ocean and noted that concentration was lower in pore waters than in sea water and that it diminishes with depth relative to the sediment concentration. The $^{234}U/^{238}U$ ratios were always greater than unity, but varied with depth in an irregular manner. They concluded that diffusion of U was limited to a short period after deposition of the sediment. By contrast, Ku *et al.* (1977) found a mean U value of 3.44 ppb and a mean U-isotopic ratio of 1.17 for the top samples in six cores. Those differences may, however, only be an artefact of the extraction procedures and of the definition of what actually constituted pore water in a sediment core.

16.2.2 Thorium and Pa

By contrast with the U isotopes which are characterized by only minor fractionation in the ocean, Th isotopes have different geochemical pathways because they do not have the same origin. Thus, it is necessary to distinguish between ^{232}Th and its granddaughter ^{228}Th in marine sediments and the Th isotopes produced as intermediate decay products in the U series. Protactinium isotopes will also be discussed here due to their similarity in geochemical behaviour to the Th isotopes. The detrital component of marine sediments has a mean ^{232}Th concentration of 10 ± 3 ppm (Heye 1969). This figure agrees well with the $CaCO_3$ free value of 8.22 ppm calculated by Ku, Broecker, and Opdyke (1968) from 75 analyses of 14 cores of different composition from around the world (range 3.1 to 18.4 ppm) and indicates that the majority of Th in marine sediments is contained in the detrital component.

Heye (1969) noted that the content of mineral biogenic component was negligible, but Holmes, Osmond, and Goodel (1968) observed that foraminifera shells had a range of Th contents from 0.2 ppm in the South Pacific to 2.1 ± 0.7 ppm in the Drake Passage. In fact, the highest contents yet documented occur in fish teeth, 63 to 230 ppm (Bernat et al. 1970), but this constituent comprises an insignificant proportion of the total biogenic component.

In metalliferous sediments, Th is depleted whereas U is enriched (as already seen in section 16.2.1). On the East Pacific Rise, for example, the average ^{232}Th content is 2.4 ppm (Boström and Rydell 1979). In three cores from the East Pacific Rise, considering that ^{232}Th is only of detritic origin, using a mean ^{232}Th concentration in the detritic phase of 10 ppm and the sedimentation rate measured in the core, Bender et al. (1971) calculated a sedimentation rate for the detritic component of a few tenth of mm/10^3 y, which is extremely low and proves that no authigenic Th exists in the deposit. In Red Sea brines, ^{232}Th is also very low, presenting a mean value of 0.25 ppm (Ku 1969).

The authigenic components are generally characterized by high and variable Th contents. In phillipsite Th content with a mean of 13 ppm is comparable to or slightly higher than in the detrital component (Bernat and Goldberg 1969; Bernat et al. 1970). As in the case of U, these authors note that Th typically decreases with depth in a sediment core and that no migration of Th occurs in the sedimentary column, although ^{228}Ra moves from the detrital fraction to the phillipsite through interstitial waters causing ^{228}Th/^{232}Th ratios in phillipsite to be as high as 9. For marine barites the average Th content is 30 ppm with a range of 2 to 190 ppm (Church and Bernat 1972). As with phillipsite, the amount of authigenic barite in pelagic ooze from the Equatorial Pacific increases with depth in the sediment column while Th decreases. The association of Th-depleted barite with Fe-Mn phases or with hydrothermal minerals such as palygorskite and sepiolite led Church and Bernat (1972) to compare these barites to magmatically derived continental barites. As with phillipsite, authigenic barites typically have ^{228}Th/^{232}Th ratios greater than unity.

In deep-sea polymetallic nodules Th concentrations are relatively high, and values of the order of 100 ppm are common. However, Th contents are quite variable often seemingly related to water depth, the highest concentrations being found in the deepest nodules (Ku and Broecker 1969), but also in some curious cases, Th decreasing with depth within a single polymetallic deposit (Lalou, Ku, Brichet, Poupeau, and Romari 1979). In shallow water Mn nodules, Th is very low, ranging from 2.31 to 5.00 ppm (Ku and Glasby 1972). In Mn-rich hydrothermal crust, Th is comparably low in TAG hydrothermal field, with a value of 2.34 to 3.95 ppm (Scott *et al.* 1974) and further depleted in the FAMOUS deposits, ranging from 0.7 to 1.5 ppm (Lalou *et al.* 1977). The same is characteristic of the Fe-rich deposits from the Northeast Pacific where Th was under the detection limit of 0.01 ppm (Piper *et al.* 1975). In sulphides presently being deposited at 21 °N, the Th is also below the detection limit, while in the oxidized parts it is somewhat higher (Lalou and Brichet 1980*b*).

In interstitial water, Th concentrations are very low, due to the low content of Th in sea water (see chapter 15). However, interesting disequilibrium is observed. The $^{228}Th/^{232}Th$ ratios in excess of a value of 14 have been documented in a sediment core from Pacific Ocean (Somayajulu and Church 1973). This disequilibrium was cited by Church and Bernat (1972) and Bernat and Goldberg (1969) to explain the excess ^{228}Th they found in authigenic barite and phillipsite.

As discussed in chapter 15, Th and Pa isotopes are produced continually from the radioactive decay of U in solution in sea water. Due to their greater insolubility they have a short residence time in solution, being rapidly adsorbed and then precipitated at the sea–sediment interface. This leads to two types of ^{230}Th and ^{231}Pa when sediment is considered as a bulk phase: the ^{230}Th (or ^{231}Pa) which constitutes an indigenous part of the detrital phase and which is in equilibrium with its parent U and that ^{230}Th (and ^{231}Pa) which may be called authigenic, adsorbed onto sediments at the sediment–water interface, and unsupported by its parent U. If this ^{230}Th and ^{231}Pa flux were strictly proportional to the U content of the water column, and the accumulation rate of sediment constant, the specific activities of the ^{230}Th or ^{231}Pa excess of a sediment would depend only on its age and on the water depth. This is not the general case because the situation is more complex. Nevertheless, this excess ^{230}Th or ^{231}Pa does allow estimation of sedimentation rates (see section 16.3.1).

Authigenic phases such as phillipsite, barite, Mn nodules or hydrothermal deposits are not uniform in their ^{230}Th or ^{231}Pa characteristics. In phillipsite and barite, the studies of Bernat and Goldberg (1969), Bernat *et al.* (1970), and Church and Bernat (1972) have shown an excess of ^{230}Th coupled with a decrease of the $^{230}Th/^{232}Th$ ratio with depth, parallel to the decrease in the bulk sediment even if this ratio is much lower in barite than in bulk sediment. Deep-sea Mn nodules also exhibit a large excess of ^{230}Th and ^{231}Pa in their outermost layers. However, shallow water Mn nodules' ^{230}Th and ^{231}Pa are depleted towards equilibrium with their parents ^{234}U and ^{235}U, the activity ratios $^{230}Th/^{234}U$ and $^{231}Pa/^{235}U$ increasing with depth within the nodules. Using the same method as the one used to date corals (see chapters 17 and 18) and making the assumptions that neither Th nor Pa was present at the time of the precipitation, and that these two nuclides result from the *in situ* decay of U in the nodules, an age may be calculated for each layer of a nodule (Ku and Glasby 1972). The same disequilibrium relationships were observed for TAG samples by Scott *et al.* (1974). In the FAMOUS Mn-rich deposit, ^{230}Th is also depleted relative to ^{234}U, but the ^{230}Th/^{234}U activity ratio, instead of increasing regularly with depth, first decreases towards the minimum value of 0.36 and then increases. These variations in ^{230}Th/^{234}U activity ratio can be attributed to the variable U content while ^{230}Th was practically constant, and thus caused Lalou *et al.* (1977) to question this method of dating such authigenic materials. Some typical results are shown in Table 16.2. Nevertheless, in sulphide and oxide deposits from East Pacific Rise, at 21 °N, ^{230}Th/^{234}U activity ratios are very low, and when interpreted in ages they are compatible with the age of the sea floor (Lalou and Brichet 1980*b*).

The ^{234}Th is a short-lived nuclide, but, as it is the direct daughter of ^{238}U, it is of importance especially in the coastal environment (see section 16.3.3). Moreover, as it is an alpha recoil nucleus from ^{238}U, it is responsible for much of the ^{234}U/^{238}U disequilibrium observed in marine sediments (Kigoshi 1971). The ^{231}Th and ^{227}Th have too short half-lives to be used in

Table 16.2 Examples of the ^{230}Th/^{234}U dating of Mn deposits

Sample	Depth interval (mm)	^{230}Th/^{234}U	'Age' (ky)	Reference
Shallow water nodules				
Loch Fyne	0.00–1.0	0.061 ± 0.004	6.5	Ku and Glasby
	2.00–4.0	0.087 ± 0.004	9.8	(1972)
Jervis Inlet	0.00–1.1	0.11 ± 0.01	12	
	1.1 –1.9	1.14 ± 0.01	16	
	3.9 –5.5	0.095 ± 0.006	11	
TAG-13–21	0.00–0.51	0.135 ± 0.010	15.7	Scott *et al.*
	0.51–0.91	0.160 ± 0.014	18.9	(1974)
	0.91–1.41	0.225 ± 0.023	27.8	
	1.41–3.01	0.233 ± 0.038	28.8	
	3.01–4.6	0.336 ± 0.021	44.5	
FAMOUS				
CYP 74 26–12	1.1– 1.5	0.88 ± 0.08	204	Lalou *et al.*
	1.5– 2.1	0.83 ± 0.09	178	(1977)
	2.1– 2.6	0.80 ± 0.09	162	
	2.6– 3.2	0.75 ± 0.07	142	
	3.2– 3.6	0.55 ± 0.06	84	
	3.6– 4.0	0.57 ± 0.09	90	
	4.0– 4.5	0.45 ± 0.05	64	
	4.5– 5.0	0.42 ± 0.05	58	
	5.0– 5.4	0.42 ± 0.09	58	
	5.4– 8.7	0.36 ± 0.02	47	
	8.7– 9.1	0.45 ± 0.05	64	
	9.1– 9.7	0.63 ± 0.06	105	
	9.7–10.5	0.75 ± 0.08	143	
	10.5–11	0.82 ± 0.09	172	

geochronology, so their behaviour in marine sediments has not been well studied. Nevertheless, Bernat and Goldberg (1969) have shown that there is a marked enrichment of ^{227}Th in phillipsites, relative to the detrital minerals which could result from a preferential uptake of ^{231}Pa or ^{227}Ac by the phillipsite.

16.2.3 Radium

Radium isotopes belong to three natural radioactive decay series: ^{226}Ra ($t_{\frac{1}{2}} = 1602$ y) in the ^{238}U series; ^{228}Ra ($t_{\frac{1}{2}} = 5.75$ y) and ^{224}Ra ($t_{\frac{1}{2}} = 3.64$ d) in the ^{232}Th series, and ^{223}Ra ($t_{\frac{1}{2}} = 11.1$ d) in the ^{235}U series. The ^{226}Ra was the first radioactive nuclide measured in deep sea sediments by Joly (1908b) and inferred by Pettersson (1937) to represent ^{230}Th. This was a simplified hypothesis because the ^{226}Ra in a core must initially increase towards equilibrium with ^{230}Th (with the half-life of ^{226}Ra) as Ra is not scavenged from sea water and thereafter decrease down the core with the ^{230}Th half-life. On a red clay core from the Central Pacific, Kröll (1953, 1954) has shown that the repartition of Ra presents a number of secondary maxima and minima, proving the migratory behaviour of Ra in marine sediments. Using auto-radiograph techniques, Arrhenius and Goldberg (1955) demonstrated that Ra and its daughter elements are bound to phillipsite. In marine barites from the Eastern Equatorial Pacific ^{226}Ra is enriched by at least an order of magnitude compared to U and ^{230}Th concentrations which indicates its uptake during precipitation (Borole and Somayajulu 1977). Due to the similarity of chemistry of Ba and Ra, this is not unexpected. If so, and if barite acts as a chemically closed system for Ra, it should be possible to use the decay of unsupported ^{226}Ra to date these deposits. Unfortunately, the Ra content is quite constant in the barites even at depth in most deep sea cores an observation which is best explained as indicating a large adsorption of pore water ^{226}Ra. The measurements of Borole and Somayajulu (1977) indicate that the rate of adsorption or exchange of ^{226}Ra by the barite is comparable to its decay rate of 5.78×10^{-3} y^{-1}. In Red Sea brines, Chung, Finkel, Kim, and Craig (1979) have shown with ^{226}Ra that each of the three hot brine layers has its own Ra content of 400, 630, and 1500 dpm/100 kg, respectively for increasing depth, but that each layer is well mixed as it is of homogeneous Ra content.

16.2.4 Radon

Due to the mobility of its progenitor ^{226}Ra and to its own mobility as a gas, ^{222}Rn is characteristically not in radioactive equilibrium, being either depleted or in excess in marine sediments. It is of more importance in the study of oceanic circulation (see chapter 15) than in sedimentology where its mobility needs to be taken into account only when measurements of U or ^{230}Th are made indirectly through ^{214}Bi, as in gamma spectrometry or through alpha tracks. Nevertheless, quite recently, it has been shown that ^{222}Rn is present in hydrothermal plumes, and that it may be used as a tracer for this injection (Klinkhammer *et al.* 1977).

16.2.5 Lead

Of the four radioactive Pb isotopes in the decay series of U and Th, only ^{210}Pb is of importance to marine sedimentology. The ^{211}Pb ($t_{\frac{1}{2}} = 36$ min) in ^{235}U series, ^{212}Pb ($t_{\frac{1}{2}} = 10.6$ h) in ^{232}Th series, and ^{214}Pb ($t_{\frac{1}{2}} = 26$ min) in ^{238}U series, have too short half-lives. By contrast, ^{210}Pb, due to its 21 y half-life, has proved to be an important nuclide for the study of sedimentary processes on the human time-scale. Moreover, because stable Pb is a potential pollutant of the atmosphere as well as estuaries and open ocean, a detailed study of its behaviour and concentration in nature through ^{210}Pb has been made, and is summarized in Robbins (1978).

Through very short half-life products (shorter than 1 h), ^{210}Pb is produced from ^{222}Rn, a gas which is at the origin of its special behaviour in the series because it has an atmospheric pathway. The ^{222}Rn continuously emanates from continental rocks containing U, enters the atmosphere and even the stratosphere. When ^{222}Rn decays into ^{210}Pb, this latter nuclide returns to the surface of the earth through rain or dust. In coastal and shallow water environments this results in a constant flux of unsupported ^{210}Pb to the surface waters, and because of the short residence time of Pb in this environment relative to its half-life, ^{210}Pb is found in excess in the coastal and shallow water sediments (Schell 1977). However, in the deep ocean sediments, this atmospheric ^{210}Pb is not important, and excess ^{210}Pb originates from the decay of ^{226}Ra in deep waters or from ^{226}Ra and ^{222}Rn migrating from the sediment towards deep water through pore waters.

In the coastal environment ^{210}Pb activity in surface sediments is often constant to a certain depth because of biological activity (see Table 16.3). Hurricanes are also considered as a possible modification agent for the ^{210}Pb repartition (for example, in Chesapeake Bay, Goldberg, Hodge, Koide, Griffin, Gamble, Bricker, Matisoff, Holdren, and Braun 1978); similarly Benninger, Aller, Cochran, and Turekian (1979) have observed in Long Island Sound that surface sediments from about 15 m depth have a ^{210}Pb excess of about 2.5 dpm/g for a sedimentation rate of about $50 \, \text{cm}/10^3$ y (as calculated from a model including mixing plus sedimentation). In San Clemente basin sediments, Koide *et al.* (1976) recorded a ^{210}Pb activity about 25 dpm/g for a deposition rate of about $6 \, \text{cm}/10^3$ y for a depth of 600 m. In two cores from the FAMOUS area, specific activity of the upper most sediment is about 37 dpm (Nozaki *et al.* 1977). In Santa Cruz basin, off California, excess ^{210}Pb activities of 40 dpm/g have been found in the uppermost part of a core sampled from 2036 m depth by Joshi and Ku (1979).

Some ^{210}Pb measurements have been carried out in other marine environments; for example, in marine barite from the central East Pacific, in a water

Table 16.3. ^{210}Pb excess in a core from
Chesapeake Bay showing a constant activity
until about 30 cm depth, due to bioturbation,
then an exponential decrease due to a sedimentation rate of 0.3 cm/y. (From Goldberg *et
al.* 1978.)

Depth (cm)	^{210}Pb (excess) (dpm/g)
0 – 0.5	7.78
0.5– 1.0	4.78
1.0– 1.5	5.18
1.5– 2.0	4.55
2.0– 2.5	4.55
2.5– 3.0	5.67
5.0– 5.5	5.09
6.5– 7.0	3.53
7.5– 8.0	3.41
8.5– 9.0	4.42
10.0–11.0	5.91
12.0–13.0	5.34
13.0–14.0	4.60
14.0–15.0	5.21
17.0–18.0	4.73
20.0–22.0	5.46
24.8–26.0	5.04
28.0–30.0	4.60
30.0–32.0	4.70
32.0–34.0	3.01
34.0–36.0	3.15
36.0–38.0	2.13
38.0–40.0	1.69
40.0–44.0	1.65
48.0–52.0	1.25
52.0–56.0	1.22
60.0–64.0	1.30
64.0–68.0	1.06
68.0–72.0	1.40

depth of 2870 m, ^{210}Pb specific activity is in the range of 20 to 66 dpm
and is depleted with respect to ^{226}Ra, which is attributed to a loss of ^{222}Rn
from the small barite crystals (Borole and Somayajulu 1977). At 21 °N on the
East Pacific Rise in the hydrothermal fluid, black pyrrhotite precipitates have
a specific activity of 750 dpm, while ^{210}Po is 1140 dpm, allowing MacDougall,
Finkel, and Chung (1979) to conclude that an effective and perhaps selective
scavenging of ^{210}Pb and ^{210}Po by the sulphides is taking place and to suggest
that the disequilibrium between ^{210}Pb and ^{210}Po might be used to date these
recent deposits.

16.3 Chronometers for sediment accumulation and mixing

From the preceeding discussion it has been shown that the content and distribution of the actinide elements in marine sediments are quite variable, and that their mobility seems to preclude their use for chronology as the closed system criterion is not generally met. However, ^{230}Th and ^{231}Pa turn out to be relatively stationary (Bernat and Goldberg 1969), and occur in the U- and Th-decay series before the more mobile nuclides so that these isotopes have been widely applied to dating Mn nodules and determining rates of deep sea sedimentation. Thus, it is especially important that direct measurement be made of the ^{230}Th and ^{231}Pa activities, because indirect techniques such as gamma spectroscopy which measures daughter activity (Osmond and Pollard 1967; Yokoyama et al. 1968; Yokoyama and Nguyen 1980), or total alpha activity such as alpha track counting (Fisher 1977a; Anderson and MacDougall 1977) can give incorrect results if substantial migration of daughter nuclides into or out of the sample has occurred.

16.3.1 Deep-sea sediment accumulation rates (^{230}Th, ^{231}Pa)

Of the radioactive elements which occur in deep-sea sediments, ^{230}Th and ^{231}Pa are in measurable excess. Formed continuously from their respective parents ^{238}U and ^{235}U dissolved in sea water, both of these nuclides are quantitatively removed (Moore and Sackett 1964), and accumulate on the sea floor as unsupported ^{230}Th and ^{231}Pa. Once in the sediment, this excess activity decreases according to the half-lives of the two nuclides, (i.e. 7.52×10^4 y for ^{230}Th, and 3.43×10^4 y for ^{231}Pa). The specific activity of a recently deposited sediment thus depends on both the nuclide flux which is roughly a function of water depth as well as the rate of sediment accumulation. In deep-sea sediments this activity is sufficiently large to be measured accurately over a range of four to five half-lives so that the ^{230}Th-excess method should have a dating range of about 3 to 4×10^5 y and the ^{231}Pa excess method a range of about 1.5×10^5 y.

The first attempt to apply the ^{230}Th-excess method to marine sediments was made by Piggot and Urry (1939, 1942) who measured ^{230}Th via its daughter nuclide ^{226}Ra and Issac and Picciotto (1953) who measured ^{230}Th directly. For Pa, the first application of the ^{231}Pa-excess method was by Sackett (1960). To apply these methods to sediments it is first necessary to determine the excess ^{230}Th or ^{231}Pa activity in a sample. As was noted in section 16.2, except for some special cases corresponding to particular environment, U is essentially linked to the detrital fraction of marine sediments in which the entire decay series should be in equilibrium. Thus, to calculate the excess of ^{230}Th (or ^{231}Pa), the activity equivalent to that required for equilibrium with the ^{238}U (or ^{235}U) present is subtracted from the total measured ^{230}Th (or ^{231}Pa) activity. To use these two methods as chronometers for sediment accumulation on the sea floor two assumptions must be made: (i) that the production of ^{230}Th and ^{231}Pa in the water column is constant, and (ii) that the sedimentation rate is constant over a particular interval of interest. These two conditions being fulfilled, the decrease of specific activity for either ^{230}Th or ^{231}Pa downwards in a sediment core should be an effective measure of the sedimentation rate, and therefore, the age of any horizon in a core is directly a function of the time elapsed since it was deposited.

As far as the first condition is concerned, the production rate of the two nuclides may be considered constant as it depends only on the U content of sea water, which has been constant over the past 400 ky (Broecker 1974). However, this does not imply a constant flux of either ^{230}Th or ^{231}Pa to the sediment because the ^{230}Th-excess (or ^{231}Pa-excess) content of a core often does not reflect the flux that may be calculated from the U content of the water column. Nevertheless, as a general rule, the ^{230}Th and ^{231}Pa flux does appear constant at any point in the deep ocean basins.

The second condition is far from being fulfilled. The first important variable in the amount of sediment settling per unit of time is the biogenic constituent. This fraction may vary, either because of climatic changes which induce major fluctuations of the planktonic production or because of a change in the depth of the lysocline which induces fluctuations in the rate of dissolution of the carbonaceous constituents of the sediments. As this component acts as a dilutent to the sediment, the ^{230}Th-excess (or ^{231}Pa-excess) has to be normalized to take this variable into account. This may be done roughly by using carbonate free values. Other methods have been proposed consisting of normalization to other elements or phases that may be considered as stable and as having the same geochemical behaviour as Th in the ocean, presuming them to be diluted or concentrated in the same way as the utilized nuclides. Baranov and Kuzmina (1958) used MnO and Fe_2O_3 for this normalization, considering that the colloidal particles of Fe and Mn hydroxides are the scavengers of ^{230}Th. A more widely used procedure proposed by Picciotto and Wilgain (1954) is the normalization to ^{232}Th. This nuclide, having a very long half-life may be considered as a stable nuclide during the time period covered by the method, and thus should have the same behaviour as ^{230}Th. Although this hypothesis has been shown to be incorrect, it does appear to work. The two Th isotopes do not have the same behaviour as they are not introduced into the ocean in the same manner. The ^{232}Th is introduced by the rivers transported on detrital particles to which it is strongly bound, and as a fraction soluble in river water which precipitates as soon as it enters the ocean, so that the ^{232}Th content of open sea water is less than 7×10^{-5} ppb (Kaufman 1969). Consequently, any ^{232}Th present in marine sediments is essentially detrital and not authigenic. By contrast, ^{230}Th is continually formed from U in the ocean water mass so that it is essentially authigenic, except for that small fraction in equilibrium with U in the detrital phase. Nevertheless, this normalization procedure apparently works for red clays which accumulate slowly in the deep ocean basins because the sedimentation rate, and thus ^{232}Th input, is constant and very low compared to ^{230}Th. For carbonate sediments the ^{232}Th normalization corresponds to a normalization to the clay fraction so that results are equivalent to calculations made on a $CaCO_3$ free basis (Ku 1976). Additionally, the normalization to ^{232}Th has an advantage relative to the $CaCO_3$ free method in that it can take into account other dilutants which are difficult to quantify such as biogenic silica as this phase is not enriched in ^{232}Th.

A second way in which chronological information can be obtained for deep-sea sediments is by means of the ^{230}Th/^{231}Pa activity ratio. As ^{230}Th and ^{231}Pa are both formed from decay of U and these nuclides can be considered to have the same behaviour, their activity ratio is not dependent on either variations in the U content of sea water or in sedimentation rate, and varies with depth in a core according to the time the sediment was deposited.

As ^{231}Pa has a half-life of 3.43×10^4 y and the ^{230}Th a half-life of 7.52×10^4 y, the ^{230}Th/^{231}Pa activity ratio increases with an apparent half-life of about 6.2×10^4 y. Although first applications of the method were encouraging (Sackett 1960; Rosholt et al. 1961), later results suggested that ^{230}Th and ^{231}Pa probably did not have exactly the same behaviour as it is frequent to find core top sediments which have a ^{230}Th/^{231}Pa activity ratio different from the calculated production ratio of 10.7 (Sackett 1964; Ku 1966). Nevertheless, the ^{230}Th/^{231}Pa ratio may be used to measure sedimentation rates if one considers that, even if the initial ratio is not the theoretical one, it has not changed at a particular site during the 1.5×10^5 y period covered by the method.

Essentially all the measurements of sedimentation rate reported in the last 15 years are based on one or more of the three dating methods: ^{230}Th-excess, ^{230}Th-excess/^{232}Th, and ^{230}Th/^{231}Pa. Other fields of application of these three methods are paleoclimatology and the study of the origin of elements in the deep-sea cores. A few examples follow.

Rosholt et al. (1961) were the first to report successful correlation of climatic events obtained by dating of different deep-sea cores, thus establishing a radiometric chronology of the last climatic cycle going back 1.5×10^5 y. Subsequently, Rona and Emiliani (1969) dated two caribbean cores using new half-life values for ^{230}Th and ^{231}Pa (7.52×10^4 y and 3.43×10^4 y respectively), and so established the best continuous climatological record available to date. However, Broecker and Ku (1969) analysed the same cores and demonstrated the sensitivity of the method to the value of the supported ^{230}Th activity and to the accuracy of the U content measurement. These authors proposed a longer time scale which agreed more closely with the Late Pleistocene paleosea-level records, and rejected the ^{230}Th/^{231}Pa dating method as an absolute dating technique on the grounds that the expected theoretical value of the ratio was not found in the samples derived from the core tops. In reply to this criticism of their approach, Emiliani and Rona (1969) used other dates for high sea level stands, demonstrating that their proposed time-scale was still acceptable. Finally, the above arguments indicate that when these dating methods are applied to deep-sea cores, only 'relative' time-scales should be expected. Nevertheless, the above work and the subsequent effort in the field have resulted in an improvement in the chronology of the late Quarternary climatic changes.

Another possible application of sedimentation rate measurements in the deep-sea environment is that of obtaining elemental budgets in cores in order to delineate the origin of different elements. Thus, Scott and Salter (1977) were able to demonstrate that in the cores from area FAMOUS Mn, Ni, Co, and Zn in one part, and Fe and Cu in another part, derive from a hydrothermal source, whereas Al, Cr, Th, and Zn derive from the detrital input. Similarly, Bender, Ku, and Broecker (1966) have shown that the deposition of Mn is nearly constant in the world oceans, by measuring Mn content in six cores previously dated by the ^{230}Th and ^{231}Pa methods. Due to a very peculiar environment in the Red Sea, another possible method for measuring sedimentation rates has been used by Ku (1969); in such deposits, ^{230}Th and ^{231}Pa are depleted with respect to U, which allows age calculation by a method similar to that used for coral dating (see chapter 17). From these ages, sedimentation rates may be deduced.

16.3.2 Manganese nodule growth rate

Manganese nodules are an authigenic formation in the deep sea, occasionally found in shallower waters (Ku and Glasby 1972). Because of their economic potential, they are one of the most extensively studied deep-sea formations. Three recent books summarize most aspects of Mn nodule research to date (Glasby 1977a; Bischoff and Piper 1979; Lalou 1979), while some papers are directly concerned with growth rates: Ku (1977); Kvenvolden and Blunt (1979); MacDougall (1979); Ku et al. (1979); Lalou, Ku, Brichet, Poupeau, and Romari (1979) Lalou, Brichet, and Jehanno (1979); Sharma and Somayajulu (1979); Ku and Knauss (1979); Schvoerer, Dantant, and Bechtel (1979); Krishnaswami et al. (1979). The growth rates of Mn nodules have been measured mainly by the ^{230}Th-excess (either normalized to ^{232}Th activity or not) and ^{231}Pa-excess methods.

Initially, as in the case of deep-sea sediments (Joly 1908b), the ^{226}Ra content of nodules was found to decrease with depth. Thus, Mn nodule growth rates of the order of 10 mm/10^3 y were measured from the unsupported ^{226}Ra (Pettersson 1943, 1955; Von Buttlar and Houtermans 1950). However, Goldberg and Arrhenius (1958) demonstrated that, as in the case of deep-sea sediments, the ^{226}Ra-excess was supported by ^{230}Th yielding an order of magnitude lower accumulation rates. Ku and Broecker (1967b, 1969) measured the growth rates of more than 20 nodules from various sites around the world using alpha spectrometry technique and reported the Mn nodule growth rate ranged from 1 to 6 mm/10^6 y. Most significantly, they found an excellent agreement between the growth rates obtained from four different U-series disequilibrium methods (^{230}Th-excess, ^{231}Pa-excess, ^{230}Th/^{232}Th, and ^{234}U/^{238}U). These results are illustrated in Fig. 16.1.

Several technical and scientific problems have been identified in this work to date. One of the technical problems derives from the fact that the nodule growth rate of the order of 1 mm/10^6 y requires a decrease of a factor of two in the ^{230}Th activity in less than one tenth of a millimetre, a thickness that cannot be resolved by the available sampling techniques. (In contrast, the deep-sea sediment accumulation rates of the order of 1 mm/10^3 y represent a factor of two decrease in the ^{230}Th activity in 10 cm making the sampling simpler.) To overcome this problem, a mean density of the material is used to evaluate the thickness of the sampled layer. Ku (1977) reports differences of about 20 per cent in the growth rates depending on the density values used in the calculation. Another consequence of the very slow growth rate of Mn nodules is that all the actinide excess is located in the outermost millimetre requiring

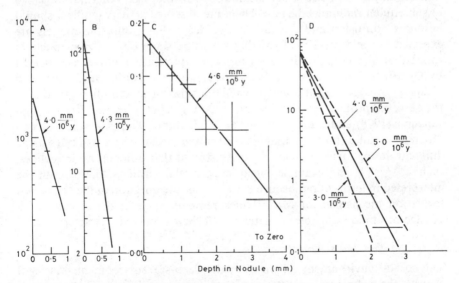

Fig. 16.1. Depth distribution of radioactivity and the accumulation rates derived from them in Mn nodule (V-21-D2) from the Pacific. (Adapted from data of Ku and Broecker 1967.) The vertical and horizontal lines denote probable error limits. Units for the y-axis: A Excess ^{230}Th (dmp/g) = dpm ^{230}Th/g − dpm ^{234}U/g; B Excess ^{231}Pa (dpm/g) = dpm ^{231}Pa/g − dpm ^{235}U/g; C Excess ^{234}U/^{238}U = (dpm ^{234}U − dpm ^{238}U)/dpm ^{238}U; D Excess ^{230}Th/^{232}Th = (dpm ^{230}Th − dpm ^{234}U)/dpm ^{232}Th. The two dashed lines are drawn to show the sensitivity in the rate estimation. (After Ku 1977, p. 253.)

high resolution sampling. This realization identified a further possible problem with the method. Since the nodule surface is very active, it could contaminate inner layers during sampling, yielding an apparent decrease in activity at the surface due to dilution and not radioactive decay. However, Heye (1975) showed, by alpha track studies on polished sections of Mn nodules, an exponential decrease of alpha emitting nuclides with depth ruling out the possible sampling artefact.

The slow growth rates of Mn nodules are well documented and accepted, raising an interesting question of how such slowly growing objects remain unburied by sediments accumulating thousands of times faster. In an effort to answer this question, different hypotheses have been put forward such as bioturbation (Paul 1976), bioturbation energy and ocean bottom current energy (Glasby 1977b), sedimentation hiatus or condensed sequences (Watkins and Kennett 1977), and special boundary layer properties allowing nodules to float (Sorem, Fewkes, McFarland, and Reinhard 1979). However, Bonatti and Nayudu (1965), Arrhenius (1967), Cherdyntsev, Kadyrov, and Novichkova (1971), and Lalou and Brichet (1972) considered the closed system assumption invalid in the case of Mn nodules, and suggested that these nodules may be formed during volcanic or hydrothermal events and subjected to subsequent radionuclidic rain. Then, the observed activity gradient should be due to diffusion and not radioactive decay. Such diffusion models are expected to yield profiles of roughly the same shape for different nuclides rather than shapes approximately proportional to their half-lives as observed by Ku (1977). Consequently, Ku et $al.$ (1979) proposed a diffusion-decay model yielding lower growth rates, and rejected the diffusion model which failed to fit the observed profiles. At the same time Lalou, Brichet, Poupeau, Romari, and Jehanno (1979) and Lalou, Brichet, and Jehanno (1979) in a study of two different nodules (i) found disagreement in the measured growth rates using different dating methods, and (ii) demonstrated that different radionuclides, such as [232]Th for example, yield exponential profiles that cannot be interpreted in terms of radioactive decay. Thus, the question of the origin and formation mechanism of Mn nodules remains open in the absence of a satisfactory model capable of matching all their observed characteristics.

16.3.3 Shallow sea sediments

The coastal environment differs from the deep-sea environment in several aspects: the sedimentation rates are much more rapid in the shallow water regime which is constantly subject to change with time. This tendency to change renders the chronology with longer-lived radionuclides, such as [230]Th and [231]Pa, inapplicable to the shallow water regime. The high sedimentation rates coupled with the fluxes of these nuclides yield low specific activities. In addition, shallow waters are subject to very active marine life which includes burrowing organisms responsible for continuous mixing of the settling particles. Whereas bioturbation is a minor effect in the deep-sea regime, affecting only the top few centimetres of pelagic sediment and yielding constant age (or activity of ^{230}Th and ^{231}Pa) in the core tops, the phenomenon is of some importance in the coastal environment, especially on the 'human time-scale', as it may yield essential information on the fate of pollutants. For this purpose, methods utilizing short-lived radionuclides such as ^{234}Th ($t_{\frac{1}{2}} = 24.1$ d), ^{228}Th ($t_{\frac{1}{2}} = 1.91$ y), ^{226}Ra ($t_{\frac{1}{2}} = 1602$ y), and ^{210}Pb ($t_{\frac{1}{2}} = 22.3$ y) have been developed.

(i) *Thorium*-234. The ^{234}Th is formed by alpha decay of ^{238}U at the theoretical rate of 3 atoms/min/l (Bhat *et al.* 1969). Thus, it is formed in sea water, and, like all other Th isotopes, is adsorbed rapidly on particulate matter after it is formed. In the shallow water environment it is scavenged rapidly, and inspite of its short half-life, ^{234}Th is found in excess in sediment.

This nuclide has been used to study particle reworking and diagenesis in shallow water sediments from Long Island Sound and New York Bight at about 20 m depth (Aller and Cochran 1976; Cochran and Aller 1979; Aller *et al.* 1980). The ^{234}Th repartition was studied in cores in which the bioturbation and physical perturbation due to climatic conditions are clearly documented by X-ray photographs. These authors have shown that ^{234}Th-excess is bound to the finest particles which may be resuspended by storms and currents, leading to a deficit of ^{234}Th total activity in sandy sediments and excess in muddy areas. This clearly shows that the distribution of reactive pollutants in such a shallow water environment is controlled by the muddy regions of the basins. Furthermore, whereas ^{234}Th excess is expected only in the first two millimetres of the sediment, as the sedimentation rate is 4 mm/y, it is actually found down to at least 12 cm. This is attributed to the activity of deposit feeders (bioturbation). The depth is subject to seasonal variations, and depends on the biological species present. Thus, these authors have calculated particle reworking rates for the depth of 0 to 5 cm in the sediment cores ranging from 1×10^{-8} cm^2/s to 1.6×10^{-6} cm^2/s, and the maximum Th residence time in the coastal environment of 1.4 d. In comparison, using ^{7}Be nuclide, Krishnaswami *et al.* (1980) report a mixing particle coefficient of the order of 10^{-7} cm^2/s for the same core from Long Island Sound.

Minagawa and Tsunogai (1980) have measured ^{234}Th on suspended and settling matter in a 60 m deep basin in Funka Bay, and conclude that the settling rate of suspended particles is related closely to the removal rate of heavy metals. Their calculated mean residence time for Th of 57 d yields a settling velocity an order of magnitude lower than the value reported by Aller and Cochran (1976) for Long Island Sound. Minagawa and Tsunogai attribute this difference in settling rates to a lower suspended particle concentration in Funka Bay.

Similar results have been obtained by Santschi *et al.* (1979) for Narragansett Bay. Subsequently, laboratory tank experiments were carried out demonstrating the analogy between Th and reactive pollutants such as Am, Pb, Po, Hg, and Cr^{3+} (Santschi, Adler, Amdurer, Li, and Bell 1980). This work has demonstrated the importance of ^{234}Th as a natural pollutant tracer in the coastal environment.

(ii) *Thorium*-228. The ^{228}Th $(t_{\frac{1}{2}} = 1.91$ y) has been used less in marine sedimentology than ^{234}Th, mainly because its geochemical behaviour and origin are less well understood. Moore and Sackett (1964) have shown that $^{228}Th/^{232}Th$ activity ratio in sea water is approximately 15, and in deep-sea sediments is often less than unity, except in some authigenic components as phillipsite and barite, while in coastal sediments ^{228}Th is in excess relative to ^{232}Th and ^{228}Ra (Koide *et al.* 1973). This is due to the migration capacity of ^{228}Ra which escapes through pore waters from the sediment towards sea water where it gives rise to ^{228}Th. Because of its short half-life this ^{228}Th component cannot return to the sediment in the open ocean, whereas in shallow coastal waters a constant 'rain' of ^{228}Th-excess is received by the sediment. This prompted Koide *et al.* (1973) to attempt to measure the sedimentation rates from the decreases of ^{228}Th-excess in HCl leachates from Santa Barbara basin and from Baja, California where annual varved sediments, previously dated with ^{210}Pb, yield the exact date of deposition. As this ^{228}Th-excess may derive from three sources: (i) the unsupported ^{228}Th coming directly from ^{228}Ra in sea water; (ii) the ^{228}Th coming from ^{232}Th which has been precipitated from sea water out of equilibrium with ^{228}Ra and ^{228}Th; and (iii) the ^{228}Th coming from ^{228}Ra arriving directly from sea water probably associated with the planktonic component, Koide *et al.* (1973) proposed a model to take into account all three components. Their conclusion was that this method could be used in relatively anoxic waters where biological activity is not sufficient to disturb the strata, and where, because of this special environment, ^{210}Pb distribution is altered. These authors also point out that the presence of ^{228}Th-excess in the top of a core may indicate that the material accumulated during the last decade has been sampled.

(iii) *Lead*-210. As seen in section 16.2.5, ^{210}Pb is delivered, essentially through an atmospheric pathway, to the surface waters of the ocean at a rate of 1 ± 0.2 dpm/cm^2/y, the river input or other potential sources being negligible (Benninger 1978). Much more work has been dedicated to this Pb isotope than to the other short lived nuclides, especially in the coastal environment, because it may be used as ^{234}Th to study the pathways of heavy metal pollutants and, because of its longer half-life, it may also be used to establish longer chronological sequences.

Being highly reactive, ^{210}Pb is rapidly scavenged from superficial coastal waters because of the high concentration of particulates in this environment. The first attempt to apply ^{210}Pb chronology to coastal sediments is due to Koide *et al.* (1973) in their study of varved sediment core from Santa Barbara basin which they used further for ^{228}Th dating purposes (see Fig. 16.2). In this

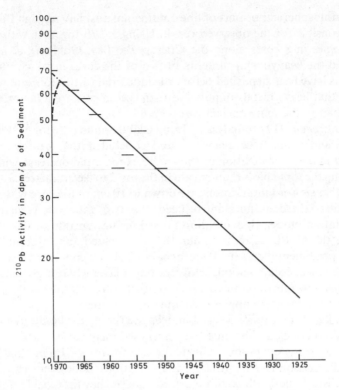

Fig. 16.2. The ^{210}Pb activity in Santa Barbara Basin core as a function of time of deposition of varve. The solid curve is drawn with the ^{210}Pb half-life. The average sedimentation rate is 0.4 cm/y. (After Koide *et al.* 1973, p. 1173.)

special environment, where no important biological activity disturbs the sediment sequence after deposition, the method works well, yielding a sedimentation rate of 0.39 cm/y, in agreement with the rate derived from counting varves. However, as a general rule, bioturbation occurs in oxygenated sediments and thus competes with the radioactive decay in the formation of the ^{210}Pb depth distribution. Nevertheless, correct sedimentation rates may be obtained from the ^{210}Pb-depth distribution provided the sedimentation rate is high, assuming bioturbation is effective only in the first few centimetres of the core, and if only core sections below this zone are analysed (Turekian and Cochran 1978).

Using ^{210}Pb, ^{228}Th, and $^{239-240}$Pu repartition in cores from Santa Barbara and Soledad basins, Koide, Bruland, and Goldberg (1975) were able to show that the $^{239-240}$Pu concentration has continued to increase from the original 1950/1954 period to 1972. This is not the repartition expected from the atmospheric fallout which should have the maximum in 1963. They concluded that an atmospheric transport of fine continental dust having high Pu content was responsible for the observed result. Using ^{210}Pb together with $^{239-240}$Pu chronologies in 8 cores along the Chesapeake Bay, Goldberg et al. (1978) compared the heavy metal contents in two of the cores in which the deeper sediments have been deposited before the industrial revolution, and were able to show that heavy metals in post industrial sediments are a factor of 2 to 3 higher than in the more ancient ones.

Benninger et al. (1979) analysed a Long Island Sound core for ^{210}Pb, ^{226}Ra, $^{239-240}$Pu, and some trace elements, and concluded that caution must be exercised in the interpretation of the ^{210}Pb exponential decrease with depth. Measuring the rapid bioturbation rate in the top 2 to 3 centimetres with ^{234}Th and another slower bioturbation rate down to 10 cm of the core with $^{239-240}$Pu distribution, these authors introduced the two rates in a mixing plus sedimentation model, thus obtaining a sedimentation rate of 0.05 cm/y. In contrast, the ^{210}Pb decrease under the bioturbated top layer indicated a 0.11 cm/y sedimentation rate. Benninger et al. (1979) concluded that in areas of slow sediment accumulation, below the upper layer where high bioturbation is evident, a low rate of bioturbation, essentially through the burrows, may be active and increases the apparent sedimentation rate.

Generally, ^{210}Pb cannot be used in deep-sea sediments because of the slow sedimentation rates, and because the present sediment surface is not preserved well during coring operations. This is the reason why ^{210}Pb method was not discussed in section 16.3.1, but, as some applications to the deep-sea cores have been published, these will be summarized here, as they may yield information about deep sea bioturbation rates and about the geochemistry of Pb in sea water.

Two cores taken by the submersible Alvin at 2500 m depth, with good preservation of the sediment–water interface, were sampled during the FAMOUS survey of the mid-Atlantic Ridge. Dating by ^{14}C and ^{210}Pb measurements was carried out in one of the cores by Nozaki, Cochran, Turekian, and Keller (1977). The ^{14}C dates are constant in the top 8 cm and increase at greater depth, yielding a sedimentation rate of 2.9 cm/10^3 y. The measured ^{210}Pb excess was attributed partly to atmospherically-derived ^{210}Pb from surface waters and partly to the in situ production from ^{226}Ra in the deep waters. The ^{210}Pb penetration down to 9 cm depth (too deep if the sedimentation rate of 2.9 cm/10^3 y is accepted), is the proof of an active bioturbation. These results lead the authors to calculate a mixing coefficient of 0.6×10^{-8} cm^2/s (see Fig. 16.3). Unfortunately, because of probable physical disruption, no conclusion could be drawn from the second core. This example illustrates well the difficulties in obtaining really undisturbed samples to which physical laws may be applied directly.

In deeper sediments from Atlantic, Pacific, and Antarctic Ocean, presenting sedimentation rates that vary by more than one order of magnitude, De Master

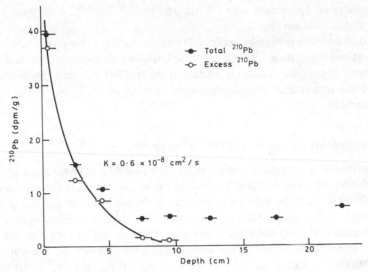

Fig. 16.3. The ^{210}Pb distribution in core 527-3 from FAMOUS area. Excess ^{210}Pb was calculated by subtracting ^{226}Ra activity from total ^{210}Pb activity. (After Nozaki *et al.* 1977, p. 170.)

and Cochran (1977) have shown that exponential decreases of ^{210}Pb are not representative of the known sedimentation rates, and have calculated mixing coefficients ranging from 1.2 to 8×10^{-9} cm^2/s. On the other hand, in a core from 2036 m depth in Santa Cruz basin off the Coast of California, ^{210}Pb exponential decrease with depth was used by Joshi and Ku (1979) to calculate a sediment flux of 0.06 g/cm^2/y, as X-ray radiography has shown that in this case the top layer had not been disturbed.

(iv) *Radium*-226. Of the four actinides used for geochronological purposes in the shallow water sedimentation studies, ^{226}Ra ($t_{\frac{1}{2}} = 1602$ y) has the longest half-life. Unfortunately, its high mobility in marine environment precludes its application to sediment geochronology. Koide *et al.* (1976) made an attempt to date sediments with unsupported ^{226}Ra in San Clemente basin off Southern California coastal region, using the fact that ^{226}Ra is concentrated by some phytoplanktonic species, specially diatoms (Shannon and Cherry 1971), and then is actively transported to the sediment in regions of high productivity. The exponential decrease of ^{226}Ra-excess activity in the first 30 cm of the core yields a sedimentation rate of 5.2 to 5.3 cm/10^3 y, subject to the hypothesis on the origin of a part of this ^{226}Ra. The ^{210}Pb chronology in the same core, gives a sedimentation rate of 100 cm/10^3 y. The authors explain this discrepancy by bioturbation which would have a larger effect on the shorter half-life nuclide than it would on the longer half-life one. Sedimentation rates, deduced from Pb, seem to be affected to a larger extent by bioturbation, while ^{14}C are in relatively good agreement with ^{226}Ra, and the ^{230}Th/^{232}Th method gives a lower sedimentation rate.

A conclusion that may be reached from the data on the geochronology of deep-sea sedimentation, as well as those on shallow water sedimentation, is that, even if progress has been made during the last 20 years, more work is needed to understand completely the behaviour of actinides in the marine environment.

16.4 Sediments as source and sink of actinides

Figure 16.4 is a schematic designed to illustrate the behaviour of different actinides at the water–sediment interface. As already seen, sediments can play the role of U isotope sink to two sources: (i) input from sea water under reducing conditions, and (ii) input from volcanic or hydrothermal events. Veeh (1967) estimated the percentage of the total world ocean surface required to be subject to reducing conditions to remove the entire stream input component. He obtained a figure of 0.4 per cent using $10 \ \mu g/cm^2/10^3$ y for the U deposition rate in shallow water anaerobic sediments, and an average stream input rate of $0.4 \ \mu g/cm^2/10^3$ y. However, as already seen in section 16.2.1, in addition to the shallow water anaerobic sediments, the metalliferous sediments and brines are also active sea water U sinks. Thus, for example, Ku (1969) estimated an U accumulation rate of $450 \ \mu g/cm^2/10^3$ y for the Red Sea brines. Other sediment components, even when they are U-enriched, like fish teeth for example, cannot be considered as the main sink, as their mass represents a minor fraction of the total sediment material. However, phosphorites are another possibly significant sink (see chapter 17).

In contrast, sediments cannot be regarded as a significant source of U with the exception of ^{234}U, formed by decay of ^{238}U in the sediments, which diffuses into the deep ocean waters via the sediment pore waters. It appears, however, that the diffusion rate of $0.3 \ dpm/cm^2/10^3$ y is insufficient to change the isotopic composition of deep waters (Ku *et al.* 1977). Another U component issues from the sea floor and is volcanogenic. Although this U fraction is not strictly sediment-derived, it is difficult to neglect it especially after the recent findings of active submarine hydrothermal vents. No estimate of the significance of this U source exists to date and no Th has been found near the vents so far.

Sediments are essentially a sink for Th and Pa nuclides. As soon as these nuclides enter the marine environment, or are formed by decay of their parents, they are efficiently scavenged. The ^{232}Th deriving from the rivers is scavenged immediately on the continental shelf, and ^{234}Th, ^{230}Th, and ^{231}Pa are continuously formed and scavenged everywhere in the oceanic water mass. Similarly, ^{210}Pb, either entering the ocean from the atmosphere or as a decay product of ^{226}Ra already in the water mass, settles rapidly at the

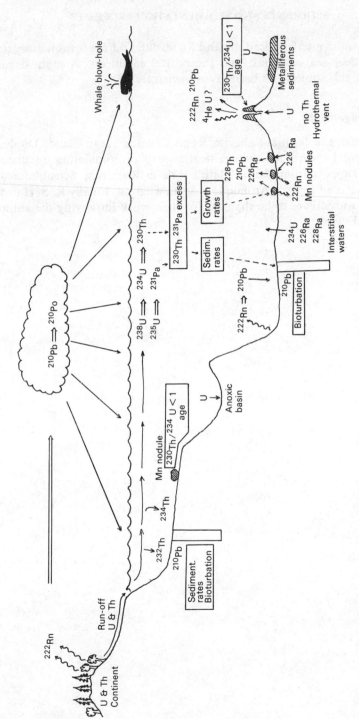

Fig. 16.4. Sketch of the behaviour of different actinides at the water–sediment interface.

sediment–water interface. Radium and Rn nuclides diffuse continuously from both the deep-sea sediments and the coastal sediments. A more detailed account of this subject has been given in chapter 15.

Acknowledgements

Thanks are due to Jacques Labeyrie, Roger Chesselet, Jean Claude Duplessy, and Laurent Labeyrie for their critical reviews or stimulating discussions. Financial support from Centre National de la Recherche Scientifique and Commissariat à l'Energie Atomique is acknowledged. Finally R. S. Harmon and M. Ivanovich are to be thanked for their work improving the author's uncertain English.

17 CARBONATE AND PHOSPHATE SEDIMENTS
H. H. Veeh and W. C. Burnett

17.1 Introduction

The applications of U-series disequilibrium methods to the age determination of marine carbonates and phosphorites are conveniently discussed together under the same chapter heading, because of similar basic models and assumptions. In both cases methods are based on the accumulation of decay products from an initially pure, or almost pure U-parent fraction. Moreover, the formation of carbonates and phosphorites in the marine environment permits the use of the known excess of ^{234}U in sea water as a convenient check on the closed system assumption underlying basic chronologic models. Finally, several of the implications of U-series ages for marine carbonates and phosphorites converge in the field of paleoclimatology and paleoceanography, with frequent cross references, and a common basis for discussion.

17.2 Uranium-series dating of marine carbonates

The applicability of the ^{230}Th growth method to marine carbonates was first successfully demonstrated by Barnes, Lang, and Potratz (1956), and Sackett and Potratz (1963) who discovered that reef building corals contain several ppm U, but are essentially free of ^{230}Th at the time of formation. Using samples from a drill core on Eniwetok, these authors found that stratigraphically older corals displayed a systematic increase in ^{230}Th towards radioactive equilibrium with ^{238}U. The subsequent discovery of a 14 to 15 per cent excess of ^{234}U activity over its parent ^{238}U in the modern ocean (Thurber 1962; Koide and Goldberg 1965) provided a convenient means of evaluating the reliability of ^{230}Th ages of marine carbonates. Both the decay of excess ^{234}U and the growth of ^{230}Th can be expressed as a function of age (Kaufman and Broecker 1965a) subject to the conditions that: (i) only U, but no Th enters the carbonate at the time of its formation; (ii) the carbonate remains a closed system with respect to U and its decay products; and (iii) the initial $^{234}U/^{238}U$ ratio can be readily assessed.

In general, marine carbonates fulfill conditions (i) and (iii). However, condition (ii) is frequently violated, and various criteria have been proposed to evaluate the suitability of a given carbonate sample for reliable age determinations (Tatsumoto and Goldberg 1959; Kaufman *et al.* 1971; Bloom, Broecker, Chappell, Mathews, and Mesolella 1974; Ku 1976; Veeh and Green 1977).

The most important criteria are:

1. There should be no evidence of recrystallization and/or deposition of void-filling cement. This can usually be ascertained by X-ray diffraction analysis and microscopic examination of the sample.
2. The U content in the sample should not differ significantly from that in a contemporary equivalent of the carbonate in question.
3. The $^{234}U/^{238}U$ ratio should be consistent with the $^{230}Th/^{234}U$ ratio of the sample, i.e. show a systematic decrease with sample age.
4. The sample should be free of ^{232}Th, since ^{232}Th would indicate the addition of common ^{230}Th (i.e. ^{230}Th not originating within the sample).
5. The $^{226}Ra/^{230}Th$ and $^{231}Pa/^{235}U$ ratios should be consistent with the $^{230}Th/^{234}U$ ages of the sample.
6. The $^{230}Th/^{234}U$ ages should be consistent with the stratigraphic position of the samples.

A sufficient number of U-series isotopic age determinations of marine carbonates, especially of corals and molluscs have been made during the last 25 years to permit a critical review of the reliability of the method's application to these materials.

17.2.1 Corals

Within the constraints listed above, it appears that reliable ages can be obtained for carefully selected coral samples. This is well illustrated in Figs 17.1 to 17.4. In Fig. 17.1, the U contents of recent (living) corals and unrecrystallized fossil corals are compared. In view of the observation that the U contents of living corals show small but significant variations at the species level (Veeh and Turekian 1968), the slight difference between the mean U contents of recent and fossil corals suggested by the data in Fig. 17.1 could be the result of a sampling bias towards the more massive coral species generally used for age determinations, rather than indicating post-depositional addition of U.

Figure 17.2 is a plot of activity ratios of $^{230}Th/^{234}U$ against $^{234}U/^{238}U$ for over 100 unrecrystallized corals of all ages, together with the expected development in time of these isotope ratios in a closed system starting with an initial $^{230}Th/^{234}U$ ratio of zero and an initial $^{234}U/^{238}U$ ratio of 1.14, the presently accepted activity ratio of U isotopes in sea water (Koide and Goldberg 1965; Ku *et al.* 1977). With a very few exceptions, the coral data do

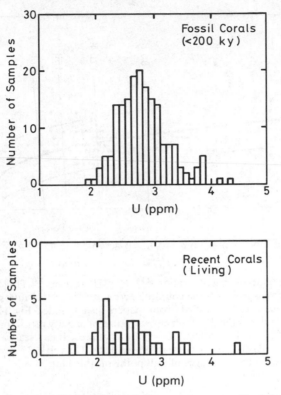

Fig. 17.1. Histograms of U contents in living and unrecrystallised fossil corals with $^{230}Th/^{234}U$ ages less than 200 ky, based on data from Broecker and Thurber (1965); Thurber *et al.* (1965); Osmond *et al.* (1965); Veeh (1966); Sakanoue *et al.* (1967); Mesolella *et al.* (1969); Veeh and Turekian (1968); Veeh and Chappell (1970); Bloom *et al.* (1974); Ku, Kimmel, Easton, and O'Neil (1974); Neef and Veeh (1977); Chappell and Veeh (1978a); Marshall and Launay (1968); Szabo (1979a, c); Veeh *et al.* (1979); Veeh (unpublished).

not differ from this model by more than 2σ, indicating that a closed system has generally been maintained. The data also implicitly confirm the assumption that $^{234}U/^{238}U$ in sea water has not varied significantly over the last 2×10^5 y, a conclusion which is consistent with the long oceanic residence time (4×10^5 y) of U (Turekian and Chan 1971).

A sensitive test of the closed system model is also provided by the determination of both $^{230}Th/^{234}U$ and $^{231}Pa/^{235}U$ in the same sample (Ku 1968). The locus of all concordant ^{230}Th and ^{231}Pa ages, a type of 'concordia curve' in analogy to the terminology of $^{206}Pb/^{238}U - ^{207}Pb/^{235}U$ dating (Wetherill 1956), is displayed in Fig. 17.3, together with actual measurements of $^{230}Th/^{234}U$ and $^{231}Pa/^{235}U$ ratios in unrecrystallized corals of different

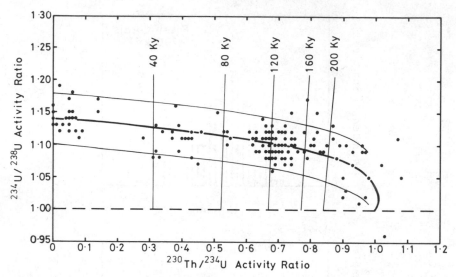

Fig. 17.2. The plot of activity ratios $^{234}U/^{238}U$ against $^{230}Th/^{234}U$ in all un-recrystallized corals in which the counting error of $^{234}U/^{238}U$ ratios is equal to, or better than ±0.02 (1σ), selected from papers listed under Fig. 17.1. The heavy horizontal curve shows the development with time of activity ratios for corals starting with $^{230}Th/^{234}U = 0$ and $^{234}U/^{238}U = 1.14$ as in present day sea water, the parallel curves on either side allowing for deviations by $\pm2\sigma$. The vertical curves are isochrons which show the $^{230}Th/^{234}U$ ages of any point on this plot.

ages. As in the case of U/Pb dating, a violation of the closed system assumption would be revealed by data points falling outside of the concordia curve. While the $^{231}Pa/^{235}U$ and $^{230}Th/^{234}U$ data in the majority of samples conform to the closed system model within the limits of experimental error, several samples show significant deviations from the concordia curve, possibly because of initial ^{231}Pa in the corals (Sakanoue *et al.* 1967), or post depositional injection of ^{231}Pa and ^{230}Th into the corals from the surrounding sedimentary matrix (see below). For these reasons, coral samples with discordant ^{231}Pa and ^{230}Th ages should be treated with caution, when a simple closed system model is used.

It is instructive to compare U-series ages of corals with ages obtained for the same material by an entirely independent method, such as ^{14}C dating (Fig. 17.4). Over the interval where both dating methods are least subject to errors caused by secondary alteration and isotope exchange processes, the age agreement is fairly good. However, there is a tendency for the data to be displaced to the same side of the line representing age concordance. This displacement increases with increasing age, so that above 10 ky, ^{14}C ages are significantly younger than ^{230}Th ages. This age difference could be caused by any of the following: (i) secular variations in atmospheric ^{14}C; (ii) contami-

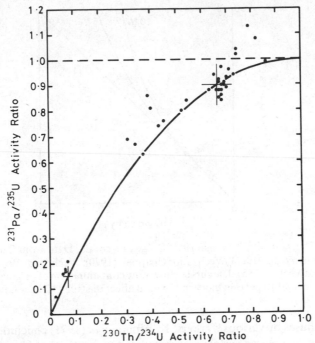

Fig. 17.3. Sympathetic variation of $^{231}Pa/^{235}U$ and $^{230}Th/^{234}U$ with time (concordia curve), using half-lives of 3.43×10^4 y and 7.52×10^4 y for ^{231}Pa and ^{230}Th, respectively, and showing path of corals starting with $^{231}Pa/^{235}U = 0$, $^{230}Th/^{234}U = 0$, and initial $^{234}U/^{238}U = 1.14$. Based on data from Sakanoue *et al.* (1967); Ku, Kimmel, Easton, and O'Neil (1974); Szabo (1979a, c). Error bars indicate 2σ of typical counting error of $^{231}Pa/^{235}U$ and $^{230}Th/^{234}U$ ratios near end points of concordia curve.

nation with modern ^{14}C; (iii) secondary leaching of U; (iv) secondary addition of ^{230}Th unaccompanied by ^{232}Th.

At present there are insufficient data to choose between these alternatives with any confidence. However, in view of the known susceptibility of 'old' samples to contamination by modern ^{14}C (Broecker and Kulp 1956), and the now well established deviation of ^{14}C ages from the absolute chronology based on tree rings (Suess 1970), there is little justification to place any more validity on coral ^{14}C ages than on U-series ages.

In summary, provided that the criteria listed above are strictly observed, it should be possible to obtain meaningful ^{230}Th ages for corals on the basis of a simple closed system model.

17.2.2 Molluscs

In contrast to corals, U-series dating of molluscs has not provided encouraging results. In a review of about 400 U-series isotopic measurements carried

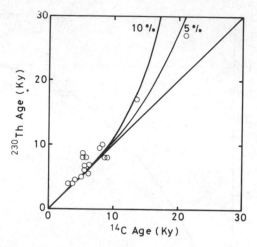

Fig. 17.4. The plot of ^{230}Th against ^{14}C ages of corals. Data from Thurber *et al.* (1965); Lalou *et al.* (1966); Veeh and Chappel (1970); Veeh and Veevers (1970); Chappell and Veeh (1978*b*). The curves show how contamination of original C in corals with 5 per cent and 10 per cent modern C would affect the true ^{14}C ages (after Broecker and Kulp 1956).

out on molluscs by various workers, Kaufman *et al* (1971) concluded that the assumption of a closed system is violated so frequently that U-series ages of mollusc samples cannot be accepted as valid unless it can be shown independently that post-depositional migration of U and/or daughter nuclides has not taken place. The evidence for migration of U and daughter nuclides includes: (i) increase of U with time; (ii) increase in ^{234}U/^{238}U ratio with time; (iii) discordant ^{230}Th/^{234}U and ^{231}Pa/^{235}U ages; and (iv) disagreement between ^{230}Th and ^{14}C ages within the range over which the latter method's dates are considered reliable (i.e. zero to 20 ky).

Some of these points are illustrated in Table 17.1, and Figs 17.5 and 17.6. The conspicuous deviation of the mollusc ^{234}U/^{238}U data in Fig. 17.5 from the closed system model is evidently due to the low initial U content in recent molluscs, making the molluscs much more susceptible to post-depositional addition of external U than corals. The U content in shells of living molluscs rarely exceeds 0.5 ppm, but is almost always much higher in fossil molluscs (Broecker 1963; Blanchard *et al.* 1967). The increase in the U content of molluscs with time is shown in Table 17.1. Although Broecker (1963) has argued that the bulk of secondary U is added early in the diagenetic history of a given fossil shell, continued addition of U with an isotopic composition typical of non-marine water is clearly indicated by the data in Fig. 17.5. It appears that shell mineralogy determines to some extent the post-depositional uptake of U. Thus calcitic mollusc shells seem to be less prone to secondary U uptake than aragonitic shells (see Table 17.1 and Fig. 17.5). However, this

Table 17.1. Uranium concentrations in marine molluscs as a function of age

Age of samples (ky)	No. of samples	% with ppm U*				
		<0.12	0.12–0.50	0.50–2.0	2.0–8.0	>8.0
Living	78	69	26	5	0	0
Fossil						
aragonitic						
<20	58	4	29	59	9	0
20–500	259	2	21	43	29	5
>500	30	0	23	37	30	10
calcitic						
>20	26	0	42	50	8	0

* Percentage of samples with U concentrations in the ppm range as shown. Based on data in Kaufman *et al.* (1971).

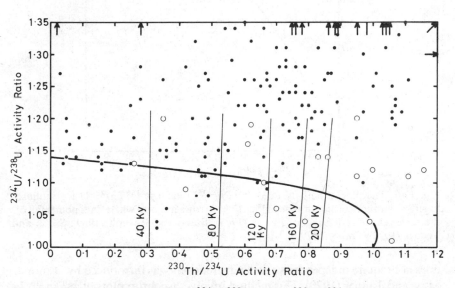

Fig. 17.5. The plot of activity ratios $^{234}U/^{238}U$ against $^{230}Th/^{234}U$ in fossil molluscs of various ages. Open circles denote calcitic mollusc shells. Arrows indicate where sample points fall off the scale. Curves as in Fig. 17.2. (Data from Kaufman *et al.* 1971.)

tendency is offset by higher contamination levels of calcitic shells with ^{232}Th, and hence by implication, with common ^{230}Th (Kaufman *et al.* 1971).

A correction for common ^{230}Th has been proposed by Kaufman and Broecker (1965a), using $^{230}Th/^{232}Th$ data from molluscs of known ^{14}C age and extrapolating to the initial $^{230}Th/^{232}Th$ ratio. Another method which

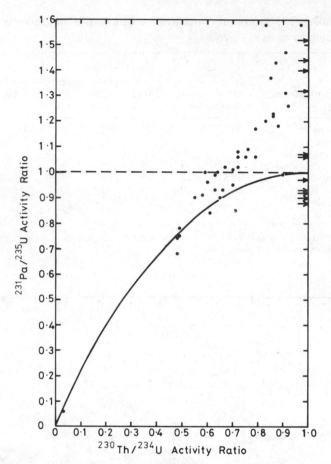

Fig. 17.6. The plot of activity ratios $^{231}Pa/^{235}U$ against $^{230}Th/^{234}U$ for molluscs together with concordia curve as in Fig. 17.3. Arrows indicate where data points fall off the scale. Data from Sakanoue *et al.* (1967); Szabo and Rosholt (1969); Szabo and Vedder (1971); Szabo (1979*a*).

does not require independent age determinations was introduced by Osmond, May, and Tanner (1970). This method involves isochron plots of $^{230}Th/^{234}U$ against $^{232}Th/^{234}U$ to represent samples of equal age but different original $^{230}Th/^{232}Th$ ratios within a given stratigraphic unit. The intercept of $^{232}Th/^{234}U = 0$ then gives the radiogenic $^{230}Th/^{234}U$ ratio from which the correct age for the array of samples forming the isochron can be derived. However, a remaining problem in both methods is the unknown time of addition of the common ^{230}Th.

That isotope exchange processes are not confined to U is clearly shown in

Fig. 17.6. On this plot, fossil molluscs having received only secondary U would tend to plot below the concordia curve. Instead, the data suggest not only secondary addition of U, but significant secondary addition of ^{231}Pa and ^{230}Th as well.

Secondary addition of ^{230}Th, ^{231}Pa, and ^{234}U need not necessarily involve migration of these nuclides via aqueous solutions. Highly energetic alpha recoil processes would be capable of injecting these nuclides from the surrounding sedimentary matrix into the shell material. Again, because of the lower original U content in molluscs, and hence, lower initial concentration levels of the radiogenic daughter nuclides, these processes would affect molluscs more severely than corals, especially when the molluscs are embedded in a sedimentary matrix of silicate minerals containing several ppm U.

It is, therefore, not surprising that the agreement between U-series ages of molluscs, and ages obtained by other dating methods is poor (Kaufman *et al.* 1971; Thurber 1965). Where U-series ages of molluscs and corals from the same stratigraphic locations have been compared, the ages are frequently discordant (Veeh and Chappell 1970; Szabo 1979*a*).

Because of these obvious violations of the closed system assumption, there have been several attempts to devise open system models in the hope that more meaningful ages can be obtained (Rosholt 1967; Szabo and Rosholt 1969; Kaufman *et al.* 1971; Hille 1979; also see chapter 3). These models take into consideration some of the migration and exchange processes discussed above. The most widely publicized open system model is that proposed by Szabo and Rosholt (1969). The model is described in chapter 3, but very briefly has the following features: (i) uranium passes into and out of the shell, while ^{234}U, ^{230}Th, and ^{231}Pa generated *in situ* remain inside the shell material; (ii) the ^{234}U, ^{230}Th, and ^{231}Pa are injected into the shell from the surrounding sedimentary matrix by recoil processes; and (iii) the chemical behaviour of ^{234}U, ^{230}Th, and ^{231}Pa subsequent to their deposition inside the shell are identical.

Although ages obtained on the basis of this model appear to be more self-consistent for molluscs (Szabo and Rosholt 1969; Szabo and Vedder 1971) as well as for corals in some instances (Szabo 1979*a*) than ages based on the closed system model, it has been critized by Kaufman *et al.* (1971) for being unrealistic, based on unproved assumptions concerning the chemical behaviour of the various isotopes concerned, and frequently yielding ages which are in conflict with known ages of the same deposit. Nevertheless, the much more universal distribution of molluscs than that of corals will continue to provide incentives for trying to improve our understanding of the migration and pathways of U-series isotopes in molluscs during and after their deposition.

17.2.3 Applications

Uranium-series dating of carbonates is most frequently applied to problems relating to the movement of sea level. Inasmuch as fossil strandlines become geological tide gauges, complete with sea level sensor (marine terraces, coral reefs), and time clock (corals, molluscs), the past position of sea level in space and time can be reconstructed. The majority of studies in the past have dealt with glacially induced sea level changes during the Quaternary, and hence with paleoclimatology (see chapter 18), but recently there has been an increasing application to problems of neotectonics as well. Most studies of this kind have concentrated on active orogenic belts and island arcs along lithosphere plate boundaries, such as New Guinea (Veeh and Chappell 1970; Chappell 1974a), the Ryukyus Island Arc (Konishi, Schlanger, and Omura 1970), the New Hebrides Islands (Neef and Veeh 1977), and the Banda Arc (Chappell and Veeh 1978a), where uplift rates range from 0.5 mm/y to more than 3 mm/y. An interesting application, so far little explored, is the quantitative evaluation of subduction processes based on uplift rates of the associated presubduction flexure of the lithosphere, as illustrated for the Loyalty Islands southwest of the New Hebrides Island Arc (Dubois, Launay, and Recy 1974; Dubois, Launay, Recy, and Marshall 1977; Benat, Launay, and Recy 1976; Marshall and Launay 1978).

In general, a systematic application of U-series dating to problems of geodynamics would require a more reliable paleo-sea level datum than is presently available (Veeh, Schwebel, van de Graaf, and Denman 1979). Considering the unique position of emerged coral reefs as geological tide gauges, future efforts in this direction could be very rewarding.

17.3 Uranium-series dating of phosphorites

17.3.1 Geochemical considerations

Phosphorites are sedimentary deposits in which the mineral apatite (normally carbonate fluorapatite) is the major component. The apatite in phosphorites is of sedimentary origin, having formed at the depositional site by replacement, or from aqueous solution at low temperature in contrast to igneous apatite which occurs as an accessory mineral in felsic and mafic rocks. A characteristic feature of sedimentary apatite is the relatively high U content (10 to 500 ppm) and the virtual absence of Th. Therefore, the theoretical basis of U-series dating of phosphorites is very similar to that already discussed for carbonates in the previous section, that is to say, the growth of daughter nuclides towards a state of radioactive equilibrium with an initially pure U parent fraction in a closed system. However, the following geochemical considerations place additional constraints on U-series dating methods as applied to phosphorites.

While corals and molluscs consist essentially of pure carbonate, marine phosphorites almost always contain small amounts of detrital phases, glauconite, biogenic opal or carbonates, as well as organic matter in addition to the apatite. Since marine apatite occurs as a cryptocrystalline variety (collophane), physical separation of the apatite from other sediment components is generally not possible. Although it has been shown by alpha track autoradiography (Altschuler, Clark, and Young 1958) and fission track distribution studies (Burnett and Veeh 1977; Kress and Veeh 1980) that most of the U in phosphorites is associated with the collophane matrix, the inclusion of other sediment components may result in the introduction of common ^{230}Th which could lead to age errors, especially in young samples.

A most important consideration is the mode of association of U with the apatite. Altschuler, Clark, and Young (1958) showed that much of the U in phosphorites is tetravalent, and proposed that it substitutes for calcium in the apatite lattice as U^{4+} which has a similar ionic radius to that of Ca^{2+}. The association of hexavalent U with phosphorites is less clear. While it has been demonstrated experimentally (Ames 1960) that $(UO_2)^{2+}$ ions can be fixed in apatite without prior reduction to U^{4+}, the large size of $(UO_2)^{2+}$ precludes a direct substitution for Ca^{2+} in the apatite lattice. The occurrence of hexavalent U in phosphorites has been variously explained by chemisorption of uranyl ions on growing apatite crystallites, or as a result of post-depositional oxidation of primary U^{4+}. Later oxidation of tetravalent U may depend on the degree of exposure to oxidizing environments (Altschuler, Clark, and Young 1958, Burnett and Gomberg 1977) and the amount of auto-oxidation of $^{238}U^{4+}$ to $^{234}U^{6+}$ during radioactive decay (see below).

Free oxides of U have been reported from fossil bones (Kochenov, Dubinchuk, and Germozenova 1973), and more recently from modern oceanic phosphorites (Baturin and Dubinchuk 1978). The principal mineral phase of U, tentatively identified by these authors, is finely disseminated uraninite. The extremely fine crystal size ($0.0x$ to $0.x$ microns) of the particles observed makes positive identification very difficult. If free oxides of U are confirmed to exist in marine phosphorites, it will become necessary to determine how U is distributed between these phases and the apatite. Based on the work of Altschuler, Clark, and Young (1958) and observations made by Veeh and Burnett (personal comm.), these authors favour apatite as the dominant host for U in marine phosphorites.

17.3.2 Marine phosphorites

Marine phosphorites typically occur as irregular masses, nodules, pellets, and sands in sediments of the outer continental shelf and upper slope in many parts of the world's oceans. The first systematic attempt to apply U-series disequilibrium methods to age determinations of marine phosphorites was reported by Kolodny (1969), and Kolodny and Kaplan (1970). These authors were guided by the high U content and by the assumption that this U is of marine origin with an initial $^{234}U/^{238}U$ activity ratio close to that of modern sea water. However, none of the phosphorites analysed by Kolodny and Kaplan (1970) contained an excess ^{234}U. In fact, many of the phosphorites showed $^{234}U/^{238}U$ ratios somewhat less than unity, suggesting selective loss of ^{234}U. This indicated that many phosphorites currently exposed on the sea floor cannot be forming at the present time, and also precluded the possibility of establishing whether or not recent phosphorites contain any ^{230}Th, and hence whether a geochronology based on $^{230}Th/^{234}U$ would be feasible.

More recently, Baturin, Merkulova, and Chalov (1972); Veeh et al. (1973); Veeh et al. (1974), and Burnett and Veeh (1977) found that phosphorite nodules occurring in organic rich diatomaceous sediments beneath areas of intense oceanic upwelling and associated high organic productivity off the coast of Peru, Chile, and Namibia have $^{234}U/\,^{238}U$ and $^{230}Th/^{234}U$ ratios close to that in modern sea water, and demonstrated that the radioactive disequilibrium of U-series isotopes in marine phosphorites can be used for meaningful age estimates of the time of phosphorite formation, subject to the following conditions: (i) the apatite in phosphorites acquired only U, but little if any ^{230}Th from sea water; (ii) the U is syngenetic with the apatite, and the time of apatite formation was short compared with the half-lives of ^{230}Th and ^{234}U; and (iii) the apatite has remained a closed system with respect to the U and Th isotopes.

Although most marine phosphorites analysed so far contain small concentrations of ^{232}Th, and hence, by implication, also common ^{230}Th, there is nothing to indicate that the common Th resides in the apatite itself. Rather, and in view of the general composition of phosphorites, it is more likely that the common ^{230}Th is associated with detrital and biogenic particulates which are known carrier phases for Th (Goldberg and Koide 1962; Scott 1968). This idea is further supported by the observation that mineralogically pure fishbone apatite coexisting with marine phosphorites are essentially free of ^{232}Th (Burnett and Veeh 1977). A correction for common ^{230}Th by the method of Kaufman and Broecker (1965a) would be possible, but depends critically on a reasonably accurate assessment of the $^{230}Th/^{232}Th$ ratio in particulate matter contained in the phosphorites concerned. Burnett and Veeh (1977) proposed a value of 4.0 as the best estimate for the average $^{230}Th/^{232}Th$ ratio of common Th in marine phosphorites based on measurements of this ratio in recent sediments closely associated with the phosphorites off Peru and Namibia. This value approaches closely the $^{230}Th/^{232}Th$ ratio in the youngest (in terms of $^{230}Th/^{234}U$ and $^{234}U/^{238}U$) phosphorite samples from Peru and Namibia. The inherent uncertainties underlying this correction are minimized by the comparatively low ^{232}Th content in marine phosphorites (1 to 10 ppm), so that the correction becomes significant only for very young samples in which the common ^{230}Th can attain a substantial fraction of the total ^{230}Th present. In spite of remaining uncertainties concerning the processes by which U becomes associated with the apatite in phosphorites, the available data are generally consistent with condition (ii) above, that is to say, that the U is syngenetic with the apatite (Altschuler, Clark, and Young 1958; Burnett and Veeh 1977; Kolodny 1980; Burnett, Veeh, and Soutar 1980). The assumption of a closed system, however, can be challenged and requires more careful examination.

Figure 17.7 is a plot of $^{234}U/^{238}U$ against $^{230}Th/^{234}U$ for marine phosphorites from a variety of locations, together with the expected change of these activity ratios with time in a closed system with an initially pure U fraction of marine origin. In contrast to a similar plot for corals which conform to a closed system model rather well (Fig. 17.2), the data in Fig. 17.7 can best be explained as the result of normal radioactive decay of excess ^{234}U in a semi-closed system from which a small fraction of ^{234}U is continuously lost. The loss of ^{234}U is most likely caused by alpha recoil and related processes (Kolodny and Kaplan 1970) and hence should affect only ^{234}U generated *in situ*, but none of the ^{234}U which was incorporated into the phosphorites originally. Apparently, these processes do not operate in carbonates to the same degree, perhaps because the U there is predominantly present in its hexavalent state which is less prone to fractionation by recoil than tetravalent U.

It is interesting to note that the initial $^{234}U/^{238}U$ activity ratio in the phosphorites tends to be somewhat higher than the $^{234}U/^{238}U$ ratio of present day sea water, in contrast to corals where there is no such difference.

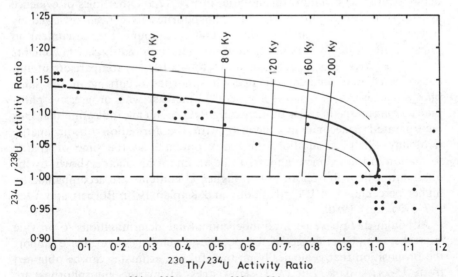

Fig. 17.7. The plot of $^{234}U/^{238}U$ against $^{230}Th/^{234}U$ in marine phosphorites with counting error in the $^{234}U/^{238}U$ ratio equal to or better than ± 0.02 (1σ). Data from Kolodny and Kaplan (1970); Veeh et al. (1974); Burnett and Veeh (1977); Kress and Veeh (1980); O'Brien and Veeh (1980). Horizontal and vertical curves as in Fig. 17.2.

This could be the result of different environments of formation of these two types of materials. Whereas corals form in well-oxygenated, open sea water, phosphorites form diagenetically within organic-rich, and frequently anoxic sediments, so that the U in the latter case is derived from pore water rather than from open sea water directly. It is, therefore, to be expected that the $^{234}U/^{228}U$ ratio in pore water should be somewhat greater than that of sea water, especially if the sediments contain older, previously formed phosphorites as well. It is quite plausible that young phosphorites growing by nucleation of apatite within organic-rich sediments incorporate U from pore water with slightly elevated $^{234}U/^{238}U$ ratios resulting from the back diffusion of ^{234}U lost from previously formed phosphorite nodules contained in the same sediments. The data in Fig. 17.7 illustrate such a process well. As the phosphorites age, the $^{234}U/^{238}U$ ratio decreases from a value initially somewhat above that of normal sea water to values below those predicted by the closed system model. However, the loss of ^{234}U appears to be a slow process as judged by the data in Fig. 17.7 and by the observation that $^{234}U/^{238}U$ and $^{230}Th/^{234}U$ ratios in old phosphorites are generally close to their equilibrium values (Kolodny and Kaplan 1970). Since the $^{230}Th/^{234}U$ method is relatively insensitive to even large variations in $^{234}U/^{238}U$ activity ratios within its normally employed range (0 to 200 ky), the leakage of ^{234}U is not likely to affect $^{230}Th/^{234}U$ ages seriously.

There is certainly no evidence in Fig. 17.7 for any secondary addition of U demonstrated so strikingly for molluscs in Fig. 17.5. Other lines of evidence against secondary addition of U to phosphorites have been reviewed by Burnett *et al.* (1980), and Kolodny (1980). Although the U content in phosphorites is highly variable, there is no indication that higher U contents reflect post-depositional addition of U to apatite. A comparison of unconsolidated and semi-consolidated phosphorites occurring in the same phosphogenic province shows a general U increase with progressive phosphorite diagenesis (in terms of increasing P_2O_5 content), indicating that U is incorporated penecontemporaneously with the formation of the apatite (Kolodny 1980). In addition, the redox potential at the time of apatite formation in the sediment appears to be an important factor, shown by the observation that phosphorites with higher U contents are accompanied by higher percentages of U^{4+} (Kolodny and Kaplan 1970; Burnett and Veeh 1977; Kolodny 1980).

Although there have been no independent age determinations to provide cross checks on U-series ages of phosphorites, several lines of evidence support the proposition that geologically meaningful age estimates can be obtained from U-series data. Firstly, U-series ages of marine phosphorites are consistent with their general macroscopic appearance. Thus phosphorites with 'young' U-series ages have dull, earthy surfaces and are in general poorly consolidated, while phosphorites with $^{230}Th/^{234}U$ and $^{234}U/^{238}U$ activity ratios near radioactive equilibrium invariably are well lithified and frequently have glazed surfaces, suggesting mechanical abrasion caused by exposure to increased wave or current action on the sea floor subsequent to their diagenetic formation within the sediment (Burnett *et al.* 1980; Kress and Veeh 1980; O'Brien and Veeh 1980). Secondly, microscopic examination (SEM) of radiometrically 'young' phosphorites reveals evidently newly formed apatite crystallites on the surface of diatom frustules (Burnett 1977), or foraminiferal tests (O'Brien and Veeh 1980), whereas the apatite in radiometrically older phosphorites frequently shows pitted surfaces, suggesting post depositional dissolution of the apatite, perhaps as a result of re-exposure to open sea water (Burnett *et al.* 1980). Thirdly, U-series ages of phosphorites and solitary corals coexisting in sediments on the upper continental slope of Eastern Australia are consistent with the depositional sequence established by sedimentological criteria (O'Brien and Veeh 1980).

An interesting application of U-series disequilibrium deals with the partition of U isotopes between different oxidation states. Chalov and Merkulova (1966) first showed that the $^{234}U/^{238}U$ ratio in the tetravelent fraction of natural U oxides is always lower than in the hexavalent fraction, indicating preferential oxidation of ^{234}U as compared to ^{238}U. Kolodny and Kaplan (1970) demonstrated that the U isotopes in marine phosphorites show a similar partition between oxidation states and proposed a quantitative model for variation of $(^{234}U/^{238}U)^{4+}$ with time. This model is based on the following assumptions: (i) a fraction of radiogenic ^{234}U in the tetravalent U in apatite is oxidized to hexavalent ^{234}U; (ii) this fraction remains constant with time; and (iii) any oxidation process which is not related to radioactive decay does not distinguish between ^{234}U and ^{238}U.

The relevant equation to describe the change in the isotopic composition of tetravalent U with time in a semi-closed valve type system is:

$$\left(\frac{^{234}U}{^{238}U}\right)^{4+} = (1-R)\left[1 - e^{(\lambda_{238}-\lambda_{234})t}\right] + \left(\frac{^{234}U}{^{238}U}\right)_0^{4+} \cdot e^{(\lambda_{238}-\lambda_{234})t} \quad (17.1)$$

where $\left(\dfrac{^{234}U}{^{238}U}\right)_0^{4+}$ = initial activity ratio at time $t = 0$

R = the fraction of radiogenic ^{234}U which is oxidized.

The data in Fig. 17.8 are consistent with Kolodny and Kaplan's (1970) model of tetravalent U in phosphorites as a 'semi-closed 'valve type' system to which no U is being added, but from which ^{234}U is able to escape'. Although, as discussed above, some limited and slow escape of ^{234}U into pore water does occur most of the mobile ^{234}U appears to be added to the hexavalent fraction of U without actually leaving the phosphorites altogether, so that the isotopic fractionation between oxidation states increases with time, while $^{234}U/^{238}U$ in total U shows only a slight negative deviation from values predicted by the closed system model.

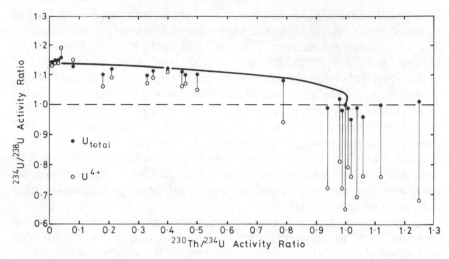

Fig. 17.8. The plot of $(^{234}U/^{238}U)$ total and $(^{234}U/^{238}U)^{4+}$ against $^{230}Th/^{234}U$ in marine phosphorites showing increasing fractionation of U isotopes between oxidation states with time. The heavy curve is drawn for initial $^{234}U/^{238}U = 1.14$ as in previous figures. Data from Kolodny and Kaplan (1970); Burnett and Veeh (1977); O'Brien and Veeh (1980).

Kolodny and Kaplan (1970) suggested that $(^{234}U/^{238}U)^{4+}$ in marine phosphorites could be used for geochronological purposes. To do this, the constant, R in eqn (17.1) must be known. Kolodny and Kaplan (1970) attempted to evaluate R in the following way. They pointed out that the right hand side of eqn (17.1) becomes equal to $(1 - R)$ for very large values of time t, and that R can, therefore, be evaluated from the experimentally determined $(^{234}U/^{238}U)^{4+}$ values in old phosphorites. Using $(^{234}U/^{238}U)^{4+}$ data in sea floor phosphorites they considered 'old' on the basis of near equilibrium $^{234}U/^{238}U$ ratios, they arrived at a mean value of 0.3 for R.

The more recent data by Burnett and Veeh (1977) and O'Brien and Veeh (1980) provide an opportunity to check this value: Figure 17.9 shows the plot of $(^{234}U/^{238}U)^{4+}$ against age as calculated from measured $^{230}Th/^{234}U$ ratios, corrected for common ^{230}Th, together with different values of R. The experimental data are in reasonable agreement with the value of 0.3 for R. It would be appropriate in this context to inquire into the physical meaning of R. Details of the various models proposed for the mechanism of $^{234}U/^{238}U$ fractionation are presented in chapter 2, but for the present purpose it is relevant to indicate that some models require the presence of an oxidizing agent in order to produce significant fractionation of the U isotopes, while other models do not. Cherdyntsev (1956) attributed the preferential release of ^{234}U from minerals to its alpha recoil induced relocation in foreign lattice sites where it is less firmly bound and more accessible to oxidation than ^{234}U atoms

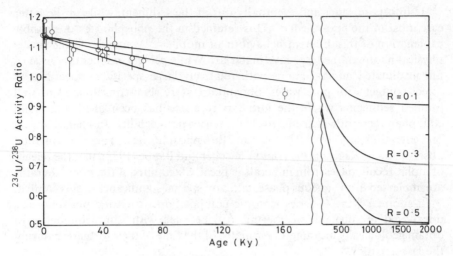

Fig. 17.9. Change of $(^{234}U/^{238}U)^{4+}$ as a function of time for different values of R according to eqn. (17.1) (see text), together with measured $(^{234}U/^{238}U)^{4+}$ in marine phosphorites of various ^{230}Th ages (corrected for common ^{230}Th). Error bars indicate $\pm 2\sigma$ (counting statistics). Note change in horizontal scale at 170 ky. Data from Burnett and Veeh (1977); O'Brien and Veeh (1980).

which were incorporated initially together with ^{238}U. Kigoshi (1971) showed by experiments with zircon that the mechanism of ^{234}U fractionation is actually due to the ejection of ^{234}Th from ^{238}U located near the surface of mineral grains, and that this mechanism is more effective in the presence of an aqueous phase in which ^{234}Th would be stopped and transformed into mobile ^{234}U. On the other hand, Rosholt, Shields, and Garner (1963) and Dooley, Granger, and Rosholt (1966) assumed that ^{234}U becomes oxidized as a result of, and not subsequent to, the radioactive transformation from ^{238}U to ^{234}U, via ^{234}Th and ^{234}Pa, perhaps because of electron stripping during this process. More recently, Fleischer and Raabe (1978) proposed that in addition to direct ejection of recoil nuclei from mineral surfaces, an important mechanism is the production of alpha-recoil tracks within the minerals, and subsequent chemical etching of the damaged regions by intergranular aqueous solutions which would release additional ^{234}U into the pore water.

Kolodny and Kaplan (1970) considered the constant R in eqn (17.1) to be the 'combined probability of a radiogenic ^{234}U atom in the tetravalent state to be displaced and then oxidized to hexavalent state'. These authors also drew attention to Ku's (1965) model for migration of ^{234}U in pelagic sediments in which 30 per cent of the *in situ* generated ^{234}U becomes mobilized, presumably by oxidation and formation of soluble uranyl carbonate complexes. The only difference between pelagic sediments and phosphorites in this regard would be that fractionation of U isotopes in the former leads to leakage of hexavalent

^{234}U into pore water and eventually out of the sediment, while in the latter case most of the hexavalent ^{234}U is retained in the phosphorites. A possible explanation of this different behaviour of mobile ^{234}U may be found in the physical nature of the respective materials. While pelagic sediments consist of unconsolidated mineral grains with relatively large specific surface area in direct contact with pore water, phosphorites are semi-consolidated to well lithified sediments where the initial pore space has been filled with solid collophane cement which effectively decreases permeability, and hence largely prevents diffusion of ^{234}U out of the phosphorites. The experimental observation by Kigoshi (1971) and Fleischer and Raabe (1978), that the release of alpha-recoil nuclei from minerals is greatly facilitated if the mineral grains are immersed in an aqueous phase, takes on special significance in this context.

In summary, the U-series isotopic composition of marine phosphorites, unexposed to subaerial weathering, can be used with some confidence to obtain geologically meaningful estimates of the time of their formation during the last 2×10^5 y.

17.3.3 Insular phosphorites

Insular phosphorites, generally believed to originate as a result of chemical interaction between bird guano and some host rock, occur on many oceanic islands which are now, or were once occupied by large bird colonies (Hutchinson 1950). Where the underlying rock is reef limestone, the phosphate mineral is predominantly apatite, but where the host rock is a silicate, the resulting phosphate minerals are a variety of Al-Fe phosphates (McKelvey 1967). Although mineralogical and geochemical arguments suggest that insular phosphorites were formed subaerially (Hutchinson 1950; Altschuler 1973), this type of materials is included in this chapter because their processes of formation are closely linked to oceanographic processes.

The first application of U-series methods to insular phosphorites was reported by Veeh and Burnett (1978). It was shown that the apatite fraction of a young phosphate deposit occurring on Ebon Atoll in the Marshall Islands contained 59 ppm of U with ^{234}U/^{238}U $= 1.15 \pm 0.02$, but no detectable ^{232}Th (less than 0.05 ppm). The age calculated from the ^{230}Th/^{234}U ratio $(4 \pm 0.3$ ky) is in excellent agreement with the ^{230}Th and ^{14}C ages of coral fragments and carbonate substrate, respectively, upon which the apatite had been deposited. These results confirm an essential prerequisite of U-series dating, that is to say, an initially pure parent fraction. The validity of the remaining assumption, such as the maintenance of a closed system over the entire dating range of the ^{230}Th/^{234}U method, has yet to be established.

17.3.4 Geological and oceanographic implications

The study of marine phosphorites serves the dual purpose of providing information about their genesis, as well as reconstructing past states of the ocean. By using modern analogues, the distribution of ancient phosphorites in space and time should offer clues to such factors as locations of former coastal upwelling centres, periods of intensified organic productivity, sea level dynamics, and the like. Thus, absolute ages of phosphorites from different parts of the world may prove extremely valuable for deciphering paleoceanographic conditions, as well as improving our understanding of the origin of economic phosphorite deposits and thus guide future exploration efforts.

The geographic distribution of all the U-series ages of marine phosphorites determined so far is presented in Fig. 17.10, together with locations of the major coastal upwelling areas of the world, and the atmospheric pressure systems which influence them. In at least three offshore areas (Eastern Australia, Peru–Chile, and Namibia) the U-series ages of sea floor phosphorites range from greater than 200 ky to Recent, suggesting that phosphorites have formed at these locations at various times throughout the Quaternary (and perhaps earlier) up to the present. Two of the areas where modern phosphorites have been identified (Peru and Namibia) are situated in the centre of well documented coastal upwelling systems, thus supporting the

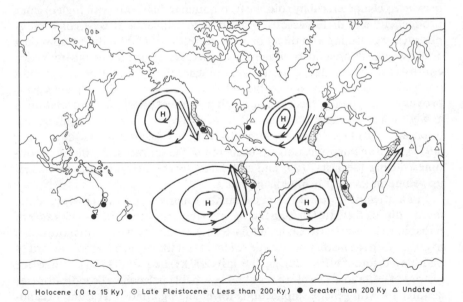

O Holocene (0 to 15 Ky) ⊙ Late Pleistocene (Less than 200 Ky) ● Greater than 200 Ky △ Undated

Fig. 17.10. Geographic distribution of marine phosphorites, together with major coastal upwelling areas (shaded) and the atmospheric pressure systems and boundary currents (double arrows) that influence them (redrawn from Thompson 1977). Each data point represents major age intervals as shown, based on several age determinations in most cases. The location of known, but as yet undated phosphorite deposits are also shown. Uranium-series age data from Kolodny and Kaplan (1970); Veeh *et al.* (1974); Burnett and Veeh (1977); Kress and Veeh (1980); O'Brien and Veeh (1980).

hypothesis that the genesis of marine phosphorites is linked to oceanic upwelling and associated organic productivity (Sheldon 1964; McKelvey 1967). However, the third area with radiometric evidence for modern phosphorite formation, Eastern Australia, differs from Peru and Namibia in several important aspects. Although limited seasonal upwelling does occur in this area, its scale is very small in comparison with coastal upwelling associated with eastern boundary currents. Moreover, the underlying cause of the East Australian upwelling is different and appears to be related to fluctuations in the East Australian Current, a complex and variable system of large, anticyclonic eddies moving southward along the coast of northern New South Wales (Godfrey, Cresswell, Golding, Pearce, and Boyd 1980), rather than being wind induced locally. In order to generalize the situation somewhat, we have taken the East Australian phosphorite deposit as an example of an 'east coast phosphogenic province' (other examples are the Agulhas Bank off southeast Africa, or the Blake Plateau east of Florida), characterized by warm, poleward flowing currents, low to moderate productivity and calcareous sediments with relatively low organic carbon content, and the Peru–Chile and Namibia phosphorite deposits as examples of 'west coast phosphogenic provinces' characterized by cold, eastern boundary currents with high organic productivity and diatomaceous, organic-rich sediments. If west coast phosphorites are forming as a direct or indirect result of wind induced upwelling and associated organic productivity, processes in the atmosphere which control eastern boundary currents will influence this type of deposit while not necessarily having any noticeable effect on an east coast phosphogenic province. One possible mechanism which may affect phosphorite formation in a west coast province such as offshore Peru or Namibia could be the intensification of the winter southeast trade winds. According to Molina-Cruz (1977), such an intensification has occurred in the southeast Pacific at various times during the last 70 ky, and is related to the development of coastal upwelling in the South American region.

In an attempt to relate the phosphorite genesis to known environmental factors during the Late Quaternary, the distribution pattern of U-series ages of phosphorites from Peru–Chile and Eastern Australia are shown in more detail together with the history of eustatic sea level and the oxygen isotopic record of planktonic foraminifera during the last 200 ky (see Fig. 17.11). While the history of eustatic sea level is a reasonably reliable paleoclimatic indicator (see chapter 18) with general implications for changing atmospheric and oceano-graphic conditions associated with the waxing and waning of continental ice sheets, the oxygen isotopic record of planktonic foraminifera in deep-sea sediment cores not only indicates changes in ocean volume, but also conveys specific information on sea surface temperatures at each site and hence, indirectly on changes in ocean circulation (Duplessy 1978).

The suggestion that phosphorite formation is largely restricted to times of

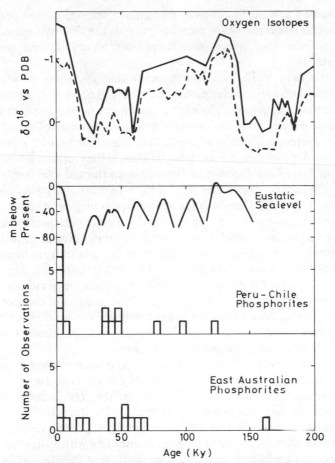

Fig. 17.11. Uranium-series ages of marine phosphorites from off Peru–Chile (Burnett and Veeh 1977) and Eastern Australia (Kress and Veeh 1980; O'Brien and Veeh 1980), compared with eustatic sea-level history (Bloom *et al.* 1974) and two typical oxygen isotope records of planktonic foraminifera during the Late Quaternary (Duplessy 1978). The ^{18}O curves based on *Globigerinoides sacculifera* from two sediment cores at opposite margins, but equal latitude, of the Indian Ocean, the top curve from off Durban, southwest Africa, the bottom curve from a core northwest of Perth, Western Australia. The major features of both curves are dominated by variations in ocean volume, while the difference in amplitude between the two curves are caused by different sea surface temperatures at the two sites.

high sea level (Burnett and Veeh 1977) is not supported by the more recent data from Eastern Australia (Kress and Veeh 1980; O'Brien and Veeh 1980). This could mean that the age groupings observed previously are simply the result of inadequate sample coverage, and that future data will fill in the gaps in the record. Alternatively, the different age patterns apparent in Fig. 17.11 could be real, indicating a basically different mechanism responsible for phosphorite genesis in these two respective areas. In view of the different oceanographic settings of the two areas, the latter interpretation deserves serious consideration.

From the data in Fig. 17.10, it appears that the areas of phosphorite formation in the Southern Hemisphere have moved north at some time during the Quaternary, suggesting a global shift of climatic belts towards the equator. Additional evidence for this is provided by the distribution of guano deposits along the western sea board of South America (Hutchinson 1950; Burnett 1980). These deposits range in age from modern in the Chincha Islands area off Peru, to at least Late Pleistocene (based on stratigraphic evidence) further south in Chile. The origin and preservation of guano deposits on oceanic islands depends on the simultaneous occurrence of high productivity and arid conditions in a given area of the ocean, the former to sustain a large bird colony, the latter to prevent dense vegetation from competing with ground nesting guano birds for space, and to prevent the guano from being washed away by frequent heavy rains (Hutchinson 1950). Although there are few radiometric ages for South American guano deposits, the geological evidence suggests that the area of maximum upwelling (assuming the guano birds follow the productivity associated with upwelling) was located further south earlier in the Quaternary than now. This agrees with the general U-series age pattern of the marine phosphorites in this area.

Minor fluctuations in climatic zonation and oceanographic conditions along the equatorial Pacific (a site of enhanced productivity due to divergence) during the last 5 ky are also indicated by the age pattern of insular phosphorites in that area (Veeh and Burnett 1978; Burnett and Lee 1980).

In conclusion, the application of U-series age determinations to phosphorites has revealed a distribution pattern in space and time which suggests a critical dependence of phosphorite genesis on one or more of such factors as sea-level dynamics, ocean circulation, sedimentation rate, continental runoff, and in the case of insular phosphorites, on atmospheric precipitation. Further refinement of the relationships between the genesis of phosphorites and specific paleoclimatic and paleoceanographic factors will have to await a more complete geochronology of marine and insular phosphorites on a world-wide basis.

Acknowledgements

A significant portion of the material presented in this chapter is based on research supported by the Australian Research Grants Committee and the U.S. National Science Foundation (grant OCE-8007047). This chapter is a contribution to Project 156 (Phosphorites) of the International Geological Correlation Program.

18 LATE PLEISTOCENE SEA-LEVEL HISTORY
W.S. Moore

18.1 Introduction

One of the remarkable successes of the application of U-series isotopes to geochronological problems is the establishment of an absolute sea-level chronology for 200 ky before present through the dating of fossil corals. Coupled with oxygen isotope stratigraphy of deep-sea sediments, these measurements have placed events of the late Pleistocene on an absolute time-scale. In this chapter the development of the U-series techniques for dating fossil corals is traced and case studies where dating and stratigraphy have provided a consistent picture of a region are presented. Finally a global curve of eustatic sea-level will be developed for the late Pleistocene and the implications of this curve for current theories of glaciations will be discussed.

A detailed knowledge of the Pleistocene sea-level record is important because it is directly coupled to the volume of ice in continental glaciers. As glaciers grow and advance, sea-level falls. The best estimates for the amount of sea-level lowering during glacial maxima are 130 to 160 m based on the apparent change in the oxygen isotope composition of benthic foraminifera (Broecker 1978). During the 19th and early 20th centuries, Quaternary geologists recognized four major glacial advances and retreats of North American and Euro-Asian ice sheets over the past 10^6 y based primarily on deposits at the terminus of the glacier called moraines (Flint 1971). However, each succeeding glacial advance largely destroyed evidence of previous glaciations making the record on land obscure and fragmentary. The best terrestrial record is preserved from the retreat of the continental ice sheets from their maxima during Late Wisconsin time, 18 ky.

The Pleistocene record in deep-sea sediments revealed a much more complex picture of global climate (Ericson and Wollin 1968). Numerous fluctuations in global ice volume and ocean temperature are recorded in the calcareous shells of foraminifera (forams) which lived in surface (planktonic) and bottom (benthic) waters. The stable isotopes of oxygen are incorporated in the shells as a function of the $^{18}O/^{16}O$ ratio of the water and temperature (Urey 1947; Epstein, Buchsbaum, Lowenstam, and Urey 1953). During glacial times a large amount of ^{18}O-depleted water is locked in glaciers, resulting in sea water and foraminiferal-CaCO$_3$ enriched in ^{18}O. Also during these cold periods decreased water temperatures cause more isotopic fractionation and result in shells enriched in ^{18}O because of the temperature effect. The relative contributions by changes in the ^{18}O/^{16}O ratio of the water and temperature to the deep-sea isotopic record have been assessed by comparing the compositions of planktonic and benthic species. Shackleton and Opdyke (1973) used this comparison to demonstrate that most of the changes in oxygen isotope composition of forams were due to changes in the ^{18}O/^{16}O ratio of the water and hence were directly related to ice volume and sea-level. Combining such deep-sea records with paleomagnetic stratigraphy revealed at least 22 so-called isotopic stages representing alternating periods of high and low Northern Hemisphere ice volumes during the past 840 ky, with 6 stages present in the last 150 ky (Emiliani 1966; Shackleton and Opdyke 1973). In addition to these main stages, numerous substages have been recognized in cores with higher sedimentation rates and hence greater resolution (Shackleton 1977; Ruddiman, McIntyre, Neifler-Hunt, and Durazzi 1980).

The sea-level record as read from fossil coral reef terraces is even more complex than the deep-sea record, with numerous short term fluctuations in level. These two records are highly complimentary as the episodic fossil coral record provides an absolute chronology for the more continuous but less reliably dated deep-sea sediment record (Shackleton and Matthews 1977).

18.2 Dating techniques for fossil corals

The details of the methodology have been discussed in chapter 3 and will not be reviewed here. However, the question of reliability of a single age or of a suite of ages bears directly on the subject of this chapter. Coral which has retained its original aragonitic composition is the only material which yields U-series ages accepted by most workers in the field. Although numerous attempts have been made to date molluscs, the open system nature of U and Th in the shells leaves the ages open to question (Kaufman et al. 1971) (see section 18.3). Field and laboratory criteria for selecting suitable corals for dating have been discussed by Bloom et al. (1974). In the field visible evidence of calcite cleavage should be enough to reject the sample. When possible, corals in their original growth position should be selected; this is often impossible, however, to assess for spherical heads. In the laboratory, X-ray diffraction analyses can be used to determine quantitatively if the sample contains calcite. The presence of high-Mg calcite indicates contamination by chemical or mechanical void filling; the presence of low-Mg calcite indicates recrystallization and/or precipitation of freshwater cement (Bloom et al. 1974). Thin section examinations are further used to verify the presence of void filling or surface cementation. The analytical data generated by the Th and U isotopic measurements are also useful in rejecting altered samples. The presence of ^{232}Th indicates secondary Th addition. High ^{234}U/^{238}U activity ratios indicate U exchange with ground water, and U contents deviating beyond the range 2.1 to 4.0 ppm indicate anomalous U content for hermatypic corals (Moore, Krishnaswami, and Bhat 1973).

The U-Series Intercomparison Project has attempted to evaluate the analytical precision and accuracy of ages obtained by all laboratories currently active in this field. The first phase of this study revealed that real isotopic differences exist among samples thought to be the same age probably because of slight differences in sample preservation (Harmon *et al.* 1979). They stress that geologists should place little confidence in single, random U-series age determinations unaccompanied by stratigraphic control.

18.3 Materials for dating

Early workers attempted to use corals, mollusc shells, and oolites for dating by the ^{230}Th/^{234}U method. It soon became obvious that molluscs were open systems with respect to U exchange. Although living molluscs usually contain very little U, the U content of the shells increases by a factor of 10 to 100 within 10^3 y of the death of the inhabitant (Broecker 1963). Some ancient samples show high ^{234}U/^{238}U activity ratios, indicative of continuous U exchange with surrounding waters (Kaufman *et al.* 1971). Numerous attempts have been made to use ^{230}Th/^{234}U dates on molluscs to establish chronologies. None of these dates can be accepted without question. In an excellent treatise on the subject, Kaufman *et al.* (1971) concluded that migration of U-series isotopes in and/or out of mollusc shells presents a serious, and at this stage insurmountable obstacle to using molluscs for dating purposes. Kaufman *et al.* (1971) reached this conclusion based on (i) a rapid (less than 10^3 y) initial increase in the U content of the shell followed by a slower but continuing (about 10^6 y) increase in U; (ii) continuing exchange of U as revealed by high ^{234}U/^{238}U activity ratios in older samples; (iii) an alarming discordance (over 50 per cent of the cases) in ^{230}Th/^{234}U and ^{231}Pa/^{235}U ages for molluscs; (iv) similar disagreements among ^{230}Th/^{234}U ages and other absolute dating techniques applied to the same sections. They considered U-series migrations as 'a serious problem which occurs in ways which can neither be reliably corrected for nor even detected, before the fact.'

Open system dating models have been applied to molluscs with various apparent degrees of success (Rosholt 1967; Szabo and Rosholt 1969; Szabo and Vedder 1971; Hille 1979), however the models have been criticized (Kaufman 1972) and the results remain equivocal. Szabo (1979a) dated five mollusc samples from a single terrace in Jamaica and another coral–mollusc pair from the Bahamas. Calculated ages based on open system (Szabo and Rosholt 1969) models reduced the ages calculated using closed system models by 10 to 20 per cent. Three of the six mollusc samples gave ages in reasonable agreement with the coral ages, two could not be determined by the open system model, and one gave an unreasonably low age. Kaufman *et al.* (1971) concluded that the open system model proposed by Szabo and Rosholt (1969) is probably invalid for the following reasons: (i) the assumptions are geochemically unlikely; (ii) the method when applied to eight samples of known age gave the correct age in only one case; (iii) the method was no more internally consistent on five sets of samples of unknown age than were simpler models based on recent gain or loss of U parents only; (iv) the model failed to explain the increase in $^{234}U/^{238}U$ with time as observed in most samples.

There has been much less work based on oolites as material for dating. The initial reports by Osmond *et al.* (1965) and Broecker and Thurber (1965) were encouraging, however, the work has not been pursued.

In this chapter any interpretations of Pleistocene sea-level record based on $^{230}Th/^{234}U$ or $^{231}Pa/^{235}U$ dating of fossil mollusc samples have been excluded. Where possible only ages of samples containing less than 5 per cent calcite have been used in the interpretations.

18.4 Ages of fossil corals

The first measurements of U and Th in corals were reported by Barnes *et al.* (1956) who recognized that since corals form with several ppm U and negligible ^{230}Th, the growth toward equilibrium of ^{230}Th with respect to ^{238}U in the coral might be an index of its age. The assumptions underlying the technique were further investigated by: Sackett (1958); Tatsumoto and Goldberg (1959); Broecker (1963); Blanchard (1963); Broecker and Thurber (1965); Thurber *et al.* (1965); Osmond *et al.* (1965). These workers recognized that the assumptions were generally valid when applied to unrecrystallized (i.e. 100 per cent aragonite) coral samples. Early investigations produced two groupings of ages: less than 6 ky and about 125 ky. By extending these studies to islands throughout the Pacific and Indian Oceans, Veeh (1966) showed that the last time sea-level was near its present position (and perhaps a bit higher) was about 120 ky. Veeh's study was critical in establishing the age of the last interglacial high sea-level because the range of sampling left no doubt that the ages reflected a global eustatic high sea-level rather than local tectonics.

During the next few years, frequent confirmations of the '120 ky' sea-level were made. Determinations of the age of a major cold to warm transition recorded in deep-sea sediments also began to converge at about 125 ky (Rosholt *et al.* 1961; Emiliani 1966; Ku 1966). Broecker and Van Donk (1970) concluded that the deep-sea oxygen isotope and fossil coral records were compatable in timing and magnitude, and published a general sea-level curve for the late Pleistocene.

A major advance in the use of fossil corals as sea-level indicators was made on the island of Barbados, West Indies. Broecker, Thurber, Goddard, Mathews, Ku, and Mesolella (1968) determined that four emerged fossil reefs had ages of 82, 104, 124, and 200 ky. By assuming that Barbados had undergone continuous, steady uplift for the past 200 ky, they deduced that these emerged fossil reefs represented previous high sea levels. The high sea stands which produced the reefs at 82 ky and 104 ky had not reached the level of the 125 ky high stand or of the present sea-level and thus could only be observed on an emerging island. These authors compared the island to a strip chart recorder with sea-level controlling the pen. When sea-level is high, a coral reef terrace is drawn on the island only to be stranded as the island continues to emerge and sea-level falls. The next high sea-level then draws another terrace on the island below the previous one. By assuming a steady uplift rate, Broecker et al. (1968) and Mesolella et al. (1969) were able to reverse the recorder and reconstruct sea-level events for the past 2×10^5 y.

Dating fossil coral terraces is a more precise technique than measuring the rates of sedimentation in the deep-sea and using these rates to determine the timing of global ice volume transitions recorded in the sediments. Broecker and Van Donk (1970) assigned an age of 127 ky to the midpoint of Termination II (the major change from cold to warm conditions) in the deep sea record, assumed constant sedimentation rates, and constructed a δ ^{18}O against time curve using data from Emiliani (1955, 1964, 1966). Other workers have followed this example and used the 120 to 127 ky datum from the fossil coral terraces or the 700 ky datum from the Brunhes–Matuyama magnetic epoch boundary reversal to establish chronologies (Shackleton and Opdyke 1973). Kominz et al. (1979) redetermined the sedimentation rate of core V28–238 and found that although the overall chronology was close to the one adopted by Shackleton and Opdyke (1973), the age of Termination II was closer to 140 ky (see section 18.5.16). The assumption of a constant sedimentation rate in studies of deep sea cores is a critical one. However, if independent ages can be established for the transitions, this assumption is not necessary. Dating fossil coral terraces provides one means of establishing this rigorous chronological framework.

18.5 Case studies

There are about a dozen areas where detailed dating of stratigraphically defined fossil coral samples has been accomplished, and numerous other areas where only a few dates with limited control are available. In this section the dates available from some of the study areas will be briefly reviewed as this is the evidence which will be used to construct a global sea-level curve for the past 160 ky. The primary sites as well as sites less intensely studied are indicated in Fig. 18.1.

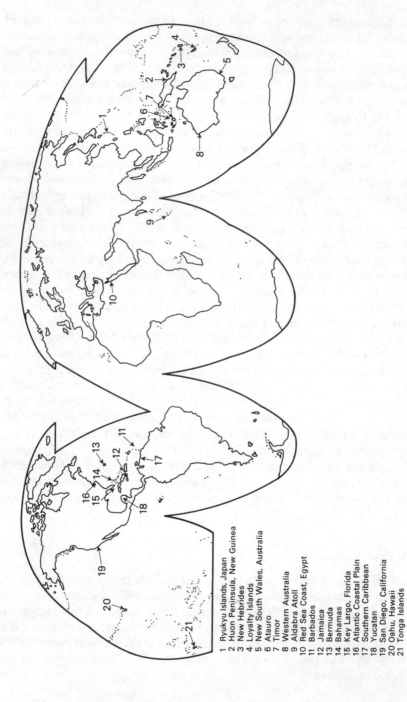

1 Ryukyu Islands, Japan
2 Huon Peninsula, New Guinea
3 New Hebrides
4 Loyalty Islands
5 New South Wales, Australia
6 Atauro
7 Timor
8 Western Australia
9 Aldabra Atoll
10 Red Sea Coast, Egypt
11 Barbados
12 Jamaica
13 Bermuda
14 Bahamas
15 Key Largo, Florida
16 Atlantic Coastal Plain
17 Southern Caribbean
18 Yucatan
19 San Diego, California
20 Oahu, Hawaii
21 Tonga Islands

Fig. 18.1. Areas referred to in the text where late Pleistocene corals have been dated using U-series techniques.

18.5.1 Barbados, West Indies

As discussed earlier the sea-level record from Barbados gives sets of discrete ages at 82, 104, 124, and 200 ky (Broecker et al. 1968). These ages have been confirmed by high precision $^{231}Pa/^{235}U$ ages on equivalent samples (Ku 1968). Subsequent to the work by Broecker et al. (1968), a younger terrace which dated 60 ± 2 ky was reported by James et al. (1971). Although it is of more limited extent than the older terraces, it lies clearly below the 82 ky terrace and should be considered to define another high sea-level event on the island. In addition to the record of high sea-levels on Barbados, evidence for an eustatic low sea-level between 125 and 105 ky has been reported by Steinen, Harrison, and Matthews (1973) based on speleothems. They estimated that sea-level fell to -71 ± 11 m relative to present during this time.

18.5.2 Huon Peninsula, New Guinea

The Barbados chart recorder model was extended to the Huon Peninsula, New Guinea, by Veeh and Chappell (1970), Bloom et al. (1974), and Chappell (1974a). More than 20 offlapping reefs are present within an 80 km stretch of coast. Variable uplift has warped these shorelines so that the altitude of a particular reef is variable along its exposure. The Huon record thus consists of a series of profiles across terraces of different elevations but correlated on the basis of stratigraphy and geomorphology.

In addition to the Barbados terraces at 60, 82, 104, 124, and 200 ky, terraces having ages of 41 and 140 ky have been documented from Huon (Bloom et al. 1974; Chappell 1974a). The 140 ky age is especially intriguing since the Milankovitch theory of orbital controls on the timing of glacial advances predicts a very cold period from 150 to 125 ky (Broecker 1966; Hays, Imbrie, and Shackleton 1976). Since much of the support for the Milankovitch theory is derived from the Pleistocene sea-level record as read in fossil coral ages, this seeming contradiction will be explored more fully in this chapter.

18.5.3 Ryukyu Islands, Japan

The Ryukyu Islands, of which Okinawa is most familiar to Westerners, lie on an active arc system, the Senkaku Gunto, where the Philippine Sea plate is being subducted beneath the Asian continental plate. This arc which is volcanically and seismically active explains the emerged limestone terraces found along the coasts of these islands. Limestone terraces which occur at elevations of 1 to 170 m have been dated by Konishi and his colleagues (Konishi et al. 1970; Konishi et al. 1974). The dating efforts were especially concentrated in the 1 to 100 ky range. Within this time interval, numerous comparisons were made among the $^{230}Th/^{234}U$, $^{231}Pa/^{235}U$, and ^{14}C dating techniques. These are summarized in Table 18.1. In most cases the $^{230}Th/^{234}U$ and ^{14}C ages agreed well in the 3 to 5 ky range but $^{231}Pa/^{235}U$ ages were

Table 18.1. U-series and ^{14}C age comparison in the Ryukyu Islands

Terrace description	Age (ky)		
	^{14}C	^{230}Th/^{234}U	^{231}Pa/^{235}U
Raised coral reef limestone, 1 to 7 m	4.76 ± 1.66 (6)*	3.73 ± 1.85 (13)	5.26 ± 2.7 (12)
Reef flat limestone and fossil reef, −6 to 1.6 m	3.24 ± 1.80 (7)	2.7 ± 1.0 (9)	7.5 ± 1.8 (6)
Raised coral reef limestone, 1.5 to 5 m	4.06 ± 2.20 (2)	1.5 ± 0.8 (2)	2.9 ± 1.0 (2)
Arik limestone, 19 to 28 m	> 37.8 (8)	40.1 ± 3.4 (2)	41.3 ± 5.5 (7)
Ryukyu limestone younger, 10 to 50 m older, 140 to 170 m	— —	52.6 ± 4.0 (5) 115 ± 14 (3)	57.7 ± 4.9 (4) 113 ± 20 (2)
Serikaku, 15 m	—	86 ± 8 (3)	92 ± 10 (2)

* numbers in brackets refer to number of samples dated, errors are standard deviations
(From Konishi et al. 1974.)

consistently several thousand years older. Samples greater than 35 ky showed good agreement between the two U-series chronometers.

High sea-level stands at 40 and 55 ky are extremely well documented on the Ryukyus; however, no ages greater than 124 to 128 ky have been reported. The Barbados II and the Huon VI terraces at 103 to 107 ky have not been reported from the Ryukyus. The Serikaku formation gives ages of 92 ky, intermediate between the 82 and 105 ky ages reported in Barbados and Huon. The only other areas which have yielded ages of 92 ky are the Red Sea coast of Egypt (Veeh and Geigengack 1970) and the U. S. Atlantic coastal plain (Cronin, Szabo, Ager, Hazel, and Owens 1981).

18.5.4 New Hebrides and Loyalty Islands

Subduction of the Australian plate beneath the New Hebridean Island arc along the New Hebrides Trench has tectonically raised the New Hebrides Islands and produced flights of emerged coral reefs (Neef and Veeh 1977). The Loyalty Islands stand between the New Hebrides and New Caledonia and consist of uplifted atolls having elevations of 50 to 140 m. The Loyalty Islands lie along a postulated lithospheric bulge; determining their uplift rate should be an aid in calculating the subduction rate of the Australian Plate (Dubois et al. 1977).

The results from the New Hebrides generally confirm the picture derived from Huon. The 140 ky Huon terrace seems to have an equivalent here which was dated at 134 ky (Neef and Veeh 1977). Other ages at 60, 81, and 120 ky are found at elevations ranging from 50 to 100 m in the New Hebrides.

The picture from the Loyalty Islands is confusing. Early studies yielded ages from 100 to 120 ky (Bernat *et al.* 1976); however Marshall and Launay (1978) have criticized this work because most of the samples were highly recrystallized. Working only with samples which were 100 per cent aragonite, and which did not appear altered in thin section examination, Marshall and Launay (1978) determined that the fossil corals found 2 m above present sea-level on Beautemps – Beaupré had ages of 200 ky. No samples measuring 120 ky were found in spite of extensive investigations. On Lifou ages of 174 and 189 ky were reported for the 3.5 m terrace. A 12 m terrace on Lifou had been dated by Bernat *et al.* (1976) at 120 ky using recrystallized samples. The only sample Marshall and Launay (1978) reported in the 125 ky range was from an elevation of 7.5 m on Ouveá. The lack of samples of this age suggests that instead of uplift, some of these islands may undergo periods of subsidence as well as emergence.

18.5.5 Timor and Atauro

These islands are located at a tectonic inflexion where the Sunda arc passes eastward into the Banda arc. Extreme tectonic activity in the area has uplifted these islands and produced flights of raised coral reef terraces. Samples from some of these terraces were dated by Chappell and Veeh (1978a). Atauro terrace 1b was dated at 103 ky and apparently correlates with the Barbados II and Huon VI terraces. Three samples from Timor similarly averaged 103 ky. Atauro terrace 2 gave a spread of ages from 111 to 160 ky. Chappell and Veeh (1978a) postulated that the lower part of the terrace which is separated by a disconformity was constructed by an early high sea-level at about 140 ky. After a regression, a major transgression caused the building of the main body of terrace 2 at about 120 ky. They emphasize that terrace 2 was almost certainly constructed by two high sea-levels about 15 ky apart rather than a single broad high at 125 ky.

On Timor a coral in growth position from the lower section of a 10 m terrace gave an age of 142 ± 8 ky while a sample in growth position from the reef crest dated 117 ± 7 ky.

18.5.6 Bahamas

In contrast to the tectonically active areas, the Bahamas offer a geologically stable area where sea-level features are most likely caused by eustatic changes. Paleo-sea level indicators include: (i) sea-cut notches; (ii) emerged reefs; (iii) emerged sea caves; and (iv) lithified coral rubble (Neumann and Moore 1975). Dates of fossil coral have yielded a broad range of ages from 100 to 145 ky with no evidence of distinct peaks that might represent specific sea-level maxima (Neumann and Moore 1975). The record on the Bahamas is thus fundamentally different from the record in the active areas. Although distinct sea-level events may be discerned from the field evidence, no strong concentrations of samples having particular ages have been observed. It is uncertain at this time if the spread in ages is due to intermittent flooding of the banks for a long period or because of statistical scatter in the data and various degrees of preservation of samples.

18.5.7 Bermuda

Stratigraphy on Bermuda is complicated by intercalated and interfingering shallow water marine carbonates and eolianites which are often but not always separated by paleosoils (Harmon et al. 1980). The interglacial Belmont formation at +1 m gives $^{230}Th/^{234}U$ ages in the range 200 to 230 ky; the Devonshire formation (+4 to +6 m) has been dated at 117 to 134 ky (Land, Mackenzie, and Gould 1967; Harmon et al. 1980). Isolated conglomerates considered to be storm deposits formed during periods of platform submergence give ages of 134, 126, 105, 97, 84, and less than 2 ky (Harmon et al. 1980). Harmon et al. (1980) stress that they see no evidence for a sea-level stand higher than the present except for the +4 to +6 m stand at 117 to 134 ky which deposited the Devonshire formation. Their evidence is based primarily on $^{230}Th/^{234}U$ dating of speleothems which indicate the absence of a fresh water table higher than present except during Devonshire time.

18.5.8 Oahu, Hawaii

Although somewhat more compressed in time, the Oahu record is quite similar to the Bahaman record with a general spread of ages from 112 to 137 ky (Ku, Kimmel, Easton, and O'Neil 1974). The mean of 23 $^{230}Th/^{234}U$ measurements, 122 ± 7 ky was supported by 15 $^{231}Pa/^{235}U$ ages which averaged 113 ± 15 ky. The conclusion reached by Ku, Kimmel, Easton, and O'Neil (1974) was that the Waimanalo shoreline on Oahu has an age of 122 ky with the expected statistical scatter about the mean. Stearns (1976) and Chappell and Veeh (1978a) pointed out that the Oahu samples fell into two groups: (i) samples of Waimanalo limestone and cemented conglomerate which averaged 133 ky and (ii) samples from deposits which lie disconformably on the Waimanalo limestone or separated from the older group by the Diamond Head tuff which averaged 119 ky.

Thus, we have two fundamentally different interpretations of the data which cannot be resolved because of the statistical spread inherent in the dating technique. What is important is that the Oahu data do not rigidly define a single 125 ky high sea-level but that they may also be interpreted as two high sea-levels of about the same magnitude but separated by about 15 ky.

18.5.9 Jamaica

The north coast of Jamaica has numerous exposures of fossil coral. The late Pleistocene Falmouth formation consists of two units separated by an unconformity (Land and Epstein 1970). Samples from the upper (+ 5 m) and lower (+ 2 m) units were dated by Moore and Somayajulu (1974). The average age of seven samples from the upper unit was 134 ± 4 ky; seven samples from the lower unit gave an average age of 123 ± 5 ky. As in the case of Oahu these ages can be interpreted as one high sea-level at about 128 ky with statistical scatter about the mean or as two high sea-level events separated by about 11 ky. The presence of the unconformity between the units strengthens the bipartite sea-level interpretation.

18.5.10 Western Australia

The western coast of Australia is considered an area of crustal stability where sea-level features should reflect eustatic changes (Veeh et al. 1979). At Rottnest Island near Perth, Szabo (1979a) obtained an average age of 132 ± 5 ky from an emerged coral reef 2 to 3 m above sea-level. Veeh (1966) had dated this deposit at 100 ± 20 ky. Field studies on the mainland at Lake MacLeod were conducted by Veeh et al. (1979) who determined an average age of 128 ky for deposits occurring between -0.5 and 1.3 m on the west shore and mean ages of 123 ky and 131 ky for deposits overlying two morphologically distinct marine erosional terraces in the Cape Cuvier area. At Cape Range a mean age of 123 ky was determined for the youngest terrace deposit at 5.5 m. Although these ages are suggestive of a bipartite sea-level, the authors do not consider the age differences as being resolved.

18.5.11 Aldabra

Corals from Aldabra Atoll in the Indian Ocean were dated by Thomson and Walton (1972). The mean of eight ages from the Upper (2 to 4 m) limestone was 127 ± 9 ky, the range was 118 to 136 ky. A lower limestone unit was present on the atoll but none of the samples from this unit met the dating criteria.

18.5.12 Southern California

The Nestor Terrace is a well developed topographic feature in the San Diego area. Ku and Kern (1974) obtained ages ranging from 109 to 131 ky (mean = 121 ± 11 ky) for three coral samples from this terrace. Two coral samples for Terrace 2 at San Nicolas Island were dated at 87 and 120 ky by Valentine and Veeh (1969) and one coral from the Cayucos Terrace gave an age of 124 ± 27 ky (Veeh and Valentine 1967). Although these ages are useful within the region for correlating terraces, the range and uncertainty of the ages is too great to resolve closely spaced sea-level events.

18.5.13 U.S. Atlantic Coastal Plain

Although local stratigraphy of the Atlantic Coastal Plain is known in great detail, the lack of absolute ages has made correlations among different sections tenuous at best. The only study which utilizes numerous $^{230}Th/^{234}U$

ages on suitable coral samples is reported by Cronin *et al.* (1981). Ages of 188, 120, 94, and 72 ky appear to correlate in sequence if not in absolute age to prominent high sea-levels documented on emergent coasts. For the 94 and 72 ky high to represent transgressive sea-levels on the Atlantic Coastal Plain, the area would have to be rising as fast as Barbados, a possibility Cronin *et al.* (1981) consider unlikely. Instead they suggest that hydroisostatic adjustment from interglacial/glacial sea-level transgressions and regressions caused ocean basin depression and Atlantic continental margin uplift. Thus uncertain tectonics obscure the eustatic picture when viewed from the Atlantic Coastal Plain.

18.5.14 Yucatan and Southern Caribbean

Ages from the tectonically stable Yucatan Peninsula cluster around 122 ± 2 ky with no evidence for other periods of sea-levels higher than present (Szabo *et al.* 1978). These samples documented a single high stage which reached a maximum elevation of + 6 m relative to present.

Limited data are available from coral terraces on Caribbean islands off the coast of Venezuela (Macsotay and Moore 1974; Schubert and Valastro 1976; Schubert and Szabo 1978). As with numerous other studies, the ages fall in the range 120 to 140 ky with no apparent relationship between the age determined and its reliability. Whether these ages represent a single broad sea-level high centred at about 130 ky or two peaks separated by about 15 ky cannot be resolved using the data available.

18.5.15 Tonga Island Arc

Ages of emerged coral reefs from three islands in the Tonga frontal arc were reported by Taylor and Bloom (1977). On Eua Island a sample at 3.5 m gave an age of 133 ky, on Tongataup a sample 5.5 m above high-tide dated 135 ky. These samples indicate that the net uplift of these Tonga Islands has been very small over the past 135 ky, and lend further support to a high sea-level at about 135 ky.

18.5.16 Synthesis of case studies

These detailed regional studies have yielded a reasonably consistent picture of sea-level changes over the past 150 ky. Certain ages which are documented in several places seem well established. Among these are high sea stands at 40, 60, 80, 105, and 120 ky. Such ages generally fit the predictions of the Milankovitch theory (Chappell 1974*b*). Recurring U-series dates for a high sea-level at 135 ky are especially troubling, however, because according to the predictions made by the Milankovitch theory, this should be a very cold period. The fossil coral ages from Timor, Atauro, the New Hebrides, Oahu, Huon, Jamaica, and Tonga strongly support a sea-level close to the modern high at about 130 to 140 ky. It is possible that these ages represent slightly altered samples which were actually deposited at 125 ky. A 7 per cent loss of U, increasing the ^{230}Th$/^{234}$U activity ratio from 0.70 to 0.75, could explain these older ages as could a 7 per cent gain of ^{230}Th. These alterations would need to occur in such a manner that the changes would not be detected by thin section or X-ray diffraction examination, because many of these older ages were obtained from samples that met criteria for reliable ages as discussed previously. To gain further insight into this problem, the δ^{18}O record in deep sea cores must be considered.

Most deep sea sediment cores with moderate to low accumulation rates record a single warm period which corresponds to a sea-level high at 125 ky (Shackleton and Opdyke 1973; Shackleton 1977). This has generally been accepted by marine geologists as the time of Termination II or the stage 5/6 boundary. The chronology for the δ^{18}O time-scale in marine sediments (and hence the age of Termination II) is based on (i) extrapolation of sedimentation rates from the Brunhes–Matuyama magnetic reversal; (ii) correlations with the 125 ky high sea-level stand from fossil corals, and (iii) limited ^{230}Th – and ^{231}Pa – based sedimentation rate determinations. Kominz et al. (1979) have redetermined sedimentation rates on core V28 – 238, the one on which Shackleton and Opdyke (1973) based their δ^{18}O time-scale. Their best estimates for the age of Termination II are 138 ky (based on an assumed constant Al accumulation) and 145 ky (based on an assumed constant rate of deposition of ^{230}Th at the site). Otherwise, the Brunhes time-scale they developed is quite similar to earlier versions.

There is also the question of whether Termination II is represented by a broad sea-level maximum or two sharper maxima. As Chappell and Veeh (1978a) have pointed out the deep-sea record could be smoothed by bioturbation so that two closely spaced warm peaks might be obscured. Recently, evidence has become available that a major meltwater event flooded the surface of the Gulf of Mexico at 138 ky (Falls and Williams 1980). This event which was separated clearly from Termination II (the 5/6 boundary, taken to be 125 ky) was identified on the basis of ^{18}O – depleted planktonic forams from four cores having accumulation rates as high as 7 cm/ky. Stable isotope analyses of a benthic foram species in the same cores recorded the same event but with less magnitude than exhibited by the near-surface dwelling species. The duration of the meltwater spike was 3 ky. Additional evidence of a meltwater spike at 135 to 140 ky has been found by J.-C. Duplessy (1980, personal comm.) in cores collected from the Labrador Sea and by A. Aksu (1980, personal comm.) in cores from Baffin Bay. These studies offer convincing confirmation of major glacial melting at about 135 ky and lend credence to the coral dates for a high sea-level stand at that time. If the time-scale of Kominz et al. (1979) which places Termination II at 140 ky is used, the age of this meltwater spike becomes 155 ky, clearly older than any coral ages from the last interglacial.

18.6 Extension of U-series techniques

As the ^{230}Th/^{234}U activity ratio approaches unity, the errors on ^{230}Th and ^{234}U determinations become increasingly important and lead to large uncertainties in age. Thus ^{230}Th/^{234}U coral ages beyond about 250 ky are usually given as minimum ages. To date older materials some workers have used He/U dating. Fanale and Schaeffer (1965) first applied He/U measurements to fossil coral with encouraging results. Bender (1973) developed the method further and Bender et al. (1979) applied it to the reef tracts of Barbados. Discordant ages for samples in the range of 200 to 400 ky based on ^{234}U/^{238}U, ^{230}Th/^{234}U, and ^{4}He/U indicated secondary ^{234}U and ^{230}Th additions to the samples. After applying corrections to the He/U data, Bender et al. (1979) obtained terrace ages of 180, 200, 220, 280, 300, 320, 460, 490, 520, 590, and 640 ky. For each interglacial isotopic stage in the deep sea sediment record, they dated at least one Barbados reef tract. Although the absolute ages are uncertain, the correlations seem reliable to plus or minus one interglacial isotopic stage.

18.7 A sea-level curve for the late Pleistocene

In Fig. 18.2 the fossil coral ages have been synthesized into a late Pleistocene sea-level curve using primarily the emergence rates and ages from Huon. Most of the features of this curve are reflected in earlier curves from Steinen et al. (1973), Bloom et al. (1974), and Chappell and Veeh (1978a). The evidence for inflexion points on the curve is cited in this section. The relative sea-level high stands follow the Roman numerical designations of the Huon reefs.

1. The high sea-level at 5 ky, which is approximately at the same level today, is recorded in Holocene reefs in many areas of the world.
2. The low sea-level at 18 ky (-130 to -160 m) represents the Late Wisconsin glacial maximum.
3. The relative high sea-level at 28 ky (-40 m) is tenuous. Huon reef complex II has been dated at 28 ky using ^{14}C from Tridacna (Chappell 1974a); however, this age has not been confirmed in other study areas.
4. The high sea-level at 40 ky (-40 m) is recorded at Huon and the Ryukyus and seems well established.
5. The high sea-level at about 50 ky (-40 m) has been dated only at Ryukyus. It has a stratigraphic expression at Huon (reef complex IIIa), but the ages of 50 ky which Veeh and Chappell (1970) reported were from 20 m below the crest of the reef and may not date the actual IIIa high (Bloom et al. 1974).
6. The high sea-levels at 60 ky (-28 m) and 84 ky (-20 m) have been dated at Huon, the Ryukyus, Barbados, and the New Hebrides. On Bermuda a sea-level above -20 m dated close to 86 ky.

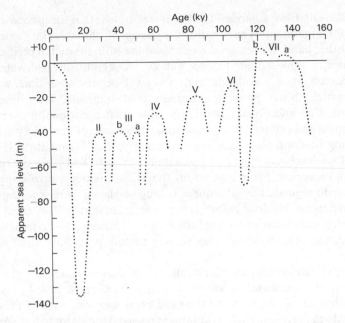

Fig. 18.2. Late Pleistocene sea-level curve relative to present sea-level. The high stands are denoted by Roman numerals following the designations of Huon, New Guinea terraces.

7. The 105 ky high sea-level (− 15 m) is based on terraces present at Barbados, Huon, Atauro, and Timor. Terraces of this age have not been reported from the Ryukyus or New Hebrides.
8. The low sea-level at 111 ky (− 70 m) is from the micritic crusts on Barbados.
9. The 120 ky high sea-level (+ 7 m) has been dated at almost all dated localities where late Pleistocene corals lie above present sea-level.
10. The 135 ky high sea-level (+ 2 m) occurs in numerous areas, often distinct stratigraphically from the 120 ky terrace. It has recently been shown that a major meltwater event flooded the Gulf of Mexico at 138 ky, close to the time of this high stand (Falls and Williams 1980).
11. Sea-level at 160 ky was about − 40 m as revealed by dating speleothems in Bahaman 'blue holes' (Gascoyne et al. 1979).

18.8 Implications for theories of the Pleistocene

There are clear correlations between ages and magnitudes of high sea-level stands deduced from fossil corals and predictions of warm northern hemisphere summers derived from the Milankovitch theory (Broecker 1966;

Mesolella *et al.* 1969; Chappell 1974*b*). It is also clear that the precessional and orbital forcing functions have a direct influence on climate (Hays *et al.* 1976). On the other hand, the sea-level record and the Milkankovitch prediction are diametrically opposed at 135 to 140 ky. A significant meltwater event accompanied by a worldwide high sea-level occurs at a time when the Milankovitch theory predicts extremely cold summers. This observation implies that an additional climatic variable must occasionally override the precessional and orbital forcing functions. In the case of the 135 ky event, this overriding function may have been responsible for Termination II since it brought sea-level near the 120 ky maximum 15 to 20 ky before the predicted warming transition. From that point, the Milankovitch controls apparently took over to regulate the subsequent timing of glacial–interglacial events as they had presumably done before. It is possible that if the 135 ky event had not occurred, precessional and orbital effects would have led to Termination II at the predicted time; however, the answer to this possibility shall never be known.

In Pleistocene geology there is no dearth of theories to explain the growth and decay of continental ice sheets. It is not the purpose here to review the theories but to call attention to the need of at least one forcing function in addition to the Milankovitch. Until this is resolved the claim to have unlocked all of the mysteries of the Pleistocene climatic cycles cannot be made.

Acknowledgments

I thank Doug Williams for reviewing the manuscript and Dee Hansen for drafting the figures and typing the text.

19 PROGRESS AND PERSPECTIVES
T.-L. Ku

19.1 Introduction

About five years ago I prepared a status report on the use of the U and Th decay series nuclides in age determination (Ku 1976). A strong impression formed then was the ever-quickening pace of this comparatively young branch of isotope geochemistry. This impression reasserts itself as one goes through the contents of this volume, noticing that over 45 per cent of the cited references are post-1976 publications. It is clear that besides geochronology, progress made in the use of the decay series nuclides as time and property tracers for a variety of geochemical processes and environments has been equally significant. The wide-ranging topics of study encompassed by these nuclides are indeed unique among many isotopic systems studied.

In their study of the deep-sea ^{210}Pb–^{226}Ra relationship, Craig *et al.* (1973) remarked euphemistically that '[The ^{238}U] family truly provides a 'Forsyte Saga' of the sea, which continues to surprise and entertain its audience of geochemists'. Reference was made to the different scavenging patterns of ^{238}U and its radiogenic descendants in sea water: Ancestral ^{238}U has an oceanic residence time of more than 10^5 years, while its offspring ^{230}Th is scavenged in only tens of years. The ^{230}Th decays to ^{226}Ra in the sediment, which wanders back to the sea; while decaying away there, ^{226}Ra produces ^{210}Pb which is eventually removed from the deep abyss in the order of fifty years. Of course, there exists a host of other parent/daughter disequilibrium relationships that are as intriguing as the above scavenging patterns. These relationships are found not just in the ^{238}U family, but in the families of ^{235}U and ^{232}Th as well. Thus, the major disequilibrium patterns among nearly all the decay-series nuclides of geochronological interest have been searched out in a rich number of natural systems.

As reviews in this volume clearly show, the field has advanced with great strides along many of its fronts. It would seem impractical as well as redundant to attempt a summary of progress and perspectives in a comprehensive way. What follows is simply a personal appraisal, relying on views which are at times impressionistic and, therefore, not entirely free from bias and omission.

19.2 Geochronology studies

The ^{230}Th and ^{231}Pa methods remain the best geochronological tools for dating in the range of 30 to 350 ky. Most of the applications have been to marine carbonates (corals and molluscs) and deep-sea deposits (sediments and Mn-nodules), although efforts to date other systems have been made, in particular cave deposits (speleothems). The major impetus for these applications comes from research into the Quaternary climates. The U-series dating of fossil coral reefs has contributed greatly to the chronology of glacio-eustatic sea-level changes (Moore, chapter 18). From Holocene back to at least 150 ky, major sea-level fluctuations with periodicities of the order of 20 ky have been identified —not a small achievement. However, on a finer scale, more work waits to be done in tying this sea-level history with the δ^{18}O record in deep-sea cores and with the timing of northern hemisphere summer insolation predicted by the Milankovitch theory (which has by and large influenced the thinking of the paleoclimatic community). There is the problem of the reported bipartite high sea stands at about 120 and 135 ky, corresponding to stage 5e of the foraminiferal δ^{18}O stratigraphy. The problem arises from the fact that the summer insolation curve shows that a cold period with large volume of continental ice existed 135 ky ago. Furthermore, not until the recent attempt of Kominz et al. (1979) has the stage 5e/6 boundary in deep-sea cores been dated at about 125 ky. As field evidence for the dual sea-levels is meager and fragmentary, one asks whether the age difference of 10 to 15 ky is resolvable from coral dating. This question becomes even more relevant in view of the appearance of coral ages around 60 ky, 72 ky, and 94 ky (see chapter 18), in addition to the 82 ky, 105 ky, and 122 ky reef sequence originally found on the island of Barbados. Clearly, to accept all these ages as representing intervals of global warming would require a major rethinking of paleoclimatic models.

Are we in a position to answer the question posed above? Listed in Table 19.1 are results of multiple ^{230}Th/^{234}U age determinations on three carbonate samples. The two coral samples are from the Barbados III terrace and the lower (6 to 10 m) terrace of Curacao. The speleothem sample 76001 is a laminated flowstone collected from Sumidero Terejapa, Chiapas, Mexico. All three were used in the U-Series Intercomparison Project (USIP). This project was initiated in 1976 as a programme of interlaboratory calibration. To date the project has entered its third phase of studies, and more than forty laboratories from nine countries have participated. During Phase I studies (Harmon et al. 1979), each participating laboratory used its own standards (spike solutions) for the measurements of U and Th isotopes.

The USIP Phase II entailed the use of a common spike as well as each laboratory's own spikes. As shown in Table 19.1, ages obtained for the terraces at Barbados and Curacao are indistinguishable within the error limits. The age values and their spreads for each terrace are about the same as those involved

Table 19.1. Multiple measurements of ^{230}Th/^{234}U ages on corals from Barbados and Curacao, and on speleothem (76001)

Barbados III	Curacao	76001
Broecker et al. (1968) 122 ± 6 ky (N = 6)	Schubert and Szabo (1978) 129 ± 7 ky (N = 5)	Gascoyne (1980) 48.2 ± 1.7 ky (N = 9)
USIP Phase I 118 ± 9 ky (N = 13)	USIP Phase I 124 ± 15 ky (N = 14)	USIP Phase II (a) 48.1 ± 8.3 ky (N = 19) (b) 49.0 ± 5.8 ky (N = 9)
USIP Phase II (a) 133 ± 16 ky (N = 19) (b) 130 ± 13 ky (N = 13)	USIP Phase II (a) 133 ± 12 ky (N = 19) (b) 132 ± 8 ky (N = 12)	

Notes: N = number of analyses. Uncertainties are one standard deviations from the mean.
(a) measured with the inter-laboratory spike.
(b) measured with each laboratory's own spike(s).

in the aforementioned 'bipartite sea stands'. Evidence showing stratigraphic separation for the samples is lacking (R. Matthews, personal comm.) nor is there indication of systematic error due to standardization (cf. Phase II results). It is important to notice that for all three samples, the statistical scatters about the means attained by a single laboratory are far better than those by different laboratories using a common standard. This could reflect some unevenness in the quality of analyses among laboratories. If true, then as far as measurement precision is concerned, the resolution of age difference of the order of 10 to 15 ky should be plausible for samples 30 to 150 ky old. What remains to be explored are the following:

1. To find out the age range that is expected for the formation of a given terrace. (This awaits tests on the among-sample variations within a single outcrop and the among-outcrop variations within the same well-defined morphostratigraphic unit of a raised terrace.)

2. To distinguish suitable specimens of coral from the unsuitable. Here we must take full advantage of the unique isotopic constraints provided by corals, besides relying on the petrologic criteria such as preservation of aragonitic and delicate internal structures (for example, columella and septae). We are dealing with a well-constrained system with three built-in clocks: ^{234}U/^{238}U, ^{230}Th/^{234}U, and ^{231}Pa/^{235}U. Strict concordancy checks among these internal clocks would ensure integrity of the sample as a closed system for the nuclides of interest.

Simultaneous measurements of ^{231}Pa/^{235}U and ^{230}Th/^{234}U ages have not been carried out as commonly as they should. The need for such a crosscheck cannot be overemphasized. Even if the ^{231}Pa method becomes highly inaccurate beyond 120 ky, the ratio ^{231}Pa/^{235}U serves as a valuable index of diagenetic disturbance (a ratio of 1.00 conveys more information than just a minimum age estimate!). In materials with U content of secondary origin such as fossil molluscs, bones, and teeth, knowing when the system becomes closed to U incorporation is of utmost importance. Means of securing this knowledge have yet to be found. However, concordance between the ^{230}Th/^{234}U and ^{231}Pa/^{235}U ages would give assurance to the fact that the U uptake at least has ceased early in the history of the sample. A hypothetical case should be examined to illustrate this point. Suppose that all the U in a fossil bone was picked up continuously at a constant rate after the death of its host; then one would expect the bone to show discordant ages such that apparent ^{230}Th age is greater than apparent ^{231}Pa age. Taking the average apparent age as $A = (^{230}$Th age $- ^{231}$Pa age)/2, A can be plotted against true age of the sample as in Fig. 19.1, for the cases of influx U having ^{234}U/^{238}U ratios of 1.00 and 3.00. It can be seen that as the sample becomes older, the difference between A and true age increases, and so does the discordancy between the ^{230}Th and ^{231}Pa ages (defined as $\Delta t(\%) = \dfrac{^{230}\text{Th age} - ^{231}\text{Pa age}}{A} \times 100$). In all prob-

ability, the situation considered here tends to be extreme; it could well be that the uptake of most of the U occurs during an earlier part of the sample's history or that it has rates which diminish with time. In any event, the example

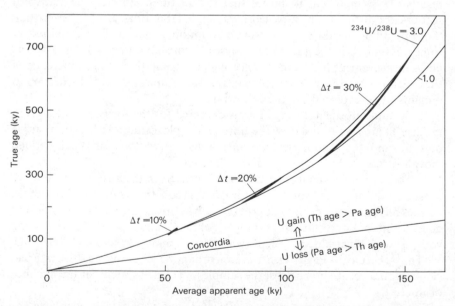

Fig. 19.1. The discordancy (denoted by Δt (%), see text for definition) between the measured ^{230}Th/^{234}U and ^{231}Pa/^{235}U ages as an indication of their deviation from the true age, for a hypothetical system in which all U is of secondary origin, and its uptake rate is constant throughout the sample's history. Two cases are shown where the taken up U has ^{234}U/^{238}U ratios initially fixed at between 1.00 and 3.00.

points to the necessity and usefulness of making the simultaneous measurements of $^{230}Th/^{234}U$ and $^{231}Pa/^{235}U$ ages as precise as possible. As interest in U-series dating of bones has grown, in view of its potential contribution to archaeology and other studies, one is especially reminded of taking advantage of the high U-content of bones for precise $^{231}Pa/^{235}U$ as well as $^{230}Th/^{234}U$ and $^{234}U/^{238}U$ analyses.

There is still room for improvement in the chronology of the late Pleistocene $\delta^{18}O$ record (hence its correlation with the coral reef and speleothem records) in deep-sea cores. A case in point is the age of $\delta^{18}O$ stage-boundary 5e/6 mentioned earlier. The ^{230}Th and ^{231}Pa methods involve the precarious assumption of constant deposition rates of sediment containing ^{230}Th and ^{231}Pa at a given point on the sea floor. The constancy of sediment influx is particularly suspect. Use of the $^{230}Th/^{231}Pa$ ratio as a direct age index is impractical. This ratio is inherently sensitive to sediment mixing, and is found to exhibit anomalous values (relative to the production ratio from U in the sea) in core tops, Mn-nodules, and particulate matter. To circumvent possible sediment-deposition-rate fluctuations, one is left with an alternative method which only assumes a uniform deposition rate of ^{230}Th or ^{231}Pa over a given area on the ocean floor—the continuous strip method. Initial applications of this technique (Sackett 1964; Kominz et al.1979) have shown reasonable success. For further research in this important area of geochronology, it would be best to include: (i) work on cores with well-preserved tops and well-known porosities; and (ii) tests for the constancy of accumulation rate of ^{230}Th and ^{231}Pa by mesurements of the U-series nuclides in strips of sediment from ^{14}C-dated cores. Perhaps comparisons between the late Wisconsin (20 to 15 ky ago) and the middle Holocene (8 to 3 ky ago) intervals would provide the most sensitive test.

U-series dating has much to contribute to problems related to the generally more variable environments on land. Research in this area still requires much ground work to enlarge the data base and to test the applicability of the method on a variety of materials. One is encouraged as one sees from chapters 6 to 8 and 10 to 13, that efforts toward fulfilling such requirements have been well started.

Significant progress is being made in the study of speleothems. Uranium-series dating coupled with stable isotopic analysis on both calcite and fluid inclusions provides paleoclimate information in continental settings, enabling detailed comparisons to be made with the marine records. Dating submerged caves from islands like Bermuda and the Bahamas has offered a unique means of determining the timing and extent of glacial sea-level lowering. Applications to problems in archaeology and geomorphic studies have been equally rewarding (see chapters 11 and 12).

The advantage of dating speleothems lies in the possibility of checking for internal consistency in a well-defined depositional sequence, even within a single specimen. Additionally, such a sequence affords the chance to test some of the concepts held regarding the behaviour of the isotopic system. For instance, the idea that the presence of ^{232}Th in calcite implies contamination of extraneous ^{230}Th in some fixed proportion to ^{232}Th on a regional basis, can be tested by analysing both clean and 'dirty' layers. Also, the ^{234}U/^{238}U ratios in the sequence may serve to index the long-term stability (or variability) of this ratio in ground-water at a given locality. Study of this aspect has already begun, but the results seem inconclusive (see chapter 11). Does the suggested fractionation between the two U isotopes indeed occur during calcite precipitation? Could some of the ^{234}U/^{238}U variations in speleothems be of diagenetic origin (for example, post-depositional migration of recoiled ^{234}U)? How firmly can the seasonal change of ground-water ^{234}U/^{238}U be taken to mean that this ratio responds to climate change on a longer-term basis? These kinds of questions are important in ground-water tracing (see chapter 9) as well as dating. They call for further research. Speleothem studies are somewhat limited by the problems of access to the deeper caves and of conservation. Most studies have been carried out in North America and Europe. Through recent contacts with our colleagues in China, there are reasons to believe that new work will soon be available from that part of the globe.

The need for dating of a large variety of impure carbonates (travertine, tufa, caliche) in the terrestrial environment, and the complications involved, have been expounded in detail in the previous chapters. Suffice it to say that fulfillment of this vital need constitutes a major challenge to workers in the U-series field for years to come. Besides the usual concern about the closed-system assumption demanded in all geochronological work, the problem involved here is the formidable task of dealing with detrital contaminants present in these mostly inorganically precipitated materials. Simple physical or chemical means of effectively isolating the pure carbonate matrix for analysis do not exist. Any correction procedures for the effects of detrital contamination require knowledge of the isotopic composition of the embedded detritus, or of the fractionation effect if selective leaching with dilute acid is employed. Acquiring such a knowledge must be a target for future research. Consider the following: carbonates can be dissolved in dilute acid, but in the process U and Th isotopes from the detrital component may also be mobilized to a different extent. One way to correct this fractionation is to compare the isotopic composition of the acid-insoluble residue with that of the detritus. Lacking information for the latter, the comparison is possible if one *assumes* that ^{238}U, ^{234}U, and ^{230}Th are in secular equilibrium with one another (Ku *et al.* 1979; Ku and Joshi 1980). The method of pseudo-isochron plots (cf. chapter 10, Fig. 10.4) obviates the need for such an assumption, but it has to *assume* no isotopic fractionation during acid leaching. Multiple analyses of total samples of coeval age circumvent this fractionation problem. Here, however, a different assumption about the detrital phase is made. This can be seen from the age equation (eqn 3.4) written in the form of:

$$
\begin{aligned}
\left[\frac{^{230}Th}{^{232}Th}\right] = & \left[\frac{^{230}Th}{^{232}Th}\right]_i (e^{-\lambda_{230}t}) \\
& + \left[\frac{^{238}U}{^{232}Th}\right]\left\{ (1 - e^{-\lambda_{230}t}) - (1 - e^{-(\lambda_{230}-\lambda_{234})t}) \times \frac{\lambda_{230}}{\lambda_{230} - \lambda_{234}} \right\} \quad (19.1) \\
& + \left[\frac{^{234}U}{^{232}Th}\right] (1 - e^{-(\lambda_{230}-\lambda_{234})t}) \cdot \frac{\lambda_{230}}{\lambda_{230} - \lambda_{234}}
\end{aligned}
$$

where the subscript i denotes initial quantity. The two unknowns in the equation are t, the time of carbonate precipitation, and $[^{230}Th/^{232}Th]_i$, the initial ratio of the two Th isotopes (presumably) in detritus. From the analyses of two or more samples of coeval ages with different U/Th ratios, t can be calculated if one *assumes* that the detrital component of the sample has the same $[^{230}Th/^{232}Th]_i$ (Ku et al. 1979). (Note that eqn (19.1) is used in the dating of igneous rocks in which $^{234}U = {^{238}U}$ generally holds; see chapter 6, eqn (6.1).) Other correction models for multiple analyses of total samples can be devised (for example, Schwarcz 1980), but all involve surmises about the isotopic makeup of the detrital phase. As it stands, work on determining the nature and regional variability of isotopic compositions of detrital materials and on selective leaching experiments seems to have fallen behind the efforts spent in modelling corrections for them—a trend that should be reversed, at least in this early stage of development.

Perhaps, mention should also be made of the importance of furthering studies in areas such as: comparison with other dating methods, particularly ^{14}C, K/A and fission-track; radioactivity distributions in samples via fission-track and alpha track analyses; and dating co-occurring materials, for example coral/molluscs and bones/cave flowstones. These investigations could provide much-needed insights into the causative aspects of age discordancy. More of these studies are anticipated.

19.3 Tracer studies

There are ten different chemical elements in the three decay series; all of them have two or more radioisotopes. These isotopes possess a wide range of decay half-lives. The usefulness of exploring the many inter-relationships among them for tracing geochemical and geophysical processes has been demonstrated in the preceding chapters. Perhaps some of the special advantages of choosing this group of nuclides as tracers are worthy of repeating. Firstly, the time element or rates of the processes being traced can be explicitly introduced. Secondly, the material balance of the system being studied can often be checked from the known distribution of the parent radionuclides.

Thirdly, compared with stable trace elements, these isotope tracers generally can be measured with greater sensitivity and less contamination problems.

Another important facet of the progress made recently has been the improvement in the analytical and sample collection capabilities. With the exception of those nuclides with short half-lives, the majority of the U-series nuclides of geochemical interest can now be routinely measured with precisions of better than 5 per cent. Improvement in alpha spectrometry is augmented by the availability of high resolution gamma spectrometry and low-level counting facilities. The U and Th alpha spectra submitted by the participating laboratories during the aforementioned USIP programme show that tail-corrections in the spectral analysis largely become unnecessary. This was certainly not the case ten to fifteen years ago; thanks are due to the advance in nuclear instrumentation and the refinement in the thin-source preparation techniques. Efforts toward intercalibration of standards have been initiated, with the first establishment of an inter-laboratory $^{232}U/^{228}Th$ standard through USIP. Tying standards of Ra and Po to this primary USIP $^{232}U/^{228}Th$ standard would seem a logical follow-up.

Sampling of water in large volumes for radioisotope analysis has been greatly facilitated by the invention of Mn-oxide coated acryllic fibres or other similar fabric (Moore and Reid 1973; Anderson 1980). Collection of sufficient quantities of oceanic suspended particulate matter for isotopic as well as other investigations has become possible in recent years. This is done by use of devices such as large volume *in situ* filtration systems (for example, Krishnaswami, Lal, Somayajulu, Weiss, and Craig 1976) and sediment traps (for example, Honjo 1978; Spencer *et al.* 1978). Large-diameter and slowly-sunk corers enable the recovery of ocean and lake bottom sediments with minimum degree of disturbance, so that water–sediment interface phenomena can be studied. Future efforts will see the collection of the so-called 'fluff material' commonly observed to exist at this interface. This material, unretrievable by coring, may hold some key information about the nature of biogeochemical transformation on the ocean and lake floors. (The current Mn-Nódule Program, MANOP, in the US has plans for sampling the fluff in deep sea using manned submersibles.) All these developments have contributed to the greater activity in this field, and they will undoubtedly continue to do so.

The field of chemical oceanography has matured considerably during the past two decades. It is fair to say that the use of U-series nuclides as marine geochemical tracers contributed handsomely to this maturation. As Cochran has observed (see chapter 15), ^{222}Rn, ^{228}Ra, and ^{226}Ra have been used for tracing water movement and mixing. Distributions of ^{210}Pb, ^{234}Th, ^{228}Th, and ^{230}Th, coupled with those of their more soluble parents ^{234}U, ^{226}Ra, and ^{228}Ra, yield information about particulate uptake rates of reactive elements (for example, Th and Pb) and settling velocities of particles in the sea. To assess the rates of bioturbation in marine deposits, tracers are provided by ^{234}Th (for estuarine sediments) and ^{210}Pb (for the more slowly depositing deep-sea sediments). In addition to determining the rate of vertical mixing in bottom waters, ^{222}Rn is also a powerful tracer for studying the gas exchange across the air–water interface and the material transfer between the pore water and overlying water. These ^{222}Rn studies apply to river, lake, and estuarine systems as well as to the ocean.

In all these applications various models have to be formulated. As data coverage improves, so does sophistication in the modelling, leading to refinement of the assumptions used. Ultimately, the accuracy of model fit offers information about the dynamics of the systems studied, or even some of the physics and chemistry involved, so that a new generation of models can be made more realistic. Such is the pattern of pursuit. In interpreting the ^{226}Ra distribution, transport equations written in terms of diffusion, advection, and *in situ* production and decay have been used to replace simple box models. Measurements of ^{222}Rn in the deep Atlantic and Pacific show that, except in major oceanic passages, the product of the apparent coefficient of vertical eddy diffusion and the density gradient is nearly constant with geographic location and height above the sea floor. This is equivalent to saying that the buoyancy flux (i.e. the rate of energy dissipation against gravity) is uniform (Sarmiento *et al.* 1976) —an observation which, if proved correct, would have important implications for the modelling of bottom-water mixing. The first-order, irreversible chemical scavenging model of reactive species in the sea is being challenged; results of simultaneous analyses of three Th isotopes (^{234}Th, ^{228}Th, ^{230}Th) on both dissolved and particulate forms suggest that desorption and regeneration may operate to return scavenged Th to solution (Bacon and Anderson 1980).

Measurements of particulate and dissolved ^{230}Th and ^{231}Pa have confirmed the fractional precipitation of the two nuclides generated by U in the sea. The initial 'surprise', caused by this seeming oceanic puzzle from earlier studies of deep-sea deposits, has given way to the contemplation of its controlling factors and implications, which remain unclear. While in the open ocean settling particles preferentially scavenge Th over Pa, nearshore they seem not to do so. Further, the efficiency of scavenging for both seems to differ between the two environments, with higher removal rates occurring in marginal basins (Anderson 1980). Factors such as concentration and nature (for example size, and chemical and mineralogical composition, etc.) of particles could all play a part, but they have yet to be identified. Success in the latter task would also aid our understanding of the geochemical behaviour of other reactive trace metals.

A first-order picture has emerged which shows that ocean boundaries (i.e. areas near the sea floor and continental margins) are active scavenging sites for the isotopes of Th, Pa, and Pb. This 'boundary scavenging' process, as mentioned in chapter 15, is worth probing further. It bears not only on removal mechanisms for reactive metals in general, but also on reactions such as cycling of Mn and Fe near the sediment–water interface. In association with this process is the transport along concentration gradients to the ocean boundaries through horizontal and vertical mixing. There is more to be learned about mixing rates in a two-or three-dimensional framework, not just through the one-dimensional models so far attempted in the majority of cases. In this regard, since Rn and the two Ra isotopes (^{226}Ra and ^{228}Ra) have inputs from the sea floor, a major task will be the determination of their bottom-source functions.

Turning from marine to terrestrial settings, one encounters systems with much more varied and complex characteristics. Tracer studies with these systems require data collection and interpretative efforts which reflect these characteristics, as seen in chapters 7, 9, and 14. Also it must be realized that some of the studies here represent avenues of research in applied earth science of a most direct nature. Determination of the U-series equilibrium state is shown to be relevant to U-ore prospecting (see chapter 14). Research efforts on Rn and ^{234}U/^{238}U in ground-water as possible earthquake precursors have stepped up considerably in recent years, with tantalizing results (for example *Geophys. Res. Lett.* **8**, no. 5, 1981 issue). The behaviour of ^{210}Pb in soils and aerosols is studied for modelling the behaviour of stable Pb or other man-made pollutants with similar properties. The removal of Th-release from mining activity and radioactive waste in ground-waters may be assessed by determination of the equilibrium relationships between ^{234}Th and ^{234}U, and between ^{228}Th and ^{228}Ra. And as a final example, there is the potential contribution of U-series disequilibrium studies to the evaluation of the nuclear waste containment problem (see chapter 7). A novel piece of research recently performed by Kaufman and Eyal (1981, personal comm.) illustrates the point. These investigators conducted leaching experiments on monazites in the laboratory. (Lanthanide orthophosphates have recently been suggested as media for disposal of actinide waste.) By comparing the nuclide relationships among ^{238}U, ^{234}U, ^{232}Th, ^{230}Th, and ^{228}Th in the leachates and the unleached samples, they were able to demonstrate that the recoil-damaged crystal regions are significantly annealled on a time-scale of 10 to 10^4 y.

Acknowledgments

I wish to acknowledge the support given to this research by the donors of the Petroleum Research Fund, administered by the American Chemical Society, and the National Science Foundation. I thank J. L. Bischoff and C. A. Huh for helpful discussions and A. Kaufman for making available his pre-publication manuscript.

APPENDIX A
DERIVATION OF THE EXPRESSIONS FOR THE ^{234}U/^{238}U AND ^{230}Th/^{234}U ACTIVITY RATIO AGE EQUATIONS

Consider the sequential decay of ^{238}U to its daughters ^{234}U and ^{230}Th.

$$^{238}\text{U} \xrightarrow{t_{\frac{1}{2}}^{238}} {}^{234}\text{U} \xrightarrow{t_{\frac{1}{2}}^{234}} {}^{230}\text{Th} \xrightarrow{t_{\frac{1}{2}}^{230}}$$

where $t_{\frac{1}{2}}$ are the respective half-lives of ^{238}U, ^{234}U, and ^{230}Th, related to their decay constants λ_{238}, λ_{234}, and λ_{230} by $\lambda = 0.693/t_{\frac{1}{2}}$.

Let ^{238}U$_0$, ^{238}U $=$ activities of ^{238}U atoms at time $t = 0$ and time t, respectively; ^{234}U$_0$, ^{234}U $=$ activities of ^{234}U atoms at time $t = 0$ and time t, respectively; ^{230}Th$_0$, ^{230}Th $=$ activities of ^{230}Th atoms at time $t = 0$ and time t, respectively.

In a multi-step decay system the first equation for the radioactive decay of ^{238}U yields $^{238}\text{U} = {}^{238}\text{U}_0 e^{-\lambda_{238}t}$ and for the second step

$$\text{d}(^{234}\text{U}/\lambda_{234})/\text{d}t = {}^{238}\text{U}_0 e^{-\lambda_{238}t} - {}^{234}\text{U}. \qquad (A.1)$$

Equation (A.1) is identical to eqn (1.15) of section 1.4, but here it is expressed in terms of activities rather than numbers of atoms. Thus, it can be solved explicitly for ^{234}U as a function of time as follows. Multiplying through by $e^{\lambda_{234}t}$, rearranging, and integrating yields

$$^{234}\text{U} e^{\lambda_{234}t} = \frac{\lambda_{234}}{\lambda_{234} - \lambda_{238}} \cdot {}^{238}\text{U}_0 \cdot e^{-(\lambda_{238}-\lambda_{234})t} + K. \qquad (A.2)$$

To find K, the initial conditions of $^{234}\text{U} = {}^{234}\text{U}_0$ at $t = 0$ will give

$$K = {}^{234}\text{U}_0 - \frac{\lambda_{234}}{\lambda_{234} - \lambda_{238}} \cdot {}^{238}\text{U}_0 \qquad (A.3)$$

and from (A.2) and (A.3)

$$^{234}\text{U} e^{\lambda_{234}t} = \frac{\lambda_{234}}{\lambda_{234} - \lambda_{238}} \cdot {}^{238}\text{U}_0 \, e^{-(\lambda_{238}-\lambda_{234})t} + {}^{234}\text{U}_0 - \frac{\lambda_{234}}{\lambda_{234} - \lambda_{238}} \cdot {}^{238}\text{U}_0$$

$$(A.4)$$

$$^{234}U = \frac{\lambda_{234}}{\lambda_{234}-\lambda_{238}} \cdot {}^{238}U_0 (e^{-\lambda_{238}t} - e^{-\lambda_{234}t}) + {}^{234}U_0 e^{-\lambda_{234}t} \qquad (A.5)$$

By remembering that at time t $^{238}U = {}^{238}U_0 e^{-\lambda_{238}t}$ and by making the appropriate substitutions one obtains

$$^{234}U = \frac{\lambda_{234}}{\lambda_{234}-\lambda_{238}} \cdot {}^{238}U [1 - e^{-(\lambda_{234}-\lambda_{238})t}] + {}^{234}U_0 e^{-\lambda_{234}t}$$

and therefore,

$$\left[\frac{^{234}U}{^{238}U} - \frac{\lambda_{234}}{\lambda_{234}-\lambda_{238}} \right] = \left[\frac{^{234}U_0}{^{238}U_0} - \frac{\lambda_{234}}{\lambda_{234}-\lambda_{238}} \right] e^{-(\lambda_{234}-\lambda_{238})t}. \qquad (A.6)$$

Considering the respective half-lives of ^{238}U and its daughter ^{234}U, it is clear that $\lambda_{234} \gg \lambda_{238}$. Thus, the approximations $e^{-\lambda_{238}t} \approx 1$ and $(\lambda_{234}-\lambda_{238}) \approx \lambda_{234}$ are appropriate, and eqn (A.6) becomes the familiar $^{234}U/^{238}U$ activity ratio age equation:

$$[(^{234}U/^{238}U)_t - 1] = [(^{234}U/^{238}U)_0 - 1]e^{-\lambda_{234}t} \qquad (A.7)$$

where $(^{234}U/^{238}U)_t$ is the activity ratio at time t and
$(^{234}U/^{238}U)_0$ is the activity ratio at time $t = 0$.

The rate of growth of ^{230}Th is given by

$$d(^{230}Th/\lambda_{230})/dt = {}^{234}U - {}^{230}Th. \qquad (A.8)$$

Equation (A.8) is identical to eqn (1.25) of section 1.4. Substituting for ^{234}U from eqn (A.5) one obtains

$$d(^{230}Th/\lambda_{230})/dt + {}^{230}Th = \frac{\lambda_{234}}{\lambda_{234}-\lambda_{238}} {}^{238}U_0 (e^{-\lambda_{238}t} - e^{-\lambda_{234}t})$$
$$+ {}^{234}U_0 e^{-\lambda_{234}t}, \qquad (A.9)$$

multiplying by $e^{\lambda_{230}t}$, rearranging and integrating gives

$$^{230}Th\, e^{\lambda_{230}t} = \frac{\lambda_{230}\lambda_{234}}{(\lambda_{234}-\lambda_{238})(\lambda_{230}-\lambda_{238})} {}^{238}U_0\, e^{-(\lambda_{234}-\lambda_{230})t}$$
$$- \frac{\lambda_{230}\lambda_{234}}{(\lambda_{234}-\lambda_{238})(\lambda_{230}-\lambda_{234})} {}^{238}U_0\, e^{-(\lambda_{234}-\lambda_{230})t} \qquad (A.10)$$
$$+ \frac{\lambda_{230}}{\lambda_{230}-\lambda_{234}} {}^{234}U_0\, e^{-(\lambda_{234}-\lambda_{230})t} + K.$$

Then, the constant, K may be found by introducing the initial condition at $t = 0$ when $^{230}Th_0 = 0$ (that is to say that there was no initial ^{230}Th activity present). Therefore,

$$K = \frac{\lambda_{230}\lambda_{234}{}^{234}U_0}{(\lambda_{234}-\lambda_{238})(\lambda_{230}-\lambda_{234})} - \frac{\lambda_{230}\lambda_{234}{}^{238}U_0}{(\lambda_{234}-\lambda_{238})(\lambda_{230}-\lambda_{238})}$$

$$- \frac{\lambda_{230}{}^{234}U_0}{(\lambda_{230}-\lambda_{234})}. \tag{A.11}$$

Substituting for K in (A.10) and rearranging yields

$$^{230}Th = \frac{\lambda_{230}\lambda_{234}{}^{238}U_0}{(\lambda_{234}-\lambda_{238})(\lambda_{230}-\lambda_{238})}(e^{-\lambda_{238}t}-e^{-\lambda_{230}t})$$

$$- \frac{\lambda_{230}\lambda_{234}{}^{238}U_0}{(\lambda_{234}-\lambda_{238})(\lambda_{230}-\lambda_{234})}(e^{-\lambda_{234}t}-e^{-\lambda_{230}t}) \tag{A.12}$$

$$+ \frac{\lambda_{230}{}^{234}U_0}{(\lambda_{230}-\lambda_{234})}(e^{-\lambda_{234}t}-e^{-\lambda_{230}t}).$$

Recognizing that $\lambda_{234} \gg \lambda_{238}$ and $\lambda_{230} \gg \lambda_{238}$, $\lambda_{234}-\lambda_{238} \approx \lambda_{234}$ and $\lambda_{230} - \lambda_{238} \approx \lambda_{230}$, and factorizing eqn, (A.12) reduces to

$$^{230}Th = {}^{238}U_0 \left[(e^{-\lambda_{238}t}-e^{-\lambda_{230}t}) + \frac{\lambda_{230}}{\lambda_{230}-\lambda_{234}}(e^{-\lambda_{234}t}-e^{-\lambda_{230}t}) \right.$$

$$\left. \times \left(\frac{{}^{234}U_0}{{}^{238}U_0}-1 \right) \right]. \tag{A.13}$$

Substituting eqn (A.7) for $({}^{234}U_0/{}^{238}U_0)$ and remembering that $^{238}U = {}^{238}U_0 \, e^{-\lambda_{238}t}$ is valid, one gets

$$\frac{^{230}Th}{^{238}U} = (1-e^{-\lambda_{230}t}) + \left(\frac{^{234}U}{^{238}U}-1 \right) \frac{\lambda_{230}}{\lambda_{230}-\lambda_{234}}(1-e^{-(\lambda_{230}-\lambda_{234})t}).$$

By multiplying throughout by $^{238}U/^{234}U$ one obtains the familiar ^{230}Th growth age relationship

$$\frac{^{230}Th}{^{234}U} = \frac{(1-e^{-\lambda_{230}t})}{^{234}U/^{238}U} + \left(1 - \frac{1}{^{234}U/^{238}U} \right) \frac{\lambda_{230}}{\lambda_{230}-\lambda_{234}}(1-e^{-(\lambda_{230}-\lambda_{234})t})$$

$$\tag{A.14}$$

APPENDIX B
AN EXAMPLE OF THE DETERMINATION
OF COUNTING ERRORS

Table B.1 contains countrates and the associated background for ^{238}U and ^{234}U, and the corresponding background corrected countrates obtained from a typical alpha spectrum. In addition two sets of counting times for the source and background mass are given.

Using eqn (5.7) for ^{238}U:

(i) $\quad ^{238}U = 8.909 \pm \sqrt{(8.909/1000 + 0.457/1000 + 0.457/1000)}$

$\qquad = 8.909 \pm 0.099$

(ii) $^{238}U = 8.909 \pm \sqrt{(8.909/100 + 0.457/100 + 0.457/1000)}$

$\qquad = 8.909 \pm 0.307$

Similarly for ^{234}U:

(i) $\quad ^{234}U = 9.448 \pm \sqrt{(9.448/1000 + 0.306/1000 + 0.306/1000)}$

$\qquad = 9.448 \pm 0.100$

(ii) $^{238}U = 9.448 \pm \sqrt{(9.448/100 + 0.306/100 + 0.306/1000)}$

$\qquad = 9.448 \pm 0.313$

Table B.1. Countrates and counting times

Countrates	^{238}U (cpm)	^{234}U (cpm)	Counting times (min)		
Raw	9.366	9.754	(i)	(ii)	
Background	0.457	0.306	1000	1000	(background)
Corrected	8.909	9.448	1000	100	(source)

To calculate error in the ^{234}U/^{238}U activity ratio one uses eqn (5.9) appropriate for determining the error in the measurement of quotients. Thus,

(i) $^{234}U/^{238}U = \dfrac{9.448}{8.909} \pm \dfrac{9.448}{8.909} \sqrt{((0.099/8.909)^2 + (0.100/9.448)^2)}$

$\qquad\qquad\quad = 1.061 \pm 0.016$

(ii) $^{234}U/^{238}U = \dfrac{9.448}{8.909} \pm \dfrac{9.448}{8.909} \sqrt{((0.307/8.909)^2 + (0.313/9.448)^2)}$

$\qquad\qquad\quad = 1.061 \pm 0.048$

Clearly, counting for longer periods decreases significantly the error associated with an activity ratio, and hence will reduce the age error limits. However, the same error in a ratio is given by high countrate samples counted for a shorter period. Thus, the error in age measurement can be reduced best by either counting high activity samples or by counting for longer periods of time.

APPENDIX C
UTAGE-3 COMPUTER PROGRAM
FOR THE ^{230}Th/^{234}U AGE CALCULATION

This program is written in FORTRAN IV and is designed to calculate ^{230}Th/^{234}U and ^{234}U/^{238}U alpha activity ratios from uncorrected ^{230}Th, ^{238}U, and ^{234}U countrates. In brief, the program routine is as follows:

1. Raw counts for each nuclide are corrected for background, allowing where necessary for natural contamination (for example, ^{228}Th and ^{232}Th).
2. The ^{228}Th activity is corrected for decay since separation from parent ^{234}U.
3. The measured ^{230}Th/^{234}U activity ratio is adjusted for chemical yield and spike ratio by the equation

$$\left[\frac{^{230}\text{Th}}{^{234}\text{U}} \right]_s = \left[\frac{^{230}\text{Th}}{^{234}\text{U}} \right]_m \cdot \left[\frac{^{228}\text{Th}}{^{232}\text{U}} \right]_{\text{spike}} \cdot \left[\frac{^{232}\text{U}}{^{228}\text{Th}} \right]_m$$

where $\left[\dfrac{^{230}\text{Th}}{^{234}\text{U}} \right]_s$ = the isotope ratio in the sample

$\left[\dfrac{^{230}\text{Th}}{^{234}\text{U}} \right]_m$ = the isotope ratio measured from the spectra

$\left[\dfrac{^{228}\text{Th}}{^{232}\text{U}} \right]_{\text{spike}}$ = the isotope ratio of the spike

$\left[\dfrac{^{232}\text{U}}{^{228}\text{Th}} \right]_m$ = the spike ratio measured from the spectra.

4. The ^{230}Th/^{234}U age is calculated using an iterative method. An estimate of the age (t_0) is used to calculate t^*, the true age of the sample by the Newton method of successive approximations. The method states that if $F(t) = 0$ then

$$t_{n+1} = t_n - \frac{F(t_n)}{F'(t_n)}$$

where t_n = the nth iterative step.

$$F(t) = -\frac{^{230}\text{Th}}{^{234}\text{U}} + \frac{(1 - e^{-\lambda_{230}t_n})}{^{234}\text{U}/^{238}\text{U}} + \frac{\lambda_{230}}{\lambda_{230} - \lambda_{234}}\left(1 - \frac{^{238}\text{U}}{^{234}\text{U}}\right)$$

$$(1 - e^{-(\lambda_{230} - \lambda_{234})t_n})$$

and $F'(t) = \lambda_{230} \cdot \dfrac{^{238}\text{U}}{^{234}\text{U}} \cdot e^{-\lambda_{230}t_n} + \lambda_{230} \cdot \left(1 - \dfrac{^{238}\text{U}}{^{234}\text{U}}\right) \cdot e^{-(\lambda_{230} - \lambda_{234})t_n}$

where $F'(t)$ is the derivative of $F(t)$ with respect to t.

The condition is that when $\dfrac{t_{n+1} - t_n}{t_{n+1}} < 0.001$, $t_{n+1} \simeq t^*$.

5. Countrate errors (1σ) are determined and then used as upper and lower limits in the subroutine to find the standard deviation of the calculated age.
6. Using this age, a value for initial ratio $(^{234}\text{U}/^{238}\text{U})_0$ is determined

$$\left(\left[\frac{^{234}\text{U}}{^{238}\text{U}}\right]_t - 1\right) = \left(\left[\frac{^{234}\text{U}}{^{238}\text{U}}\right]_0 - 1\right)e^{-\lambda_{234}t}$$

in order to establish the consistency of the results in relation to an initial condition.
7. $^{230}\text{Th}/^{232}\text{Th}$ ratio, U concentration, U and Th chemical yields are determined and printed out along with the age data. A copy of a listing of the program is given below together with a typical output.

```
C   UTAGE3 : THIS PROGRAMME PROCESSES RAW INTEGRAL COUNTS
C   TO GIVE CORRECTED COUNT RATES IN CPM AND U/TH ISOTOPIC
C   RATIOS.
C   A SAMPLE AGE IS CALCULATED BY AN ITERATIVE PROCEDURE
C   INPUT VARIABLES :
C       U8 AND BU8 = INTEGRAL COUNTS FOR SAMPLE AND
            BACKGROUND FOR U-238
C       U4 AND BU4 = DITTO FOR U-234
C       U2 AND BU2 = DITTO FOR U-232
C       TH2 AND BTH2 = DITTO FOR TH-232
C       TH0 AND BTH0 = DITTO FOR TH-230
C       TH8 AND BTH8 = DITTO FOR TH-228
C       VORW = SAMPLE VOLUME OR MASS (G OR L)
C       UT, THT, BT = COUNT TIMES FOR U SAMPLE, TH SAMPLE AND
C       BACKGROUND (SECS)
C       SPAM = AMOUNT OF SPIKE USED (MG)
C       THUD 1 = DELAY TWIXT LAST SEP. AND COUNTING OF U
            SAMPLE (DAYS)
C       THUD2 = DELAY TWIXT LAST SEP. AND COUNTING OF TH
C       SAMPLE (DAYS)
C   ROUTING CODES:
C       I SL = 0 FOR LIQUID SAMPLES
C            = 1 FOR SOLID SAMPLES
C       I ERT = 0 FOR 0.05 LEVEL OF CONFIDENCE (COUNT RATE ERROR)
C            = 1 FOR S.D. (COUNT RATE ERROR)
```

```
C        ILDS = 1 FOR LAST DATA SET
C
         REAL*8 NAME, U238, U234, U232, TH232, TH230, TH228
5        READ(5, 10)I SL, IERT, ILDS
10       FORMAT(3I 1)
         WRITE(6, 15) ISL, IERT, ILDS
15       FORMAT(1H1, 3I 1)
         READ(5, 20)NAME, VORW, UT, THT, BT, SPAM, THUD1, THUD2,
       1 CHEFU, CHEFTH
20       FORMAT(A8, F8.3, 3E8.0, F7.4, 2E4.1/2F6.2)
         WRITE(6, 25)NAME, VORW, UT, THT, BT, SPAM, THUD1, THUD2
       1, CHEFU, CHEFTH
25       FORMAT(1H, A8, F8.3, 3F8.0, F5.2, 2F5.1, 2F6.2)
         READ(5, 30)U8, U4, U2, TH2, TH0, TH8
30       FORMAT(6E8.0)
         WRITE(6, 35)U8, U4, U2, TH2, TH0, TH8
35       FORMAT(1H , 6F8.0)
         READ(5, 40) BU8, BU4, BU2, BTH2, BTH0, BTH8
40       FORMAT(6F8.0)
         WRITE(6, 45)BU8, BU4, BU2, BTH2, BTH0, BTH8
45       FORMAT(1H , 6F8.0)
         WRITE(6, 140)
C
C  DATA STATEMENT SYMBOLS :
C        SPACT AND SPACTE = SPIKE ACTIVITY AND ERROR OF U-232
         (CPM)
C        SPAR AND SPARE = SPIKE TH-228/U-232 ACTIVITY RATIO AND
         ERROR
C        UFACT = FACTOR TO CALC, U CONCS. FROM COUNT RATES
C        CHEF = CHAMBER EFFICIENCY (PERCENT)
C        BU8 = INTEGRAL COUNT FOR REAGENT BLANK U-238
C        SIMILARLY FOR BU4-BTH8
C        RUT AND RTHT = COUNT TIMES FOR U AND TH REAGENT
C        BLANK SAMPLES (SECS)
C
         DATA SPACT, SPACTE, SPAR, SPARE, UFACT/12.322, .104,
       1 1.007, .013, .747/
         DATA RU8, RU4, RU2, RTH2, RTH0, RTH8, RUT, RTHT/0.0, 0.0, 0.0, 0.0,
       1 0.0, 0.0, 1., 1./
         DATA U238, U234, U232, TH232, TH230, TH228/'U-238',
       1 'U-234', 'U-232', 'TH-232', 'TH-230', 'TH-228'/
C
C  UR8 = RAW COUNT RATE (CPM) FOR U-238 ETC. ETC.
C
         UR8 = U8*60./UT
         UR4 = U4*60./UT
         UR2 = U2*60./UT
         THR2 = TH2*60./THT
         THR0 = TH0*60./THT
         THR8 = TH8*60./THT
```

```
C
C   BUR8 = BACKGROUND COUNT RATE (CPM) FOR U-238 ETC. ETC.
C
        BUR8 = BU8*60./BT
        BUR4 = BU4*60./BT
        BUR2 = BU2*60./BT
        BTHR2 = BTH2*60./BT
        BTHR0 = BTH0*60./BT
        BTHR8 = BTH8*60./BT
C
C   BUR8 = REAGENT BLANK COUNT RATE (CPM) FOR U-238 ETC. ETC.
C
        RUR8 = RU8*60./RUT
        RUR4 = RU4*60./RUT
        RUR2 = RU2*60./RUT
        RTHR2 = RTH2*60./RTHT
        RTHR0 = RTH0*60./RTHT
        RTHR8 = RTH8*60./RTHT
C
C   CUR8 = CORRECTED COUNT RATE (CPM) FOR U-238 ETC. ETC.
C
        CUR8 = UR8-BUR8-RUR8
        CUR4 = UR4-BUR4-RUR4
        CUR2 = UR2-BUR2-RUR2
        CTHR2 = THR2-BTHR2-RTHR2
        CTHR0 = THR0-BTHR0-RTHR0
        CTHR8 = THR8-BTHR8-RTNR8-THR2+BTHR2+RTHR2
C
C   PCU8 = % CORRECTION DUE TO BACKGROUND AND REAGENT
C     BLANK FOR U-238 ETC.
C
        PCU8 = 100.*(BUR8 + RUR8)/UR8
        PCU4 = 100.*(BUR4 + RUR4)/UR4
        PCU2 = 100.*(BUR2 + RUR2)/UR2
        I F(THR2.EQ.0) GOTO 331
        PCTH2 = 100.*(BTHR2 + RTHR2)/THR2
331     PCTH2 = 0
        PCTH0 = 100.*(BTHR0 + RTHR0)/THR0
C       PCTH8 = 100.*(BTHR8 + RTHR8)/THR8
C
C   CALC. OF CORRECTION IN U-232 AND TH-228 COUNT RATES DUE TO
C     DECAY AND GROWTH OF DAUGHTERS
C   COR2 = CORRECTION IN U-232 COUNT RATE DUE TO GROWTH OF
C     TH-228
C   COR8 1 = CORRECTION IN TH-228 COUNT RATE DUE TO DECAY OF
C     TH-228
C   COR8 2 = CORRECTION IN TH-228 COUNT RATE DUE TO GROWTH
C     OF RA-224
C
        DEC2 = 2.637E-5
```

```
          DEC8 = 9.940E-4
          DEC4 = 0.19
          B1 = 1. − EXP( − DEC8 ∗ THUD1)
          COR2 = 28.∗CUR2∗B1/100.
          CUR2 = CUR2 − COR2
          COR8 1 = (EXP(DEC8∗THUD2) − 1.)∗CTHR8
          B2 = DEC4/(DEC4 − DEC8)
          B3 = EXP( − DEC8∗THUD2) − EXP( − DEC4∗THUD2)
          COR8 2 = 5.5∗B2∗CTHR8∗B3/100.
          CTHR8 = CTHR8 + COR8 1 − COR8 2
C
C   EU8 = ERROR IN U-238 COUNT RATE ECT. ETC.
C
          A = 1.96
          I F(IERT.EQ.1) A = 1.
          EU8 = A∗7.75∗SQRT(UR8/UT + BUR8/BT + RUR8/RUT)
          EU4 = A∗7.75∗SQRT(UR4/UT + BUR4/BT + RUR4/RUT)
          EU2 = A∗7.75∗SQRT((UR2 − COR2)/UT + BUR2/BT + RUR2/RUT)
          ETH2 = A∗7.75∗SQRT(THR2/THT + BTHR2/BT + RTHR2/RTHT)
          ETH0 = A∗7.75∗SQRT(THR0/THT + BTHR0/BT + RTHR0/RTHT)
          ETH8 = A∗7.75∗SQRT((THR8 + COR8 1 − COR8 2)/THT)
        1 + BTHR8/BT + RTHR8/RTHT)
C
C   CONCU  AND  CONCUE = CONCENTRATION  AND  ERROR  OF
        URANIUM IN SAMPLE
C
          CONCU = CUR8∗SPACT∗SPAM/(CUR2∗UFACT∗VORW)
          CONCUE = CONCU∗SQRT(EU8∗EU8/(CUR8∗CUR8) + SPACTE
        1∗SPACTE/(SPACT∗SPACT) + EU2∗EU2/(CUR2∗CUR2))
C
C   PEXU AND PEXTH = % EXTRACTION (CHEMICAL) OF U AND TH
C
          PEXU = CUR2∗10000./(SPACT∗SPAM∗CHEFU)
          PEXTH = CTHR8∗10000./(SPACT∗SPAR∗SPAM∗CHEFTH)
C
C   CALC. OF ACTIVITY RATIOS AND ERRORS
C   RAT48 AND ER48 = U-234/U-238 AR AND ERROR
C   RAT04 AND ER04 = TH-230/U-234 AR AND ERROR
C   RAT02 = TH-230/TH-232 AR
C
          RAT48 = CUR4/CUR8
          RAT04 = CTHR0∗CUR2∗SPAR/(CUR4∗CTHR8)
          I F(CTHR2.LE. .0004) GO TO 332
          RAT 02 = CTHR0/CTHR2
          GO TO 333
332       RAT02 = 1050
333       CONTINUE
          ER48 = RAT48∗SQRT(EU4∗EU4/(CUR4∗CUR4)
        1 + EU8∗EU8/(CUR8∗CUR8))
          ER04 = RAT04∗SQRT(ETH0∗ETH0/(CTHR0∗CTHR0)
        1 + EU4∗EU4/(CUR4∗CUR4) + EU2∗EU2/(CUR2∗CUR2) + ETH8∗
        1 ETH8/(CTHR8∗CTHR8) + SPARE∗SPARE/(SPARE∗SPAR))
```

```
        I F(I SL.EQ.0) GO TO 50
        I F(RAT 04.GT.1.) GO TO 50
        SUB04 = RAT04
        CALL THAGE (SUB04, RAT48, 5000., T)
C
C   SUBROUTINE THAGE CALCS. SAMPLE AGE BY AN ITERATIVE
C   METHOD

        SUB04 = RAT04 + ER04
        CALL THAGE(SUB04, RAT48, T, TU)
        SUB04 = RAT04 − ER04
        CALL THAGE(SUB04, RAT48, T, TL)
        TER = ABS((TU-TL)/2.0)
        TU = ABS(TU − T)
        TL = ABS(T − TL)
C
C   IT AND ITER = SAMPLE AGE AND ERROR (INTEGRAL YEARS)
C
        IT = INT(T + .5)
        ITER = INT(TER + .5)
        SUB48 = RAT48
        CALL UAGE(SUB48, RIN48, IT)
C
C   SUBROUTINE UAGE CALCS. INITIAL U-234/U-238 AR FROM SAMPLE
        AGE
C
        SUB48 = RAT48 + ER48
        CALL UAGE(SUB48, RIN48U, IT)
        SUB48 = RAT48 − ER48
        CALL UAGE(SUB48, RIN48L, IT)
        RIN48E = ABS((RIN48U-RIN48L)/2.)
50      WRITE(6, 135)
        WRITE(6, 60) NAME
60      FORMAT(1H, 20X, 'SAMPLE NUMBER = ', A8)
        WRITE(6, 70) UT
70      FORMAT(21X, 'U COUNT TIME = ', F8.0, 'SECONDS')
        WRITE(6, 80) THT
80      FORMAT(21X, 'TH COUNT TIME =', F8.0, 'SECONDS')
        WRITE(6, 90) BT
90      FORMAT(21X, 'BG COUNT TIME = ', F8.0, 'SECONDS')
        IF(ISL.EQ.1) GOTO 110
        WRITE(6, 100) VORW
100     FORMAT(21X, 'SAMPLE VOLUME =', F6.2, 'LITRES')
        GOTO 125
110     WRITE(6, 120) VORW
120     FORMAT(21X, 'SAMPLE MASS =', F8.4, 'GRAMS')
125     WRITE(6, 128) SPAM
128     FORMAT(21X, 'SPIKE ADDED =', F6.2, 'GM')
130     WRITE(6, 135)
135     FORMAT(19X, '********************************')
        WRITE(6, 140)
```

```
140       FORMAT(1H)
          WRITE(6, 150)
150       FORMAT(1H, 10X, 'COUNTS', 6X, 'COUNTS', 6X, 'RAW COUNT', 2X,
          1 'BACKGROUND', 3X, 'PERCENT', 4X, 'CORRECTED')
          WRITE(6, 160)
160       FORMAT(1H, 13X, 'TOTAL', 3X, 'BACKGROUND', 5X, 'RATE', 5X, 1
          'COUNT RATE', 2X, 'CORRECTION', 2X, 'COUNT RATE', 2X, 'ERROR')
          WRITE(6, 170)
170       FORMAT(1H, 8X, '------------------------------------------------------------------------')
          WRITE(6, 175)
175       FORMAT(1H +, 48X, '----------------------------------------------------------------')
          WRITE(6, 180)U238, U8, BU8, UR8, BUR8, PCU8, CUR8, EU8
          WRITE(6, 140)
          WRITE(6, 180)U234, U4, BU4, UR4, BUR4, PCU4, CUR4, EU4
          WRITE(6, 140)
          WRITE(6, 180)U232, U2, BU2, UR2, BUR2, PCU2, CUR2, EU2
          WRITE(6, 140)
          WRITE(6, 180) TH232, TH2, BTH2, THR2, BTHR2, PCTH2, CTHR2, ETH2
          WRITE(6, 140)
          WRITE(6, 180) TH230, TH0, BTH0, THR0, BTHR0, PCTH0, CTHR0, ETH0
          WRITE(6, 140)
          WRITE(6, 180) TH228, TH8, BTH8, THR8, BTHR8, PCTH8, CTHR8, ETH8
180       FORMAT(1H, A8, F10.0, F12.0, 2F12.3, F12.2, F12.3, F7.3)
          WRITE(6, 170)
          WRITE(6, 175)
          WRITE(6, 140)
          WRITE(6, 190)
190       FORMAT('INFORMATION CALC. FROM ABOVE DATA :')
          WRITE(6, 200)
200       FORMAT(1H, 55X, 'ERROR')
          WRITE(6, 210) PEXU
210       FORMAT(1H, 20X, 'CHEMICAL YIELD OF U =', F7.2, '%')
          WRITE(6, 220) PEXTH
220       FORMAT(1H, 20X, 'CHEMICAL YIELD OF TH =', F7.2, '%')
          IF(I SL.EQ.0) GOTO235
          WRITE(6, 230) CONCU, CONCUE
230       FORMAT(1H, 20X, 'URANIUM CONCENTRATION =', F8.3, 2X, F5.3,
          1 'PPM')
          GOTO239
235       WRITE(6, 238) CONCU, CONCUE
238       FORMAT(1H, 20X, 'URANIUM CONCENTRATION =', F8.3, 2X, F5.3,
          1 'MI CRO G/L')
239       WRITE(6, 240) RAT48, ER48
240       FORMAT (1H, 20X, 'PRESENT U-234/U-238 AR = ', F8.3, 2X, F5.3)
          I F(ABS(RAT02).GT.1000.)GOTO270
          WRITE(6, 250) RAT02
250       FORMAT (1H, 20X, 'PRESENT TH-230/TH-232 AR = ', F6.1)
          IF(ABS(RAT 02).GT.50.) GOTO 290
          WRITE(6, 260)
260       FORMAT(1H, 40X, 'HENCE SAMPLE AGE UNRELIABLE')
          GOTO 290
```

```
270     WRITE(6, 280)
280     FORMAT(1H, 20X, 'PRESENT TH-230/TH-232 AR = GT 1000')
290     WRITE(6, 300) RAT04, ER04
300     FORMAT(1H, 20X, 'PRESENT TH-230/U-234 AR = ', F8.3, 2X, F5.3)
        IF(RAT04.LT.1.) GOTO320
        WRITE(6, 310)
310     FORMAT(1H, 35X, 'HENCE SAMPLE AGE GT OR EQ 300000 YEARS
      1 B.P.')
        GOTO 360
320     IF(I SL.EQ.0) GOTO 360
        WRITE(6, 330)
330     FORMAT(1H, 20X, 'INFERRED')
        WRITE(6, 340) RIN48, RIN48E
340     FORMAT(1H, 22X, 'INITIAL U-234/U-238 AR =', F8.3, 2X, F5.3)
        WRITE(6, 350) IT
350     FORMAT(1H, 20X, 'INFERRED SAMPLE AGE = ', I8,
      1 'YEARS')
        WRITE(6, 355) TU
355     FORMAT(1H, 24X, 'UPPER AGE ERROR =', F6.0, 'YEARS')
        WRITE(6, 356) TL
356     FORMAT(1H, 24X, 'LOWER AGE ERROR = ', F6.0, 'YEARS')
360     WRITE(6, 370)
370     FORMAT(19X, 52('*'))
        I F(ILDS.EQ.1) GOTO 380
        GOTO5
380     STOP
        END
        SUBROUTINE THAGE(SUB04, RAT48, TS, TSR)
        TSR = TS
        DC4 = 2.794E − 6
        DC0 = 9.215E − 6
        C1 = DC0/(DC0-DC4)
15      C2 = EXP(− DC0*TSR)
        C3 = EXP((DC4 − DC0)*TSR)
        FUNC = (1. − C2)/RAT48 + C1*(1. − 1./RAT48)*(1. − C3) − SUB04
        DERIV = DC0*C2/RAT48 + DC0*(1. − 1./RAT48)*C3
        D = FUNC/DERIV
        TSR = TSR − D
        I F(ABS(D/TSR).GT. .001) GOTO 15
        RETURN
        END
        SUBROUTINE UAGE (SUB48, RSR48, I T)
        DC4 = 2.794E − 6
        RSR48 = (SUB48 − 1.)*EXP(DC4*FLOAT(I T)) + 1.
        RETURN
        END
```

```
111
125.19 P3   8.470   60000.   60000.   60000.   1.08   1.0   1.0   37.02   15.85
2140.        2376.    3385.      8.      768.   1668.
41.           27.      22.       2.        3.     44.
```

****

SAMPLE NUMBER = 125.19 P3
U COUNT TIME = 60000. SECONDS
TH COUNT TIME = 60000. SECONDS
BG COUNT TIME = 60000. SECONDS
SAMPLE MASS   = 8.4700 GRAMS
SPIKE ADDED   = 1.08 GM

****

	COUNTS TOTAL	COUNTS BACKGROUND	RAW COUNT RATE	BACKGROUND COUNT RATE	PERCENT CORRECTION	CORRECTED COUNT RATE	ERROR
U-238	2140.	41.	2.140	0.041	1.92	2.099	0.047
U-234	2376.	27.	2.376	0.027	1.14	2.349	0.049
U-232	3385.	32.	3.385	0.032	0.95	3.352	0.058
TH-232	8.	2.	0.008	0.002	0.0	0.006	0.003
TH-230	768.	1.	0.768	0.001	0.13	0.767	0.028
TH-228	1668.	44.	1.668	0.044	2.64	1.604	0.041

INFORMATION CALCULATED FROM ABOVE DATA: ERROR

CHEMICAL YIELD OF U   = 48.24%
CHEMICAL YIELD OF TH   = 53.02%
URANIUM CONCENTRATION   =   1.858   0.070 PPM
PRESENT U-234/U-238 AR   =   1.119   0.034
PRESENT TH-230/TH-232 AR = 127.8
PRESENT TH-230/U-234 AR   =   0.694   0.044
INFERRED
   INITIAL U-234/U-238 AR =   1.169   0.048
INFERRED SAMPLE AGE   = 124484 YEARS
   UPPER AGE ERROR   = 15277. YEARS
   LOWER AGE ERROR   = 13482. YEARS

****

APPENDIX D
COMPUTER PROGRAM FOR THE ^{231}Pa/^{235}U AGE CALCULATION USING THE ^{227}Th MEASUREMENT METHOD

The program for the ^{231}Pa/^{235}U age calculation using the ^{227}Th measurement method was written by Mel Gascoyne of McMaster University and the method in this Appendix derives from his Ph.D. thesis work (Gascoyne 1980). As already indicated in section 5.5.2, the program carries out the following functions.

1. It needs the raw countrates, delay times, count times, background, blank and sample activities.
2. The countrates are corrected for background and reagent blank (using some previously determined values for the laboratory). These are TN1 to TN7 corresponding to the corrected countrates of ^{238}U, ^{234}U, ^{232}U, ^{232}Th, ^{230}Th, ^{228}Th, and ^{227}Th.
3. The TN7 is then corrected for (i) ingrowth of ^{224}Ra which is present either as detector background, or as daughter of ^{228}Th which grows in after plating out (from detrital Th or spike contamination)—this correction is usually negligible but may become important if sample contains detrital Th; (ii) ^{223}Ra ingrowth after plating out—this may be negligible if the source is counted immediately after plating because the rate of decay of ^{227}Th is approximately equal to the rate of ^{223}Ra ingrowth, but if there is a delay in counting the correction must be applied; (iii) decay of ^{227}Th while unsupported after separation from its ^{227}Ac parent somewhere on the ion exchange column.
4. Counting errors are then calculated from the sample background, and reagent blank activities.
5. The sample age is calculated using the iterative method already described in Appendix C except that the function $F(t)$ is defined by the following age equation (see eqn (3.8))

$$\frac{^{231}\text{Pa}}{^{230}\text{Th}} = \frac{1 - e^{-\lambda_{231}t}}{21.7(1 - e^{-\lambda_{230}t}) + \left\{ \left(\left(\frac{^{234}\text{U}}{^{238}\text{U}} \right)_t - 1 \right) \cdot (\lambda_{230}/(\lambda_{230} - \lambda_{234})) \cdot (1 - e^{-(\lambda_{230} - \lambda_{234})t}) \right\}}$$

$$\tag{D.1}$$

which, as shown in chapter 3, is derived from the division of

$$\left(\frac{^{231}\text{Pa}}{^{235}\text{U}}\right)_t = \left(\frac{^{227}\text{Th}}{^{235}\text{U}}\right)_t = 1 - e^{-\lambda_{231}t} \qquad (D.2)$$

by the $^{230}\text{Th}/^{234}\text{U}$ age equation (see eqn (3.4) of section 3.3.1), taking into account the fact that the $^{238}\text{U}/^{235}\text{U}$ activity ratio in the environment is a constant 21.7.

```
C       PROGRAM TST (INPUT LINE 5, OUTPUT LINE 6)
C       PROGRAM CORRECTS RAW COUNT RATE DATA AND COMPUTES
C       PA-231/TH-230, U-234/U-238, TH-230/TH-232 AND TH-228/TH-232
C       ISOTOPE RATIOS AND THE AGE
C       N1 = U-238 COUNT RATE
C       N2 = U-234 COUNT RATE
C       N3 = U-232 COUNT RATE
C       N4 = TH-232 COUNT RATE
C       N5 = TH-230 COUNT RATE
C       N6 = TH-228 COUNT RATE
C       N7 = TH-227 COUNT RATE
C       CNTU = ACCUMULATION TIME FOR U SPECTRUM (MINS)
C       CNTT = ACCUMULATION TIME FOR TH SPECTRUM (MINS)
C       CNBD = ACCUMULATION TIME FOR BKGD SPECTRUM (MINS)
C       DCPT = DELAY BETWEEN PLATING OUT AND COUNTING TH
C       (MINS)
C       DSPT = TIME BETWEEN PLATING OUT AND SEPARATION FROM
C       AC-227 (HRS)
C       WEIGHT = WEIGHT OF SAMPLE IN GRAMS
C       B1 = COUNTER BACKGROUND CORRECTION FOR U-238 AND
C       TH-232
C       B2 = COUNTER BACKGROUND CORRECTION FOR U-234 AND
C       TH-230
C       B3 = COUNTER BACKGROUND CORRECTION FOR U-232 AND
C       TH-228
C       B4 = COUNTER BACKGROUND CORRECTION FOR TH-232
C       B5 = COUNTER BACKGROUND CORRECTION FOR TH-230
C       B6 = COUNTER BACKGROUND CORRECTION FOR TH-228
C       B7 = COUNTER BACKGROUND CORRECTION FOR TH-227
        REAL*8 U238, U234, U232, TH232, TH230, TH228, TH227, CM, ER
        REAL N1, N2, N3, N4, N5, N6, N7, LAM3, LAM7, NT75, N21, NT54, LAM4
        DATA CM, ER/'C/MIN', 'ERROR ='/
        DATA U238, U234, U232, TH232, TH230, TH228, TH227/ 'U238 = ',
     1  'U234 = ', 'U232 = ', 'TH232 = ', 'TH230 = ', 'TH228 = ',
     2  'TH227 = '/
        GO TO 160
100 IF (I STOP.EQ.0) GO TO 799
160 READ (5, 101) T11, T12, T13, TGS, CNTU, CNTT, CNBD, DCPT, DSPT
101 FORMAT (A4, A4, A2, 6F10.0)
        READ(5, 170) DA1, DA2, DA3
```

```
170 FORMAT (A4, A4, A2)
    IF (TGS.EQ.0.) GO TO 799
997 CONTINUE
102 READ (5, 103) N1, N2, N3, N4, N5, N6, N7
103 FORMAT (7E10.0)
104 READ(5, 105) B1, B2, B3, B4, B5, B6, B7, WEIGHT
105 FORMAT (8F7.3)
    READ (5, 129) I STOP
129 FORMAT(I1)
    TAPPX = 0.0
    NLOOP = 0
C
C     CALCULATION  OF  COUNT  RATES  CORRECTED  FOR
C     BACKGROUND AND REAGENT BLANK.
C
C     FOR RB CORRECTION ASSUME U YIELD = 50%
C     TH REAGENT BLANK IS ZERO
C
C
    RB1 = 0.008
    RB2 = 0.025
    TN1 = N1 − B1 − RB1
    TN2 = N2 − B2 − RB2
    TN3 = N3 − B3
    TN4 = N4 − B4
    TN5 = N5 − B5
    TN6 = N6 − B6
    TN7 = N7 − B7
C
C *************************************************************************
C
C     CORRECTION TO TN7 FOR RA-224 INGROWTH FROM TH-228
C     DETERMINATION OF TOTAL RA-224
C
    LAM4 = 1.3224E − 4
    RA224 = TN6∗(1. + ((EXP(− LAM4∗(CNTT + DCPT)) −
  1 EXP(− LAM4∗DCPT)))/(LAM4∗CNTT))
C     CORRECT TH-227 FOR RA-224 INGROWTH
C
    TN7 = TN7 − (0.945∗RA224)
C
C     CORRECTION TO TN7 FOR TH-227 DECAY AND RA-223 GROWTH
C     AFTER PLATING, TPN 7 IS TH-227 COUNT RATE AT TIME
C     OF PLATING OUT
C
    LAM3 = 4.2113E − 5
    LAM7 = 2.6492E − 5
    XA = (1 + LAM3/(LAM3 − LAM7)))/(− LAM7)
    XB = (1/(LAM3 − LAM7))
    TPN7 = TN7∗CNTT)/((XA∗(EXP(− LAM7∗(CNTT + DCPT)) −
```

```
    1 (EXP(-LAM7*DCPT)))) + (XB*(EXP(-LAM3*(CNTT + DCPT))-
    2 EXP(-LAM3*DCPT))))
C
C      CORRECTION TO TPN7 FOR TH227 DECAY AFTER SEPARATION
C      FROM AC-227 ON TH COLUMN
C      TSN7 IS TH-227 COUNT RATE AT TIME OF COLUMN SEPARATION
C
       TSN7 = TPN7/EXP(-LAM7*DSPT*60.)
       NT75 = TSN7/TN5
       N21 = TN2/TN1
       NT54 = TN5/TN4
C
C      ER = TOTAL ERROR IN U, TH COUNT RATES DERIVED FROM:-
C         1) U, TH COUNT RATE
C         2) BLANK DISC COUNT RATE FOR U, TH COUNT TIMES
C         3) BLANK DISC COUNT RATE FOR BLANK DISC COUNT TIME
C         4) REAGENT BLANK.
C
C
C
       BR1 = 0.004
       BR2 = 0.022
       BR4 = 0.000
       BR5 = 0.000
       ER1 = SQRT((TN1/CNTU) + (B1/CNTU) + (B1/CNBD) + (BR1*BR1))
       ER2 = SQRT((TN2/CNTU) + (B2/CNTU) + (B2/CNBD) + (BR2*BR2))
       ER4 = SQRT((TN4/CNTT) + (B1/CNTT) + (B1/CNBD) + (BR4))
       ER5 = SQRT((TN5/CNTT) + (B2/CNTT) + (B2/CNBD) + (BR5))
       ER7 = SQRT((TN7/CNTT) + (B4/CNTT) + (B4/CNBD))
       ER21 = (TN2/TN1)*SQRT(ER1*ER1/(TN1*TN1)
    1 + ER2*ER2/(TN2*TN2))
       ER54 = (TN5/TN4)*SQRT(ER5*ER5/(TN5*TN5)
    1 + ER4*ER4/(TN4*TN4))
       ERS7 = ER7*TSN7/TN7
       ER75 = (TSN7/TN5)*SQRT(ERS7*ERS7/(TSN7*TSN7)
    1 + ER5*ER5/(TN5*TN5))
       R75 = NT75
       T = TGS
       CALL PADATE(R75, N21, T, TN1, TN2)
C
C      SUBROUTINE PA DATE SOLVES AGE EQUATION BY AN
C      ITERATIVE METHOD
C
       IF (TGS.GE.350000.) GO TO 106
       TU = TGS
       R75 = NT75 - ER75
       CALL PADATE(R75, N21, TU, TN1, TN2)
       TL = TGS
       R75 = NT75 + ER75
       CALL PADATE (R75, N21, TL, TN1, TN2)
       I = (T + 50.)/100.
```

```
        T = I*100
        RTL = T - TL
        RTU = TU - T
        I = (RTU + 50.)/100.
        RTU = I*100
        I = (RTL + 50.)/100.
        RTL = I*100
        CALL UDATE (N21, RUO, T)
C
C       SUBROUTINE UDATE COMPUTES INITIAL U-234/U-238 RATIOS
C       FROM MEASURED U-234/U-238 AND CALCULATED AGE (T)

        R21 = N21 + ER21
        CALL UDATE (R21, RUOU, TU)
        R21 = N21 - ER21
        CALL UDATE (R21, RUOL, TL)
        RUOA = ABS((RUOU - RUOL)*2.0)
106 CONTINUE
107 WRITE(6, 108)
108 FORMAT(1H1)
        WRITE (6, 109) TI1, TI2, TI3, DA1, DA2, DA3
109 FORMAT (1H, 'SAMPLE NO. = ', A4, A4, A2, 5X, 'DATE RUN', A4, A4, A2)
        WRITE(6, 133)
133 FORMAT(1H, 48 ('-'))
        WRITE(6, 134)
        WRITE (6, 110) CNBD
110 FORMAT (1H, 'BLANK DISC COUNT TIME =', F6.0, 1X, 'MINS')
        WRITE (6, 111) DCPT
111 FORMAT(1H, 'DELAY BETWEEN PLATING & COUNTING TH = ',
    1 F6.0, 'MINS')
        WRITE (6, 112) DSPT
112 FORMAT (1H, 'TIME BETWEEN PLATING AND SEPARATION
    1 FROM AC-', '227 = ', F6.0, 'HRS')
        WRITE(6, 134)
        WRITE (6, 113)
113 FORMAT (1H, 'RAW DATA')
        WRITE (6, 114)CNTU, CNTT
114 FORMAT(1H, 'U COUNT TIME = ', F6.0, 'MINS', 5X, 'TH COUNT'
    1, 'TIME = ', F6.0, 'MINS')
        WRITE(6, 115) U238, N1, CM, TH232, N4, CM
115 FORMAT(1H, A8, F8.4, 2X, A3, 5X, A8, F8.4, 2X, A3)
        WRITE(6, 115) U234, N2, CM, TH230, N5, CM
        WRITE(6, 115) U232, N3, CM, TH228, N6, CM
        WRITE(6, 116) TH227, N7, CM
116 FORMAT (27X, A8, F8.4, 2X, A3)
        WRITE (6, 117)
117 FORMAT (1H, 'BACKGROUND')
        WRITE (6, 115) U238, B1, CM, TH232, B1, CM
        WRITE (6, 115) U234, B2, CM, TH230, B2, CM
        WRITE (6, 115) U232, B3, CM, TH228, B3, CM
```

```
      WRITE (6, 116) TH227, B4, CM
      WRITE (6, 118)
      WRITE (6, 134)
  118 FORMAT (1H, 'CORRECTED DATA')
      WRITE (6, 119)
  119 FORMAT (1H, 18X, 'ERROR', 27X, 'ERROR')
      WRITE (6, 120) U238, TN1, ER1, CM, TH232, TN4, CM
  120 FORMAT (1H, A8, F8.3, 2X, F5.3, 1X, A3, 5X, A8, F8.4, 8X, A3)
      WRITE (6, 121) U234, TN2, ER2, CM, TH230, TN5, ER5, CM
  121 FORMAT (1H, A8, F8.3, 2X, F5.3, 1X, A3, 5X, A8, F8.4, 2X, F5.3, 1X, A3)
      WRITE (6, 122) U232, TN3, CM, TH228, TN6, CM
  122 FORMAT (1H, A8, F8.3, 8X, A3, 5X, A8, F8.4, 8X, A3)
      WRITE (6, 123) TH227, TSN7, ERS7, CM
  123 FORMAT (1H, 32X, A8, F8.4, 2X, F5.3, 1X, A3,/)
      WRITE (6, 133)
      WRITE (6, 134)
      IF (NT54. GT.999.) GO TO 1245
      WRITE (6, 125) NT54, ER54
  125 FORMAT ('TH-230/TH-232 = ', F7.3, 2X, F7.3,/)
      GO TO 1265
 1245 WRITE (6, 1255)
 1255 FORMAT ('TH-230/TH-232 = GT 1000')
 1265 WRITE (6, 126) NT 75, ER 75
  126 FORMAT ('TH-227/TH-230 =', F5.3, 2X, F5.3,/)
      IF (T.GE.350000.) GO TO 131
      WRITE (6, 127)
  127 FORMAT (1H, 19X, 'ERROR', 24X, 'ERROR')
      WRITE (6, 128) N21, ER21, RUO, RUOA
  128 FORMAT (1H, 'U234/U-238 =', F5.3, 2X, F5.3, 2X, '(U-234/U-238)0 ='
     1, F5.3, 2X, F5.3)
  144 WRITE (6, 120)
  145 WRITE (6, 146) RTU
  146 FORMAT (1H, 27X, '+', 1X, F7.0)
  147 WRITE (6, 148)T
  148 FORMAT (1H, 'PROTACTINIUM AGE = ', F7.0, 12X, 'YEARS')
  149 WRITE (6, 150) RTL
  150 FORMAT (28X, '-', 1X, F7.0)
      WRITE (6, 133)
  134 FORMAT (1H)
      GO TO 100
  131 WRITE (6, 124) N21, ER21
  124 FORMAT (1H, 'U-234/U-238 = ', F5.3, 2X, F5.3,/)
      WRITE (6, 132)
  132 FORMAT (1H, 'AGE IS GREATER THAN 350,000 YEARS B.P.')
      GO TO 100
  799 STOP
      END
      SUBROUTINE PADATE (NT75, N21, TGS, TN1, TN2)
      REAL NT75, N21, TN1, TN2, LAM0, LAM1, LAM4
      WRITE (6, 333) NT75, N21, TGS, TN1, TN2
      LAM0 = 9.2173E-6
```

```
      LAM1 = 2.0208E-5
      LAM4 = 2.7950E-6
      TN = TN2/TN1
  10  SAM1 = EXP (− LAM1∗TGS)
      SAM0 = EXP (− LAM0∗TGS)
      SAM04 = EXP (− (LAM0 − LAM4)∗TGS)
      ROY = (TN − 1)∗(LAM0/(LAM0 − LAM4))
      FUNC = ((1 − SAM1)/(21.7∗((1 − SAM0) + ROY∗(1 − SAM04)))) − NT75
      DERIV = ((LAM1∗SAM1)∗((1 − SAM0) + ROY∗(1 − SAM04)) −
  1   (1 − SAM1)∗(LAM0∗SAM0) + (TN − 1)∗LAM0∗SAM04))/
  2   (21.7∗((1 − SAM0) + (ROY∗(1 − SAM0)))∗∗2.)
      ALF = FUNC/DERIV
      TGS = TGS − ALF
      WRITE (6, 333) ROY, FUNC, DERIV, ALF, TGS
 333  FORMAT (5F18.7)
      IF (TGS.GE.350000.) GO TO 11
      IF (ABS (ALF/TGS).GT.0.001) GO TO 10
  11  CONTINUE
      RETURN
      END
      SUBROUTINE UDATE (N21, RUO, T)
      REAL N21
      THALF 4 = 2.7950E-6
      RUO = ((N21) − 1.0)∗EXP (THALF4∗T) + 1.0
      RETURN
      END
```

SAMPLE NO. = 76121-3   DATE RUN THE DATE‖

BLANK DISC COUNT TIME = 1261. MINS
DELAY BETWEEN PLATING & COUNTING TH = 2911. MINS
TIME BETWEEN PLATING AND SEPARATION FROM AC-227 = 21. HRS

RAW DATA

U COUNT TIME = 592.   MINS   TH COUNT TIME = 1286. MINS
U238 = 15.7000   C/M   TH232 = 0.0400   C/M
U234 = 17.5760   C/M   TH230 = 6.4070   C/M
U232 = 0.0300   C/M   TH228 = 0.0700   C/M
                                     TH227 = 0.4076   C/M

BACKGROUND
U238 = 0.0   C/M   TH232 = 0.0   C/M
U234 = 0.0   C/M   TH230 = 0.0   C/M
U232 = 0.0   C/M   TH228 = 0.0   C/M
                                     TH227 = 0.0   C/M

CORRECTED DATA
                  ERROR   ERROR
U238 = 15.692   0.163 C/M   TH232 = 0.0400   C/M
U234 = 17.551   0.174 C/M   TH230 = 6.4070   0.071 C/M
U232 = 0.030   C/M   TH228 = 0.0700   C/M
                                     TH227 = 0.3789   0.017 C/M

                         ERROR
TH-230/TH-232 = 160.175 22.402
TH-227/TH-230 = 0.059   0.003

                 ERROR   ERROR
U234/U-238 = 1.118 0.016   (U-234/U-238)0 = 1.158   0.111
                                           + 15200.
PROTACTINIUM AGE = 101900.   YEARS
                        −13100.

APPENDIX E
EQUATIONS FOR THE U-TREND MODEL

The model is based on the assumptions that

$$(^{234}U - {}^{238}U) = \frac{\lambda_{234}}{(\lambda_{234} - \lambda_0)} e^{-\lambda_0 t} - \frac{\lambda_{234}}{(\lambda_{234} - \lambda_0)} e^{-\lambda_{234} t} + e^{-\lambda_{234} t} \quad (E.1)$$

and

$$(^{234}U - {}^{230}Th) = \frac{-3\lambda_{234}\lambda_{230}}{(\lambda_{234} - \lambda_0)(\lambda_{230} - \lambda_0)} e^{-\lambda_0 t}$$

$$- \frac{3\lambda_{234}\lambda_{230}}{(\lambda_{230} - \lambda_0)(\lambda_{230} - \lambda_{234})} e^{-\lambda_{234} t}$$

$$- \frac{3\lambda_{234}\lambda_{230}}{(\lambda_0 - \lambda_{230})(\lambda_{234} - \lambda_{230})} e^{-\lambda_{230} t} - 2e^{-\lambda_{234} t} + e^{-\lambda_{230} t}, \quad (E.2)$$

where λ_0 is the decay constant of $F(0)$ or $\lambda_0 = \ln 2/\text{half-period of } F(0)$, λ_{234} is the decay constant of ^{234}U ($0.280 \times 10^{-5}\,\text{y}^{-1}$), λ_{230} is the decay constant of ^{230}Th ($0.922 \times 10^{-5}\,\text{y}^{-1}$) and t is time. For samples of a single age, the rate of change between parent–daughter activities would be represented by the first derivatives of eqns (E.1) and (E.2).

$$\Delta(^{234}U - {}^{238}U) = \frac{-\lambda_{234}\lambda_0}{\lambda_{234} - \lambda_0} e^{-\lambda_0 t} + \frac{\lambda_{234}\lambda_{234}}{\lambda_{234} - \lambda_0} e^{-\lambda_{234} t} - \lambda_{234} e^{-\lambda_{234} t} \quad (E.3)$$

and

$$\Delta(^{234}U - {}^{230}Th) = \frac{3\lambda_0\lambda_{234}\lambda_{230}}{(\lambda_{234} - \lambda_0)(\lambda_{230} - \lambda_0)} e^{-\lambda_0 t}$$

$$+ \frac{3\lambda_{234}\lambda_{234}\lambda_{230}}{(\lambda_0 - \lambda_{234})(\lambda_{230} - \lambda_{234})} e^{-\lambda_{234} t}$$

$$+ \frac{3\lambda_{234}\lambda_{230}\lambda_{230}}{(\lambda_0 - \lambda_{230})(\lambda_{234} - \lambda_{230})} e^{-\lambda_{230} t} + 2\lambda_{234} e^{-\lambda_{234} t} - \lambda_{234} e^{-\lambda_{230} t} \quad (E.4)$$

The slope of the line represented by the quotient of eqns (E.3) and (E.4) is used for the isochron. To accommodate measured isotopic data, the isotopic variations were normalized to ^{238}U and a U-trend model written in the form of a Bateman equation (see section 1.4):

$$\frac{Y}{X} = \frac{(^{234}\text{U} - {}^{238}\text{U})/^{238}\text{U}}{(^{234}\text{U} - {}^{230}\text{Th})/^{238}\text{U}} = \frac{C_1 e^{-\lambda_0 t} + C_2 e^{-\lambda_{234} t}}{C_3 e^{-\lambda_0 t} + C_4 e^{-\lambda_{234} t} + C_5 e^{-\lambda_{230} t}} \quad \text{(E.5)}$$

where $\quad C_1 = \dfrac{-\lambda_0 \lambda_{234}}{\lambda_{234} - \lambda_0}, C_2 = \dfrac{\lambda_{234} \lambda_{234}}{\lambda_{234} - \lambda_0} - \lambda_{234}, C_3 = \dfrac{3\lambda_0 \lambda_{234} \lambda_{230}}{(\lambda_{234} - \lambda_0)(\lambda_{230} - \lambda_0)},$

$$C_4 = \frac{3\lambda_{234} \lambda_{234} \lambda_{230}}{(\lambda_0 - \lambda_{234})(\lambda_{230} - \lambda_{234})} + 2\lambda_{234},$$

$$\text{and } C_5 = \frac{3\lambda_{234} \lambda_{230} \lambda_{230}}{(\lambda_0 - \lambda_{230})(\lambda_{234} - \lambda_{230})} - \lambda_{230}.$$

Equation (E.5) is an empirical model equation; the numerical constants in the coefficients preceding the exponential terms were determined by computer synthesis to provide a model with the best fits for deposits of known age. The alternative U-trend isochron is represented by

$$\frac{Y}{X - Y} = \frac{(^{234}\text{U} - {}^{238}\text{U})/^{238}\text{U}}{(^{238}\text{U} - {}^{230}\text{Th})/^{238}\text{U}} \quad \text{(E.6)}$$

and this form is usually used for computer solution of the age.

REFERENCES

Adams, J. A. S. and Gasparini, P. (1970). *Gamma-ray spectrometry of rocks*, Elsevier Science, New York. [14.1, 14.4.7]
—— Richardson, K. A. (1960). *Econ. Geol.* **55**, 1673. [7.3.2]
—— Weaver, C. E. (1958). *Bull. Am. Ass. Petrol. Geol.* **42**, 387. [2.2.2]
—— Osmond, J. K., and Rogers, J. J. W. (1959). *Phys. Chem. Earth* **3**, 299. [6.2]
Aitken, M. J. (1974). *Physics and archaeology.* Oxford University Press. [3.1]
—— (1978). *J. Phys. Rep.* **40**, 281. [3.1]
Alberti, G., Bettanili, C., Salvetti, F., and Santoli, S. (1959). *Ann. Chim., Roma.* **49**, 199. [4.2.2]
Alekin, D. A. and Brazhnikova, L. V. (1961). *Gidrokhim. Mater.* **32**, 12. [15.2.1]
Alekseev, F. A., Gottikh, R. P., Zverev, V. L., Spiridonov, A. I., and Grombkov, A. P. (1973). *Unpublished MS.*, IAEA Panel on Application of uranium isotope disequilibrium in hydrology, Vienna. [9.3.4]
Allegre, C. J. (1968). *Earth planet. Sci. Lett.* **5**, 209. [6.4.3]
—— Condomines, M. (1976). *Earth planet. Sci. Lett.* **28**, 395. [4.3.1, 4.4, 6.4.2, 6.4.3, 6.5.2, 6.5.3, 10.3.2]
Aller, R. C. (1977). *The influence of macrobeuthos on chemical diagenesis of marine sediments*, Ph.D. Thesis, Yale University, 599pp. [15.3.3]
—— Cochran, J. K. (1976). *Earth planet. Sci. Lett.* **29**, 37. [2.5.3, 13.3.3, 15.2.2, 15.2.5, 15.3.3]
—— Benninger, L. K., and Cochran, J. K. (1980). *Earth planet. Sci. Lett.* **47**, 161. [15.3.3, 16.3.3]
Allison, S. K. and Warshaw, S. D. (1953). *Rev. Mod. Phys.* **25**. [5.1]
Aly, H. F., Herpes, U., and Herr, W. (1975). *J. radioanal. Chem.* **26**, 279. [5.3.4]
Allred, A. L. and Rochow, E. G. (1958). *J. inorg. nucl. Chem.* **5**, 264. [2.1.1]
Altschuler, Z. S. (1973). In: *Environmental phosphorus handbook*, ed. E. S. Griffith, A. Beeton, J. M. Spencer, and D. T. Mitchell, p. 33. Wiley–Interscience, New York. [17.3.3]
—— Clark, R. S. Jr., and Young, E. J. (1958). *U. S. Geological Survey Prof. Paper 314-D*, 45. [6.3, 17.3.1, 17.3.2]
Ames, L. L. Jr. (1960). *Econ. Geol.* **55**, 354. [17.3.1]
Amiel, S. (1961). *Analytical application of delayed neutron emission in fissionable elements.* Israel AEC Rep. IA-621, Tel Aviv, Israel. [5.3.4]
—— (1962). *Anal. Chem.* **34**, 1683. [5.3.4]
Amin, B. S., Krishnaswami, S., and Somayajulu, B. L. K. (1974). *Earth planet. Sci. Lett.* **21**, 342. [15.3.3]
Amin, B. S., Lal, D., and Somayajulu, B. L. K. (1975). *Geochim. Cosmochim. Acta* **39**, 1187. [3.1]

Anderson, M. E. and MacDougall, J. D. (1977). *EOS Trans. Am. geophys. Un.* **58**, 420. [16.3]

Anderson, R. F. (1980). Ph.D. dissertation, Woods Hole Oceanographic Institution—MIT [9.3]

Anderson, R. F., Bacon, M. P., and Brewer, P. G. (1980a). *Trans. Am. geophys. Un.* **61**, 268. [15.3.1, 15.3.2]

—— —— —— (1980b). *Trans. Am. Geophys. Un.* **61**, 1012. [15.3.1, 15.3.2]

Andjelkovic, M. and Rajkovic, D. (1959). In: *2nd U.N. Int. Conf. Peaceful Uses At. En., Geneva.* Report A/CONF. 15/P/474. [4.2.2]

Andrews, J. N. and Kay, R. L. F. (1978). *Short Pap. Fourth Int. Conf. Geochron. Cosmochron. Isotope Geol*, ed. R. E. Zartman. U. S. Geological Survey Open File Report 78–701, 11. [9.3.4]

—— Lee, D. J. (1979). *J. Hydrol.* **41**, 233. [2.3.3, 3.1, 3.3.3, 9.3.2]

—— Wood, D. F. (1972). *Trans. Inst. Min. Metall.* **81**, B197. [9.3.2]

—— Giles, I. S., Kay, R. L. F., Lee, D. J., Osmond, J. K., Cowart, J. B., Fritz, P., Barker, J. F., and Gale, J. (In Prep.). [9.3.4]

Anestad-Fruth, E. (1963). *Uranium series disequilibrium in recent volcanic rocks.* MA Thesis, Columbia University, New York, 22pp. [3.4.4]

Apte, B. G., Subbaraman, P. R., and Gupta, J. (1965). *Anal. Chem.* **37**, 106. [4.2.1]

Armands, G. and Landergreen, S. (1960). *Intl. Geogr. Congr. Rept.* **21**, (Sess. Norden, Pt XV), 51. [2.3.1]

Armi, L. and D'Asaro, E. (1980). *J. geophys. Res.* **85**, 469. [15.5.2]

Arrhenius, G. (1967). *Trans. Am. geophys. Un.* **48**, 604. [16.3.2]

—— Goldberg, E. D. (1955). *Tellus* **7**, 226. [15.4.1, 16.2.3]

Artree, R. W., Cabell, M. J., Cushing, R. L., and Pieroni, J. J. (1962). *Can. J. Phys.* **40**, 194. [1.5]

Asaro, F. and Perlman, I. (1955). *Phys. Rev.* **99**, 37. [1.5]

—— Stephens Jr., F., and Perlman, I. (1953). *Phys. Rev.* **92**, 1495. [1.5]

Asikainen, M. and Kahlos, H. (1979). *Geochim. Cosmochim. Acta* **43**, 1681. [4.5.2, 9.2.3, 14.4.4]

Atkinson, T. C., Harmon, R. S., Smart, P. L., and Waltham, A. C. (1978). *Nature* **272**, 24. [11.3.4, 13.3.3]

Aumento, F. (1971). *Earth planet. Sci. Lett.* **11**, 90. [7.1, 15.2.5]

—— Mitchell, W. S., and Fratta, M. (1976). *Can. Mineralogist* **14**, 269. [15.2.5]

Austin, S. R. (1975). In: *Radon in uranium mining*, p. 151. International Atomic Energy Agency Publication STI/PUB/391, Vienna. [14.4.5]

Bacon, M. P. (1976). *Applications of Pb-210/Ra-226 and Po-210/Pb-210 disequilibria in the study of marine geochemical processes*, Ph.D. Thesis, Woods Hole Oceanographic Institution, Massachusetts, 165 pp. [15.6.3]

—— (1978). *Earth planet. Sci. Lett.* **39**, 250. [6.5.3, 7.1]

—— (1979). *EOS Trans. Am. geophys. Un.* **60**, 283. [16.2.1]

—— Anderson, R. F. (1980). *EOS Trans. Amer. Geophys. Un.* **61**, 1011 [19.3]

—— Elzerman, A. W. (1980). *Nature* **284**, 332. [15.7]

—— Spencer, D. W., and Brewer, P. G. (1976). *Earth planet. Sci. Lett.* **32**, 277. [2.5.3, 4.4, 15.3.3, 15.6.2, 15.6.3, 15.7]

—— Brewer, P. G., Spencer, D. W., Murray, J. W., and Goddard, J. (1980). *Deep Sea Res.* **27**, 119. [2.5.3]

Bada, J. L. (1972). *Earth planet. Sci. Lett.* **15**, 223. [13.2.2]

Balustrieri, L. and Murray, J. W. (1979). In: *Am. Chem. Soc. Symp. Ser. No. 93: Chemical Modelling in Aqueous Systems* p. 214 American Chemical Society, New York. [8.2.2]

Baranov, S. A., Zelenkov, A. G., Schepkin, G. I., Beruchlov, V., and Malov, A. F. (1959). *Izv. Akad. Nauk. SSSR* **23**, 1402. [5.3.1]

Baranov, V. I. and Kuzmina, L. A. (1958). *Geochem.* **2**, 131. [16.3.1]

—— Surkov, Yu. A., and Vilinskii, V. D. (1958). *Geochem. Intl.* **5**, 591. [2.5.1]

Baranowski, J. and Harmon, R. S. (1978). In: *Short Papers 4th Int. Conf. Geochron., Cosmochron., Isotope Geol.*, ed. R. E. Zartman. U. S. Geol. Survey Open File Report 78–701, 22. [6.4.3]

Barker, F. B. and Johnson, J. O. (1964). *US Geol. Surv. Water Supp. Pap. 1696B*. [2.5.3]

Barnes, F. Q. and Ruzicka, V. (1972). *Int. Geol. Congr. Montreal, Section 4 (Mineral Deposits)*, **159**. [14.2]

Barnes, J. W., Lang, E. J., and Potratz, H. A. (1956). *Science* **124**, 175. [17.2, 18.4]

Barretto, P. M. C. (1975). In: *Radon in uranium mining*, p. 129. International Atomic Energy Agency Publication STI/PUB/391. [14.4.5]

Barth, K. W. F. (1952). *Theoretical petrology*, John Wiley and Sons, New York. [15.1.1]

Bastin, B. (1978) *Annls. Soc. géol. Belg.* **101**, 13. [13.2.2, 13.3.3]

Bateman, H. (1910). *Proc. Camb. phil. Soc. math. phys. Sci.* **15**, 423. [1.4]

Baturin, G. N. (1973). *Geochim. Int.*, 1031. [16.2.1]

—— Dubinchuk, V. T. (1978). *Okeanologiya* **18**, 1036. [17.3.1]

—— Kochenov, A. V. (1973). *Geokhimya* **10**, 1529. [15.2.4, 16.2, 16.2.1]

—— Merkulova, K. L., and Chalov, P. I. (1972). *Mar. Geol.* **13**, M37. [17.3.2]

Bayer, E. and Möllinger, H. (1959). *Agnew. Chem.* **71**, 426. [4.2.1]

Bayer, G. (1969). *Handbook of geochemistry*, ed. K. H. Wedepohl. Springer-Verlag, Berlin. [2.1.2]

Bell, K. G. (1954). *Uranium and thorium in sedimentary rocks*. In: *Nuclear geology*, ed. H. Faul, J. Wiley and sons, New York. [2.4.3]

Bender, M. L. (1973). *Geochim. Cosmochim. Acta* **37**, 1229. [3.3.3, 4.4, 18.6]

—— Ku, T. L., and Broecker, W. S. (1966). *Science* **151**, 325. [16.3.1]

—— Fairbanks, R. G., Taylor, F. W., Matthews, R. K., Goddard, J. G., and Broecker, W. S. (1979). *Geol. Soc. Am. Bull.* **90**, 577. [3.3.3, 12.2.2, 18.6]

—— Broecker, W. S., Gornitz, V., Middler, U., Kay, R., Sun, S. S., and Biscaye, P. (1971). *Earth planet. Sci. Lett.* **12**, 425. [15.2.5, 16.2.1, 16.2.2]

Benninger, L. K. (1976). *The uranium series radionuclides as tracers of geochemical processes in Long Island Sound*. Ph.D. Thesis, Yale Univ. [4.6, 5.5.4]

—— (1978). *Geochim. Cosmochim. Acta* **42**, 1165. [15.6.2, 16.3.3]

—— Aller, R. C., Cochran, J. K., and Turekian, K. K. (1979). *Earth planet. Sci. Lett.* **43**, 241. [16.2.5, 16.3.3]

—— Lewis, D. M., and Turekian, K. K. (1975). In: *Marine chemistry in the coastal environment*, ed. T. Church. A.C.S. Symp. Series, **18**. American Chemical Society, New York. [8.5, 15.6.1]

Bernat, M. and Goldberg, E. D. (1969). *Earth planet. Sci. Lett.* **5**, 308. [16.2, 16.2.2, 16.3]

—— Launay, J., and Recy, J. (1976). *C. r. hebel. Séanc. Acad. Sci., Paris* **282**, 9. [17.2.3, 18.5]

—— Bieri, R. H., Koide, M., Griffin, J. J., and Goldberg, E. D. (1970). *Geochim. Cosmochim. Acta* **34**, 1053. [16.2, 16.2.1, 16.2.2]

Bertine, K. K., Chan, L. H., and Turekian, K. K. (1970). *Geochim. Cosmochim. Acta* **34**, 641. [8.1, 8.2.1, 8.2.3, 15.2.1, 16.2, 16.2.1]

Bertolini, G. and Coche, A. (1968). *Semiconductor detectors*. North Holland Publ. Co., Amsterdam. [5.2.1]

—— Rota, A. (1968). *Beta and electron spectroscopy in semiconductor detectors*, ed. G. Bertolini and A. Coche. North Holland Publ. Co., Amsterdam. [5.3.2]

Berzina, I. G., Yeliseyeva, O. P., and Popenko, D. P. (1974). *Int. Geol. Rev.* **16**, 1191. [6.2]

Bethe, H. A. (1937). *Rev. mod. Phys.* **9**, 161. [1.3.1]

—— Ashwin, J. (1953). *Passage of radiations through matter in experimental nuclear physics*, Vol. I. ed. Segrè. Wiley and Sons, New York. [5.1]

Beus, A. A. and Grigorian, S. V. (1977). *Geochemical exploration methods for mineral deposits*. Applied Publishing Ltd, Wilmette, Illinois. [14.3]

Bevington, P. R. (1969). *Data reduction and error analysis for the physical sciences*. McGraw-Hill Book Co., New York. [5.4]

Bhat, S. G. and Krishnaswami, S. (1969). *Proc. Indian Acad. Sci. A* **LXX**, [8.1, 8.2.1, 8.4, 15.2.1, 15.2.6, 15.4.1]

—— —— Lal, D., Rama, and Moore, W. S. (1969). *Earth planet. Sci. Lett.* **5**, 483. [2.5.1, 2.5.3, 4.4, 4.6, 5.5.3, 15.2.2, 15.3.3, 16.3.3]

Bievre, de P. (1976). *Advances in mass spectrometry. 7A.* In: *Proc. 7th Int. Mass Spect. Conf., Florence, It.*, ed. N. R. Daly. Heyden and Son Ltd., on behalf of the Institute of Petroleum, London. [5.3.4]

Bischoff, J. L. and Fyfe, W. S. (1968). *Am. J. Sci.* **266**, 65. [11.1]

—— Piper, D. Z. (1979). *Marine geology and oceanography of the Pacific Manganese Nodule Province. Marine Science Series. No. 9*. Plenum Press, New York. [16.3.2]

Blackwell, B. (1980). unpub. M.Sc. Thesis, McMaster Univ., Hamilton. [12.4.1]

Blanchard, R. L. (1963). *U decay series disequilibrium in age determination of marine calcium carbonates*, Washington Univ. Ph.D. thesis, St. Louis, Mo. [18.4]

Blanchard, R. L. and Oakes, D. (1965). *J. geophys. Res.* **70**, 2911. [15.2.2]

—— Cheng, M. H., and Potratz, H. A. (1967). *J. geophys. Res.* **70**, 4055. [17.2, 17.2.2]

Blank, H. R. and Tynes, E. W. (1965). *Geol. Soc. Am. Bull.* **76**, 1387. [10.2.1]

Bloch, S. (1980a). *Geol. Soc. Am. (Abs. Prog.)*, 388, [9.2.3]

—— (1980b). *Geochim. Cosmochim. Acta* **44**, 373. [7.1, 8.2.1, 15.2.1, 15.2.5]

—— Bischoff, J. L. (1979). *Geology* **7**, 193. [15.2.5]

—— Key, R. M. (1980). *Am. Assoc. Petr. Geol.* **65**, 154. [9.2.3]

Bloom, A. L., Broecker, W. S., Chappell, J. M. A., Mathews, R. K., and Mesolella, K. J. (1974). *Quatern. Res.* **4**, 185. [17.2, 17.2.1, 17.3.4, 18.2, 18.5, 18.7]

Bock, R. and Bock E. (1950). *Zhur anorg. Chem.* **263**, 146. [4.2.2]

Boltwood, B. B. (1904a). *Am. J. Sci.* **168**, 97. [14.1]

—— (1904b). *Am. J. Sci.* **168**, 378. [14.1]

Bonatti, E. and Nayudu, Y. R. (1965). *Am. J. Sci.* **263**, 17. [16.3.2]

—— Fisher, D. E., Joensuu, O., and Rydell, H. S. (1971). *Geochim. Cosmochim. Acta* **35**, 189. [16.2.1]

Bonner, O. D. and Smith, L. L. (1957). *J. Phys. Chem.* **61**, 326. [4.2.6]

Bordes, F. (1972). *A tale of two caves*. Harper and Row, New York. [12.2.1, 13.2.2]

—— Thibault, C. (1977). *Quatern. Res.* **8**, 115. [13.2.2]

Borole, D. V. and Somayajulu, B. L. K. (1977). *Mar. Chem.* **5**, 291. [16.2.3, 16.2.5]

—— Krishnaswami, S., and Somayajulu, B. L. K. (1977). *Estuar coastal mar. Sci.* **5**, 743. [8.1, 8.2.1, 8.2.2, 15.2.2]

—— —— —— (1981). (In press). [8.2.1, 8.2.2]

—— —— —— (1982). *Geochim. Cosmochim. Acta* **46**, 125. [15.2.1, 15.2.6]

Boström, K. and Peterson, M. N. A. (1969). *Mar. Geol.* **7**, 427. [15.2.5]

—— Rydell, H. (1979). In: *La genèse des nodules de manganèse*, ed. C. Lalou, p. 151. C.N.R.S. Paris. [16.2, 16.2.1, 16.2.2]

Bothe, W. (1932). *Handb. Phys.* **22** (2), 1. [5.1]

Boulad, A. P. and Michard, G. (1976). *Earth planet. Sci. Lett.* **32**, 77. [15.2.4]

Bowen, H. J. M. (1975). *Activation analysis in radiochemical methods in analysis*, ed. D. I. Coomber. Plenum Press, New York [5.3.4]

Bowie, S. H. V., Simpson, P. R., and Rice, C. M. (1973). In: *Proc. 4th Int. Geochem. Explor. Symp.*, *17–20 Apr. 1972*, ed. M. J. Jones, p. 458. Institute of Mining and Metallurgy, London. [6.2]

Boyle, R. W. (1974). *Geol. Surv. Canada Paper 74–45*. [14.2]

Brain, C. K. (1970). *Nature* **225**, 1112. [12.2.1]

Breger, I. A. and Deul, M. (1956). *US. Geol. Surv. Prof. Pap.* **300**, 505. [2.2.3, 2.2.5, 2.3.1]

Brewer, P. G., Nozaki, Y., Spencer, D. W., and Fleer, A. P. (1980). *J. Mar. Res.* **38**, 703. [15.2.5, 15.3.1]

Briel, L. I. (1976). *Investigation of the $^{234}U/^{238}U$ disequilibrium in the natural waters of the Santa Fe river basin in North Central Florida*, Florida State Univ., Ph.D. Thesis. [4.3.2, 4.4, 9.2.2]

Briggs, G. H. (1927). *Proc. R. Soc. A.* **114**, 341. [5.1]

Bril, K. and Holzer, S. (1961). *Anal. Chem.* **33**, 55. [4.2.1]

Brits, R. J. (1979). *Chem. Geol.* **25**, 347. [4.4]

—— Dias, H. A. (1978). *Radiochem. Radioanal. Lett.* **35**, 313. [4.4]

Broecker, W. S. (1963). *J. geophys. Res.* **68**, 2817. [3.3, 17.2.2, 18.3, 18.4]

—— (1965). In: *Symp. Diffus. Oceans Fresh Wat.*, ed. T. Ichiye. Lamont Doherty Geological Observatory, Columbia University, Palisades, N. Y. [5.5.3, 15.4.1, 15.5]

—— (1966). *Science* **151**, 299. [18.5, 18.8]

—— (1974). *Chemical oceanography*. Harcourt-Brace-Jovanovich, New York. [8.2.1, 15.2.1, 15.2.3, 16.3.1]

—— (1978). *The cause of glacial to interglacial climatic change. In Evolution of planetary atmosphere and climatology of the earth.* p. 165. Centre National D'Etudes Statales, France. [18.1]

—— Bender, M. L. (1972). *Age determinations on marine sediments.* In: *Proc. Symp. Calib. Hominoid Evol., 1971*, ed. W. W. Bishop and J. A. Miller. Scottish Academic Press, Edingburgh. [3.1]

—— Kaufman, A. (1965), *Geol. Soc Amer. Bull.* **76**, 535. [11.5.3, 13.2.2]

—— —— (1970). *J. geophys. Res.* **75**, 7679, [15.5.1, 15.5.2]

—— Ku, T. -L. (1969). *Science* **166**, 404. [3.4.5, 13.2.1, 16.3.1]

—— Kulp, J. L. (1956). *Am. Antiq.* **22**, 1. [17.2.1]

—— Peng, T. H. (1971). *Earth planet. Sci. Lett.* **11**, 99. [15.5.1]

—— —— (1974). *Tellus* **26**, 21. [15.5.1]

—— Thurber (1965). *Science* **149**, 58. [17.2.1, 18.3, 18.4]

—— Van Donk, J. (1970). *Rev. geophys. space Phys.* **8**, 169. [13.2.1, 18.4]

—— Cromwell, J., and Li, Y. H. (1968). *Earth planet. Sci. Lett.* **5**, 101. [15.5.2]

—— Goddard, J., and Sarmiento, J. (1976). *Earth planet Sci. Lett.* **32**, 220. [8.4, 15.4.1]

—— Kaufman, A., and Trier, R. (1973). *Earth planet. Sci. Lett.* **20**, 35. [4.3.2, 4.4, 15.3.3]

—— Li, Y. H., and Cromwell, J. (1967). *Science* **158**, 1307. [15.4.1, 15.5.1, 15.5.2]

—— Kaufman, A., Ku, T, -L., Chung, Y. C., and Craig, H. (1970). *J. geophys. Res.* **75**, 7682. [15.4.1]

—— Thurber, D. L., Goddard, J., Matthews, R. K., Ku, T.-L., and Mesolella, K. J. (1968). *Science* **159**, 297. [18.4, 18.5, 19.2]

Brown, H. and Silver, L. T. (1956). *U.S. Geol. Survey Prof. Paper 300. 91.* [7.2]

Bruland, K. W. (1974). *Pb-210 geochronology in the coastal marine environment*, Ph.D. Thesis, Univ. of California, San Diego, 106 pp. [15.2.5]

—— Koide, M., and Goldberg, E. D. (1974). *J. geophys. Res.* **79**, 3083. [2.4.3, 15.6.2]

Budo, V. and Skoflek, I. (1964). In: *Tata, eine mittelpalaolithische Travertin—Siedlung in Ungarn*, ed. L. Vertes. *Archaeol. Hungar., S. N.* **43**, 51. [13.3.5]

Bumstead, H. A. and Wheeler, L. P. (1903). *Am. J. Sci.* **166**, 328. [14.1]

—— —— (1904). *Am. J. Sci.* **167**, 97. [14.1]

Bunker, C. M., Bush, C. A., and Forbes, R. B. (1973). *J. Res., U.S. Geol. Survey*, **659**, [6.1]

Bunney, L. R., Ballon, N. E., Pascual, J., and Foti, S. (1959). *Anal. Chem.* **31**, 324. [4.2.1]

Burba, P. and Lieser, K. H. (1979). *Z. analyt. Chem.* **298**, 373. [4.4]

Burcham, W. E. (1963). *Nuclear physics, an introduction.* Longmans, London. [1.2, 1.3.2, 5.1, 5.2]

Burnett, W. C. (1977). *Geol. Soc. Am. Bull.* **88**, 813. [17.3.1, 17.3.2]

—— (1980). *Bull Geol. Soc., Lond.*, (in press). [17.3.4]

—— Gomberg, D. N. (1977). *Sedimentol.* **24**, 291. [17.3.1, 17.3.2]

—— Lee, A. I. N. (1980) *Geo. J.* **4**, 423. [17.3.4]

—— Veeh, H. H. (1977). *Geochim. Cosmochim. Acta* **41**, 755. [4.4, 17.3.1, 17.3.2, 17.3.4]

—— —— Soutar, A. (1980). *SEPM Special Publication* **29**, 61. Society of Economic Palaeontologists and Mineralogists, Tulsa, Oklahoma. [17.3.2]

Burton, J. D. (1975). In: *Chemical oceanography* V. 3, 2nd edition, ed. V. P. Riley and G. Skirrow, p. 91. Academic Press, London. [15.2.3]

Burton, W. M. and Stewart, N. G. (1960). *Nature* **186**, 584. [15.7]

Busigin, A., van der Vooren, A. W., Lin Pai, H., and Phillips, C. R. (1979). *Miner. Sci. Eng.* **11**, 193. [14.5]

Butler, F. E. (1965). *Anal. Chem.* **37**, 340. [4.2.2]

Butzer, K. W. (1961). *Climatic changes in arid regions since the Pliocene.* In: *A History of land use in arid regions*, ed. L. D. Stamp. UNESCO Arid Zone Research **17**. [11.5]

—— (1965). *Science* **153**, 1718. [12.2.1]

—— (1975). In: *After the Australopitchecines*, ed. K. Butzer and G. L. Issac. Mouton, The Hague [12.2.2, 12.4.2]

—— Brown, F. H., and Thurber, D. L. (1969). *Quaternaria* **11**, 15. [12.4.2]

Cadigan, R. A. and Felmlee, J. K. (1977). *J. geochem. Explor.* **8**, 381. [14.4.4]

—— —— (1979). In: *Geochemical Exploration 1978*, ed. J. R. Watterson and P. K. Theobald, p. 401. Association of Exploration Geochemists. Elsevier, Amsterdam. [14.4.4]

Callahan, C. M. (1961). *Anal. Chem.* **33**, 1160. [4.2.1]

Campbell, E. E. and Moss, W. D. (1965). *Health Phys.* **11**, 737. [4.5.1]

Cappelani, F. and Restelli, G. (1968). *X-ray and gamma-ray spectroscopy in semiconductor detectors*, ed. G. Bertolini and A. Coche. North-Holland Publ. Co., Amsterdam. [5.3.3]

Catanzero, E. J. (1968). *Radiometric dating for geologists*, ed. E. I. Hamilton and R. M. Farquhar, p. 225. Wiley and Sons, London. [3.1]

—— Kulp, J. L. (1964). *Geochim. Cosmochim. Acta.* **28**, 87. [3.1]

Cerrai, E., Dugnani Lonati, R., Gazzarini, F., and Tongiorgi, E. (1965). *Rc. Soc. miner. ital.* **21**, 47. [3.4.4, 6.4.3]

Chakarvarti, S. K. and Nagpaul, K. K. (1979). *Nucl. Tracks* **3**, 85. [5.3.4]

Chalov, P. I. and Merkulova, K. I. (1966). *Dokl. Akad. Nauk. SSSR* **167**, 146. [17.3.2]

—— —— Tuzova, T. V. (1966a). *Dokl. Akad. Nauk SSSR* **169**, 89. [3.4.1]

—— —— —— (1966b). *Geochem. Intl.* **3**, 1149. [13.2.2]

—— Svetlichnaya, N. A., and Tuzova, T. V. (1970a). *Dokl. Akad. Nauk SSSR* **195**, 190. [3.4.1]

—— —— —— (1970b). *Geokhimiya*, Issue 7, 848. [3.4.1]

—— Tuzova, T. V., and Musin, Ye. A. (1964). *Geokhimiya* **5**, 404. [3.4.1]

Chan, L. H., Edmond, J. M., Stallard, R. F., Broecker, W. S., Chung, Y. C., Weiss, R. F., and Ku, T.-L. (1976). *Earth planet. Sci. Lett.* **32**, 258. [15.4.1]

Chappell, J. M. A. (1974a). *Geol. Soc. Am. Bull.* **85**, 553. [17.2.3, 18.5, 18.7]

—— (1974b). *Nature* **252**, 199. [18.5, 18.8]

—— Veeh, H. H. (1978a). *Geol. Soc. Am. Bull.* **89**, 356. [17.2.1, 17.2.3, 18.5, 18.7]

—— —— (1978b) *Nature* **276**, 602. [17.2.1]

Cheeseman, R. V., and Wilson, A. L. (1973). *U. K. Water Research Association Report TM 78*. [4.3.2]

Cherdyntsev, V. V. (with P. I. Chalov *et al.*) (1955). *Trans 3rd Session Commiss. Determ. Abs. Age Geol. Formns*, p. 175. Izd. Akad. Nauk SSSR. [2.5.1, 3.3, 3.4.1, 8.1, 15.2.3]

—— (1956). *Abundance of chemical elements*. University of Chicago Press, Chicago. [17.3.2]

—— (1970). *Geothermics (Spec. Issue 2)* **2**, [9.2.3]

—— (1971). *Uranium-234*, Israel Program for Scientific Translations, Jerusalem. [3.1, 3.3, 9.1.3, 9.1.4, 9.2.2, 9.3.3, 10.3.1, 11.3.3, 11.4, 11.5.3, 12.2.2, 12.4.1, 12.4.3, 12.5, 13.2.2, 13.3.2, 13.3.5, 14.4.2, 14.4.3]

—— Kadyrov, N. B., and Novichkova, N. (1971). *Geochem. Int.* **8**, 211. [16.3.2]

—— Kazachevskii, I. V., and Kuz'mina, E. A. (1963). *Geochem. Int.* **3**, 271. [3.4.1]

—— —— —— (1965). *Geochem. Int.* **2**, 749. [5.5.2, 11.3.3]

—— Kislitsina, G. I., Kuptsov, V. M., Kuz'mina, E. A., and Zverev, V. L. (1967). *Geochem. Int.* **4**, 639. [3.4.4, 6.4.3, 6.5.1]

—— Kuptsov, V. M., Kuz'mina, E. A., and Zverev, V. L. (1968). *Geochem. Int.* **5**, 56. [3.4.4, 6.4.3]

—— Senina, N., and Kuz'mina E. A. (1975). *Abh. Zent. Geol. Inst. Palaont.* **23**, 7. [11.4, 12.2.1, 12.4.1, 13.3.2]

Chetham-Strode, A., Tarrant, J. R., and Silva, R. J. (1961). *IRE Trans. nucl. Sci.* **NS8**, 59. [5.3.1]

Chow, S. N. and Carswell, D. J. (1963). *Aust. J. appl. Sci.* **14**, 193. [4.2.2]

Chow, T. J. and Goldberg, E. D. (1960). *Geochim. Cosmochim. Acta* **20**, 192. [15.4.1]

Chung, Y. C. (1973). *Earth planet. Sci. Lett.* **17**, 319. [15.5.2]

—— (1974a). *Earth planet. Sci. Lett.* **23**, 125. [15.4.1]

— (1974b). *Earth planet. Sci. Lett.* **21**, 295. [15.5.2]

—— (1976). *Earth planet. Sci. Lett.* **32**, 249. [15.4.1]

—— Craig, H. (1972). *Earth planet. Sci. Lett.* **14**, 55. [15.5.2]

—— —— (1973). *Earth planet. Sci. Lett.* **17**, 306. [15.4.1]

—— —— (1980). *Earth planet. Sci. Lett.* **49**, 267. [15.4.1]

—— Finkel, R., Kim, K., and Craig, H. (1979). *EOS Trans. Am. geophys. Un.* **60**, 857. [16.2.3]

—— Craig, H., Ku, T.-L., Golder, J., and Broecker, W. S. (1974). *Earth planet Sci. Lett.* **23**, 116. [15.4.1]

Church, S. E. (1975). *Rev. Geophys.* **13**, 98. [5.3.4]

Church, T. M. and Bernat, M. (1972). *Earth planet. Sci. Lett.* **14**, 139. [16.2, 16.2.1, 16.2.2]

Clark, S. P. Jr., Peterman, Z. E., and Heier, K. S. (1966). In: *Handbook of physical constants*, ed. S. P. Clark, Jr. *Geol. Soc. of America Memoir* **97**, 521. Geological Society of America, Boulder, Colorado. [6.1]

Clarke, R. M. (1975). *Antiquity* **49**, 251. [12.1]

Clarke, W. B. and Kugler, G. (1973). *Econ. Geol.* **68**, 243. [4.4, 9.2.3]

Cliff, K. D. (1978a). *Phys. Med. Biol.* **23**, 55. [14.4.5, 14.5]

—— (1978b). *Phys. Med. Biol.* **23**, 696. [14.4.5, 14.5]

Cochran, J. K. (1979). *The Geochemistry of* ^{226}Ra *and* ^{228}Ra *in Marine Deposits*, Ph.D. Thesis, Yale University, New Haven, Conn., 260 pp. [15.3.2, 15.4.1, 16.2.3]

Cochran, J. K. (1980a). In: *Ecosystem processes in the deep oceans*, Rubey Colloquium, Vol. 2. Prentice-Hall Inc., Englewood Cliffs, New Jersey. (In press). [15.1.1]

—— (1980b). *Earth planet. Sci. Lett.* **49**, 381. [15.4.1]

—— Aller, R. C. (1979). *Estuar. Coast. Mar. Sci.* **9**, 739. [15.3.3, 16.3.3]

—— Krishnaswami, S. (1980). *Am. J. Sci.* **280**, 849. [15.2.4, 15.2.6, 15.4.1]

—— Osmond, J. K. (1976). *Deep Sea Res.* **23**, 193. [15.3.2]

—— Rye, D. M., and Landman, N. H. (1980). *Trans. Am. geophys. Un.* **61**, 275. [15.7]

Compton, A. H. (1922). *Bull. Nat. Res. Coun.* **4** (20), 10. [5.1]

Compston, W. and Jeffery, P. M. (1959). *Nature* **184**, 1792. [3.1]

Condomines, M. and Allegre, C. J. (1980). *Nature* **288**, 354. [6.4.3]

—— Berna't, M., and Allegre, C. J. (1976). *Earth planet. Sci. Lett.* **33**, 122. [6.5.2]

Coope, R. G. (1977). *Phil. Trans. R. Soc. Lond.* B **280**, 313. [13.2.2]

Cotton, F. A. and Wilkinson, G. (1972). *Advanced inorganic chemistry*, (3rd ed.) Wiley-Interscience, New York. [2.1.1]

Covell, D. F. (1975). *Gamma-ray spectrometry in radiochemical methods in analysis*, ed. D. I. Coomber, Plenum Press, New York. [5.3.2]

Cowan, C. L., Reines, F., Harrison, F. B., Kruse, H. W., and McGuire, A. D. (1956). *Science* **124**, 103. [See also: Reines, F. and Cowan, C. L. (1953). *Phys. Rev.* **92**, 830; and Reines, F. and Cowan, C. L. (1959). *Phys. Rev.* **113**, 273.] [1.3.2]

Cowart, J. B. (1974). ^{234}U *and* ^{238}U *in the Carrizo Sandstone Aquifer of South Texas*, Ph.D. Thesis. Florida State Univ., Talahassee. [4.3.2, 4.4]

—— (1979). *Geol. Soc. Am. (Abs. Prog.)* **11**, 405. [9.2.2]

—— (1980a). *Earth planet Sci. Lett.* **48**, 277. [9.2.3]

—— (1980b). In: *Natural radiation environment III*, ed. T. F. Gesell and W. M. Lowder, p. 711. U.S. Department of Energy. [9.2.2]

—— Osmond, J. K. (1974). In: *Isotope techniques in groundwater hydrology 2* (Proc. Symp. IAEA, Vienna), 131. [2.2.4, 9.1.5, 9.2.3, 9.3.4]

—— —— (1977). *J. geochem. Explor.* **8**, 365. [2.5.1, 2.5.2, 14.4.2]

—— —— (1980). *U.S. Dept. Energy Report GJBX 119*. [9.2.3, 14.4.2]

—— Kaufman, M. J., and Osmond, J. K. (1978). *J. Hydrol.* **36**, 161. [9.2.2, 9.2.3]

Craig, H., Krishnaswami, S., and Somayajulu, B. L. K. (1973). *Earth planet. Sci. Lett.* **17**, 295. [2.5.3, 4.4, 4.6, 5.5.4, 15.6.3, 19.1]

Cronin, T. M., Szabo, B. J., Ager, T. A., Hazel, J. E., and Owens, J. P. (1981). *Science* **211**, 233. [18.5]

Culler, F. L. (1956). *Proc. Int. Conf. Peaceful Uses At. En., Geneva. Paper P/822*. U. N., New York. [4.2.1]

Cuninghame, J. G. (1975). Methods of detection and measurement of radioactive nuclides. *In Radiochemical methods in analysis*, ed. D. I. Coomber. Plenum Press, New York. [5.2.1]

Curie, P., Curie, M., and Bémont, G. (1898). *C.r. hebd. Séanc. Acad. Sci., Paris* **127**, 1215. [15.4.1]

Curtis, G. H. (1975). *World Archaeol.* **7**, 198. [3.1]

Dall 'Aglio, M., Gragnani, M., and Locardi, E. (1974). *Formation of uranium ore deposits* (Proc. Symp. IAEA), 183. IAEA, Vienna. [2.2.5]

Dalrymple, G. B. and Lanphere, M. A. (1969). *Potassium argon dating: principles, techniques and applications to geochronology*, Freeman, San Francisco. [3.1]

D'Amore, F. (1975). *Geothermics* **4**, 96. [9.2.3]

Danis, A. and Valjin, V. (1979). *Nucl. Tracks* **3**, 83. [5.3.4]

Davis, R. and Schaeffer, O. A. (1955). *Ann. N. Y. Acad. Sci.* **62**, 105. [3.1]

Dearnaley, G. and Whitehead, A. B. (1961). *Nucl. Instr. Meth.* **12**, 205. [5.2.1]

Débenath, A. (1976). *J. Human Evol.* **6**, 297. [12.1, 12.4.1]

Deines, P., Langmuir, D., and Harmon, R. S. (1974). *Geochim. Cosmochim. Acta* **38**, 1147. [13.2.2]

Delaney, M. (1980). *Trans. Am. geophys. Un.* **61**, 259. [15.2.5]

DeMaster, D. J. (1979). *The marine budgets of silica and ^{32}Si*, Ph.D. Thesis, Yale University, New Haven, Conn., 308 pp. [15.2.5, 15.3.2, 15.4.1]

—— Cochran, J. K. (1977). *EOS Trans. Am. geophys. Un.* **58**, 1154. [16.3.3]

Denham, G. R., Anderson, R. F., and Bacon, M. P. (1977). *Earth planet. Sci. Lett.* **35**, 384. [4.4]

De Voto, R. H., Mead, R. H., Martin, J. P., and Bergquest, L. E. (1980). *New Mex. Bur. Mines Min. Resour. Mem. 38 (in press).* [14.4.8]

Diamond, R. M. (1955). *J. Am. chem. Soc.* **77**, 2978. [4.2.4]

Dion, E. P. (1979). *Trans. Am. Geophys. Un.* **60**, 851. [15.4.1]

Discendenti, A., Nicoletti, M., and Taddeucci, A. (1970). *Period. Min.* **39**, 3. [6.4.3]

Doe, B. R. (1979). In: *Geochemical Exploration 1978*, ed. J. R. Watterson and P. K. Theobald, p. 227. Association of Exploration Geochemists. [14.4.8]

Doi, K., Hirono, S., and Sakamaki, Y. (1975). *Econ. Geol.* **70**, 628. [2.2.3, 2.3.1, 8.2.2]

Dokiya, Y., Yamazaki, S., and Fuwa, K. (1974). *Env. Lett.* **7**, 1. [4.3.2]

Dooley, J. R., Granger, H. C., and Rosholt, J. N. (1966). *Econ. Geol.* **61**, 1326. [17.3.2]

Dubois, J., Launay, J., and Recy, J. (1974). *Tectonophys.* **24**, 133. [17.2.3]

—— —— —— Marshall, J. (1977). *Can. J. earth Sci.* **14**, 250. [17.2.3, 18.5]

Duplessy, J. C. (1978). In: *Climatic change*, ed. J. Gribbin, p. 46. Cambridge University Press, New York. [17.3.4]

—— Labeyrie, J., Lalou, C., and Nguyen, H. V. (1970). *Nature* **226**, 631. [11.3.3, 13.3.3]

Duschner, H., Born, H. J., and Kun, J. I. (1973). *Int. J. appl. Rad. Isotopes* **24**, 433. [4.5.1]

Duyckaerto, G. and Lejeune, R. (1960). *J. Chromat* **3**, 58. [4.2.4]

Dyck, W. (1978). In *Short course in uranium deposits: their mineralogy and origin*, ed. M. M. Kimberley, p. 57. Mineral. Assoc. of Canada. [14.4.5]

—— (1976). *J. geochem. Explor.* **5**, 3. [14.4.8]

—— (1979). In: *Geophysics and geochemistry in the search for metallic ores*, ed. P. J. Hood. Geol. Surv. Canada Econ. Geol. Report No. 31, p. 489. [14.4.8]

—— Jonasson, I. R. (1977). *Water Res.* **11**, 705. [14.4.8]

—— Tan, B. (1978). *J. geochem. Explor.* **10**, 153. [9.2.3]

Dyer, F. F., Emergy, J. F., and Leddicotte, G. W. (1962). *A comprehensive study of the neutron activation analysis of uranium by delayed neutron counting*. ORNL-3342. Oak Ridge National Laboratories, Oak Ridge, Tennessee. [5.3.4]

Dymond, J. and Veeh, H. H. (1975). *Earth planet Sci. Lett.* **28**, 13. [15.3.2]

Dysard, J. E. and Osmond, J. K. (1975). *EOS Trans. Am. geophys. Un.* **56**, 1006. [16.2.1]

Eakins, J. D. and Morrison, R. T. (1978). *Int. J. appl. radiat. Isotopes* **29**, 531. [4.2.7, 4.4]

Eardley, A. J., Shuey, R. T., Gvosdetsky, V., Nash, W. P., Dane Pritchard, M., Gray, D. C., and Kukla, G. J. (1973). *Geol. Soc. Am. Bull.* **84**, 211. [13.3.1]

Edmond, J. M., Measures, C., McDuff, R. E., Chan, L. H., Collier, R., Grant,

B., Gordon, L. I., and Corliss, J. B. (1979). *Earth planet Sci. Lett.* **46**, 1. [15.2.5]

—— —— Mangum, B., Grant, B., Selater, F. R., Collier, R., Hudson, A., Gordon, L. I., and Corliss, J. B. (1979). *Earth planet Sci. Lett.* **46**, 19. [15.2.5]

Eisenbud, M. (1977). In: *Int. Symp. Areas High Nat. Radioact.* (Proc. of Conf. held in Pocos de Caldas, Brazil, June 16–20, 1975), p. 167. Brazil Acad. Sciences, Rio de Janeiro. [14.4.4]

Elmore, D., Anoutareman, N., Fulbright, H. W., Gove, H. E., Hans, H. S., Nishiizumi, K., Murrell, M. T., and Honda, M. (1980). *Phys. Rev. Lett.* **45**, 589. [3.1]

—— Fulton, B. R., Clover, M. R., Marsden, J. R., Gove, H. E., Naylor, H., Purser, K. H., Kilius, L. R., Beukens, R. P., and Litherland, A. E. (1979). *Nature* **277**, 22. [3.1]

—— Gove, H. E., Ferraro, R., Kilius, L. R., Lee, H. W., Chang, K. H., Beukens, R. P., Litherland, A. E., Russo, C. J., Purser, K. H., Murrell, M. T., and Finkel, R. C. (1980). *Nature* **286**, 138. [3.1]

Emiliani, C. (1955). *J. Geol.* **63**, 538. [12.1, 13.1, 13.2.1, 18.4]

—— (1964). *Geol. Soc. Am. Bull.* **75**, 129. [18.4]

—— (1966). *J. Geol.* **74**, 109. [18.1, 18.4]

—— Rona, E. (1969). *Science* **166**, 1551. [16.3.1]

Engel-Kemeir, D. (1967). *Nucl. Instr. Meth.* **48**, 335. [5.2.1]

England, J. B. A. (1974). *Techniques in nuclear structure physics, Part I*. Macmillan Press Ltd, London. [5.2.1]

Epstein, S., Buchsbaum, R., Lowenstam, H., and Urey, H. C. (1953). *Geol. Soc. Am. Bull.* **64**, 1315. [18.1]

Ericson, D. B. and Wollin, G. (1968). *Science* **162**, 1227. [18.1]

Evans Jr., H. T. (1963). *Science* **141**, 154. [2.1.1]

Evans, R. D. (1955). *The atomic nucleus*, McGraw-Hill, New York. [1.2, 1.3.1, 5.1, 5.4]

—— Kip, A. F. (1938). *Am. J. Sci.* **36**, 321. [15.4.1]

—— —— Moberg, E. G. (1938). *Am. J. Sci.* **36**, 241. [15.4.1]

Eve, A. S. (1907). *Phil. Mag.* **13**, 248. [15.4.1]

—— (1909). *Phil. Mag.* **18**, 102. [15.4.1]

Ewing, T. E. (1979). *Econ. Geol.* **74**, 678. [3.1]

Fairbanks, R. G. and Matthews, R. K. (1978). *Quat. Res.* **10**, 181. [13.2.1]

Falls, W. F. and Williams, D. F. (1980). *EOS Trans. Am. geophys. Un.* **61**, 258. [18.5, 18.7]

Fanale, F. P. and Schaeffer, O. A. (1965). *Science* **149**, 312. [3.1, 3.3.3, 18.6]

Fano, U. (1953). *Nucleonics* **11**, (8), 8; **11**, (9), 55. [5.1]

Farquhar, R. M., Bregman, N., Badone, E., and Beebe, B. (1978). *Proc. 1978 Symp. Archaeom. Archaeol. Prospect.*, Bonn (abs.). [12.2.2]

Feth, J., and Barnes, I. (1979). *U.S. Geol. Surv. Water Res. Open File Rep. 79–35*, (map). [11.4]

Fisher, D. E. (1972). *Earth planet. Sci. Lett.* **14**, 255. [4.4]

—— (1977a). *Nature* **265**, 227. [5.3.4, 5.5, 16.3]

—— (1977b). *J. radioanal. Chem.* **38**, 477. [5.3.4, 5.5, 6.2]

—— (1978). *Geophys. res. Lett.* **5**, (4), 233. [5.3.4, 5.5]

—— Boström, K. (1969). *Nature* **224**, 64. [15.2.5, 16.2, 16.2.1]

Fitch, F. J., Foster, S. C., and Miller, J. A. (1974). *Rep. Prog. Phys.* **37**, 1433. [3.1]

—— Hooker, P. J., and Miller, J. A. (1976). *Nature* **263**, 740. [3.1]

Fleischer, R. L. (1975). *World Archaeol.* **7**, 136. [3.1]

—— (1980). *Science* **207**, 979. [7.3.1, 10.3.1]

—— Mogro–Campero, A. (1978). *J. geophys. Res.* **18**, (B7), 3539. [14.4.5]

—— —— (1979). *Geophys.* **44**, 1541. [14.4.5]

—— Price, P. B. (1964a). *Geochim. Cosmochim. Acta* **28**, 1705. [5.3.4]

—— —— (1964b). *Geochim. Cosmochim. Acta* **28**, 775. [5.3.4]

—— Raabe, O. G. (1978). *Geochim. Cosmochim. Acta* **42**, 973. [7.3.1, 10.3.1, 15.2.1, 15.4.1, 17.3.2]

—— Price, P. B., and Walker, R. M. (1965). *Ann. rev. nucl. Sci.* **15**, 1. [5.3.4]

—— —— —— (1975). *Nuclear tracks in solids*, University of California Press, Berkeley. [2.2.1, 6.2]

Flint, R. F. (1971). *Glacial and Quaternary geology.* J. Wiley and Sons, New York. [11.5, 13.2.2, 18.1]

Flynn, W. W. (1968). *Anal. Chim. Acta.* **43**, 221. [4.5.1, 5.5.4]

Folk, R. L. and Assereto, R. (1976). *J. sedim. Petrol.* **46**, 486. [12.2.1]

Ford, D. C., Schwarcz, H. P., Drake, J. J., Gascoyne, M., Harmon, R. S., and Latham, A. G. (1981). *J. Arct. Alp. Res.* **13**, 1. [11.3.4]

Fornaca–Rinaldi, G. (1968a). *Boll. Geofis. teor. appl.* **10**, 3. [11.3.3]

—— (1968b). *Earth planet. Sci. Lett.* **5**, 120. [12.4.1]

Føyn, E., Karlik, B., Pettersson, H., and Rona, E. (1939). *Nature* **143**, 275. [5.5.3]

Francis, C. W., Chester, G., and Hoskin, L. A. (1970). *Envir. Sci. Techn.* **4**, 586. [3.4.8]

Franke, H. W. and Geyh, M. (1970). *Umschau* **3**, 91. [13.3.3]

Freiser, H. and Morrison, G. H. (1959). *Ann. rev. nuc. Sci.* **9**, 221. [4.2.6]

Friedlander, G., Kennedy, J. W., and Miller, J. M. (1964). *Nuclear and radiochemistry* (2nd edition) Wiley and Sons, New York. [1.3.3]

Frisch, O. R. (1944). *Isotopic analysis of uranium samples by means of their alpha-ray groups.* BR-49 (unpublished report). [5.2.1]

Fukuoka, T. (1974). *Geochem. Jour.* **8**, 109. [6.4.3]

Gäggler, H., Von Gunten, H. R., and Nyffler, U. (1976). *Earth planet. Sci. Lett.* **33**, 119. [4.4, 5.5.4]

Gale, N. H. (1967). *IAEA Symp. Radioact. Dating Meth Low-Level Count., Mexico 1967. (Paper SM 87/38)*, 431. IAEA, Vienna. [4.4, 5.3.4]

Gamow, G. (1928). *Z. f Physik.* **51**, 204. [1.3.1]

—— Critchfield, C. L. (1949). *Theory of atomic nucleus and nuclear energy sources,* Oxford University Press, London. [1.3.1]

—— Hantermans, F. (1928). *Z. f Physik.* **52**, 496. [1.3]

Garlick, G. F. (1952). *Prog. nucl. Phys.* **2**, 51. [5.2.1]

Gascoyne, M. (1977). *U series dating of speleothems: An investigation of technique, data processing and precision.* Tech. Memo 77–4, Dept. of Geology, McMaster Univ. [5.5.1, 5.5.2]

—— (1977). *Trace elements in calcite—the only cause of speleothem color?* Proc. Nat. Speleo. Soc. Conv. Alpena, Mich., p. 39, National Speleological Society, Care House, Huntsville, Alabama. [11.3.1]

—— (1980). *Pleistocene climates determined from stable isotopic and geochronological studies of speleothem.* UnPub. Ph.D. Thesis, McMaster University, [2.3.1, 5.5.2, 11.3.4, 12.2.1, 13.3.3, 19.2]

—— (1981). *J. Geochem. Explor.* **14**, 199. [11.3.4]

—— (1981). *J. Hydrol.* (in press). [13.3.3]

—— Ford, D. C., and Schwarcz, H. P. (1981). *Can. J. Earth Sci.* (in press). [11.3.4]

—— Schwarcz, H. P., and Ford, D. C. (1978). *Trans. Brit. Cave Res. Assoc.* **5**, 91. [3.5, 13.3.3]

—— —— —— (1980). *Nature* **285**, 474. [13.3.3]

—— Benjamin, G. J. B., Schwarcz, H. P., and Ford, D. C. (1979). *Science* **205**, 806. [11.3.4, 13.2.1, 18.7]

—— Currant, A. P., and Lord, T. (1981). (In preparation.) [13.3.5]

Gibson, W. M. (1961). *The radiochemistry of lead NAS-NS 3040.* National Academy of Sciences, U.S.A. [4.2.6]

Glasby, G. P. (1977a). *Marine manganese deposits*. Oceanography series No. **15**, Elsevier, New York. [16.3.2]

—— (1979b). *N. Z. J. Sci.* **20**, 187. [16.3.2]

Gleadow, A. J. W. (1980). *Nature* **284**, 225. [12.1]

Glover, K. M. (1956). *Performance of a six position gridded ionization chamber for alpha energy measurement*. AERE C/R 2091. [5.3.1]

—— (1971). *Alpha spectra and associated techniques and measurements*. Paper 26 in Proc. Int. Conf. Chem. Nucl. Data Meas. Appl., University of Kent, Canterbury. British Nuclear Energy Society. [5.3.1]

Godfrey, J. S., Cresswell, G. R., Golding, T. J., Pearce, A. F., and Boyd, R. (1980). *J. phys. Oceanogr.* **10**, 430. [17.3.4]

Goede, A., Green, D. C., Harmon, R. S., and Hitchman, M. A. (1981). *Nature*, (in press). [13.3.3]

Gofman, J. W. and Seaborg, G. T. (1949). *Paper No. 19.14: The Transuranium Elements. Nat. Nuc. Energy. Ser., Div. IV, 14B.* McGraw-Hill, New York. [1.5]

Goldberg, E. D. (1963). *Geochronology with ^{210}Pb in radioactive dating*, IAEA, Vienna, 121. [3.4.8]

—— Arrhenius, G. (1958). *Geochim. Cosmochim. Acta* **13**, 153. [15.1.1, 16.3.2]

—— Koide, M. (1958). *Science* **218**, 1003. [15.4.1]

—— —— (1962). *Geochim. Cosmochim. Acta* **26**, 417. [4.2.2, 4.4, 7.1, 17.3.2]

—— Gamble, E., Griffin, J. J., and Koide, M. (1977). *Estuarine Coast. Marine Sci.* **5**, 549. [3.4.8]

—— Hodge, V., Koide, M., Griffin, J., Gamble, E., Bricker, O. P., Matisoff, G., Holdren Jr., G. R., and Braun, R. (1978). *Geochim. Cosmochim. Acta* **42**, 1413. [16.2.5, 16.3.3]

Golden, J. and Maddock, A. G. (1956). *J. inorg. nucl. Chem.* **2**, 46. [4.2.3]

Golding, A. S. (1961). *Anal. Chem.* **33**, 406. [4.2.4]

Goldich, S. S. and Mudrey, M. G. Jr., (1969). *Geol. Soc. Am. Abst. Prog.* **7**, 80. [7.2]

—— —— (1972). In: *Contributions to recent geochemistry and analytical chemistry*, (A. P. Vinogradov Volume) ed. A. I. Turaginov, p. 415. Nauka Publ. Office, Moscow. [7.2]

Goldschmidt, V. M. (1954). *Geochemistry*. Clarendon Press, Oxford [2.1.1, 2.1.2]

Goudie, A. (1973). *Prog. Geog.* **5**,77. [10.2.1]

Green, H. S., Stringer, C. B., Collcutt, S. N., Currant, A. P., Huxtable, J., Schwarcz, H. P., Debenham, N., Emblton, C., Bull, P., Molleson, T. I., and Bevins, R. E. (1981). *Nature* **294**, 707. [12.4.1]

Gurney, R. W. and Condon, E. U. (1928). *Nature* **122**, 439. [1.3.1]

—— —— (1929). *Phys. Rev.* **33**, 127. [1.3.1]

Gustavsson, J. E. and Högberg, S. A. C. (1973). *Boreas* **1**, 247. [2.4.2]

Hagemann, F. (1950). *J. Am. chem. Soc.* **72**, 768. [4.2.4, 5.5.3]

Halbach, P., Von Borstel, D., and Gunderman, K-D. (1980) *Chem. Geol* **29**, 117. [2.3.1]

Hambleton-Jones, B. B. (1978). *Miner. Sci. Eng.* **10**, 182. [7.1, 14.2, 14.3, 14.4.1, 14.4.4, 14.4.6, 14.6]

—— (1979). In: *Training course on radiometric prospecting techniques. Lecture 3* South Africa Atomic Energy Board, Pretoria. [14.1]

—— Smit, M. C. B. (1979). In: *Training course on radiometric prospecting techniques*. Lecture 8. South Africa Atomic Energy Board. Pretoria. [14.4.5]

Hamilton, E. I. (1965). *Applied geochronology*. Academic Press, New York. [3.1]

Van der Hammen, T., Wijmstra, T. A. and Zagwijn, W. H. (1971). *The floral record of the Late Cenozoic in Europe*. In: *The Late Cenozoic Glacial Ages*, ed. K. K. Turekian, Yale University Press, New Haven, Connecticut. [13.2.2]

Hammer, C. U., Clausen, H. B., Dansgaard, W., Gundestup, N., Johnson, S. J., and Reeh, N. (1978). *J. Glaciol.* **20**, 3. [13.2.2]

Hammond, D. E., Simpson, H. J., and Mathieu, G. G. (1977). *J. geophys. Res.* **82**, 3913. [15.5.2]

Hanna, G. C. (1959). *Alpha-Radioactivity*. In: *Experimental nuclear physics Vol III* ed. E. Segré, Wiley and Son, New York. [1.3.1]

Hanor, J. S. and Chan, L. H. (1977). *Earth planet. Sci. Lett.* **37**, 242. [8.1, 8.4]

Hansen, R. O. (1965). *Isotopic distribution of uranium and thorium in soils weathered from granite and alluvium*, Ph.D. Thesis, Univ. of Calif., Berkeley, 135p. [7.3.2, 10.3.2]

Hansen, R. O. (1970). *Soil Sci.* **110**, 31. [10.3.2]

—— Huntington, G. L. (1969). *Soil Sci.* **108**, 257. [7.3.2, 10.3.2]

—— Stout, P. R. (1968). *Soil Sci.* **105**, 44. [7.3.2, 10.2.1, 10.3.2]

Hanshaw, B. and Hallet, B. (1978). *Science* **200**, 1267. [13.2.2]

Harley, J. (1974). Ed. *H.A.S.L. Procedures Mannual HASL-300*. [4.2.5]

Harmon, R. S. (1975). *Late Pleistocene paleoclimate in North America as inferred from isotopic variations in speleothems*. Unpublished Ph.D. Thesis, McMaster University, Hamilton, Ontario. [11.3.4]

—— Schwarcz, H. P. (1981). *Nature* **290**, 125. [11.3.4, 13.3.3]

—— Ford, D. C., and Schwarcz, H. P. (1977). *Can. J. Earth Sci.* **14**, 2543. [11.3.4, 13.3.3]

—— Glazek, J., and Nowak, K. (1980). *Nature* **284**, 132. [10.2.2, 12.4.1, 13.3.2]

—— Land, L. S., and Mitterer, R. M., (1981). *Science* (in press). [18.5]

—— Schwarcz, H. P., and Ford, D. C. (1978a). *J. Geol.* **86**, 373. [11.3.4, 13.3.3]

—— —— —— (1978b). *Quatern. Res.* **9**, 205. [11.3.4, 13.2.1, 13.2.2]

—— —— O'Neil, J. R. (1979). *Earth planet. Sci. Lett.* **42**, 254. [11.3.4, 13.3.3]

—— Ku, T.-L., Matthews, R. K., and Smart, P. L. (1979). *Geology* **7**, 405. [18.2, 19.2]

—— Schwarcz, H. P., Ford, D. C., and Koch, D. L. (1979). *Geology* **7**, 430. [11.3.4]

—— —— Thompson, P., and Ford, D. C. (1978). *Geochim. Cosmochim. Acta* **42**, 433. [11.3.4, 12.5]

—— Thompson, P., Schwarcz, H. P., and Ford, D. C. (1975). *Nat. Speleo. Soc. Bull.* **37**, 21. [2.1.2, 4.4]

—— —— —— —— (1978). *Quatern. Res.* **9**, 54. [3.4.1, 11.3.4, 13.3.3]

—— Land, L. S., Mitterer, R., Garrett, P., Schwarcz, H. P., and Larson, G. (1981). *Nature* **289**, 481. [11.3.4, 13.2.1]

Harshman, E. N. (1972). *U.S. Geological Survey Prof. Paper 745*, 82. [7.2]

Hart, R. (1973). *Can. J. earth. Sci.* **10**, 799. [15.2.5]

Hathaway, L. R. and James, G. W. (1975). *Anal. Chem.* **47**, 2035. [4.4]

Haynes, C. V., Jr. (1967). In: *Pleistocene extinctions – The search for a cause*, ed. P. S. Martin and H. E. Wright. INQUA Congress 6. Yale University Press., New Haven, Connecticut. [12.2.2]

—— (1968). *Southern Arizona Guidebook III*. Ariz. Geol. Soc. **79**. [12.2.2]

—— Agogino, G. A. (1960). *Denver Mus. Nat. Hist. Proc.* **9**, 5. [12.2.2]

—— Doberenz, A. R., and Allen, J. A. (1966). *Am. Antiq.* **31**, 517. [12.2.2]

Hays, J. D., Imbrie, J., and Shackleton, N. J. (1976). *Science* **194**, 1121. [13.2.1, 18.5, 18.8]

Hedges, R. E. M. (1979). *Nature* **281**, 19. [3.1]

Heier, K. S. and Adams, J. A. S. (1965). *Geochim. Cosmochim. Acta* **29**, 53. [2.2.2]

Henderson, G. H. (1922). *Proc. R. Soc. A* **102**, 496. [5.1]

—— (1925). *Proc. R. Soc. A* **109**, 157. [5.1]

Hendy, C. H. (1971). *Geochim. Cosmochim. Acta* **35**, 801. [13.3.3]

—— Wilson, A. T. (1968). *Nature* **216**, 48. [13.2.2, 13.3.3]

—— Healy, T. R., Rayner, E. M., Shaw, J., and Wilson, A. T. (1979). *Quatern. Res.* **11**, 172. [10.2.2, 11.5.2, 11.5.3, 13.3.1]

Hennig, G. J. (1979). *Unpublished Ph.D. Thesis*, Universität zü Köln, Köln, West Germany. [11.3.4]

Hernegger, F. and Karlik, B. (1935). *Göteborgs K. Vetensk. − o. VitterhSamh. Handl., B* **4** (12), 1. [15.4.1]

Herr, W., Wolff, R., Eberhardt, P., and Kopp, E. (1967). *Proc. Symp. Radioact. Dating Meth. Low Level Counting, Monaco.* IAEA, Vienna. [3.1]

Hesford, E., McKay, H. A. C., and Scargill, D. (1957). *J. inorg. nucl. Chem.* **4**, 321. [4.2.2]

Heye, D. (1969). *Earth planet. Sci. Lett.* **6**, 112. [16.2, 16.2.2]

—— (1975). *Geol. Jb.* **E-5**, 3. [16.3.2]

Heyrand, M. and Cherry, R. D. (1979). *Mar. Biol.* **52**, 227. [15.7]

Higashi, S. (1958). *Japan Analyst* **7**, 441. [4.2.2]

Hilaire-Marcel, C., Soucy, J. M., and Cailleux, A. (1979). *Can. J. Earth Sci.* **16**, 1494. [13.2.2]

Hille, P. (1979). *Earth planet. Sci. Lett.* **42**, 138. [3.7, 12.3.2, 17.2.2, 18.3]

Holland, H. D. (1978). *The chemistry of the atmosphere and oceans.* John Wiley and Sons, New York. [8.2.1]

—— Kulp, J. L. (1954). *Geochim. Cosmochim. Acta* **5**, 214. [2.3.2]

Hollander, J. M., Perlman, I., and Seaborg, G. T. (1953). *Rev. Mod. Phys.* **25**, 469. [1.5]

Hollin, J. (1980). *Nature* **283**, 629. [13.3.1]

Honjo, S. (1978). *J. mar. Res.* **36**, 469. [19.3]

Holmes, A. (1967). (ed.) *The determination of radionuclides in materials of biological origin.* R. 5474. AERE, Harwell. [4.5.1]

—— Holmes, D. C. (1978). *Principles of physical geology*, John Wiley and Sons, New York. [3.1]

Holmes, C. W., Osmond, J. K., and Goodel, H. G. (1968). *Earth planet. Sci. Lett.* **4**, 308. [16.2, 16.2.1, 16.2.2]

Hostetler, P. B. and Garrels, R. M. (1962). *Econ. Geol.* **57**, 137. [2.3.1, 2.4.1]

Howell, F. C., Cole, G. H., Kleindienst, M. R., and Szabo, B. J. (1972). *Nature* **237**, 51. [12.2.2, 12.4.3]

Hurford, A. J., Gleadow, J. W., and Naeser, C. W. (1976). *Nature* **263**, 738. [3.1]

Hurley, P. M. (1950). *Geol. Soc. Am. Bull.* **61**, 1. [7.2]

Hussain, N. and Krishnaswami, S. (1980). *Geochim. Cosmochim. Acta* **44**, 1287. [3.4.6, 4.6, 9.1.3]

Hutchinson, G. E. (1950). *Bull. Am. Mus. nat. Hist.* **96**, 1. [17.3.3, 17.3.4]

Hyde, E. K., Perlman, I., and Seaborg, G. T. (1964). *The nuclear properties of the heavy elements, Vol II.* Prentice Hall, New Jersey. [1.5]

Ikeya, M. (1978). *Archaeometry* **20**, 147. [12.5, 13.2.2]

Immel, R. and Osmond, J. K. (1976). *Chem. Geol.* **18**, 263. [15.2.4]

Inoue, T. and Tanaka, S. (1979). *Nature* **277**, 209. [3.1]

Irlweck, K. and Sopantin, H. (1975). *Electrolytic fabrication of calibration sources in alpha spectrometry.* UCRL-Trans-10983. Livermore Laboratory, University of California. [4.5.1]

Irving, A. J. (1978). *Geochim. cosmochim. Acta* **42**, 743. [2.2.1]

Ishii, D. and Takenchi, T. (1961). *Japan Analyst* **10**, 1125. [4.2.2]

Ishimori, T. (1955). *Bull. chem. Soc. Japan* **28**, 432. [4.2.7]

—— Okimo, H. (1956). *Bull. chem. Soc. Japan* **29**, 78. [4.2.1]

—— Watanabe, K., and Kimura, K. (1960). *J. atom. Energy Soc. Japan* **21**, 750. [4.2.3]

Issac, N. and Picciotto, E. (1953). *Nature* **171**, 742. [15.4.1, 16.1, 16.3.1]

Ivanovich, M., Ku, T. L., Harmon, R. S., and Smart, P. (1981). (In preparation.)
[5.5.1]

Izett, G. A. and Naeser, C. W. (1976). *Geology* **4**, 597. [7.3]

—— Wilcox, R. E., Powers, H. A., and Desborough, G. A. (1970). *Quatern. Res.* **1**, 121. [7.3]

Jackson, N. (1960). *J. Sci. Inst.* **37**, 139. [4.5.1]

James, N. P., Mountjoy, E. W., and Omura, A. (1971). *Geol. Soc. Am. Bull.* **82**, 2011. [3.7, 18.5]

Jelinek, A., Farrand, W., Haas, G., Horowitz, A., and Goldberg, P. (1973). *Paleorient*, **1**, 151. [13.2.1]

Jenkins, I. L. and McKay, H. A. C. (1954). *Trans. Faraday Soc.* **50**, 107. [4.2.2]

Jenkins, W. J., Edmond, J. M., and Corliss, J. B. (1978). *Nature* **272**, 156. [15.2.5]

Johnson, J. O. (1971). *Determination of radium-228 in natural waters.* US Geol. Survey Water Supply Paper 1696-G, 26 pp. [5.5.3]

Johnson, J. (1979). *Investigation of Mg and Sr distributions in speleothem.* Unpub. B.Sc. thesis, Dept. of Geology, McMaster Univ. Hamilton, Ont. Canada. [13.3.3]

Joly, J. (1908*a*). *Phil. Mag.* **15**, 385. [15.4.1]

—— (1908*b*). *Phil. Mag.* **16**, 190. [15.4.1, 16.1, 16.2.3, 16.3.2]

—— (1909). *Phil. Mag.* **18**, 396. [14.1, 15.4.1]

Joshi, L. U. and Ganguly, A. K. (1970). *Proc. Chem. Symp. IIT. Madras, Nov. 1970.* 137. Indian Department of Atomic Energy. [4.4]

—— —— (1976). *Geochim. Cosmochim. Acta* **40**, 1491. [4.4]

—— —— (1977). *J. radioanalyt. Chem.* **41**, 15. [4.4]

—— Ku, T. L. (1979). *J. radioanalyt. Chem.* **52**, 329. [16.2.5, 16.3.3]

Kadko, D. (1980*a*). *Earth planet. Sci. Lett.* **49**, 360. [15.3.2, 15.5.2]

—— (1980*b*). *Earth planet. Sci. Lett.* **51**, 115. [15.3.2]

Kanwisher, J. (1963). *Deep Sea Res.* **10**, 195. [15.5.1]

Kapitza, P. (1924). *Proc. R. Soc. A* **106**, 602. [5.1]

Katz, J. J. and Seaborg, G. T. (1957). *The chemistry of the actinide elements.* Methuen, London. [4.2.3]

Katzin, L. I. and Stoughton, R. W. (1956). *J. inorg. nucl. Chem.* **3**, 229. [4.2.3]

Kaufman, A. (1964). $^{230}Th-^{234}U$ *dating of carbonates from Lakes Lahontan and Bonneville*, Lamont Geol. Labs., Ph.D. Thesis, Columbia University. [3.3, 4.5.1, 11.5.2, 11.5.3]

—— (1969). *Geochim. Cosmochim. Acta* **33**, 717. [2.2.4, 15.3.1, 16.3.1]

—— (1971). *Geochim. Cosmochim. Acta* **35**, 1269. [3.6, 10.2.1, 10.2.2, 11.5.1, 11.5.3, 12.2.1, 13.3.1]

—— (1972). *Earth planet Sci. Lett.* **14**, 447. [12.3.2, 18.3]

—— Broecker, W. S. (1965). *J. geophys. Res.* **70**, 4039. [2.5.3, 3.6, 17.2, 17.2.2, 17.3.2]

—— Li, Y. H., and Turekian, K. K. (1981). *Earth planet. Sci. Lett.* **54**, 385. [15.3.3]

—— Broecker, W. S., Ku, T-L., and Thurber, D. L. (1971). *Geochim. Cosmochim. Acta* **35**, 1155. [2.1.2, 2.4.1, 3.7, 12.2.3, 12.3.2, 13.2.2, 13.3.1, 13.3.5, 17.2, 17.2.2, 18.2, 18.3]

—— Trier, R. M., Broecker, W. S., and Feely, H. W. (1973). *J. geophys. Res.* **78**, 8827. [15.4.1, 15.4.2]

Kazachevskii, I. V., Cherdyntsev, V. V., Kuzmina, E. A., Sulevzhitskii, L. D., Mochalova, V. G., and Kyuregyan, T. N. (1964). *Geokhimiya* **11**, 1116. [8.1, 8.2.1]

Keepin, G. R., Wimitt, T. F., and Zeigler, R. K. (1957). *Phys. Rev.* **107**, 1044; *J. nucl. Energy* **6**, 1. [5.3.4]

Keith, M. L. and Anderson, G. M. (1963). *Science* **141**, 634. [13.2.2]

Kelley, L., Spiker, E., and Rubin, M. (1978). *Radiocarbon* **20**, 283. [12.2.2]

Kendall, A. C. and Broughton, P. L. (1978). *J. sedim. Petrol.* **48**, 519. [11.3.1]

Kennedy, J., Davis, R. V., and Robinson, B. K. (1956). *A.E.R.E. Report C/R. 1896.* [4.2.1]

Key, R. M., Guinasso JR. N. L., and Schink, D. R. (1979). *Mar. Chem.* **7**, 221. [5.5.3, 15.5.2]

—— Brewer, R. L., Stockwell, J. H., Guinasso, N. L., and Schink, D. R. (1979). *Mar. Chem.* **7**, 251. [4.4, 4.6, 5.5.3]

Kharkar, D. P., Turekian, K. K., and Scott, M. R. (1969). *Earth planet. Sci. Lett.* **6**, 61. [3.1]

—— Thomson, J., Turekian, K. K., and Forster, W. O. (1976). *Limnol. Oceanogr.* **21**, 294. [15.7]

Khlapin, V. G. (1926). *Dokl. Akad. Nauk SSSR*, 178. [3.3]

Khopar, S. M. and De, A. K. (1960). *Analyst* **85**, 376. [4.2.1]

Kigoshi, K. (1967). *Science* **156**, 1932. [3.4.4, 6.4.3]

—— (1971). *Science* **173**, 47. [2.5.1, 7.3.1, 9.1.2, 10.3.1, 15.2.1, 15.4.1, 16.2.2, 17.3.2]

—— (1973). *Unpublished MS.*, IAEA Panel on Application of Uranium Isotope Disequilibrium in Hydrology, Vienna. [9.3.4]

Killeen, P. G. (1979). In: *Geophysics and geochemistry in the search for metallic ores*, ed. P. J. Hood. Geol. Surv. Canada Econ. Geol. Report 31, p. 163. [14.4.6]

—— Heier, K. S. (1975). *Geochim. Cosmochim. Acta.* **39**, 1515. [6.3]

Kimura, K. (1960). *J. atom. Energy Soc. Japan.* **2**, 585. [4.2.3]

King, C-Y. (1978). *EOS, Trans Am. geophys. Un.* **58**, 434. [9.2.3]

Kirby, H. W. (1959). *The radiochemistry of protactinium NAS-NS-3016.* [4.2.3]

—— (1961). *J. inorg. nucl. Chem.* **18**, 8. [1.5, 11.3.4]

—— Brodbeck, R. M. (1954). *USAEC Rep. MLM 1003.* [4.2.4]

—— Salutsky, M. L. (1964). *The radiochemistry of radium NAS-NS-3057.* [4.2.4, 5.5.3]

Klinkhamer, G., Bender, M. L., and Weiss, R. F. (1977). *Nature* **269**, 319. [15.3.2, 16.2.4]

Kluge, E. and Lieser, H. (1980). *Radiochim. Acta* **27**, 161. [4.4]

Knauss, K. G. and Ku, T.-L. (1980). *J. Geol.* **88**, 95. [10.3.2]

—— —— Moore, W. S. (1978). *Earth planet. Sci. Lett.* **39**, 235. [4.4, 4.6, 5.5.3, 15.3.3, 15.4.2]

Knight, G. B. and Machlin, R. L. (1948). *Phys. Rev.* **74**, 1540. [1.5]

Ko, R. and Weiller, M. R. (1962). *Anal. Chem.* **34**, 85. [4.2.2]

Kochenov, A. V., Dubinchuk, V. T., and Germozenova, E. V. (1973). *Sov. Geol.* **3**, 69. [17.3.1]

Koczy, F. F. (1958). In: *Proc. Int. Conf. Peaceful Uses At. En., 2nd Geneva* **18**, 351. [15.4.1]

—— (1963). In: *The Sea, Vol. 3.*, ed. M. N. Hill, p. 816 Interscience, New York. [15.2.5]

—— Szabo, B. J. (1962). *J. ocean. Soc. Jap.*, (20th anniv. vol.) 590. [15.4.1]

—— Tomic, E., and Hecht, F. (1957). *Geochim. Cosmochim. Acta* **11**, 86. [8.2.1, 15.2.5, 15.4.2]

—— Picciotto, E., Poulaert, G., and Wilgain, S. (1957). *Geochim. Cosmochim. Acta* **11**, 103. [3.4.7]

Kohman, T. P. (1947). *Am. J. Phys.* **15**, 356. [1.1]

Kohout, F. A. (1965). *Trans. N. Y. Acad. Sci. Ser. II.* **28**, 249. [9.2.2]

Koide, M. and Bruland, K. W. (1975). *Anal. chim. Acta* **75**, 1. [4.4, 5.5.3]

—— Goldberg, E. D. (1964). In: *Progress in oceanography, Vol. 3*, ed. M. Sears, p. 172. Pergamon Press, Oxford. [15.2.3]

—— —— (1965). In: *Progress in oceanography, Vol. 3*, ed. M. Sears, p. 173. Pergamon Press, Oxford. [17.2, 17.2.1]

—— Bruland, K. W., and Goldberg, E. D. (1973). *Geochim. Cosmochim. Acta* **37**, 1171. [2.5.3, 3.4.7, 16.3.3]

—— —— —— (1976). *Earth planet. Sci. Lett.* **31**, 31. [15.4.1, 16.2.5, 16.3.3]
—— Griffin, J. J., and Goldberg, E. D. (1975). *J. geophys. Res.* **80**, 4153. [16.3.3]
Kolodny, Y. (1969). *Nature* **224**, 1017. [17.3.2]
—— (1980). In: *The sea* **7**, (in press). [17.3.2]
—— Kaplan, I. R. (1970). *Geochim. Cosmochim. Acta* **34**, 3. [12.3.3, 17.3.2, 17.3.4]
—— —— (1973). In: *Proc. Symp. Hydrogeochem. Biogeochem., 1: Hydrogeochem.*
   418. Clarke Co., Washington, D. C. [15.2.5]
Kominz, M. A., Heath, G. R., Ku, T.-L., and Pisias, N. G. (1979). *Earth planet. Sci.
   Lett.* **45**, 394. [3.4.3, 12.1, 13.2.1, 18.4, 18.5, 19.2]
Komura, K. and Sakanoue, M. (1967). *Sci. Rep. Kanazawa Univ.* **12**, 21. [3.4.1, 3.5,
   11.3.3, 12.3.2]
—— —— Konishi, K. (1978). *Proc. Japan Acad.* **54**, 505. [5.5]
Konishi, K., Omura, A., and Nakamichi, O. (1974). In: *Proc. 2nd Int. Coral Reef Symp.*
   2, 595. Great Barrier Reef Committee, Brisbane. [3.7, 18.5]
—— Schlanger, S. O., and Omura, A. (1970). *Mar. Geol.* **9**, 225. [17.2.3, 18.5]
Korkisch, J. (1969). *Modern methods for the separation of rarer metal ions.* Pergamon
   Press, Oxford. [4.2.1, 4.2.7]
—— Antal, P. (1960). *Zhur. analyt. Chem.* **173**, 126. [4.2.2]
—— Steffan, I., Arrhenius, G., Fisk, M., and Fraser, J. (1977). *Anal. Chim. Acta.* **90**,
   151. [4.3.1]
Kraemer, T. F. (1981). *Earth planet. Sci. Lett.* **56**, 210. [9.1.3, 9.1.4, 9.2.3]
Kraus, K. A. and Moore, G. E. (1955). *J. Am. chem. Soc.* **77**, 1383. [4.2.3]
—— Nelson, F. (1956). *Proc. Int. Conf. Peaceful Uses At. En., Geneva 1955.* **7**, 113.
   United Nations, New York. [4.2.1]
Kress, A. G. and Veeh, H. H. (1980). *Mar. Geol.* **36**, 143. [17.3.1, 17.3.2, 17.3.4]
Kröll, V. (1953). *Nature* **171**, 742. [2.4.2, 15.4.1, 16.2.3]
—— (1954). *Deep Sea Res.* **1**, 211. [15.4.1, 16.2.3]
—— (1955). *Repts. Swedish Deep Sea Expol. 10*, ed. H. Pettersson, p. 3. [15.4.1]
Kronfeld, J. (1972). *Hydrologic investigations and the significance of supported*
   $^{234}U/^{238}U$ *disequilibrium in the groundwaters of Central Texas*, Ph.D. Thesis, Rice
   University, Houston, Texas. [4.3.2, 4.4]
—— (1974). *Earth planet. Sci. Lett.* **21**, 327. [2.5.1]
—— Adams, J. A. S. (1974). *J. Hydrol.* **22**, 77. [9.1.5, 9.3.4]
—— Gradsztajn, E., Muller, H. W., Radin, J., Yaniv, A., and Zach, R. (1975). *Earth
   planet Sci. Lett.* **27**, 342. [9.1.4, 9.2.3]
Krishnaswami, S. (1976). *Geochim. Cosmochim. Acta* **40**, 425. [15.2.5, 15.3.2]
—— Cochran, J. K. (1978). *Earth planet. Sci. Lett.* **40**, 45. [15.5.2]
—— Lal, D., and Somayajulu, B. L. K. (1976). *Earth planet. Sci. Lett.* **32**, 403. [15.3.1]
—— Somayajulu, B. L. K., and Chung, Y. (1975). *Earth planet. Sci. Lett.* **27**, 388.
   [15.6.2]
—— Benninger, L. K., Aller, R. C., and Von Damm, K. L. (1980). *Earth planet. Sci.
   Lett.* **47**, 307. [5.5.4, 16.3.3]
—— Cochran, J. R., Turekian, K. K., and Sarin, N. M. (1979). In: *La genese des
   modules de manganese*, ed. C. Lalou, p. 251. CNRS, Paris. [16.3.2]
—— Lal, D., Martin, J. M., and Meybeck, M. (1971). *Earth planet. Sci. Lett.* **11**, 407.
   [3.4.8]
—— —— Somayajulu, B. L. K., Weiss, R. F., and Craig, H. (1976). *Earth planet. Sci.
   Lett.* **32**, 420. [15.3.1, 19.3]
—— —— —— Dixon, F. S., Stonecipher, S. A. and Craig, H. (1972). *Earth planet. Sci.
   Lett.* **16**, 84. [4.3.2, 4.4, 5.5.3, 5.5.4, 15.3.1]
Krolopp, E. (1964). In: *Tata, eine mittelpäläolithische Travertin—Siedlung in Ungarn*,
   ed. L. Vertes. *Archaeol. Hungar., S. N.* **43**, 87. [13.3.5]

Krough, T., (1973). *Geochim. Cosmochim. Acta* **37**, 485. [4.3.1, 4.4]
Krylov, A. Y. and Atrashenok. L. Y. (1959). *Geochem.* (English Trans.) **3**, 307. [7.2]
Ku, T-L. (1965). *J. geophys. Res.* **70**, 3457. [2.5.2, 3.4.2, 4.3.1, 4.4, 4.5.1, 4.6, 8.2.1, 15.2.5, 15.2.6, 16.2.1, 17.3.2]
—— (1966). *Uranium series disequilibrium in deep-sea sediments*, Ph.D. Thesis, Columbia University, New York. [5.5.2, 15.2.6, 15.3.1, 15.3.2, 16.2, 16.2.1, 16.2.2, 16.3.1, 18.4]
—— (1968). *J. geophys. Res.* **73**, 2271. [3.7, 4.4, 5.5.2, 11.3.4, 12.2.2, 17.2.1, 18.5]
—— (1969). In: *Hot brines and recent heavy metal deposits in the Red Sea*, ed. E. T. Degans and D. A. Ross, p. 512. Springer Verlag, New York. [16.2.1, 16.4]
—— (1976). *Annu. Rev. earth planet. Sci.* **4**, 347. [2.3.2, 2.4.2, 2.5.3, 3.1, 3.4.1, 3.4.2, 3.4.3, 3.4.4, 3.4.5, 3.4.7, 3.6, 16.1, 16.3.1, 17.2]
—— (1977). In: *Marine manganese deposits*, ed. G. P. Glasby, Elsevier Oceanography Series 15, **8**, 249. Elsevier Scentific Publishing, New York. [16.3.2, 16.4]
—— Broecker, W. S. (1966). *Science* **151**, 448. [13.2.1]
—— —— (1967*a*). *Prog. Oceanogr.* **4**, 95. [3.4.5]
—— —— (1967*b*). *Earth planet. Sci. Lett.* **2**, 317. [16.3.2]
—— —— (1969). *Deep Sea Res.* **16**, 625. [15.2.5, 16.2, 16.2.1, 16.3.2]
—— Glasby, G. P. (1972). *Geochim. Cosmochim. Acta* **36**, 699. [16.2, 16.2.1, 16.2.2, 16.3.2]
Ku, T.-L. and Joshi, L. U. (1980). In: *Management of environment* (ed. B. Patel) p. 551. Wiley Eastern, Bombay.[19.2]
—— Kern, J. P. (1974). *Geol. Soc. Am. Bull.* **85**, 1713. [18.5]
—— Knauss, K. G. (1979). In: *La genese des nodules de manganese*. Ed. C. Lalou, CNRS Paris. [3.4.3, 16.3.2]
—— Lin, M. C. (1976). *Earth planet. Sci. Lett.* **32**, 236. [15.4.1]
—— Bischoff, J. L., and Boersma, A. (1972). *Deep Sea Res.* **19**, 233. [15.3.2]
—— Broecker, W. S., and Opdyke, N. (1968). *Earth planet. Sci. Lett.* **4**, 1. [16.2, 16.2.2]
—— Huh, C. A., and Chan, P. S. (1980). *Earth planet. Sci. Lett.* **49**, 293. [15.4.1]
—— Knauss, K. G., and Mathieu, G. G. (1974). *EOS Trans. Am. geophys. Un.* **55**, 314. [2.2.4, 3.4.1, 9.2.2]
—— —— —— (1977). *Deep Sea Res.* **24**, 1005. [8.2.1, 8.2.2, 10.3.1, 15.2.3, 15.2.4, 15.2.6, 16.2, 16.4, 17.2.1]
—— Omura, A., and Chen, P. S. (1979). In: *Marine geology and oceanography of the Pacific Manganese Nodule Province*, ed. J. L. Bischoff and D. Z. Piper, p. 791. Marine Science Series No. 9. Plenum Press, New York. [3.4.3, 16.3.2]
—— Bull, W. E., Freeman, S. T., and Knauss, K. G. (1979). *Bull. Geol. Soc. Am.* **90**, 1063. [3.6, 10.2.1, 10.2.2, 12.2.1, 12.3.1, 13.2.2, 13.3.4, 19.2]
—— Kimmel, M. A., Easton, W. H., and O'Neil, T. J. (1974). *Science* **183**, 959. [3.7, 17.2.1, 18.5]
—— Li, Y. H., Mathieu, G. G., and Wong, H. K. (1970). *J. geophys. Res.* **75**, 5286. [15.4.1]
Kukla, G. L. (1977). *Earth Sci. Rev.* **13**, 307. [13.1, 13.2.2, 13.4]
—— Lozek, L. (1958). *Ceskoslovensky Kras* **XI**, 59. [13.2.2]
Kulmatov, R. A. and Kist, A. A. (1978). *Zav. Lab.* **44** (12), 1482. [5.3.4]
Kunzendorf, H. and Friedrich, G. H. W. (1976). *Geochim. Cosmochim. Acta* **40**, 849. [2.1.2, 2.2.3]
Kuptsov, V. M. and Cherdyntsev, V. V. (1969). *Geochem. Int.* **6**, 532. [16.2.1]
Kuroda, R. and Seki, T. (1980). *Z. analyt. Chem.* **300**, 107. [4.4]
Kutschera, W., Henning, W., Paul, M., Smither, R. K., Stephenson, E. J., Yutema,

I. L., Alberger, D. E., Cumming, J. B., and Harbottle, G. (1980). *Phys. Rev. Lett.* **45**, 592. [3.1]

Kuznetsov, Y. V., Simonyak, Z. N., Elizarova, A. N., and Lisitsyan, A. P. (1966). *Radiokhimiya* **8**, 455. [15.3.1]

Kvenvolden, K. A. and Blunt, D. J. (1979). In: *Marine geology and oceanography of the Pacific Manganese Nodule Province*, ed. J. L. Bischoff and D. Z. Piper, p. 763. Marine Science Series No. 9, Plenum Press, New York. [16.3.2]

Lachenbruch, A. H. (1968). *J. geophys. Res.* **73**, 6977. [6.1]

Eahoud, J. A., Miller, D. S., and Friedman, G. M. (1966). *J. sedim. Petrol.* **36**, 541. [12.3.2]

Lal, D. and Schink, D. R. (1960). *Rev. sci. Instr.* **31**, 395. [5.5.4]

Lally, A. E. and Eakins, J. D. (1978). *Symp. determ. radionuclides environ. bio. mats.* CEGB London (1978). [4.5.1]

Lalou, C. (1979). (ed.) *La genese des nodules de manganese*. CNRS Paris. [16.3.2]

——Brichet, E. (1972). *C. r. hebd. Séanc. Acad. Sci., Paris* **275**, 815. [16.3.2]

—— —— (1980a). *Nature* **284**, 251. [16.2, 16.2.1]

—— —— (1980b). *C. r. hebd. Séanc. Acad. Sci., Paris*, **290**, 819. [16.2, 16.2.1, 16.2.2]

—— —— Jehanno, C. (1979). In: *La genèse des nodules de manganese*, ed. C. Lalou, p. 271. CNRS, Paris. [16.3.2]

—— Labeyrie, J., and Delibrias, G. (1966). *C. r. hebd. Séanc. Acad. Sci., Paris, ser. D*, **263**, 1946. [17.2.1]

—— Brichet, E., Ku, T. L., and Jehanno, C. (1977). *Mar. Geol.* **24**, 245. [16.2.1, 16.2.2]

—— —— Poupeau, G., Romari, P., and Jehanno, C. (1979). In: *Geology and oceanography of the Pacific Manganese Nodule Province*, ed. J. L. Bischoff and D. Z. Piper, p. 815. Marine Science Series No. 9, Plenum Press, New York. [16.3.2]

—— Ku, T. L., Brichet, E., Poupeau, G., and Romari, P. (1979). In: *La genèse des nodules de manganese* ed. C. Lalou, p. 261. CNRS, Paris. [16.2.2, 16.3.2]

Lambert, G. and Nezami, M. (1965). *Nature* **206**, 1343. [15.7]

—— Sanak, J., and Ardouin, B. J. (1974). *J. Rech. Atmos.* **8**, 647. [15.7]

Lambert, R. S. J. (1971). *The Phanerozoic time scale Part 1 (Suppl.).* Geological Society, London. [3.1]

Lancelot, J. R., Vitrac, A., and Allegre, C. J. (1975). *Earth planet Sci. Lett.* **25**, 189. [1.5, 2.5.1, 14.1]

Land, L. S. and Epstein, S. (1970). *Sedimentol.* **14**, 187. [18.5]

—— Mackenzie, G. T., and Gould, S. J. (1967). *Geol. Soc. Am. Bull.* **78**, 993. [18.5]

Langmuir, D. (1971). *Geochim. Cosmochim. Acta* **35**, 1023. [11.1]

—— (1978). *Geochim. Cosmochim. Acta* **42**, 547. (See also **43**, 1991.) [2.2.3, 2.2.4, 2.3.1, 2.4.1, 8.2.2, 8.2.3, 9.1.2, 14.4.1]

Larsen, E. S. III (1957). *U.S. Geol. Survey TEI-700*, 249. [7.2]

—— Gottfried, D. (1960). *Am. J. Sci.* **258**, 151. [6.3, 7.2]

Latham, A. G., Schwarcz, H. P., Ford, D. C., and Pearce, G. W. (1979). *Nature* **280**, 383. [13.2.2]

Lauritzen, S. E. and Gascoyne, M. (1980). *Norsk geogr. Tidsskr.* **34**, 77. [11.3.4]

Lederer, C. M. and Shirley, V. (1978). *Table of isotopes* (7th edition), Wiley-Interscience, New York. [1.5]

Lee, C., Bada, J. L., and Peterson, E. (1976). *Nature* **259**, 183. [13.2.2]

—— Kim, N. B., Lee, I. C., and Chung, K. S. (1977). *Talanka* **24**, 241. [4.4]

Leonova, L. L. and Tauson, L. V. (1958). *Geochem.* (English trans.) **7**, 815. [7.2]

Lester, O. C. (1918). *Am. J. Sci.* **196**, 621. [14.1]

Levinson, A. A. (1980). *Introduction to exploration geochemistry* (2nd edn.), Applied Publishing, Wilmette, Illinois. [14.2, 14.4.4, 14.4.5]

—— Bland, C. J. (1978). *Can J. earth Sci.* **15**, 1867. [14.4.5]

—— Coetzee, G. L. (1978). *Miner. Sci. Eng.* **10**, 19. [14.4.5]

—— Bland, C. J., and Parslow, G. R. (1978). *Can. Inst. Min. Met. (CIM) Bull.* **71**, No. 796, 59. [14.4.6]

Levorsen, A. I. (1967). *Geology of petroleum*, W. H. Freeman, San Fransisco. [3.3.3]

Lewis, D. M. (1976). *The geochemistry of manganese, iron, uranium, Pb-210 and major ions in the Susquehanna River*, Ph.D. Thesis, Yale University. [8.1, 8.2.1, 8.2.3, 8.5]

—— (1977). *Geochim. Cosmochim. Acta* **41**, 1557. [8.5, 15.6.1]

Li, Y. H. and Chan, L. H. (1979). *Earth planet. Sci. Lett.* **43**, 343. [8.4]

—— Feely, H. W., and Santschi, P. H. (1979). *Earth planet. Sci. Lett.* **42**, 13. [5.5.3, 15.3.3]

—— Ku, T-L., Mathieu, G. G., and Wolgemuth, K. (1973). *Earth planet. Sci. Lett.* **19**, 352. [2.5.3, 15.4.1]

—— Mathieu, G. G., Biscaye, P., and Simpson, H. J. (1977). *Earth planet. Sci. Lett.* **37**, 237. [2.5.3, 8.1, 8.4, 15.4.1]

Libby, W. F. (1952). *Radiocarbon dating*. The University of Chicago Press, Chicago. [16.1]

Liddicoat, J. C., Opdyke, N. D., and Smith, G. I. (1980). *Nature* **286**, 22. [11.5.3, 13.3.1]

Lietzke, T. A. and Lerman, A. (1975). *Earth planet. Sci. Lett.* **24**, 337. [15.5.2]

Lively, R. S., Harmon, R. S., Levinson, A. A., and Bland, C. J. (1979). *J. geochem. Explor.* **12**, 57. [14.4.3]

Livingstone, D. A. (1963). *Prof. Pap. U.S. Geol. Surv. 440-G.*, 64. [15.2.1]

Lopatkina, A. P. (1964). *Geochem. Intl.* **4-6**, 788. [2.2.4]

Lowdon, J. A. and Blake, W., Jr. (1968). *Radiocarbon* **10**, 207. [12.2.2]

Ludwig, K. R. (1978). *Econ. Geol.* **73**, 29. [4.4]

—— (1980). *Earth planet. Sci. Lett.* **46**, 212. [3.1]

—— Lindsey, D. A., Zielinski, R. A., and Simmons, K. R. (1980). *Earth planet. Sci. Lett.* **46**, 221. [3.1]

Lyons, J. B. (1964). *U.S. Geol. Survey Bull.* **1144-F**, 43. [6.2]

MacDougall, J. D. (1973). *Trans. Am. geophys. Un.* **54**, 988. [7.1]

—— (1977). *Earth planet. Sci. Lett.* **35**, 65. [7.1, 15.2.5]

—— (1979). In: *Marine geology and oceanography of the Pacific Manganese Nodule Province*, ed. J. L. Bischoff and D. Z. Piper, p. 775. Marine Science Series No. 9, Plenum Press, New York. [16.3.2]

—— Finkel, B., and Chung, Y. (1979). *EOS Trans. Am. geophys. Un.* **60**, 864. [16.2.1, 16.2.5]

—— Maier, R., and Sutherland-Hawkes, P. (1980). *Nature* **284**, 230. [12.1]

—— Finkel, R. C., Carlson, I., and Krishnaswami, S. (1979). *Earth planet. Sci. Lett.* **42**, 27. [6.5.3]

McKelvey, V. E. (1967). *U. S. Geol. Survey Bull. 1252-D*, 1. [17.3.3, 17.3.4]

—— Everhart, D. L., and Garrels, R. M. (1956). *Proc. Intl. Conf. Peaceful Uses Atom. Energy, Geol. Uranium and Thorium* **6**, U.N., Geneva. [2.4.1]

Mackenzie, A. B., Baxter, M. S., McKinley, I. G., Swan, D. S., and Jack, W. (1979). *J. radioanal. Chem.* **48**, 29. [4.3.2, 4.4, 4.6]

McKinney, C. R. (1980). *Geol. Soc. Am. Ann. Mtg*, (abs.), 481. [12.4.3]

McMillan, R. H. (1978). In: *Short course in uranium deposits: their mineralogy and origin*, ed. M. M. Kimberley, p. 187. Mineralogical Association of Canada. [14.2, 14.3]

Macsotay, O. and Moore, W. S. (1974). *Publicaciones de la Comision Organizadora de*

la 111 Conferencia de las Niciones Unidas Sobre Dereche Del Mar, Caracas, (p. 63).
[18.5]

Madansky, L. and Rosetti, F. (1956). *Phys. Rev.* **102**, 464. [1.5]

Mallory, E. C., Johnson, J. O., and Scott, R. C. (1969). *USGS Water Supply Paper 1535–0*, 31 pp. [8.1, 8.2.1]

Mangini, A. and Dominik, J. (1979). *Sedim. Geol.* **23**, 113. [16.2, 16.2.1]

—— Sonntag, G. (1977). *Earth planet. Sci. Lett.* **37**, 251. [4.4, 5.5.2, 15.3.2]

—— —— Bertsch, G., and Müller, E. (1979). *Nature* **278**, 337. [8.1, 8.2.1, 15.2.1]

Mankinen, E. A. and Dalrynple, G. B. (1979). *J. geophys. Res.* **84**, 615. [13.2.1]

Manskaya, S. M. and Drozdova, T. V. (1968). *Geochemistry of organic substances.* Pergamon Press, Oxford. [2.3.1]

Manton, W. I. (1973). *Earth planet. Sci. Lett.* **19**, 83. [4.4]

Marine, I. W. (1976). *DP-1356*. E. I. DuPont de Nemours and Co., Savannah River Laboratory, NTIS. [9.3.2]

—— (1979). *Water Resour. Res.* **15**, 1130. [3.1, 3.3.3, 9.3.2]

Marks, A. E. (1977). *Prehistory and Paleoenvironments in the Central Negev, Israel. Vol. 1, Part. I.* Southern Methodist University Press, Dallas. [12.2.2, 12.4.1]

Marmier, P. and Sheldon, E. (1969). *Physics of nuclei and particles. Vol. I.* Academic Press, New York. [1.3.3, 5.1]

Marshall, R. R. and Hess, D. C. (1960). *Anal. Chem.* **32**, 960. [3.1]

Marshall, J. F. and Launay, J. (1978). *Quatern. Res.* **9**, 186. [17.2.1, 17.2.3, 18.5]

Marteus, C. S., Kipphut, G. W., and Klump, V. J. (1980). *Science* **208**, 285. [15.5.2]

Martin, J. M. and Meybeck, M. (1979). *Mar. Chem.* **7**, 173. [8.2.1, 8.3]

—— —— Pusset, M. (1978). *Neth. J. sea Res.* **12**, 338. [8.1, 8.2.1, 8.2.2, 8.2.3]

—— Nijampurkar, V. N., and Salvadori, F. (1978). In: *Biogeochemistry of estuarine sediments*, ed. E. D. Goldberg, p. 111. UNESCO publication. [8.1, 8.2.2, 8.2.3, 8.3, 15.2.2]

Martinez, P. and Senftle, F. E. (1960). *Rev. sci. Instr.* **31**, 974. [5.3]

Mathieu, G. G. (1977). $^{222}Rn-^{226}Ra$ *technique of analysis.* Ann. Tech. Report COO-2185-0 to ERDA, Lamont-Doherty Geological Observatory, Palisades, New York. [5.5.3]

Matsumoto, E. (1975). *Geochim. Cosmochim. Acta* **39**, 205. [15.3.3]

May, S. (1973). *Ind. At. Spat.* **2**, 31. [5.3.4]

Mazor, E. (1962). *Geochim. Cosmochim. Acta* **26**, 765. [2.3.3]

—— (1978). *Pure appl. Geophys.* **17**, 262. [9.2.3]

Megumi, K. (1979). *J. geophys. Res.* **84**, 3677. [4.4, 5.5, 7.3.2, 10.3.2, 12.2.2]

Meinke, W. W. (1964). *USAEC Report AECD. 2739.* [4.2.7]

Mcnard, H. W. and Smith, S. M. (1966). *J. geophys. Res.* **71**, 4305. [15.2.5]

Mesolella, K. J., Mathews, R. K., Broecker, W. S., and Thurber, D. L. (1969). *J. Geol.* **77**, 250. [11.3.4, 17.2.1, 18.4, 18.8]

Meyer, S. and Urlich, C. (1923). *Sber. Akad. Wiss. Wien. IIa* **132**, 279. [1.5]

Michel, J., and Moore, W. S. (1980). *Sources and behaviour of natural radioactivity in Line Aquifers near Leesville, South Carolina.* Clemson University Water Resources Res. Inst. Tech. Report *No. 83*, 73 pp. [5.5.3]

—— —— (1980). *Health Phys.* **38**, 663. [8.4, 9.2.1, 9.2.2, 9.3.3]

Millard, H. T. (1963). *Anal. Chem.* **35**, 1017. [4.2.7]

Millard, Jr. H. D. (1976). *US Geol. Survey Paper 840*, 61. [5.3.4]

Miller, J. A. (1972). *Dating Pliocene and Pleistocene using the K/Ar and* $^{40}Ar/^{39}Ar$ *methods.* In: *Proc. Symp. Calib. Hominoid Evol.*, 1971, ed. W. W. Bishop and J. A. Miller, p. 63. Scottish Academic Press, Edingburgh. [3.1]

Miller, H. W. and Brouns, R. J. (1952). *Anal. Chem.* **24**, 536. [4.5.1]

Minagawa, M. and Tsunogai, S. (1980). *Earth planet. Sci. Lett.* **47**, 51. [16.3.3]

Mitchell, R. F. (1960). *Anal. Chem.* **32**, 326. [4.5.1]

Mitterer, R. M. (1975). *Earth planet. Sci. Lett.* **28**, 275. [13.2.2]

Miyake, V. and Sugimura, Y. (1961). *Science* **133**, 1823. [4.2.2]

—— —— Tsubota, H. (1964). *The national radiation environment*, ed. J. Adams and W. M. Lowder, p. 39. University of Chicago Press, Chicago. [8.1, 8.2.1, 8.4]

—— —— Uchida, T. (1966). *J. geophys. Res.* **71**, 3083. [15.2.3]

—— —— Yasujima, T. (1970). *J. oceanogr. Soc. Japan* **26**, 130. [2.2.4, 15.3.1]

—— Saruhashi, K., Katsuragi, Y., Kanazawa, T., and Sugimura, Y. (1964). In: *Recent researches in the fields of hydrosphere, atmosphere and nuclear geochemistry*, ed. Y. Miyake and T. Koyama, p. 127. Maruzen Co. Ltd., Tokyo. [15.2.2]

Mo, T., Suttle, A. D., and Sackett, W. M. (1973). *Geochim. Cosmochim. Acta* **37**, 35. [15.2.5, 16.2, 16.2.1]

Molina-Cruz, A. (1977). *Quatern. Res.* **8**, 324. [17.3.4]

Moore, G. W. (1952). *Nat. Speleo. Soc. News* **10**, 2. [11.3]

Moore, H. E., Poet, S. E., and Martell, E. A. (1974). *J. geophys. Res.* **79**, 5019. [15.7]

Moore, W. S. (1967). *Earth planet. Sci. Lett.* **2**, 231. [8.1, 8.2.1, 8.2.3, 8.3, 8.4, 15.2.1, 15.4.1]

—— (1969). *Oceanic concentrations of radium-228 and a model for its supply.*, Ph.D. Thesis, SUNY-Stony Brook, 135 pp. [15.4.2]

—— (1969a). *J. geophys. Res.* **74**, 694. [4.4, 4.5.1, 4.6, 5.5.3, 8.4, 15.4.1, 15.4.2]

—— (1969b). *Earth planet. Sci. Lett* **6**, 437. [2.5.3, 3.4.7, 5.5.3, 15.4.1, 15.4.2]

—— (1972). *Earth planet. Sci. Lett.* **16**, 421. [15.4.2]

—— (1976). *Deep Sea Res.* **23**, 647. [4.3.2, 4.4, 5.5.3, 15.4.2]

—— Reid, D. (1973). *J. geophys. Res.* **78**, 8880. [4.4, 5.5.3, 19.3]

—— Sackett, W. M. (1964). *J. geophys. Res.* **69**, 5401. [15.3.1, 15.4.2, 16.3.1, 16.3.3]

—— Somayajulu, B. L. K. (1974). *J. geophys. Res.* **79**, 5065. [5.5.2, 18.5]

—— Krishnaswami, S. and Bhat, S. G. (1973). *Bull. Mar. Sci.* **23**, 157. [18.2]

—— Ching, J. H., Talwani, P., and Stevenson, D. A. (1977). *EOS, Trans. Am. geophys. Un.* **58**, 434. [9.2.3]

Moreira-Nordemann, L. C. (1980). *Geochim. Cosmochim. Acta* **44**, 103. [7.3.2, 8.1, 8.2.1, 10.3.2]

Morgan, J. W. (1971). *Radiochim. Acta.* **15**, 190. [4.5.1]

Morrison, G. H. (1973). *J. radioanal. Chem.* **18**, 9. [5.3.4]

Morrison, R. B. (1965). In: *Means and correlation of Quaternary succession*, ed. R. B. Morrison and H. E. Wright, Jnr., Proc. 7th Congr. Int. Assoc. Quatern. Res., University of Utah Press, Salt Lake City. [11.5]

Mott, N. F. and Massey, H. S. W. (1948). *Theory of atomic collisions*. Oxford University Press, London. [5.1]

Müller, P. J. and Mangini, A. (1980). *Earth planet. Sci. Lett.* **51**, 94. [15.3.2]

Murray, S. M. (1975). *Accumulation rates of sediments and metals off Southern California as determined by ^{210}Pb method*, Ph.D. Thesis, University Southern California, Los Angeles, pp. 146. [3.4.7]

Muto, T., Hirono, S., and Kurata, H. (1968). *Japan At. En. Res. Inst. Rep.*, NSJ Transl. No. 91, from *Mining Geol. (Japan) 1965* **15**, 289. [8.2.2]

Myers, W. A. and Lindner, M. (1971). *J. inorg. nucl. Chem.* **33**, 3233. [2.1.1]

Naeser, C. W., Izett, G. A., and Wilcox, R. E. (1973). *Geology* **1**, 187. [7.3]

Neef, G. and Veeh, H. H. (1977). *Nature* **269**, 682. [17.2.1, 17.2.3, 18.5]

Nelson, F. and Kraus, H. (1954). *J. Am. chem. Soc.* **76**, 5916. [4.2.6]

—— Kraus, K. (1955). *J. Am. chem. Soc.* **77**, 801. [4.2.4]

—— Murase, T., and Kraus, K. (1964). *J. Chromat.* **13**, 503. [4.2.3]

Nemodruck, A. A. and Verotinitskaya, I. E. (1962). *Zhur. Anal. Chim.* **17**, 481. [4.2.1]

Nettenberg, F. (1978). *Trans. geol. Soc. S. Afr.* **81**, 379. [10.2.1, 12.2.1]

Neuerburg, G. J. (1956). *U.S. Geol. Survey Prof. Paper 300*, **55**, [6.2]

—— Antweiler, J. C., and Bieler, B. H. (1956). *Resumenes de los Trabajos Presentados*, 20th Congreso Geologico Internacional U.S. Geological Survey. [7.2]

Neumann, A. C. and Moore, W. S. (1975). *Quatern. Res.* **5**, 215. [18.5]

Nicolaysen, L. O. (1961). *Ann. N. Y. Acad. Sci.* **91**, 198. [3.1]

Nikolaev, D. S. Lazarev, K. F., Korn, O. P., and Drozhzhin, V. M. (1966). *Radiokhimiya* **8**, 469. [15.2.5]

Ninkovitch, D. and Shackleton, N. J. (1975). *Earth planet Sci. Lett.* **27**, 20. [13.3.3]

Nishimura S. (1970). *Earth planet. Sci. Lett.* **8**, 293. [6.5.1]

Nishizumi, K., Arnold, J. R., Elmore, D., Ferraro, R. D., Gove, H. E., Finkel, R. C., Beukens, R. P., Chang, K. H., and Kilius, L. R. (1979). *Earth planet. Sci. Lett.* **45**, 285. [3.1]

Nkomo, I. T., Rosholt, J. N., and Dooley, J. R. Jr. (1979). *Wyo. Geol. Assn. Earth Sci. Bull.* **12**, 1. [7.3.1]

—— Stuckless, J. S., Thaden, R. E., and Rosholt, J. N. (1978). In: *Resources of the Wind River Basin*, ed. R. G. Boyd, p. 335. 30th Annual Field Conf. Guidebook, Wyoming Geological Association. [7.3.1, 14.3]

Noakes, J. E., Kim, S. M., and Supernaw, I. R. (1967). *Trans. Am. geophys. Un.* **48**, 237. [15.2.2]

Nozaki, Y. and Tsunogai, S. (1973a). *Anal. Chim. Acta* **64**, 209. [4.3.2, 4.4]

—— —— (1973b). *Earth planet. Sci. Lett.* **20**, 88. [15.6.2]

—— —— (1976). *Earth planet. Sci. Lett.* **32**, 313. [15.6.3]

—— Thomson, J., and Turekian, K. K. (1976). *Earth planet. Sci. Lett.* **32**, 304. [15.6.2, 15.7]

—— Turekian, K. K., and von Damm, K. (1980). *Earth planet. Sci. Lett.* **49**, 393. [15.6.3]

—— Cochran, J. K., Turekian, K. K., and Keller, G. (1977). *Earth planet. Sci. Lett.* **34**, 167. [16.2.5, 16.3.3, 16.4.2]

Oakley, K. P. (1969). In: *Science in Archaeology 2nd ed.* ed. D. Brothwell and E. Higgs. Thames and Hudson, London. [12.2.2]

—— (1980). *Bull. Br. Mus. nat. Hist. (Geol.)* **34**, 1. [12.2.2]

O'Brien, G. and Veeh, H. H. (1980). *Nature* **288**, 690. [17.3.2, 17.3.4]

Oldfield, F., Appleby, P. G., and Butterbee, R. W. (1978). *Nature* **271**, 339. [3.4.8]

O'Riordan, M. C. (1978). *Geological aspects of uranium in the environment*, ed. P. McL. Duff, p. 37. Geological Society, London. [14.5]

Osmond, J. K. and Cowart, J. B. (1976a). *Atomic Energy Review* **14**, 621. [2.5.1, 2.5.2, 3.4.1, 7.2, 7.3.1, 8.1, 8.2.1, 9.1.1, 9.1.2, 9.2.1, 10.3.1, 11.3.4, 15.2.1, 15.2.2]

—— —— (1976b). *Geol. Soc. Am. Abst. Prog.* **8**, 823. [9.2.3]

—— Pollard, L. D. (1967). *Earth planet. Sci. Lett.* **3**, 476. [5.5.1, 16.1, 16.3]

—— Carpenter, J. R., and Windom, H. L. (1965). *J. geophys. Res.* **70**, 1843. [17.2.1, 18.3, 18.4]

—— Kaufman, M. I., and Cowart, J. B. (1974). *Geochim. Cosmochim. Acta* **38**, 1083. [9.2.2]

—— May, J. P., and Tanner, W. F. (1970). *J. geophys. Res.* **75**, 469. [17.2.2]

Ostrihansky, L. (1976). *Radioactive disequilibrium investigations, Elliot Lake area, Ontario.* Geol. Surv. Canada Paper 75–38. Part 2, 21. [5.5, 14.1, 14.4.2]

Oversby, V. M. and Gast, P. W. (1968). *Earth planet. Sci. Lett.* **5**, 199. [3.4.4, 5.5.4, 6.5.1, 6.5.3, 14.1]

Owers, M. J. and Parker, A. (1964). *AERE Report R-4466* [4.2.4]

Pankhurst, R. J. (1970). *Scot. J. Geol.* **6**, 83. [3.1]

Parker, W., Blidstein, H., and Getoff, N. (1964). *Nucl. Instr. Meth.* **26**, 55. [4.5.1]

Patterson, C. C. (1951). *AEC Rep. AECD-3180.* [3.1]

Paul, A. Z. (1976). *Nature* **263**, 50. [16.3.2]

Pécsi, M. (1973). *Magyar tudom. Akad., Földr. Kutato Intéz., Földr. Köz 2*, 109. [11.4, 12.4.1]

Peirson, D. H., Cambray, R. S., and Spicer, G. S. (1966). *J. geophys. Res.* **77**, 6515. [3.4.8]

Penck, A. and Bruckner, E. (1909). *Die Alpen im Eiszeitalter*, Leipzig. [13.4]

Peng, T.-H., Goddard, J. G., and Broecker, W. S. (1978). *Quatern. Res.* **9**, 319. [10.2.2, 11.5.2, 11.5.3, 11.6, 13.2.2, 13.3.1]

—— Takahashi, T., and Broecker, W. S. (1974). *J. geophys. Res.* **79**, 1777. [15.5.1]

—— Broecker, W. S., Mathieu, G. G., Li, Y. H., and Bainbridge, A. E. (1979). *J. geophys. Res.* **84**, 2471. [15.5.1]

Pennington, W., Cambrey, R. S., Eakins, J., and Harkness, D. D. (1976). *Freshwater Biol.* **6**, 317. [3.4.8]

Perel'man, A. I. (1977). *Geochemistry of elements in the supergene Zone.* Keter Publishing, Jerusalem. [14.4.4]

Perlman, I. and Rasmunssen, J. O. (1957). In: *Handbuch der Physik. Vol 42*, Springer-Verlag, Berlin. [1.3.1]

—— Ghiorso, A., and Seaborg, G. T. (1950). *Phys. Rev.* **77**, 26. [1.3.1]

Pertlik, F. (1969). *Handbook of geochemistry*, ed. K. H. Wedepohl, Springer-Verlag, Berlin. [2.1.2]

Petrow, H. G. and Lindström, R. (1961). *Anal. Chem.* **33**, 313. [4.2.4]

—— Neitzel, O. A., and DeSosa, M. A. (1960). *Anal. Chem.* **32**, 926. [4.2.4, 4.2.6]

Pettersson, H. (1937). *Anz. Akad. Wiss. Wien K* **1**, 127. [15.4.1, 16.1, 16.2.3]

—— (1943). *Medd Oceanogr. Inst., Göteborg* **6**, 1. [16.3.2]

—— (1951). *Nature* **167**, 942. [15.4.1]

—— (1953). *Am. Scient.* **41**, 245. [15.4.1]

—— (1955). *Deep Sea Res. Supp.* **3**, 335. [15.4.1, 16.3.2]

Pfeifer, V. and Hecht, F. (1960). *Mikrochim. Acta* **3**, 378. [4.2.1]

Picciotto, E. and Wilgain, S. (1954). *Nature* **173**, 632. [3.4.4, 16.3.1]

Piggot, G. S. and Urry, W. D. (1939). *J. Wash. Acad. Sci.* **29**, 405. [3.4.2, 16.1, 16.3.1]

—— —— (1941). *Am. J. Sci.* **239**, 81. [15.4.1]

—— —— (1942). *Bull. geol. Soc. Am.* **53**, 1187. [3.4.2, 15.4.1, 16.1, 16.3.1]

Piper, D. Z., Veeh, H. H., Bertrand, W. G., and Chase, R. L. (1975). *Earth planet. Sci. Lett.* **26**, 114. [16.2, 16.2.1, and 16.2.2]

Pliler, R. and Adams, J. A. S. (1962). *Geochim. Cosmochim. Acta* **26**, 1137. [7.2, 7.3.2]

Plummer, L. N., Vacher, H., Mackenzie, F. T., Bricker, O. P., and Land, L. S. (1976). *Geol. Soc. Am. Bull.* **87**, 1301. [11.1]

Poet, S. E., Moore, H. E., and Martell, E. A. (1972). *J. geophys. Res.* **77**, 6515. [8.5]

Porter, S. C. (1979). *Geol. Soc. Am. Bull.* **90**, 609. [13.2.2]

Price, P. B. and Walker, R. M. (1963). *Appl. phys. Letts.* **2** (2), 23. [5.3.4]

Preston, M. A. (1947). *Phys. Rev.* **71**, 865. [1.3.1]

—— (1962). *The physics of the nucleus.* Addison-Wesley, New York. [1.3.1]

Puphal, K. W. and Olsen, D. R. (1972). *Anal. Chem.* **44**, 284. [4.5.1]

Puri, H. S. and Winston, G. O. (1974). *Spec. Pub.* **20**, Bureau of Geology, State of Florida. [9.2.2]

Raisbeck, G. and Yion, F. (1979). *Nature* **277**, 42. [3.1]

—— —— Stephen, C. (1979). *J. Phys. Lett.* **40**, 241. [3.1]

Rama, Koide, M., and Goldberg, E. D. (1961). *Science* **134**, 98. [2.5.3, 4.4, 5.5.4, 8.5, 15.6.1, 15.6.2, 15.6.3]

Ramsay, W. and Travers, M. W. (1898). *Proc. R. Soc. Lond.* **62**, 325. [14.1]

Rankama, K. and Sahama, T. G. (1950). *Geochemistry*, p. 636. University of Chicago Press, Chicago. [7.1]

Rayleigh (Lord). (1896). *Proc. R. Soc. Lond.* **60**, 56. [14.1]

Raynolds, J. H. (1960). *Geochim. Cosmochim. Acta* **20**, 101. [3.1]

Redfield, A. C. (1967). *Science* **157**, 687. [13.2.2]

Reeves, C. C. Jr. (1976). *Caliche*. Estacado Books, Lubbock. [12.2.1]

Reid, D. F., Key, R. M., and Schink, D. R. (1979). *Earth planet. Sci. Lett.* **43**, 223. [5.5.3]

―― Moore, W. S., and Sackett, W. M. (1979). *Earth planet. Sci. Lett.* **43**, 227. [4.3, 8.4, 15.4.2]

Reimer, G. M., Denton, E. H., Friedman, I., and Otton, J. K. (1979). *J. Geochem.* **11**, 1. [9.2.3]

Reynolds, R. L. and Goldhaber, M. B. (1978). *Econ. Geol.* **73**, 1677. [9.2.3]

Reyss, J. L., Yokoyama, Y., and Duplessy, J. C. (1978). *Deep Sea Res.* **25**, 491. [5.5.1]

Richardson, K. A. (1964). In: *Natural radiation environment*, ed. J. A. S. Adams and W. M. Lowder, p. 39. University of Chicago Press, Chicago. [6.2]

Riss, W. (1924). *Sber. Akad. Wiss. Wien. IIa* **132**, 91. [1.5]

Robbins, J. A. (1978). In: *The biogeochemistry of lead in the environment*, ed. J. O. Nriagu, p. 285. Elsevier–North Holland Biomedical Press, New York. [16.3.1]

―― Edgington, D. N. (1975). *Geochim. Cosmochim. Acta* **39**, 285. [3.4.8]

Roberge, J. and Gascoyne, M. (1978). *Geog. phys. Quart.* **32**, 281. [11.3.4]

Rogers, J. J. W. and Adams, J. A. S. (1969). *Handbook of geochemistry*, ed. K. H. Wedepohl. Springer-Verlag, Berlin. [2.1.2, 2.2.1, 2.2.4, 2.5.1, 6.1, 6.2]

Rona, E. (1964). *Science* **144**, 1595. [7.2]

―― Emiliani, C. (1969). *Science* **163**, 66. [13.2.1, 16.3.1]

―― Urry, E. D. (1952). *Am. J. Sci.* **250**, 242. [8.1, 8.2.1, 8.4, 15.4.1]

―― Gilpatrick, L. O., and Jeffrey, L. M. (1956). *Trans. Am. geophys. Un.* **37**, 697. [5.3.4, 8.1, 8.2.1, 15.2.3]

Rona, P., and Joeusuu, O. (1974). *Mem. Am. Assoc. petrol. Geol.* **20**, 570. [16.2.1]

Rose, A. W. and Korner, L. A. (1979). In: *Geochemical exploration 1978*, ed. J. R. Watterson and P. K. Theobald, p. 65. Assoc. Exploration Geochemists, Rexdale, Canada. [14.4.4]

Rosholt, J. N. (1957). *Anal. Chem.* **29**, 1398. [4.4]

―― (1959). *U.S. Geol. Surv. Bull. 1084-A*. [14.1]

―― (1967). *Open-system model for uranium-series dating of Pleistocene samples*, In: *Proc. Symp. Radioact. Dat. Meth. Low-Level Count., Monaco*, p. 299. IAEA, Vienna [3.7, 5.5.2, 17.2.2, 18.3]

―― (1976). *Geol. Soc. Am. Bull. Abs. Prog.* **8**, 1076. [10.2.1]

―― (1980a). *Nucl. Technol*, **51**, 143. [7.3, 10.3.1]

―― (1980b). *Uranium-trend dating of Quaternary sediments*. US Geol. Surv. Open-File Rep. 80–1087. [3.7, 4.4, 10.2.1, 10.3.1, 10.3.2]

――Antal, P. S. (1962). *US Geological Survey Prof. Paper 450-E*, E 108. [5.5.2, 11.3.3, 13.3.2]

―― Bartel, A. J. (1969). *Earth planet. Sci. Lett.* **7**, 141. [7.3.1, 14.3]

―― Butler, A. P., Garner, E. L., and Shields, W. R. (1965). *Econ. Geol.* **60**, 199. [2.5.2]

―― McKinney, C. R. (1980). *Uranium series disequilibrium investigations related to the WIPP Site, New Mexico, Part II*. US Geol. Survey Open file Report 80–879. [8.2.3]

―― Szabo, B. J. (1968). *Proc. Int. Conf. Mod. Trends in Activ. Anal.*, p. 327. NBS Gaithersburg, Maryland. [5.3.4, 5.5]

—— Doe, B. R., and Tatsumoto, M. (1966). *Geol. Soc. Am. Bull.* **77**, 987. [7.3.2, 8.2.3, 10.2.1, 10.3.2]

—— Prijana, and Noble, D. C. (1971). *Econ. Geol.* **66**, 1061. [6.3]

—— Shields, W. R., and Garner, E. L. (1963). *Science* **13**, 224. [17.3.2]

—— Tatsumoto, M., and Dooley, J. R. (1965). *Econ. Geol.* **60**, 477. [14.4.3]

—— Zartman, R. E., and Nkomo, I. T. (1973). *Geol. Soc. Am. Bull.* **84**, 989. [6.3, 14.3]

—— Emiliani, C., Geiss, J., Koczy, F. F., and Wangersky, P. J. (1961). *J. Geol.* **69**, 162. [3.4.5, 13.2.1, 15.3.2, 16.3.1, 18.4]

—— —— —— —— (1962). *J. geophys. Res.* **67**, 2907. [13.2.1]

Roubet, C. (1969). *L'Anthropologie* **73**, 503. [12.4.2]

Ruddiman, W. F., McIntyre, A., Neifler-Hunt, V., and Durazzi, J. T. (1980). *Quatern. Res.* **13**, 33. [18.1]

Rudran, K. (1969). *AERE Report R-5987.* [4.5.1]

Rundo, J. (1959). *AERE Report HP/R-627.* [4.5.1]

Rutherford, E. (1924). *Phil. Mag.* **XLVII**, 277. [5.1]

—— (1927). *Phil. Mag.* **IV**, Series 7, 580. [1.3.1]

—— Soddy, F. (1902). *Phil. Mag.* **IV**, Series 6, 370. [1.4]

—— —— (1903). *Phil. Mag.* **V**, Series 6, 576. [1.4]

Ruzicka, V. (1970). *Geol. Surv. Can. Paper 70-48.* [14.2]

Rydell, H. S. (1969). *Ph.D. Diss.*, Florida State Univ., Tallahassee. [9.2.1]

—— Bonatti, E. (1973). *Geochim. Cosmochim. Acta* **37**, 2557. [16.2.1]

—— Kreamer, T., Boström, K., and Joeusuu, O. (1974). *Mar. Geol.* **17**, 151. [15.2.5]

Sackett, W. M. (1958). *Ionium-uranium ratios in marine deposited calcium carbonates and related materials*, Washington Univ. Ph.D. thesis, St. Louis, Mo. [18.4]

—— (1960). *Science* **132**, 1761. [3.4.5, 4.4, 5.5.2, 15.3.2, 16.3.1]

—— (1964). *Ann. N. Y. Acad. Sci.* **119**, 339. [3.4.5, 15.3.1, 15.3.2, 16.3.1, 19.2]

—— (1965). In: *Symp. Mar. Geochem.*, ed. D. R. Schink and J. T. Corless, p. 29. *University of Rhode Island.* [3.4.3]

—— (1966). *Science* **154**, 646. [15.3.2]

—— Cook, G. (1969). *Trans. Gulf-Cst Ass. geol. Socs* **19**, 233. [8.1, 8.2.1, 8.2.3, 8.3, 15.2.2, 15.2.3, 16.2, 16.2.1]

—— Potratz, H. A. (1963). *U.S. Geol. Survey Prof. Paper 260-BB.* [17.2]

—— Broecker, W. S., and Thurber, D. L. (1965). *The geochemistry of* ^{231}*Pa and the dating of pelagic sediments.* Ann. Rep. AEC Contract AT (30–1). USAEC, Washington, DC. [3.4.3]

—— Potratz, H. A., and Goldberg, E. D. (1958). *Science* **128**, 204. [4.2.2]

—— Mo, T., Spalding, R. F., and Exner, M. E. (1973). In: *Symp. Interact. Radioact. Contam. Const. Mar. Env.*, Seattle, Washington. IAEA, Vienna. [8.2.1, 15.2.1, 15.2.5]

Sakanoue, M. and Yoshioka, M. (1974). *Quatern. Res. (Tokyo)* **13**, 220. [12.4.3]

—— Konishi, K., and Komura, K. (1967). *Stepwise determinations of thorium, protactinium, and uranium isotopes and their application in geochronological studies.* In: *Proc. Symp. Radioact. Dat. Meth. Low-Level Count., Monaco*, p. 313. IAEA, Vienna. [3.7, 17.2.1, 17.2.2]

—— Yoneda, S., Onishi, K., Koyama, K., Komura, K. and Nakanishi, T. (1968). *Geochem. J.* **2**, 71. [14.4.2]

Santschi, P. H., Li, Y-H., and Bell, J. (1979). *Earth planet. Sci. Lett.* **45**, 201. [2.5.3, 5.5.4, 15.3.3, 16.3.3]

—— —— Carson, S. R. (1980). *Est. Coast. Mar. Sci.* **10**, 635. [15.3.3]

—— Adler, D., Amdurer, M., Li, Y. H., and Bell, J. J. (1980). *Earth planet. Sci. Lett.* **47**, 327. [15.3.3, 16.3.3]

Sargent, B. W. (1933). *Proc. R. Soc. Lond. A* **139**, 659. [1.3.2]

Sarmiento, J. L. (1978). *A study of mixing in the deep sea based on STD, radon-222 and radium-228 measurements*, Ph.D. Thesis, Columbia University, New York, 238 pp. [15.3.3, 15.4.2, 15.5.2]

—— Broecker, W. S., and Biscaye, P. E. (1978). *J. geophys. Res.* **83**, 5068. [15.5.2]

—— Feely, H. W., Moore, W. S., Bainbridge, A. E., and Broecker, W. S. (1976). *Earth planet. Sci. Lett.* **32**, 357. [15.4.2, 15.5.2, 19.3]

Satterly, J. (1911). *Proc. Camb. phil. Soc. math. phys.* **16**, 360. [15.4.1]

Schaeffer, O. A. (1967). *Direct dating of fossils by the He-U method*, In: *Proc. Symp. Radioact. Dat. Meth. Low-Level Count., Monaco*, p. 395. IAEA, Vienna. [3.1, 3.3.3, 4.4]

—— Thompson, S. O., and Lark, N. L. (1960). *J. geophys. Res.* **65**, 4013. [3.1]

Schell, W. R. (1977). *Geochim. Cosmochim. Acta.* **41**, 1019. [4.3.2, 4.4, 16.2.5]

Schreiner, G. D. (1958). *Proc. R. Soc. Lond. A* **245**, 112. [3.1]

Schroeder, J. H., Miller, D. S., and Friedman, G. M. (1970). *J. sedim. Petrol.* **40**, 672. [2.4.1]

Schubert, C. and Valastro, S. (1976). *Geol. Soc. Am. Bull.* **87**, 1131. [18.5]

—— Szabo, B. J. (1978). *Geologie Mijnb.* **57**, 325. [18.5, 19.2]

Schvoerer, M., Dautant, A., and Bechtel, F. (1979). In: *La genèse des nodules de manganèse*, ed. C. Lalou, p. 295. CNRS, Paris. [16.3.2]

Schwarcz, H. P. (1979). *Uranium series dating of contaminated travertines: A two component model.* McMaster University Techn. Memo 79–1, 14p. [3.3.1, 3.6]

—— (1980). *Archaeometry* **22**, 3. [3.1, 3.5, 3.6, 4.4, 11.4, 12.3.1, 12.4.1, 19.2]

—— Débenath, A. (1979). *C. r. hebd. Séanc. Acad. Sci., Parls* **288**, 1155. [12.4.1]

—— Mohamad, D. (1981). *Determination of uranium isotope ratios in groundwater by track counting methods* (in preparation). [5.3.4]

—— Richmond, G. (1981). (research in progress). [11.4]

—— Suttcliffe, A. (1981). (manuscript in prep.) [12.2.1]

—— Goldberg, P., and Blackwell, B. (1980). *Israel J. earth Sci.* **29**, 157. [12.4.1]

—— Blackwell, B., Goldberg, P., and Marks, A. E. (1979). *Nature* **277**, 558. [10.2.2, 12.4.1, 13.2.2, 13.3.2]

—— Harmon, R. S., Thompson, P., and Ford, D. C. (1976). *Geochim. Cosmochim. Acta* **40**, 657. [11.3.4, 13.3.3]

Scott, M. R. (1968). *Earth planet. Sci. Lett.* **4**, 245. [8.2.3, 8.3, 15.3.1, 17.3.2]

Scott, R. C., and Barker, F. B. (1962). *U.S. Geol. Survey Prof. Paper 426*, 115. [9.2.2]

—— Salter, P. F. (1978). *Trans. Am. geophys. Un.* **59**, 118. [8.5, 16.3.1]

—— Osmond, J. K., and Cochran, J. K. (1972). In: *Antarctic oceanology II: The Australian–New Zealand sector*, ed. D. E. Hayer, p. 317. American Geophysical Union, Washington. [15.2.5]

—— Scott, R. B., Rona, P. A., Butler, L. W., and Nalwalk, A. J. (1974). *Geophys. Res. Lett.* **1**, 355. [15.2.5, 16.2, 16.2.1, 16.2.2]

Segré, E. (1959). *Experimental nuclear physics Vol III.* Wiley and Sons, New York. [1.2, 1.3.1]

Seuter, F. (1953). In: *Kosmische Strahlung*, ed. W. Heisenberg, Berlin. [5.1]

Severne, B. C. (1978). *J. geochem. Explor.* **9**, 1. [9.2.3]

Shackleton, N. J. (1967). *Nature* **215**, 15. [13.2.1]

—— (1977). *Phil. Trans. R. Soc., Lond.* **280**, 169. [18.1, 18.5]

—— Matthews, R. K. (1977). *Nature* **268**, 618. [18.1]

—— Opdyke, N. D. (1973). *Quatern. Res.* **3**, 39. [11.3.4, 12.1, 13.2.1, 18.1, 18.4, 18.5]

Shacklette, H. T. and Connor, J. J. (1973). *U.S. Geol. Surv. Prof. Paper 574-E.* [14.4.7]

Shannon, L. V. and Cherry, R. D. (1971). *Earth planet. Sci. lett.* **11**, 339. [16.3.3]

—— —— Orren, M. J. (1970). *Geochim. Cosmochim. Acta* **34**, 701. [15.7]

Shapiro, M. H., Melvin, J., Tombrello, T. A., and Whitcomb, J. H. (1977). *EOS. Trans. Am. geophys. Un.* **58**, 434. [9.2.3]

Sharma, P. and Somayajulu, B. L. K. (1979). In: *La genèse des nodules de manganèse* ed. C. Lalou, p. 281. CNRS, Paris. [16.3.2]

Sharpe, J. (1960). *Nuclear radiation detectors.* Methuen, London. [5.2]

Sheldon, R. P. (1964). *U.S. Geol. Survey Prof. Paper 501-C*, C106. [17.3.4]

Sheppard, R. A. and Gude, A. J. III (1968). *U.S. Geol. Survey Prof. Paper 597*, 38. [7.3]

Sholkovitz, E. R. (1976). *Geochim. Cosmochim. Acta* **40**, 831. [8.2.2]

Siegbahn, K. (1965). *Alpha-, beta- and gamma-ray spectroscopy. Vol. 1.* North Holland Publishing Co., Amsterdam. [5.1, 5.2, 5.3.2]

Siffert, P. (1966). *Ph.D. Thesis.* University of Strasbourg. (unpublished). [5.3.1]

—— Coche, A., and Cappellani, F. (1968). *Heavy charged particle spectrometry in semiconductor detectors*, ed. G. Bertolini and A. Coche, p. 306. North-Holland Publishing Co., Amsterdam. [5.3.1]

Sill, C. W. (1976). *NBS Special Publication* **422**, National Bureau of Standards, Washington. [4.3.2]

—— (1979). *Anal. Chem.* **51**, 1307. [4.3.2, 4.4]

Sinha, A. K. (1972). *Earth planet. Sci. Lett.* **16**, 219. [4.4]

Smales, A. A., Airey, L., Woodward, J., and Mapper, D. (1957). *AERE Report CR-2223.* [4.5.1]

Smith, G. I. (1979). *U.S. Geol. Survey Prof. Pap. 1043*, 130 pp. [11.5.3, 13.3.1]

Smith, A. R., Wollenberg, H. A., and Mosier, D. F. (1980). In: *Natural radiation environment III*, ed. T. F. Gesell and W. M. Lowder, p. 154. U.S. Dept. Energy. [9.2.3]

Smith, A. Y., Barretto, P. M. C., and Pournis, S. (1976). In: *Exploration for uranium ore deposits*, p. 185. International Atomic Energy Agency Pub. STI/PUB/434. [14.4.5]

Smith, B. N. and Epstein, S. (1971). *Plant Physiol.* **47**, 380. [13.2.2]

Smith, K. A. and Mercer, E. R. (1970). *J. radioanal. Chem.* **5**, 303. [5.5.3]

Somayajulu, B. L. K. and Church, T. M. (1973). *J. geophys. Res.* **78**, 4529. [15.2.4, 15.4.1, 16.2.1, 16.2.2]

—— Craig, H. (1976). *Earth planet. Sci. Lett.* **32**, 268. [2.5.3, 15.6.3]

—— Goldberg, E. D. (1966). *Earth planet. Sci. Lett.* **1**, [15.3.1, 15.4.2]

—— Lal, D., and Craig, H. (1973). *Earth planet. Sci. Lett.* **18**, 181. [3.1]

—— Tatsumoto, M., Rosholt, J. N., and Knight, R. J. (1966). *Earth planet. Sci. Lett.* **1**, 387. [3.4.4, 6.4.1, 6.4.2, 6.5.1, 6.5.3]

Sorem, R. K., Fewkes, R. H., McFarland, W. D., and Reinhard, W. R. (1979). In: *La genese des nodules de manganese*, ed. C. Lalou, p. 61. CNRS. Paris. [16.3.2]

Spalding, R. F. and Matthews, T. D. (1972). *Quatern. Res.* **2**, 470. [13.2.1]

Spalding, R. G. and Sackett, W. M. (1972). *Science* **175**, 629. [8.1, 8.2.1, 8.4, 15.2.1]

Spencer, D. W., Brewer, P. G., Fleer, A., Honjo, S., Krishnaswami, S., and Nozaki, Y. (1978). *J. Mar. Res.* **36**, 493. [15.3.1, 19.3]

Spiridonov, A. I. and Tyminiskii, V. G. (1971). *Izv. Akad. Nauk. Fiz. Zemli*, 214. [9.2.3]

—— Sultankhodzhagv, A. N., Surganova, N. A., and Tyminskii, V. G., (1969). *Uzbek. geol. Zh.* **4**, 82. [4.4]

Staub, H. (1953). *Detection methods in experimental nuclear physics, Vol. I*, ed. E. Segrè. Wiley and Sons, New York. [5.2]

Stearns, C. E. (1976). *Quatern. Res.* **6**, 445. [18.5]

—— Thurber, D. L. (1965). *Quaternaria* **7**, 29. [12.2.2, 12.4.2, 17.2]

—— —— (1967). $^{230}Th/^{234}U$ *dates of Late Pleistocene marine fossils from the Mediterranean and Moroccan littorals.* In: *Progress in Oceanography.* Pergamon, Oxford. [13.3.5]

Steen-McIntyre, V., Fryxell, R., and Malde, H. E. (1975). *Proc. Southwestern Anthro. Assoc. Soc. Mexicana de Antropl. Joint Mtg.*, Los Alamos Sci. Lab., New Mexico. Ed. D. Snow. [12.4.3]

Steinen, R. P., Harrison, R. S., and Matthews, R. K. (1973). *Geol. Soc. Am. Bull. 84*, 63. [18.5, 18.7]

Stephens, F. S., Asaro, F., and Perlman, I. (1957). *Phys. Rev.* **107**, 1091. [1.5]

Strehlow, F. W. E. (1960). *Anal. chem.* **32**, 1185. [4.2.1, 4.2.2, 4.2.6]

—— Rethemeyer, R., and Bothma, C. J. C. (1965). *Anal. Chem.* **37**, 106. [4.2.1, 4.2.2, 4.2.6]

Strutt, R. J. (1904). *Proc. R. Soc. Lond.* **73**, 191. [14.1]

Stuckless, J. S. (1979). *Contrib. Geol.* **17**, 173. [14.3]

—— Ferreira, C. P. (1976). In: *Exploration for uranium ore deposits.* IAEA-SM-208/17, 717. [6.3, 7.3.1, 14.3]

—— Nkomo, I. T. (1978). *Econ. Geol.* **73**, 427. [6.3, 14.3]

—— —— (1980). *Econ. Geol.* **75**, 289. [14.3]

—— Peterman, Z. E. (1977). *Wyo. Geol. Assn. Earth Sci. Bull.* **10**, 3. [6.1]

—— Bunker, C. M., Bush, C. A., Doering, W. P., and Scott, J. H. (1977). *US Geol. Survey. J. Res.* **5**, 61. [5.3.4]

Stuiver, M., Heusser, C. J., and Yang, I. C. (1978). *Science* **200**, 16. [12.1]

Suess, H. E. (1965). *J. Geophys. Res.* **70**, 5937. [13.2.2]

—— (1970). In: *Radiocarbon variations and absolute chronology*, ed. I. U. Olsson, p. 595. Wiley–Interscience,New York. [17.2.1]

Sugimura, Y. and Tsubota, H. (1963). *J. mar. Res.* **21**, 71. [4.2.4]

Sykes, L. R. and Raleigh, C. B. (1975). *EOS, Trans. Am. geophys. Un.* **56**, 838. [9.2.3]

Syromyatnikov, N. G. (1965). *Atomnaya Energiya* **19**, No. 2, 169. [14.4.2]

—— Ivanova, E. T. (1968). *Geochem. Int.* **3**, 299. [7.3.1]

—— Ibrayev, R. A., and Mukashev, F. A. (1967). *Geokhimiya*, No. 7, 834. [English abstract only on p. 697 of *Geochem. Int.*] [14.4.2]

Szabo, B. J. (1967). *Geochim. Cosmochim. Acta* **31**, 1321. [15.4.1]

—— (1969). *Un. Int. Quatern. Stud.* VIII INQUA Conf., Paris, 941. [10.3.1]

—— (1979a). *J. geophys. Res.* **84**, 4927. [3.7, 12.2.2, 12.4.3, 17.2.1, 17.2.2, 18.3, 18.5]

—— (1979b). *J. archaeol. Sci.* **6**, 201. [12.2.2, 12.4.3]

—— (1979c). *Mar. Geol.* **29**, M11. [17.2.1]

—— (1980). *Arctic Alpine Res.* **12**, 95. [12.2.2, 12.3.3]

—— Butzer, K. W., (1979). *Quatern. Res.* **11**, 257. [10.2.1, 11.5.3, 12.2.1, 12.4.1]

—— Collins, D. (1975). *Nature* **254**, 680. [12.2.2, 12.4.3, 13.3.5]

—— Rosholt, J. N. (1969). *J. geophys. Res.* **74**, 3253. [3.7, 5.3.4, 12.2.2, 12.3.2, 13.2.2, 13.3.5, 17.2.2, 18.3]

—— Sterr, H. (1978). *In Short Papers of the Fourth International Conference, Geochronology, Cosmochronology, Isotope Geology*, ed. R. E. Zartman. *U.S. Geol. Survey Open-File Report 78–701*, 416. [10.2.1]

—— Tracey, J. I. (1977). *Uranium-series and radiocarbon dating of coral samples from Eniwetok Atoll cores reveal hiatus in coral growth between 8000 and 132 000 years BP.* In: *Vol. Abs. 10th Int. Un. Quatern. Cong., Birmingham, UK*, p. 456. [3.7]

—— Vedder, J. G. (1971). *Earth planet. Sci. Lett.* **11**, 283. [3.7, 12.2.2, 17.2.2, 18.3]

—— Carr, W. J., and Gottschall, W. C. (1981). *U.S. Geol. Survey Open File Report, 81–119* [10.2.1, 10.2.2]

—— Koczy, F. F., and Östlund, G. (1967). *Earth planet. Sci. Lett.* **3**, 51. [15.4.1]

—— Malde, H. E., and Irwin-Williams, C. (1969). *Earth planet. Sci. Lett.* **6**, 237. [3.7, 12.2.2, 12.4.3]

—— Stalker, A. Macs., and Churcher, C. S. (1973). *Can. J. earth Sci.* **10**, 1464. [12.2.2, 13.2.2, 13.3.5]

—— Dooley, J. R., Taylor, R. B., and Rosholt, J. N. (1970). *US Geol. Surv. Prof. Paper 700–B*, B90. [3.7, 12.2.2]

—— Gottschall, W. C., Rosholt, J. N., and McKinney, C. R. (1980). *U. S. Geol. Survey Open-File Report 80–879*, 22 p. [10.2.2]

—— Ward, W. C., Ewidie, A. E., and Brady, M. J. (1978). *Geology* **6**, 713. [18.5]

Szalay, A. (1958). In: *Proc. Int. Conf. Peaceful Uses At. En., Surv. Raw Mats Res. 2*, p. 182. U.N. Geneva. [2.2.3, 2.3.1]

—— Samsoni, Z. (1969). *Geochem. Int.* **6**, 613. [2.2.1, 7.2, 8.2.2, 14.3]

Szoghy, I. M. and Kish, L. (1978). *Can. J. earth Sci.* **15**, 35. [14.4.2]

Tabushi, I. (1979). *Nature* **280**, 665. [4.4]

Taddeucci, A., Broecker, W. S., and Thurber, D. L. (1967). *Earth planet. Sci. Lett.* **3**, 338. [6.4.3]

Talvitie, N. A. (1972). *Anal. Chem.* **44**, 280. [4.5.1]

Tanner, A. B. (1980). *Natural radiation environment III*, ed. T. F. Gesell and W. M. Lowder, p. 5. U.S. Department of Energy. [9.2.3, 9.3.3]

Tatsumoto, M. (1966). *Science* **153**, 1097. [6.5.3]

—— Goldberg, E. D. (1959). *Geochim. Cosmochim. Acta* **17**, 201. [17.2, 18.4]

—— Hedge, C. E., and Engel, A. E. F. (1965). *Science* **150**, 886. [2.2.1]

Tauson, L. V. (1956). *Geochem.* [English trans.] **3**, 236. [7.2]

Taylor, A. E. (1952). *Rep. Prog. Phys.* **15**, 1. [5.1]

Taylor, F. W. and Bloom, A. L. (1977). In: *Proc. 3rd Int. Coral. Reef Symp.*, ed. D. L. Taylor, p. 275. University of Miami, Miami. [18.5]

Teng, T.-L. (1980). *J. geophys. Res.* **85**, 3089. [9.2.3]

—— Ku, T.-L., and McIlreath, R. P. (1975). *EOS, Trans. Am. geophys. Un.* **56**, 1019. [9.2.3]

Thompson, G. (1973). *Trans. Am. geophys. Un.* **54**, 1015. [7.1]

—— Lumsden, D. N., Walker, R. L., and Carter, J. A. (1975). *Geochim. Cosmochim. Acta* **39**, 1211. [11.3.4, 12.5, 13.3.2]

Thompson, J. D. (1977). In: *Desertification* ed. M. N. Glantz, p. 103. Westview Press, Boulder. [17.3.4]

—— Walton, A. (1972). *Nature* **240**, 145. [18.4]

—— Turekian, K. K., and McCaffre, R. J. (1975). In: *Estuarine Research, Vol. 1*, p. 28. Academic Press, New York. [15.2.5]

Thomson, J. and Turekian, K. K. (1976). *Earth planet. Sci. Lett.* **32**, 297. [4.4, 4.6, 15.6.2, 15.6.3]

Thompson, P. (1973). *Speleochronology and late Pleistocene climates*. Unpublished Ph.D. Thesis, Dept. of Geology, McMaster Univ. [5.5.1, 5.5.2, 11.3.4]

—— (1973). *Procedure for extraction and isotopic analysis of uranium and thorium from speleothem*. Tech. Memo 73–9 Geology Dept., McMaster University, Hamilton, Ontario, Canada. [11.3.4]

—— Ford, D. C., and Schwarcz, H. P. (1975). *Geochim. Cosmochim. Acta* **39**, 661. [3.4.1, 11.3.4]

—— Schwarcz, H. P., and Ford, D. C. (1976). *Geol. Soc. Am. Bull.* **87**, 1730. [11.3.4, 13.3.3]

Thurber, D. L. (1962). *J. geophys. Res.* **67**, 4518. [2.5.1, 7.2, 15.2.3, 17.2]

—— (1963). *Anomalous $^{234}U/^{238}U$ and an investigation of the potential of ^{234}U for Pleistocene chronology*, Ph.D. Thesis, Columbia University, New York. [3.4.1]

—— (1965). In: *Symp. Mar. Geochem.*, ed. D. R. Schink and J. T. Corless. Graduate School of Oceanography, University of Rhode Island, Occasional Publication No. 3. Kingston, Rhode Island. [17.2.2]

—— Broecker, W. S., Blanchard, R. L., and Potratz, H. A. (1965). *Science* **149**, 55. [3.1, 3.4.1, 3.5, 17.2, 17.2.1, 18.4]

Tieh, T. T., Ledger, E. B., and Rowe, M. W. (1980). *Chem. Geol.* **29**, 227. [6.2, 7.3.2]

Tilling, R. I. and Gottfried, D. (1969). *U.S. Geol. Survey Prof. Paper 614-E*, 29. [6.1, 6.3]

—— —— Dodge, F. C. W. (1970). *Geol. Soc. Am. Bull.* **81**, 1447. [6.1, 6.2]

Tilton, G. R. (1951). *AEC Rep. AECD-3182*. [3.1]

—— Patterson, C. C., and Davis, G. L. (1954). *Geol. Soc. Am. Bull.* **65**, 1314. [3.1]

—— —— Brown, H., Inghram, M. G., Hayden, R., Hess, D., and Larsen, E. S., Jr., (1955). *Geol. Soc. Am. Bull.* **66**, 1131. [7.2]

Titayeva, N. A. and Veksler, T. L. (1977). *Geochem. Int.* **14**, 99. [7.2, 7.3.1]

—— Filonov, V. A., Ovchenkov, V. Ya., Veksler, T. I., Orlova, A. V., and Tyrina, A. S. (1973). *Geochem. Int.* **10**, 1146. [14.4.3]

—— Taskayev, A. I., Ovchenkov, V. Ya., Aleksakhin, R. M., and Shuktomova, I. I. (1977). *Geochem. Int.* **5**, 57. [2.5.3]

Torgersen, T. (1980). *J. geochem. Explor.* **13**, 57. [9.2.3, 9.3.2]

Travers, M. W. (1899). *Proc. R. Soc. Lond.* **64**, 130. [14.1]

Trier, R. M., Broecker, W. S., and Feely, H. W. (1972). *Earth planet. Sci. Lett.* **16**, 141. [15.4.2]

Tripathi, V. S. (1979). *Geochim. Cosmochim. Acta* **43**, 1979. [2.3.1]

Tsunogai, S. and Nozaki, Y. (1971). *Geochem. J.* **5**, 165. [15.6.2]

Turekian, K. K. (1971). *Impingement of man on the ocean*, ed. D. W. Hood, p. 9. John Wiley, New York. [8.4, 15.4.1]

—— (1973). *U and Th decay series nuclides abundances in marine plankton, Yale Univ. AD773-626*. [4.5.1]

—— (1976). *Oceans*. Prentice-Hall Inc., Englewood Cliffs, New Jersey. [15.2.5]

—— Bertine, K. K. (1971). *Nature* **229**, 250. [16.2]

—— Chan, L. H. (1971). In: *Activation analysis in geochemistry and cosmochemistry*, ed. A. O. Brunfeldt and E. Steinnes, p. 311. Universitetsforlaget, Oslo. [8.1, 8.2.1, 15.2.1, 15.2.3, 15.3.2, 17.2.1]

—— Cochran, J. K. (1978). In: *Chemical oceanography, 2nd ed., Vol. 7*, ed. J. P. Riley and R. Chester, p. 313. Academic Press, New York. [8.2.1, 8.2.2, 8.4, 15.2.1, 15.2.6, 15.4.1, 15.6.3, 16.3.3]

—— Johnson, D. G. (1966). *Geochim. Cosmochim. Acta* **30**, 1153. [15.4.1]

—— Nelson, E. (1976). *Uranium series dating of the travertines of Conna De L'Arago (France)*. Colloque I, Datations Absolues et Analyses Isotopiques en Prehistoire, Methodes et Limites. Union Des Sci. Prehist. IX Congres, Nice, p. 172. [3.6, 12.4.1]

—— Nozaki, Y. (1980). In: *Isotope marine chemistry*, p. 157. Uchida Rokakuho Pub. Co. Ltd., Tokyo. [15.6.2, 15.6.3, 15.7]

—— Katz, A., and Chan, L. (1973). *Limn. Oceanogr.* **18**, 240. [15.3.1]

—— Kharkar, D. P., and Thomson, J. (1973). *Uranium and thorium decay series nuclide abundances in marine plankton*. Final report to Advanced Research Projects Agency, ARPA Order 1793, Contract N00014-67-A-0097-0022. [5.5.4]

—— Kharkar, D. P., and Thomson, J. (1974). *J. Rech. Atmos.* **8**, 639. [15.7]

—— Nozaki, Y., and Benninger, L. K. (1977). *Ann. Rev. earth planet. Sci.* **5**, 227. [15.6.1]

Ulomov, V. I. and Mavashev, B. Z. (1967). *Dokl. Akad. Nauk. SSSR* **176**, 319. [9.2.3]

Urey, H. C. (1947). *J. chem. Soc.* 562. [18.1]

Urry, W. D. (1941). *Am. J. Sci.* **239**, 191. [4.2.1]

—— Piggot, C. S. (1941). *Am. J. Sci.* **239**, 633. [5.5.3]

Valentine, J. W. and Veeh, H. H. (1969). *Geol. Soc. Am. Bull.* **80**, 1415. [18.5]

Valentine, K. W. G. and Dalrymple, J. B. (1976). *Quatern. Res.* **6**, 209. [13.2.2]

Van, N. H. and Lalou, C. (1969). *Radiochimica Acta* **12**, 156. [4.4]

Van der Weidjen, C. H. and Langmuir, D. (1976). *Geol. Soc. Am. Abs. Prog., 1976*

Meet., Denver, Colarado. Geological Society of America, Boulder, Colorado. [8.2.2]

Veeh, H. H. (1966). *J. geophys. Res.* **71**, 3379. [3.4.1, 17.2.1, 18.4, 18.5]

—— (1967). *Earth planet. Sci. Lett.* **3**, 145. [7.1, 8.2.1, 15.2.5, 15.2.6, 16.2, 16.2.1, 16.4]

—— (1968). *Geochim. Cosmochim. Acta* **32**, 117. [15.2.3]

—— Boström, K. (1971). *Earth planet. Sci. Lett.* **10**, 372. [15.2.5]

—— Burnett, W. C. (1978). *Nature* **275**, 460. [17.3.3, 17.3.4]

—— Chappell, J. (1970). *Science* **167**, 862. [17.2.1, 17.2.2, 17.2.3, 18.5, 18.7]

—— Geigengack, R. (1970). *Nature* **226**, 155. [18.5]

—— Green, D. C. (1977). In: *Biology and geology of coral reefs*, **4**, 183, ed. O. A. Jones and R. Endean. Academic Press, New York. [17.2]

—— Turekian, K. K. (1968). *Limnol. Oceanogr.* **13**, 304. [17.2.1]

—— Valentine, J. W. (1967). *Geol. Soc. Am. Bull.* **78**, 547. [18.5]

—— Veevers, J. J. (1970). *Nature* **226**, 536. [17.2.1]

—— Burnett, W. C., and Soutar, A. (1973). *Science* **181**, 844. [15.2.5, 17.3.2]

—— Calvert, S. E. and Price, N. B. (1974). *Mar. Chem.* **2**, 189. [4.4, 15.2.5, 17.3.2, 17.3.4]

—— Schwebel, D., van de Graaff, W. J. E., and Denman, P. D. (1979). *J. geol. Soc. Aust.* **26**, 285. [17.2.1, 17.2.3, 18.5]

Veselesky. J. (1974). *Radiochim. Acta.* **21**, 151. [4.3.2, 4.4]

Vine, D. J., Swanson, V. E., and Bell, K. G. (1958). In: *Proc. Int. Conf. Peaceful Uses At. En., Survey Raw Mats. Res.* **2**, 187 UN, Geneva. [2.2.3]

Vlcek, E. and Mania, D. (1977). *Anthropologie* **15**, 159. [12.4.1]

Volchok, H. L. and Kulp, J. L. (1957). *Geochim. Cosmochim. Acta* **11**, 219. [15.4.1]

Von Buttlar, H. and Houtermans, G. (1950). *Maturwissenchaften* **37**, 400. [16.3.2]

Vorob'ev, A. A., Komer, A. P., and Korolev, V. A. (1962). *Zh. eksp. Teor. Fiz.* **43**, 426. [5.3.1]

Wagner, G. A. (1966). *Z. Naturf. A.* **21**, 733. [5.3.4]

Walton, A. J. and Debenham, N. C. (1980). *Nature* **284**, 42. [12.2.1]

Wapstra, A. H. and Nijgh, G. J. (1959). *Nuclear spectroscopy tables*, North-Holland, Amsterdam. [1.3.2]

Wasserburg, G. J. and Steiger, R. H. (1967). *Systematics in the Pb–U–Th systems and multiphase assemblages*, In: *Proc. Symp. Radioact. Dat. Meth. Low-Level Count., Monaco*, p. 331. IAEA, Vienna. [3.1]

Wasson, J. T., Adler, B., and Oeschger, H. (1967). *Science* **155**, 447. [3.1]

Watkins, N. D. and Kennett, J. P. (1977). *Mar. Geol.* **23**, 103. [16.3.2]

Webb, T., III and Bryson, T. A. (1972). *Quatern. Res.* **2**, 70. [13.2.2]

Weiss, R. F. (1977). *Earth planet. Sci. Lett.* **37**, 257. [15.3.2]

Wendorf, F., Laury, R. L., Albritton, C. C., Schild, R., Haynes, C. V., Damon, P., Shafiqullah, M., and Scarborough, R. (1975). *Science* **187**, 740. [12.2.1]

Wetherill, G. W. (1956). *Trans. Am. geophys. Un.* **37**, 320. [3.1, 17.2.1]

White, J. C. and Ross, W. J. (1961). *U.S.A.E.C. Report. NAS-NS 3102.* [4.2.1]

White, W. B. and Van Gundy, J. J. (1974). *Nat. Speleo. Soc. Bull.* **36**, 5. [11.3.1]

Whitfield, J. M., Rogers, J. J. W., and Adams, J. A. S. (1959). *Geochem. Cosmochim. Acta.* **17**, 248. [6.3]

Wigley, T. M. L. (1975). *Wat. Resour. Res.* **11**, 324. [13.2.2]

—— Plummer, L. N., and Pearson, F. J. (1978). *Geochim. Cosmochim. Acta* **42**, 1117. [13.2.2]

Williams, G. E. and Polach, H. A. (1971). *Geol. Soc. Am. Bull.* **82**, 3069. [13.2.2, 13.3.4]

Wilkening, M. H. and Clements, W. E. (1975). *J. geophys. Res.* **80**, 3828. [15.6.1]

—— —— Stanley, D. (1975). In: *The natural radiation environment 2*. USERFA, Oak Ridge. [8.5]

Wilkinson, D. H. (1950). *Ionization chambers and counters.* Cambridge University Press, London . [5.2.1]

Wilson, A. T., Hendy, C. H., and Reynolds, C. P. (1979). *Nature* **279**, 315. [13.4]

Wilson, J. D., Webster, R. K., Milner, G. W. C., Barnett, G. A., and Smales A. A. (1960). *Anal. Chim. Acta* **23**, 505. [15.2.3]

Winograd, I. J., and Friedman, W. (1975). *U.S. Geol. Survey Prof. Paper 712—C*, 126. [9.2.2]

—— Pearson, F. J. (1976). *Wat. Resour. Res.* **12**, 1125. [9.2.2]

Wintle, A. G. (1978). *Can. J. Earth Sci. Lett.* **15**, 1977. [12.5, 13.2.2]

—— Aitkin, M. J. (1977). *Archaeometry* **19**, 111. [13.2.2]

Woillard, G. M. (1978). *Quatern. Res.* **9**, 1. [13.2.2]

Wolery, T. J. and Sleep, N. H. (1976). *J. Geol.* **84**, 249. [15.2.5]

Wollenberg, H. A. (1975). *Second U.N. Symp. dev. use geotherm. res.*, San Francisco. Abstracts III—95. University of California, Berkeley, California. [9.2.3, 9.3.3]

—— Smith, A. R. (1968). *J. geophys. Res.* **73**, 1481. [6.2]

Wood, D. F. and McKenna, R. H. (1962). *Anal. Chim. Acta.* **27**, 446. [4.2.1]

Yamamoto, M. and Schwarcz, H. P. (1981). (In preparation). [13.3.3]

Yapp, C. J. and Epstein, S. (1977). *Earth planet. Sci. Lett.* **34**, 333. [13.2.2]

Yokoyama, Y. and Nguyen, H. V. (1980). In: *Isotope marine chemistry*, ed. E. D. Goldberg, Y. Horibe, and K. Saruhashi. **14**, 235. Uchida—Rokakuha, Tokyo. [5.5.1, 5.5.2, 5.5.3, 5.5.4, 16.1, 16.3]

—— Tobailem, J., Grjebine, T., and Labeyrie, J. (1968). *Geochim. Cosmochim. Acta* **32**, 347. [5.5.1, 16.1, 16.3]

York, D. (1969). *Earth planet. Sci. Lett.* **5**, 320. [10.2.1]

—— Farquhar, R. M. (1972). *The earth's age and geochronology.* Pergamon Press, Oxford. [3.1]

Zartman, P. R. (1964). *J. Petrol.* **5**, 359. [3.1]

Zeuner, F. E. (1959). *The Pleistocene Period.* Hutchison, London. [13.2.2]

Ziegler, V. (1974). *Formation of uranium ore deposits.* IAEA, Public. STI/PUB/374, 661. [14.2]

Zielinski, R. A. (1978). *Geol. Soc. Am. Bull.* **89**, 409. [6.1, 6.2, 6.3]

—— (1979). *Chem. Geol.* **27**, 47. [7.2]

—— (1980). *Am. assoc. petro. geol. Bull.* (In press). [7.2]

—— Rosholt, J. N. (1978). *J. Res. U.S. Geol. Survey* **6**, 489. [9.2.1]

—— Linsey, D. A., and Rosholt, J. N. (1980). *Chem. Geol.* **29**, 139. [7.3, 10.3.1]

—— Peterman, Z. E., Stuckless, J. S., Rosholt, J. N., and Nkomo, I. T. (1981). (in press). *Contrib. Mineral Petrol.* [6.2, 7.2]

INDEX